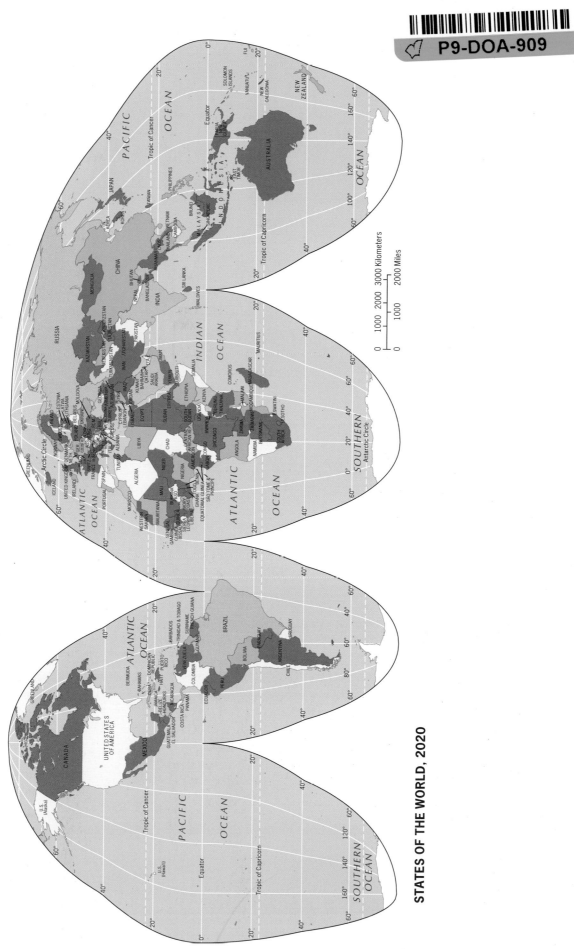

STATES OF THE WORLD, 2020

WileyPLUS

WileyPLUS gives you the freedom of mobility and provides a clear path to your course material and assignments, helping you stay engaged and on track.

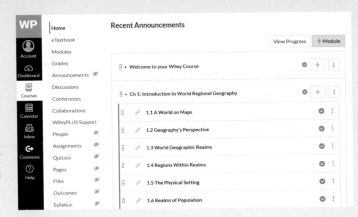

When course materials are presented in an organized way, you are more likely to stay focused, develop mastery, and participate in class. WileyPLUS provides you a clear path through the course material.

Starting with Wiley's quality curated content and adaptive practice questions, you can quickly access material most relevant to your course and understand topics that are most challenging to you. This easy-to-use, intuitive interface saves you time, helps you on track, and keeps you motivated throughout your course.

 Customized Content

Your course has been customized with videos, documents, pages, and relevant links to keep you motivated.

 Interactive eTextbook

You can easily search content, highlight and take notes, access instructor's notes and highlights, and read offline.

Adaptive Practice Questions

Quickly identify areas that are most challenging for you and then focus on material most relevant to your needs.

 Linear Design and Organization

Course modules organized by learning objective include eTextbook content, videos, animations, interactives, and practice questions.

 Calendar

The course calendar syncs with other features—like assignments and syllabus—so any updates from your instructor will immediately appear.

 Student App

You can view due dates, submit assignments, read material, and communicate with your instructor all from your phone.

WILEY

The World Today

Concepts and Regions in Geography

Eighth Edition

About the Cover

A small fishing boat slowly makes its way through the melting ice in Disko Bay, off Greenland's west coast, in the summer of 2019. We are well inside the Arctic Circle. The Ilulissat Glacier, one of the largest in the world, deposits about 35 billion tons of ice into this arm of the North Atlantic Ocean every year. As the glacier reaches the bay, the ice fragments into pieces (a process called 'calving'). Some of the biggest icebergs are grounded in the Bay for years, until they are sufficiently melted and light enough to float, and then drift southward. In the process, all this melting ice contributes to the rising of global sea level. This is an important site for research on climate change and global warming. In the late spring and early summer of 2019, scientists here reported historical record high temperatures. The rate of melting from the Greenland Ice Sheet is estimated to have quadrupled since 2003.

The World Today

Concepts and Regions in Geography

Eighth Edition

JAN NIJMAN

Distinguished University Professor, Geosciences,
Georgia State University
Professor of Geography, University of Amsterdam

MICHAEL SHIN

Professor of Geography, University of
California, Los Angeles

PETER O. MULLER

Professor Emeritus of Geography and
Regional Studies, University of Miami

WITH CONTRIBUTIONS FROM

ANN-MARGARET ESNARD

Professor of Regional and Urban Planning,
Andrew Young School of Policy Studies,
Georgia State University

RICHARD GRANT

Professor of Geography and Regional Studies,
University of Miami

WILEY

VICE PRESIDENT AND DIRECTOR	Veronica Visentin
EXECUTIVE EDITOR	Glenn Wilson
ASSISTANT EDITOR	Monica Rogers
INSTRUCTIONAL DESIGN LEAD	Karen Staudinger
ASSISTANT INSTRUCTIONAL DESIGNER	Megan Garvin
SENIOR MARKETING MANAGER	Carolyn Wells
SENIOR PRODUCTION EDITOR	Trish McFadden
SENIOR DESIGNER	Wendy Lai
ART DIRECTOR	Jonathon Boylan

Front cover image: Cover Photography: © J. Nijman

This book was set in 9.5 Source Sans Pro by Lumina Datamatics and printed and bound by Quad Graphics. The cover was printed by Quad Graphics.

ePUB ISBN: 978-1-119-57761-4

Printed in the United States of America

The inside back cover will contain printing identification and country of origin if omitted from this page. In addition, if the ISBN on the back cover differs from the ISBN on this page, the one on the back cover is correct.

V10017753_022620

This book is dedicated to the current generation of college students who are raising awareness about the challenges and consequences of global climate change and who are claiming their rights to the future world.

Preface

This is the Eighth Edition of a book that, since its first appearance in 2002, has undergone several significant changes while maintaining its essential strengths. We first wrote this book because so many of you, teachers and students alike, expressed the need for a text that provided an introduction to concepts and regions in world geography—but that did so in a concise manner and in the context of current world affairs. That is still the thrust of the book today, and this Eighth Edition, we believe, meets that challenge better than ever.

The approach to the integrated treatment of regions and concepts is a constant throughout all editions, but over the years we have adapted to the changing world around us. The evolving regional content of the chapters, and their organization, reflect the dynamic nature of the world's geography; the changing (and growing number of) concepts mirror the progress of the discipline; and the ongoing introduction of new (digital) features reflects the instructional possibilities of new technologies.

As always, we put a premium on top-quality, original maps to accompany our regional analyses, and we rely, in large part, on our own field observations. As geography changes over time, so do many of our maps. Much of our time and research for this book involves field visits and investigations in many of the world's realms. This is expressed most concretely in the numerous field notes, photographs, and on-location videos supplied by the authors throughout the book. For the geographer, there is no substitute for *being there*. Since the 7th edition, the author team has revisited key sites in all of the world realms discussed in this book, from the Arctic to subtropical Australia, from Easter Island to the Gobi Desert, from Cape Town to the Peruvian Andes.

Geographic Knowledge, Global Awareness

The content before you is an information highway to geographic knowledge and global awareness. Universities and colleges prepare their students to become critical thinkers, creative problem solvers, and, increasingly, global citizens. If we are living in a "global village," then students must get to know the world as they know their own local areas. But while the influence and interests of the United States reach across the globe, it is no secret that geographic illiteracy at home remains widespread. Much of what students encounter in this content is of immediate, practical value to them—as citizens, as consumers, as voters, as travelers, as job-seekers. North America is a geographic realm marked by intense global connections and relationships. Those involvements require countless, often instantaneous decisions, whether they relate to international business, media, affairs of state, environmental management, disaster relief, and many other spheres. Such decisions must be based on the best possible knowledge of the world beyond our borders. That understanding can be gained by studying the layout of our world, its environments, societies, traditions, resources, policies, development strategies, and other properties—in short, its regional geography.

The New Author Team

The World Today carries the imprint of its founding author, Harm J. de Blij. His passing in 2014 brought some important and ongoing changes, but this book will always radiate his passionate belief we all share: that geography is fundamental to our understanding of the world and to our efforts to shape and sustain it. The previous edition already marked a transition toward a new author team and that transition is continued in the present 8th edition.

Long-time co-authors Jan Nijman and Peter Muller are joined by Michael Shin and by contributing authors Ann-Margaret Esnard and Richard Grant. Michael Shin, Professor of Geography at UCLA, contributes essential digital interactive content to every chapter. He also authors innovative features on *Technology & Geography* and *Geographic Information Analysis*. Michael's combined expertise in online education, GIS, and regional geography are instrumental in taking this book to the cutting edge of digital learning. Ann-Margaret Esnard is a Distinguished University Professor at the Andrew Young School of Policy Studies at Georgia State University. She contributes several features boxes pertaining to regional planning projects around the world. Richard Grant is responsible for the chapter on Subsaharan Africa. Professor of Geography and Regional Studies at the University of Miami, he is one of the discipline's foremost authorities on the geography of contemporary Subsaharan Africa. Richard has more than two decades of field research experience in Southern and West Africa and is affiliated with prominent universities across the realm.

Chapter Organization

For convenience, and in response to feedback from our users, we decided to renumber the chapters of the book: in the past, the introductory chapter stood by itself and was not numbered, then the successive chapters on the various realms were numbered 1 to 12. In this new edition, the Introduction is Chapter 1, and the successive realm chapters are numbered 2 to 13.

The internal organization of the chapters, one that has proven so useful over the years to instructors and students, remains the same: aside from the introductory chapter, all chapters are divided roughly in two halves. The first part, *Defining the Realm*, concerns itself with realm-wide geographies (e.g., South America or Europe). The second part, *Regions of the Realm*, focuses on finer-scale and more richly detailed regional and sub-regional geographies within the realm (e.g., the Andean West, or Mediterranean Europe).

We know that many instructors cope with the challenge of covering all of the world in a single semester course. If instructors feel the need to cut material, the structure of our chapters allows for the option to assign only the first half of selected chapters. In that case, students will not be exposed to the detailed and finer-scale issues but they still get a concise and coherent overview at the scale of the realm.

New and Established Features

In this edition, we present two new features and further develop several features that were first introduced in the 7th edition. Some are especially designed for digital interactive platforms, and all have counterparts in print and e-text.

1. ***From the Field Notes***
   One of the signature features of our book, From the Field Notes, provides first-hand observations and reflections from the field by the authors. For every edition, we include new or updated field notes from around the world.

2. ***Geographic Information Analysis***
   The *Geographic Information Analysis (GIA)* features are guided interactive exercises that foster geographical thinking, spatial awareness, and data literacy. For this edition, they are updated and revised, and new features were created. The 24 *GIAs* in this edition are carefully designed to interactively engage students through the use of compelling maps and other spatial data visualizations, with the purpose to highlight and reinforce key concepts, issues, and locations in world regional geography. From examining the pattern of carbon dioxide emissions at the global scale in the Chapter 1, to mapping population projections across China's Pearl River Delta in Chapter 10, the variety in content and form of these features will stimulate both student curiosity and learning. Moreover, these features can and should be used as the basis for in-class or online discussions, or for independent or group-based student assignments.

3. ***Interactive Maps***
   This new edition also offers a new set of interactive web maps. Accessed through WileyPLUS, the interactive maps give students the freedom to navigate and explore the map by panning and zooming into areas and places, thus getting a better sense of the spatial dimensions, at varying scales, of the phenomenon in question. From examining regional variations in population density to identifying which cities of the world are most prominent when viewed using night-time satellite imagery, giving students the ability to explore and examine the world in detail, independently, across various themes, encourages engagement with the text and enhances student learning. The interactive maps can also be used by instructors to design their own student exercises or assignments.

4. ***On-Location Video Material***
   We are gradually integrating videos from our field visits into the various chapters in WileyPLUS, often accompanying Field Notes. These videos allow students a more effective and dynamic "place experience." We are selective with these on-location videos, including them where we think they have the greatest pedagogical value, either in the illustration or emphasis of a particular concept or problem, or where we feel the students would most benefit from a "feel" for a place that may otherwise seem distant to them. In this edition, we start with eight new videos and we will be expanding their number over time.

5. ***Regional Impacts of Global Climate Change***
   We created another new feature in this 8th edition to cover the ever-increasing importance of global climate change. The book includes ten new feature boxes, spread across the various chapters, that focus on the impact of climate change in a particular realm. For instance, in the chapter on Europe the feature covers extreme weather events; in South America, it focuses on the drying of lakes in Bolivia; in East Asia, it is concerned with flooding in the Pearl River Delta; and in Subsaharan Africa the feature concentrates on the increasing frequency and severity of cyclones. Taken together, this new collection of features underscores the reality of global climate change in local contexts around the world.

6. ***Technology & Geography***
   We continue this important features that was first introduced in the 7th edition. The *Technology & Geography* feature, repeated in every chapter, illustrates the impact of evolving technologies (digital or otherwise) on geography and geographic regions; it underscores the dynamic nature of the subject matter in this book: regions change, and the role or meaning of geography can change as a result of new technologies.

Highlights of Newly Added Content

The entire text has been thoroughly updated in terms of statistics and recent or even current developments. Additionally, we have revised and replaced text, maps, and features in more substantial ways, wherever we observe critical shifts in the geographic makeup of realms and regions.

Since the last edition appeared in early 2016, the world has undergone some momentous changes, all of which have important geographic dimensions. Throughout the text, we emphasize three general trends that are affecting many different realms and regions around the world. The first pertains to global climate change—it has been underway for some time, of course, but the acceleration and deepening of climate change impacts, locally and globally, demand more systematic examination (hence the new feature described above). The second more general trend refers to international migration: a testament to increased global information flows and mobility, but also a reflection of the global disparities in development, standards of living, and political freedom and safety. Furthermore, increased international migration has been a source of tension and reactionary politics, including populism, in receiving countries, particularly in the United States and various European countries. Third, this edition pays close attention to China's growing global footprint, in almost all realms, be it through infrastructure investment, trade, diplomacy, or military involvements.

No summary can adequately encapsulate everything this new edition contains, but here are some chapter-by-chapter highlights of the new content.

Chapter 1, the **Introduction**, as usual, provides the foundation and sets the stage for the regional chapters, discussing a broad range of basic geographic concepts along with a set of cornerstone world maps. It includes new sections and text on the importance of global climate change and global refugee and migration flows. Look for a new Geographic Information Analysis (GIA) on global migration flows and the first feature box on Global Climate Change that focuses on the Anthropocene; also direct your attention to the first set of interactive maps, one on the concept of scale and the other depicting the world's population distribution.

Chapter 2, **North America**, has new sections on immigration and shifting demographics, and another on the growing urban–rural divide in the United States. Also, look for new materials on growing urban and regional inequalities, and for updated GIAs on the U.S. presidential election. The climate change feature box (including on-location video material) refers to the recent wildfires in California and flooding events in South Florida.

Chapter 3, **Middle America**, contains updated and expanded text on the realm's northern boundary with the United States and on the impact of the expanded Panama Canal. The chapter also provides the latest available geographic information on Middle America's challenges related to the narcotics trade and concomitant drug wars. Also, see a new GIA on migrant remittances from the U.S. to Middle American countries.

Chapter 4, **South America**, provides an interesting new section comparing race relations in Brazil and the United States and has a new section on the Venezuelan crisis. In addition, look for new material on China's economic footprint in South America and a new feature box on global climate change and vanishing Lake Poopó in Bolivia. The GIAs on income in-

equalities and wine-producing regions have been updated with increased interactivity.

Chapter 5, **Europe**, discusses the latest developments (as of January 2020) regarding Brexit, and contains new sections on geopolitics in the Baltics and on extreme weather events (e.g., record-setting heatwaves in 2018 and 2019) induced by climate change. It also presents new text on the challenges of international migration and its political ramifications in various European countries. Also, check out the new interactive maps (on population distribution and climate), on-location video material from Iceland, and updated GIAs on immigration and demography and on the Russian presence in Ukraine.

Chapter 6, **Russia/Central Asia**, emphasizes the enormous challenges for Moscow to exercise control over a country that stretches across nine time zones and the realms and regions that surround it. The chapter provides important updates on Russia's demographic east–west imbalance, the geopolitical situation vis-à-vis Ukraine, and relations with the Central Asian republics and China. In addition, there is a new section on Siberia's thawing permafrost; look for new field notes from Lake Baikal, Kamchatka, and Moscow.

Chapter 7, **North Africa/Southwest Asia (NASWA)**, presents important updates and new material on this highly volatile realm. This includes new sections and maps on the territorial demise of ISIS, the refugee problems resulting from the ongoing war in Syria, and the devastating conflict in Yemen. There is also new text on the steady rise of the Gulf States (with new field notes), on evolving women's rights in Tunisia and Saudi Arabia, and a shift of jihadism toward the African Transition Zone. Look for interactive features on the geography of wealth and democracy and on the realm's growing water crisis.

Chapter 8, **Subsaharan Africa**, offers new sections on the impacts of climate change in the realm, along with a new field note from Cape Town on "Day Zero" (the day when the city's water supply was predicted to run dry). New interactive features and GIAs focus on the ethnic map of Subsaharan Africa, geographies of colonization and decolonization, population distribution, and the geography of Chinese investments across the realm. The chapter contains major updates on China's foothold in the realm and the divergent economic trajectories of various countries.

Chapter 9, **South Asia**, provides substantial new text on Afghanistan and the complicated triangular geopolitical relationship with Pakistan and India, including an update on the situation in Kashmir. Look for various new field notes (including on-location video material) from Nepal, Bhutan, and India; for a new feature box on climate change and extreme weather events; for new maps of Nepal/Bhutan, and Sri Lanka; and for interactive materials on India's demographic gender imbalance and a comparison of megacity Dhaka with Los Angeles.

Chapter 10, **East Asia**, introduces new sections on China's massive and ongoing Belt and Road Initiative, Mongolia's development challenges, Japan's aging population, the gender pay gap in this realm, and the Demilitarized Zone that divides the two Koreas. Look for new GIAs on Korean missile ranges and on the increasing threats of flooding in the Pearl River Delta, one of China's megacity regions. There are several new maps, as well as new field notes, with photographic and on-location video from Lhasa (Tibet), the Korean DMZ, Mongolia, and Japan. This is perhaps the most dynamic realm in the world, much of it driven by China, and is reflected in the breadth and currency of this chapter.

Chapter 11, **Southeast Asia**, introduces the concept of "hybrid warfare" (multifaceted conflict short of all-out military war) as applied to the South China Sea. Hybrid warfare has become an important issue in geopolitical studies, as it is thought to be proliferating in different parts of the world (another example is the conflict between Russia and Ukraine). The chapter also offers new material on the precarious position of Islamic minority communities in Myanmar, Thailand, and the Philippines. The climate change feature focuses on one of the world's most exposed countries, Myanmar, and one of the new GIAs examines the realm's trade patterns, especially in relation to China. Also, look for new field notes and photographic and video material from Laos, Vietnam, Cambodia, and Myanmar.

Chapter 12, **The Austral Realm**, includes expanded material on matters of immigration and the challenges of climate change in Australia. It also offers new interactive features on the realm's population distribution and Aboriginal geographies. And look for a new map of the world as seen from Australia and New Zealand.

Chapter 13, has been slightly retitled as **The Pacific Realm and Polar Regions**, where we treat the Arctic and Antarctic as individual regions aside from the Pacific Realm. Note a new figure in this chapter that explains the latitudinal positioning of the Arctic and Antarctic Circles and expanded material on the impacts of global warming. The GIA on competition in the Arctic incorporates the latest available information on the extent of melting of the polar ice cap. New material on the Pacific Realm engages with the exposure of low-lying islands to rising sea levels, and it draws comparisons with small and isolated Caribbean islands in terms of obstacles to economic development.

WileyPLUS Instructor Resources

A broad spectrum of electronic resources are available to support instructors:

Test Bank—Prepared by Travis Bradshaw of Liberty University. Includes over 1200 multiple-choice, true/false, and fill-in-the-blank short-answer questions.

PowerPoint Lecture Slides—Prepared by Michele Barnaby of Pittsburg State University. These slides highlight key chapter topics to help reinforce students' grasp of essential concepts.

Clicker Questions—Also prepared by Michele Barnaby of Pittsburg State University. A bank of questions is available for instructors who utilize personal-response-system technology in their courses.

Image Gallery—all figures in the text can be extracted as part of a comprehensive image gallery that is available for instructors to isolate and use at their convenience.

WileyPLUS Student Resources

Students also have access to a variety of tools to help gain a better understanding of course concepts and provide further insight to the material:

Geographic Information Analysis Quizzes—Guided assignments use this powerful new tool to engage students through the use of compelling maps and other spatial data visualizations, and is designed to highlight and reinforce key concepts, issues, and places in world regional geography.

Interactive Graphics—These features present processes, relationships, data, and layers in a dynamic fashion—all controllable by the student. Instead of skipping quickly over static images, students explore Interactive Graphics in ways that allow them to see how the parts relate to the whole, visualize data, build processes, and relate elements to each other.

On-Location Videos—Eight new videos, filmed by the author team during their field work, help to immerse the students and trigger their imagination about distant parts of the world.

Adaptive Practice—Every student has a different learning starting point, and Adaptive Practice provides new opportunities for practice to effectively prepare for class or quizzes and exams. Adaptive Practice creates a personalized study plan that is focused on an individual's needs which makes study time more efficient. Students begin with a quick, chapter-level diagnostic to determine their initial level of understanding, and they can use the dashboard and quick reports to see what topics they know and don't know. From there students are able to focus on the concepts they need to master.

Acknowledgments

Since the publication of the first edition of *The World Today* (and of *Geography: Realms, Regions, and Concepts*, its predecessor and companion text), we have been fortunate to receive advice and assistance from hundreds of people. One of the rewards associated with the publication of a book of this kind is the steady stream of correspondence and other feedback it generates. Geographers, economists, political scientists, education specialists, and others have written us, often with fascinating enclosures. We still make it a point to respond personally to every such email, and our editors have communicated with many of our correspondents as well. Moreover, we have considered every suggestion made, and many who wrote or transmitted their reactions through other channels will see their recommendations in print in this edition.

Student Response

A major part of the correspondence we receive comes from student readers. We would like to take this opportunity to extend our deep appreciation to the several million students around the world who have studied from our world regional geography books. In particular, we thank the students from more than 150 different colleges across the United States who took the time to send us their opinions. Students told us they found the maps and graphics attractive and functional. We have not only enhanced the map program with exhaustive updating but have added a number of new maps to this 8th edition as well as making significant changes in many others. Generally, students have told us that they found the pedagogical devices quite useful. We continue to develop the study aids the students cited as effective: a list of each chapter's key concepts, ideas, and terms (numbered for quick reference in the text itself); a box that summarizes each realm's major geographic qualities; a chapter-end box of thought questions entitled Points to Ponder: and an extensive and still-expanding glossary.

Faculty Feedback

In developing the 8th edition, we are indebted to the following people, for advising us on numerous matters:

Randy Bertolas, *Wayne State College (Nebraska)*
Sean Cannon, *BYU (Idaho)*
Davis Harris, *NC AT&T State University (Greensboro)*
John Havir, *Ashland Community & Technical College, Kentucky*
Carry Holeski, *Southern State Community College (Alabama)*
Joseph Pasztor, *University of Lethbridge (Canada)*
Scott Philips, *Clovis Community College (Fresno, CA)*
Roberg Smith, *Colorado State University*
Teresa York, *Temple University College (Texas)*

In addition, many colleagues and scholars from around the world assisted us with earlier editions, and their contributions continue to grace the pages of this book. Among them are:

James P. Allen, *California State University, Northridge*
Thomas L. Bell, *University of Tennessee–Knoxville*
Stephen S. Birdsall, *University of North Carolina*
Bilal Butt, *University of Wisconsin–Madison*
Leo Dillon, *U.S. Department of State*
J. Douglas Eyre, *University of North Carolina*
Charles Fahrer, *Georgia College and State University*
Fang Yong-Ming, *Shanghai, China*
Edward J. Fernald, *Florida State University*
Jason B. Greenberg, *Sullivan University (Kentucky)*
Margaret Gripshover, *Western Kentucky University*
Chris Hanson, *University of Miami*
Ray Henkel, *Arizona State University*
Richard C. Jones, *University of Texas at San Antonio*
J. Miguel Kanai, *University of Sheffield (UK)*
Gil Latz, *Indiana University–Purdue University, Indianapolis*
Max Lu, *Kansas State University*
Ian MacLachlan, *University of Lethbridge (Alberta)*
Melinda S. Meade, *University of North Carolina*
Henry N. Michael, *Temple University (Pennsylvania)*
David B. Miller, *National Geographic Society*
Imelda Moïse, *University of Miami*
Veronica Mormino, *William Rainey Harper College (Illinois)*
Eugene J. Palka, *U.S. Military Academy (New York)*
Clifton W. Pannell, *University of Georgia*
Bimal K. Paul, *Kansas State University*
Charles Pirtle, Georgetown University (D.C.)
J. R. Victor Prescott, *University of Melbourne (Australia)*
Elvira Maria Restrepo, *George Washington University*
Mika Roinila, *SUNY College at New Paltz*
Rinku Roy Chowdhury, *Clark University (Massachusetts)*
Shouraseni Sen Roy, *University of Miami*
Ira Sheskin, *University of Miami*
John D. Stephens, *University of Washington*
Justin Stoler, *University of Miami*
George E. Stuart, *North Carolina*
Diana Ter-Ghazaryan, *University of Miami*
Canute Vander Meer, *University of Vermont*
Jack Weatherford, *Macalester College*
Henry T. Wright, *University of Michigan*

Faculty members from a large number of North American colleges and universities continue to supply us with vital feedback and much-appreciated advice. Our publishers arranged several feedback sessions, and we are most grateful to the following professors for showing us where the text could be strengthened and made more precise:

Martin Arford, *Saginaw Valley State University (Michigan)*
Donna Arkowski, *Pikes Peak Community College (Colorado)*
Greg Atkinson, *Tarleton State University (Texas)*
Christopher Badurek, *Appalachian State University (North Carolina)*
Denis Bekaert, *Middle Tennessee State University*
Donald J. Berg, *South Dakota State University*
Jill (Alice) Black, *Missouri State University*
Kathleen Braden, *Seattle Pacific University*
David Cochran, *University of Southern Mississippi*
Joseph Cook, *Wayne County Community College (Michigan)*
Deborah Corcoran, *Southwest Missouri State University*
Marcelo Cruz, *University of Wisconsin at Green Bay*
William V. Davidson, *Louisiana State University*

Larry Scott Deaner, *Kansas State University*

Jeff de Grave, *University of Wisconsin–Eau Claire*

Jason Dittmer, *Georgia Southern University*

James Doerner, *University of Northern Colorado*

Steven Driever, *University of Missouri–Kansas City*

Elizabeth Dudley-Murphy, *University of Utah*

Dennis Ehrhardt, *University of Louisiana–Lafayette*

Bryant Evans, *Houston Community College*

William Flynn, *Oklahoma State University*

William Forbes, *Stephen F. Austin State University (Texas)*

Bill Foreman, *Oklahoma City Community College*

Eric Fournier, *Samford University (Alabama)*

Gary A. Fuller, *University of Hawai'i*

Randy Gabrys Alexson, *University of Wisconsin–Superior*

William Garbarino, *Community College of Allegheny County (Pennsylvania)*

Hari Garbharran, *Middle Tennessee State University*

Chad Garick, *Jones County Junior College (Mississippi)*

Jon Goss, *University of Hawai'i*

Debra Graham, *Messiah College (Pennsylvania)*

Sara Harris, *Neosho County Community College (Kansas)*

John Havir, *Ashland Community & Technical College (Kentucky)*

John Hickey, *Inver Hills Community College (Minnesota)*

Shirlena Huang, *National University of Singapore*

Ingrid Johnson, *Towson State University (Maryland)*

Kris Jones, *Saddleback College (California)*

Uwe Kackstaetter, *Front Range Community College, Westminster (Colorado)*

Cub Kahn, *Oregon State University*

Thomas Karwoski, *Anne Arundel Community College (Maryland)*

Robert Kerr, *University of Central Oklahoma*

Eric Keys, *University of Florida*

Jack Kinworthy, *Concordia University–Nebraska*

Marti Klein, *Saddleback College (California)*

Christopher Laingen, *Kansas State University*

Heidi Lannon, *Santa Fe College (Florida)*

Unna Lassiter, *Stephen F. Austin State University (Texas)*

Richard Lisichenko, *Fort Hays State University (Kansas)*

Catherine Lockwood, *Chadron State College (Nebraska)*

George Lonberger, *Georgia Perimeter College*

Claude Major, *Stratford Career Institute (Quebec)*

Steve Matchak, *Salem State College (Massachusetts)*

Chris Mayda, *Eastern Michigan University*

Trina Medley, *Oklahoma City Community College*

Dalton W. Miller, Jr., *Mississippi State University*

Ernando F. Minghine, *Wayne State University (Michigan)*

Tom Mueller, *California University (Pennsylvania)*

Irene Naesse, *Orange Coast College (California)*

Valiant C. Norman, *Lexington Community College (Kentucky)*

Richard Olmo, *University of Guam*

J. L. Pasztor, *Delta College (Michigan)*

Iwona Petruczynik, *Mercyhurst College (Pennsylvania)*

Paul E. Phillips, *Fort Hays State University (Kansas)*

Rosann Poltrone, *Arapahoe Community College (Colorado)*

Jeff Popke, *East Carolina University*

David Privette, *Central Piedmont Community College (North Carolina)*

Joel Quam, *College of DuPage (Illinois)*

Rhonda Reagan, *Blinn College (Texas)*

Paul Rollinson, *Missouri State University*

A. L. Rydant, *Keene State College (New Hampshire)*

Justin Scheidt, *Ferris State University (Michigan)*

Kathleen Schroeder, *Appalachian State University (North Carolina)*

Nancy Shirley, *Southern Connecticut State University*

Dmitrii Sidorov, *California State University–Long Beach*

Dean Sinclair, *Northwestern State University (Louisiana)*

Richard Slease, *Oakland, North Carolina*

Susan Slowey, *Blinn College (Texas)*

Dean B. Stone, *Scott Community College (Iowa)*

Jamie Strickland, *University of North Carolina at Charlotte*

Ruthine Tidwell, *Florida Community College at Jacksonville*

Irina Vakulenko, *Collin County Community College (Texas)*

Cathy Weidman, *Austin, Texas*

Kirk White, *York College of Pennsylvania*

Thomas Whitmore, *University of North Carolina*

Keith Yearman, *College of DuPage (Illinois)*

Laura Zeeman, *Red Rocks Community College (Colorado)*

Yu Zho, *Vassar College (New York)*

Personal Appreciation   We are privileged to work with an excellent team of professionals at John Wiley & Sons: Glenn Wilson, Executive Editor; Carolyn Wells, Marketing Manager; Trish McFadden, Production Editor; Wendy Lai, Senior Designer; Megan Garvin, Assistant Instructional Designer; and Instructional Design Lead Karen Staudinger. We also acknowledge the support and encouragement from others colleagues at Wiley, most notably Vice President and Director Veronica Visentin. Others who played key roles in this process are Katy Gabel of Lumina Datamatics and, last but definitely not least, cartographers Don Larson and Terry Bush of Mapping Specialists, Ltd. in Fitchburg, Wisconsin.

Finally, above all else, we express our gratitude to our families for yet again seeing us through the constantly challenging schedule of creating this latest edition of *The World Today*.

JAN NIJMAN
Atlanta, Georgia

MICHAEL SHIN
Los Angeles, California

PETER O. MULLER
Coral Gables, Florida
January, 2020

Brief Contents

Contents

7 The North African/ Southwest Asian Realm   221

8 The Subsaharan African Realm   265

9   The South Asian Realm   309

10   The East Asian Realm   351

11   The Southeast Asian Realm   **395**

12   The Austral Realm   **433**

13 The Pacific Realm and Polar Regions 451

FIGURE 1-1

© J. Nijman, P.O. Muller and John Wiley & Sons, Inc

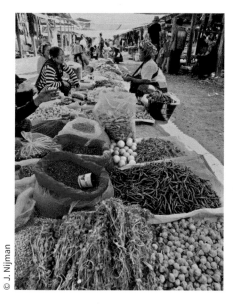

CHAPTER 1

© J. Nijman

Pa'o women are trading vegetables, fruits, spices, and herbs in the small market town of Nyaungshwe in central Myanmar. The *Pa'o* are one of the country's larger ethnic groups (estimated near 2 million), and the women are recognizable by their colorful, turban-like, headdress. Myanmar has only been open to foreign visitors since the early 2010s. It is a country with a highly strategic location in Southeast Asia, a rich culture, and a dizzying array of ethnic groups. You can read more about Myanmar in Chapter 11 of this book.

Introduction: World Regional Geography

IN THIS CHAPTER

- The power of maps
- The spatial order of the world
- Persistent cultural diversity
- Geopolitics and the specter of terrorism
- Technology and geography
- The challenge of sustainable development
- Widening global inequalities
- The threats of global climate change
- Migrants and refugees in an unequal world

CONCEPTS, IDEAS, AND TERMS

Mental maps **[1]**

Cartography **[2]**

Geographic information system (GIS) **[3]**

Geospatial data **[4]**

Spatial perspective **[5]**

Scale **[6]**

Geographic realm **[7]**

Transition zone **[8]**

Region **[9]**

Absolute location **[10]**

Relative location **[11]**

Formal region **[12]**

Spatial system **[13]**

Hinterland **[14]**

Functional region **[15]**

Natural landscape **[16]**

Continental drift **[17]**

Tectonic plate **[18]**

Pacific Ring of Fire **[19]**

Climate **[20]**

Global climate change **[21]**

Anthropocene **[22]**

Holocene **[23]**

Population distribution **[24]**

Urbanization **[25]**

Megacity **[26]**

Cultural landscape **[27]**

State **[28]**

Sovereignty **[29]**

European state model **[30]**

Geopolitics **[31]**

Development **[32]**

Sustainable development **[33]**

Inclusive development **[34]**

Global core **[35]**

Global periphery **[36]**

Core–periphery relationship **[37]**

Core area **[38]**

Uneven development **[39]**

Globalization **[40]**

Migration **[41]**

What are your expectations as you open this book? You have signed up for a course that will take you around the world to try to understand how it functions today. You will discover how interesting and challenging the discipline of geography is. We hope that this course, and this book, will open new vistas, bring new perspectives, raise your awareness of place, and help you navigate our increasingly complex and often daunting world.

You could not have chosen a better time to study geography. The world is changing on many fronts and is doing so ever faster in response to the rapid advancement of communication technologies. The world truly is "shrinking," and the likelihood has never been greater that your professional career will be taking you to places far from home.

Moreover, the United States is still the world's biggest economy, with political influence to match. Intentionally or not, the United States is affecting nations and peoples, lives and livelihoods, from pole to pole. That power confers on Americans a responsibility to learn as much as they can about those places, nations, and livelihoods, so that the decisions of their government representatives are well informed. But in this respect, the United States is no superpower. Geographic literacy is a measure of international comprehension and awareness, and Americans' geographic literacy ranks low. That is a liability, for both the United States and the rest of the world, because such geographic fogginess afflicts not only voters but also the representatives they elect, from local school boards to the federal government.

A World on Maps

Just a casual glance at the pages that follow reveals a difference between this and other textbooks: there are almost as many maps as there are pages. Geography is more closely identified with maps than any other discipline, and we urge you to give as much (or more!) attention to the maps in this book as you do to the text. It is often said that a picture is worth a thousand words, and the same applies to maps. When we write "see Figure XX," we really mean it—and we hope that you will get into the habit. We humans are territorial creatures, and the boundaries that fence off our 200 or so countries reflect our divisive ways. Other kinds of borders—between religions, languages, wealth, and poverty—partition our planet as well. When political and cultural boundaries are at odds, there is nothing like a map to summarize the circumstances. Just look, for example, at the map of the African Transition Zone in Chapter 7: this corridor's turbulence and challenges are steeped in geography.

Maps in Our Minds

All of us carry in our minds maps of the world around us: the apartment building or house we live in, the streets nearby, the way to school or workplace, the general layout of our hometown or city. You will know what lane to use when you turn into a shopping mall, or where to park at the movie theater. You can probably draw from memory a pretty good map of your hometown.

These **mental maps [1]** allow you to navigate your activity space with efficiency, predictability, and safety. When you arrived as a first-year student on a college or university campus, a new mental map would have started forming. At first you needed the navigation tool on your smartphone or a hard-copy map to find your way around, but soon you dispensed with that because your mental map became sufficient. And it will continue to improve as your activity space expands.

If a well-formed mental map is useful for decisions in daily life, then an adequate mental map is surely indispensable when it comes to decision making in the wider world. You can give yourself an interesting test. Choose some part of the world beyond North America in which you have an interest or about which you have a strong opinion—for example, Israel, Iran, Russia, North Korea, or China. On a blank piece of paper, draw a map that reflects your impression of the regional layout there: the country, its neighbors, its internal divisions, major cities, water bodies, and so forth. That is your mental map of the place. Put it away for future reference, and try it again at the end of this course. You will have proof of your improved mental-map inventory.

Mapping Revolutions

The maps in this book show larger and smaller parts of the world in various contexts. Some depict political configurations; others display ethnic, cultural, economic, or environmental features. **Cartography [2]** (the making of maps) has undergone a dramatic and continuing technological revolution. Earth-orbiting satellites equipped with remote-sensing technology (special on-board sensors and imaging instruments) transmit remotely sensed information to computers on the surface, recording the expansion of deserts, the shrinking of glaciers, the depletion of forests, the growth of cities, and myriad other geographic phenomena. Earthbound computers possess ever-expanding capabilities not only to organize this information but also to display it graphically. This allows geographers to develop a **geographic information system** or **GIS [3]**, bringing geospatial data to a monitor that would have taken months to assemble just three decades ago (see **Box 1-1**).

A parallel map revolution is embodied in the astounding proliferation of navigation systems in cars and on mobile phones. Smartphones enable the use of maps on the go, and many of us, in the developed world at least, have already

Box 1-1 Technology & Geography

Geospatial Data: Understanding Our World

It is estimated that approximately 80 percent of all data contain geographic information that can be mapped. Moreover, the technologies that we use daily, from our automobiles to our cell phones to the Internet, are also aware of our location, and sometimes share it with others without our knowledge. Advances in mapping technology that make use of **geospatial data [4]** are driving innovations such as self-driving cars and trucks; enabling smart cities to improve sustainability; and permitting scientists to monitor the causes and consequences of global environmental change.

The National Aeronautics and Space Administration—NASA—builds instruments and launches satellites to collect and analyze data about the Earth and beyond. By coupling locational information to data on wind speed, pollution, and atmospheric conditions, a more complete picture of the Earth's atmospheric dynamics can be obtained. Computer simulations using such big geospatial datasets reveal temporal and regional variations in the amount of carbon dioxide (CO_2) present in the atmosphere, as well as how pollutants and other emissions (e.g., smoke) travel across North America, China, and Europe.

The image below is a single frame taken from a movie that shows how CO_2 emissions travel around the world. The image shows the situation in January. Emissions of CO_2 (increasing intensities shown from yellow to orange to red) are most visible in the winter months of the Northern Hemisphere. The integration of geospatial data, technology, and geography increases our understanding of climate change, and can also inform how we respond to this critical issue.

NASA's Goddard Space Flight Center

Interactive

become dependent on them to traverse cities, find a store or restaurant, even to move around shopping malls. While the personalized maps on our smartphones allow us to navigate more efficiently, the maps in this book are aimed at better understanding the world and its constituent parts.

Satellites—even spy satellites—cannot record everything that occurs on the Earth's surface. Sometimes the borders between ethnic groups or cultural sectors can be discerned by satellites—for example, in changing house types or religious shrines—but this kind of information tends to require on-the-ground verification through field research and reporting. No satellite view of Iraq could show you the distribution of Sunni and Shi'ite Muslim adherents. Many of the boundaries you see on the maps in this book cannot be observed from space because long stretches are not even marked on the ground. So the maps you will be "reading" here have their continued uses: they summarize complex situations and allow us to begin forming durable mental maps of the areas they represent.

There is one other point we should make that is especially important when it comes to world maps: never forget that the world is a sphere, and to project it onto a two-dimensional flat surface must necessarily entail some very significant distortions. Try peeling an orange and flattening the entire peel on a surface—you will have to tear it up and try to stretch it in places to get the job done. Take a look at **Figure 1-1** and note how the Atlantic Ocean and other segments of the planetary surface are interrupted. You can produce a map like this in many different ways, but you will always end up distorting things. When studying world maps, there is nothing like having a globe at hand to remind you of our three-dimensional reality.

Geography's Perspective

Geography is sometimes described as the most interdisciplinary of disciplines. That is a testimonial to geography's historic linkages to many other fields, ranging from geology to economics and from sociology to environmental science. And, in some ways, geography is in the lead because interdisciplinary studies and research are now more prevalent than ever.

Most disciplines focus on one key theme: economics is about money; political science is about power; psychology is about the mind; biology is about life forms. Geography is about the explanation of space on the Earth's surface, also referred to as terrestrial space. Social space (cities, buildings, political boundaries, etc.) as well as natural space (climates, terrain, vegetation, etc.) are not randomly configured. Instead, there generally prevails a particular order, regularity, even predictability about the ways in which space is organized. Sometimes it is the deliberate work of human beings, and sometimes it is the work of nature, but both produce specific patterns. Geographers consider these spatial patterns and processes as not only interesting but also crucial to how we live and how we organize our societies. This **spatial perspective [5]** has defined geography from its beginning.

Environment and Society

There is another connection that binds geography and has done so for centuries: an interest in the relationships between human societies and the natural environment. Geography lies at the intersection of the social and natural sciences and integrates perspectives from both, being the only discipline to do so explicitly. This perspective comes into play frequently: environmental modification is in the news on a daily basis in the form of worldwide climate change, but this current surge of global warming is only the latest phase of endless atmospheric and ecological fluctuation. Geographers are involved in current environmental issues not only by considering climate change in the context of the past, but also by looking carefully at the implications of global warming for human societies. Geographers are acutely aware that

human beings will always be part of nature, no matter how far technology advances.

More generally, think of this relationship between humans and their environment as a two-way street. On one hand, human beings have always had a transformative effect on their natural surroundings, from the burning of forests to the creation of settlements. On the other hand, humans have always been heavily dependent on the natural environment, their individual and collective behaviors very much a product of it. There are so many examples that it is hard to know where to begin or when to end: we eat what nature provides, and traditional diets vary regionally; rivers allow us to navigate and connect with other peoples—or they serve as natural boundaries like the Rio Grande; wars are fought over access to water or seaports; landlocked countries seem to have different global connections than islands; and so forth.

Spatial Patterns

Geographers, therefore, need to be conversant with the location and distribution of salient features on the Earth's surface (**Fig. 1-2**). This includes the natural (physical) world, simplified in Figure 1-1, as well as the human world; our inquiry will view these in temporal (historical) as well as spatial perspective. The spatial structure of cities, the layout of farms and fields, the networks of transportation, the configurations of rivers, the patterns of climate—all these form part of our investigation. As you will find, geography employs a comprehensive spatial vocabulary with meaningful terms such as *area*, *distance*, *direction*, *clustering*, *proximity*, *accessibility*, and many others that we will encounter in the pages ahead. For geographers, some of these terms have more specific definitions than is generally assumed. There is a difference, for example, between *area* (surface) and *region*, between *boundary* and *frontier*, and between *place* and *location*. Sometimes, what at first may seem to be simple ideas turn out to be rather complex concepts.

Scale and Scope

One very prominent term in the geographic vocabulary is **scale [6]**. Whenever a map is created, it represents all or part of the Earth's surface at a certain level of detail. Obviously, Figure 1-1 displays a very low level of detail; it is little more than a general impression of the distribution of land and water as well as lower and higher elevations on our planet's surface. A limited number of prominent features such as the Himalayas and the Sahara are named, but not the Pyrenees Mountains or the Nile Delta. At the bottom of the map you can see that one inch at this scale must represent about 1650 miles of the real world, leaving the cartographer little scope to insert information.

A map such as Figure 1-1 is called a *small-scale* map because the ratio between map distance and real-world distance,

© John Wiley & Sons, Inc.

FIGURE 1-2 **Interactive** **This map is an assemblage of nighttime satellite images that show the dominance of electric lighting in certain parts of the world and darkness in others. What does this spatial distribution suggest about the geography of development?**

expressed as a fraction, is very small at 1:103,750,000. Increase that fraction (i.e., zoom in), and you can represent less territory—but also enhance the amount of detail the map can exhibit. In **Figure 1-3**, note how the fraction increases from the smallest (1:14,600,000) to the largest (1:14,500). San Francisco is just a dot on Map A but an urban area on Map C. Does this mean that world maps like Figure 1-1 are less useful than larger-scale maps? It all depends on the purpose of the map. In this chapter, we often use world maps to show global distributions as we set the stage for the more detailed discussions to follow. In later chapters, the scale tends to become larger as we focus on smaller areas, even on individual countries and cities. But whenever you read a map, be aware of the scale because it is a guide to its utility.

The importance of the scale concept is not confined to maps. Scale plays a fundamental role in geographic research and in the ways we think about geographic problems—scale in terms of *level of analysis*. This is sometimes referred to as *operational scale*, the scale at which social or natural processes operate or play out. For instance, if you want to investigate the geographic concentration of wealth in the United States, you can do so at a range of scales: within a neighborhood, a city, a county, a State,* or at the national level. You choose the scale

that is the most appropriate for your purpose, but it is not always that straightforward. Suppose you had to study patterns of ethnic segregation: what do you think would be the most relevant scale(s)?

In this book, our main purpose is to understand the geography of the world at large and how it works, and so, inevitably, we must deal with broad spatial entities. Our focus is on the world's realms and on the main regions within those realms, and in most cases we will have to forgo analyses at a finer scale. For our purposes, it is the big picture that matters most.

World Geographic Realms

Ours is a globalized, interconnected world, a world of international trade and travel, migration and movement, tourism and television, financial flows and Internet traffic. It is a world that, in some contexts, has taken on the properties of a "global village"—but that village still has its neighborhoods. Their names are Europe, South America, Southeast Asia, and others familiar to us all. Like the neighborhoods of a city or town, these global neighborhoods may not have sharply defined borders, but their persistence, after tens of thousands of years of human dispersal, is beyond doubt. We call such global neighborhoods **geographic realms [7]**. Each of the realms possesses a particular combination of environmental, cultural, and organizational properties.

*Throughout this book we will capitalize State when this term refers to an administrative division of a country: for example, the U.S. State of Ohio or the Australian State of New South Wales. Since this term is also synonymous with country (e.g., the state of Brazil), we use the lower case when referring to such a national state.

EFFECTS OF SCALE

A) Scale 1:14,600,000 (regional)

B) Scale 1:1,800,000 (metropolitan area)

C) Scale 1:225,000 (city)

D) Scale 1:14,500 (neighborhood)

FIGURE 1-3 Interactive

These blended, characteristic qualities are imprinted on the landscape, giving each realm its own traditional attributes and social settings. As we come to understand the human and environmental makeup of these geographic realms, we learn not only where they are located but also why they are located where they are (a central question in geography), how they are constituted, and what their future is likely to be in our fast-changing world. **Figure 1-4**, therefore, forms the overall framework for our investigation in this book.

Criteria for Geographic Realms

The existence and identification of world geographic realms depend on a combination of factors. Our world consists of a highly complex and variable environment of large and small continents, enormous oceans and countless waterways, innumerable islands, diverse habitats and cultures, and intricate political geographies. What constitutes a realm depends on the circumstances, but we can still identify three main sets of criteria that apply to all realms:

- *Physical and Human* Geographic realms are based on sets of spatial criteria. They are the largest units into which the inhabited world can be divided. The criteria on which such a broad regionalization is based include both physical (i.e., natural) and human (or social) yardsticks. For instance, South America is a geographic realm because physically it is a continent and culturally it comprises comparable societies. The realm called South Asia, on the other hand, lies on a Eurasian landmass shared by several other geographic realms; high mountains, wide

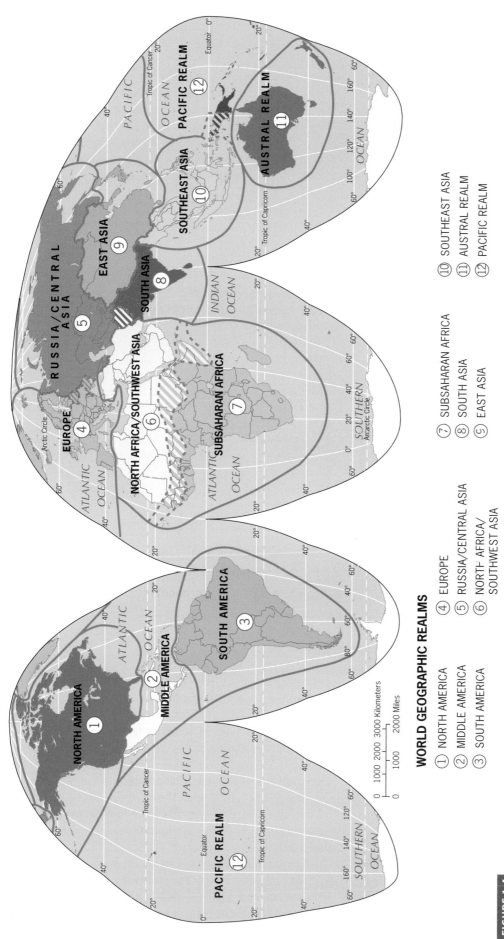

FIGURE 1-4

WORLD GEOGRAPHIC REALMS

- ① NORTH AMERICA
- ② MIDDLE AMERICA
- ③ SOUTH AMERICA
- ④ EUROPE
- ⑤ RUSSIA/CENTRAL ASIA
- ⑥ NORTH AFRICA/ SOUTHWEST ASIA
- ⑦ SUBSAHARAN AFRICA
- ⑧ SOUTH ASIA
- ⑨ EAST ASIA
- ⑩ SOUTHEAST ASIA
- ⑪ AUSTRAL REALM
- ⑫ PACIFIC REALM

© J. Nijman, P. O. Muller, and John Wiley & Sons, Inc.

deserts, and dense forests combine with a distinctive social fabric to create this well-defined realm centered on India.

- **Functional** Geographic realms are the result of the interaction of human societies and natural environments, a *functional* interaction revealed by farms, mines, fishing ports, transport routes, dams, bridges, villages, and countless other features that mark the landscape. According to this criterion, Antarctica is a continent but not a geographic realm.

- **Historical** Geographic realms represent the most comprehensive and encompassing definition of the great clusters of humankind in the world today. China lies at the heart of such a cluster, as does India. Most of Africa constitutes a geographic realm from the southern margin of the Sahara (an Arabic word for desert) to the Cape of Good Hope and from its Atlantic to its Indian Ocean shores. Geographic realms are generally defined on the basis of physical features, but they are also the product of historical evolution.

Figure 1-4 displays the twelve world geographic realms based on these criteria. As we will show in greater detail later, waters, deserts, and mountains as well as cultural and political shifts mark the borders of these realms. We shall discuss the positioning of these boundaries as we examine each realm.

Delineating Realms: Boundaries and Transition Zones

Oceans and seas are the most common natural boundaries of the world's realms, such as the South Atlantic to Subsaharan Africa's west or the North Atlantic to North America's east. But where two geographic realms meet, **transition zones [8]**, not sharp boundaries, often mark their contacts.

We need only remind ourselves of the border zone between the geographic realm in which most of us live, North America, and the adjacent realm of Middle America. The line in Figure 1-4 coincides with the boundary between Mexico and the United States, crosses the Gulf of Mexico, and then separates Florida from Cuba and the Bahamas. But Hispanic influences are strong in North America north of this boundary, and the U.S. economic influence is strong south of it. The line, therefore, represents an ever-changing zone of regional interaction. Again, there are many ties between South Florida and the Bahamas, but the Bahamas resemble a Caribbean more than a North American society. On the other hand, metropolitan Miami has so many ethnic Cuban residents that it is sometimes referred to as the second-largest Cuban city after Havana.

In Africa, the transition zone from Subsaharan to North Africa is so wide and well defined that we have put it on the world map; elsewhere, transition zones tend to be narrower and less easily represented. In the third decade of this century, such countries as Belarus (between Europe and Russia/Central Asia) and Afghanistan (between Southwest Asia and South Asia) lie in inter-realm transition zones. Remember, over much (though not all) of their length, borders between realms are zones of regional change.

Transition zones are fascinating spaces: it is almost as if they rebel against a clear ordering of the world's geography. They remind us that the world is a restless and contested place with shifting boundaries and changing geographic fortunes. They challenge the geographer, and, above all, they underscore just how complex the study of geography is.

Geographic Realms: Dynamic Entities

Had we drawn Figure 1-4 before Columbus made his voyages beginning in 1492 (and assuming we had the relevant geographical knowledge), the map would have looked different: indigenous states and peoples would have determined the boundaries in the Americas; Australia and New Guinea would have constituted a single realm, and New Zealand would have been part of the Pacific Realm. The colonization, Europeanization, and Westernization of the world changed that map dramatically. Since World War II, the world map has been redrawn as a result of decolonization and the rise and then demise of the Cold War. That Cold War division between western and eastern Europe has now given way to far-reaching European integration across that geographic realm. Realms and regions are dynamic entities, and their geographies are always subject to change.

Two Types of Realms

The world's geographic realms can be subdivided into two categories. The first are *monocentric* realms that are dominated by a single major political entity, in terms of territory and/or population. North America (United States), Middle America (Mexico), East Asia (China), South Asia (India), Russia/Central Asia (Russia), and the Austral Realm (Australia) are all monocentric realms. They are, in their entirety, heavily influenced by the presence of that single country. It is as if the realm is organized around them.

The second type of realm is *polycentric* in nature. In these, the appearance, functioning, and organization of the realm are dispersed among a number of more or less equally influential regions or countries. Europe, North Africa/Southwest Asia, Subsaharan Africa, and the Pacific Realm all fall into this category. Polycentric realms can be more volatile in some ways, their development determined by the sum of many different parts.

Two of the world's realms are a bit more difficult to categorize. Southeast Asia is a dynamic realm that contains about a dozen countries, some of them regarded as emerging economies. Arguably, Indonesia is becoming the most influential power, but it would be premature to label this a monocentric realm. The other realm that seems to fall in between is South America. Here it is Brazil that has the biggest population and the largest and most influential economy—but it does not dominate its realm the way the U.S. does.

Of course, some of the dominant powers in the monocentric realms influence events beyond their realm and demonstrate a truly global reach. The United States has dominated world events (though not unchallenged) since World War II, but it soon had to make way for newly emergent powers such as Japan and Germany. Nowadays, China exercises a major influence not just across East and Southeast Asia but around the globe. Our discussion of the various realms will give due consideration to the influence of global trends and outside powers.

Regions Within Realms

The compartmentalization of the world into geographic realms establishes a broad global framework, but for our purposes a more refined level of spatial classification is needed. This brings us to an important organizing concept in geography: the **region [9]**. To establish regions within geographic realms, we need more specific criteria.

Let us use the North American realm to demonstrate the regional idea. When we refer to a part of the United States or Canada (e.g., the South, the Midwest, or the Prairie Provinces), we employ a regional concept—not scientifically but as part of everyday communication. We reveal our perception of local or distant space as well as our mental image of the region we are describing.

But what exactly is, for example, the Midwest? How would you draw this region on the North American map? Regions seem easy to imagine and describe, but they can be difficult to outline with precision on a map. You might suggest that, broadly, the Midwest lies between the Ohio River and the Rocky Mountains. Or you might use the borders of States: certain States are part of this region, others are not. Alternatively, you could use agriculture as the principal criterion: the Midwest is where corn and/or soybeans occupy a certain percentage of the farmland. Each method results in a different delimitation; a Midwest based on physical boundaries is different from a Midwest based on State borders or farm production. Therein lies an important principle: regions are devices that allow us to make spatial generalizations, and they are based on criteria to help us construct them. If you were studying the geography of presidential-election politics, then a Midwest region defined by State boundaries would make sense. If you were studying agricultural distributions, you would need a different definition.

Criteria for Regions

Given these different dimensions of the same region, we can identify properties that all regions have in common:

- **Area** To begin with, all regions have *area*. This observation would seem obvious, but there is more to this idea than meets the eye. Regions may be intellectual constructs, but they are not abstractions: they exist in the real world, and they occupy space on the Earth's surface.

- **Boundaries** It follows that regions have *boundaries*. Occasionally, nature itself draws sharp dividing lines, for instance, along the crest of a mountain range or along a coast. More often, regional boundaries are not self-evident, and we must determine them using criteria that we establish for that purpose. For example, to define a cultural region, we may decide that only areas where more than 50 percent of the population belong to a particular religion or speak a certain language qualify to be part of that region.

- **Location** All regions also possess *location*. Often the name of a region contains a locational clue, as in Amazon Basin or Indochina (a region of Southeast Asia lying between India and China). Geographers refer to the **absolute location [10]** of a place or region by providing the latitudinal and longitudinal extent of the region with respect to the Earth's grid coordinates. A more useful measure is a region's **relative location [11]**, that is, its location with reference to other regions. Again, the names of certain regions reveal aspects of their relative locations, as in *Mainland* Southeast Asia and *Equatorial* Africa.

- **Homogeneity** Many regions are marked by a certain *homogeneity* or sameness. Homogeneity may lie in a region's cultural properties, its physical characteristics, or both. Siberia, a vast region of northeastern Russia, is marked by a sparse human population that resides in widely scattered small settlements of similar form, frigid climates, extensive areas of permafrost (permanently frozen subsoil), and cold-adapted vegetation. This dominant uniformity makes it one of Russia's natural and cultural regions, extending from the Ural Mountains in the west to the Pacific Ocean in the east. When regions display a measurable and often visible internal homogeneity, they are called **formal regions [12]**. But not all formal regions are visibly uniform. For instance, a region may be delimited by the area in which, say, 90 percent or more of the people speak a particular language. This cannot be seen in the landscape, but the region is a reality, and we can use that criterion to draw its boundaries accurately. It, too, is a formal region.

- **Regions as Systems** Other regions are marked not by their internal sameness but by their functional integration—that is, by the way they work. These regions are defined as **spatial systems [13]** and are formed by the areal extent of the activities that define them. Take the case of a large city with its surrounding zone of suburbs, urban-fringe countryside, satellite towns, and farmlands. The city supplies goods and services to this encircling zone, and it buys farm products and other commodities from it. The city is the heart, the *core* of this region, and we call the surrounding zone of interaction the city's **hinterland** [14]. But the city's influence wanes along the outer *periphery* of that hinterland, and there lies the boundary of the functional region of which the city is the focus. A **functional region** [15], therefore, is usually forged by a structured, urban-centered system of interaction. It has a core and a periphery.

Interconnections

Even if we can easily demarcate a particular region and even if its boundaries are sharp, that does not mean it is isolated from other parts of the realm or even the world. All human-geographic regions are more or less interconnected, being linked to other regions. As we shall see, globalization is forging ongoing integration and connections among regions around the world. Trade, migration, education, television, computer linkages, and other interactions sometimes blur regional identities. Interestingly, globalization tends to have a seemingly paradoxical effect: in some ways, regions and places become more alike, more homogeneous (think of certain consumption patterns), but in other respects the contrasts can become stronger (for example, a reassertion of ethnic or religious identities).

The Physical Setting

Natural Landscapes

The landmasses of Planet Earth present a jumble of **natural landscapes** [16] ranging from rugged mountain chains to smooth coastal plains (Fig. 1-1). Certain continents are readily linked with a dominant physical feature—for instance, North America and its Rocky Mountains, South America with its Andes and Amazon Basin, Europe with its Alps, Asia with its Himalaya Mountains and massive river basins, and Africa with its Sahara and Rift Valley complex. Physical features have long influenced human activity and movement. Mountain ranges form barriers to movement but have also channeled the spread of agricultural and technological innovations around them. Large deserts similarly form barriers, as do rivers, although rivers also facilitate accessibility and connectivity between people. River basins in Asia still contain several of the planet's largest population concentrations: the advantages of fertile soils and ample water supplies that first enabled clustered human settlement now sustain hundreds of millions in crowded South and East Asia.

As we study each of the world's geographic realms, we will find that physical landscapes continue to play significant roles in this modern world. That is one reason why the study of world regional geography is so important: it puts the human map in environmental as well as regional perspective (see, for example, **Box 1-2**).

Box 1-2 From the Field Notes …

"On the descent into Tibet's Lhasa Gongga Airport, I have a great view of the Yarlung Zangbo Valley, its braided stream channels gently flowing toward the distant east. The Yarlung Zangbo is the highest major river on Earth, running from the Tibetan Plateau into northeastern India where it joins the mighty Brahmaputra River that continues on to Bangladesh where it empties into the Indian Ocean. It was mid-October and the water levels were low. The landing strip of the airport can be seen in the center-right of the photo, on the south bank. The airport is quite far from Lhasa, the Tibetan capital, located about 62 kilometers (40 mi) to its southwest. Despite major road and tunnel construction, it is still more than an hour's drive. The airport had to be built away from the city and in this widest part of the valley because it allows the easiest landings and takeoffs in this especially rugged terrain. It lies at 3700 meters above sea level (12,100 ft), one of the highest airports in the world."

© J. Nijman

Geology and Natural Hazards

Our planet may be 4.5 billion years old, but it is far from placid. As you read this chapter, Earth tremors are shaking the still-thin crust on which we live, volcanoes are erupting, storms are raging. Even the continents themselves are moving measurably, pulling apart in some areas, colliding in others. Hundreds of thousands of human lives are lost to natural calamities of this sort in almost every decade.

More than a century ago, the German geographer Alfred Wegener used spatial analysis to explain something that is obvious even from a small-scale map like Figure 1-1: the apparent jigsaw-like fit of the landmasses, especially across the South Atlantic Ocean. His theory of **continental drift** [17] held that the landmasses on the map were actually pieces of a supercontinent, ***Pangaea***, that existed hundreds of millions of years ago. Today we know that the continents are "rafts" of relatively light rock that rest on slabs of heavier rock known as **tectonic plates** [18] whose movement is propelled by giant circulation cells in the red-hot magma below (when this molten magma reaches the surface through volcanic vents, it is called lava).

When moving tectonic plates collide, earthquakes and volcanic eruptions result, and the physical landscape is thrown into spectacular relief. **Figure 1-5** shows the outlines of the tectonic plates and the coincidence of earthquakes and volcanic eruptions. The Earth's largest ocean is almost completely encircled by active volcanoes and earthquake epicenters. Appropriately, this geologic frame is known as the **Pacific Ring of Fire** [19]. In comparison, Russia, Europe,

Africa, and Australia are relatively safe; in other realms, the risks are far greater in one sector than in others (western as opposed to eastern North and South America, for instance). As we shall discover, for certain parts of the world the activity mapped in Figure 1-5 presents a clear and present danger. Some of the world's largest cities, such as Tokyo and Mexico City, lie in zones that are highly vulnerable to sudden disaster—as indeed occurred with Japan's huge 2011 earthquake and tsunamis not very far north of Tokyo.

Climate

The prevailing **climate** [20] constitutes a key factor in the geography of realms and regions (in fact, some regions are essentially defined by climate). But climates change: those dominating in certain regions today may not have prevailed there several thousand years ago. Thus any map of climate, including the maps in this chapter, is but a frozen snapshot of our constantly changing world.

Global Climate Change In the 2020s, we are living in an era of **global climate change** [21], particularly natural global warming that has been accelerated by anthropogenic (human-source) causes. Since the Industrial Revolution, we have been emitting gases that have enhanced nature's ***greenhouse effect*** whereby the sun's radiation becomes trapped in the Earth's atmosphere. This is leading to a series of atmospheric changes, especially the overall warming of the planet. One important international organization of experts,

FIGURE 1-5

the Intergovernmental Panel on Climate Change (IPCC), predicts an increase of about 2°C (3.6°F) during the course of the twenty-first century for the world at large, but with significant regional variability (e.g., more at higher latitudes, less at lower latitudes). Precipitation patterns are predicted to become more variable, particularly in regions where they are already seasonal. And without massive intervention through, for example, reductions of CO_2 emissions, the rise in temperature may well exceed 2°C.

This change in mean temperature may seem small but is expected to have significant impacts on global climatic patterns, agricultural zones, and the quality of human lives. With a rise of 2°C, for example, more than half of the world's coral reefs are expected to disappear. The full ramifications are not known, but scenarios are being modeled so that societies can confront the changes that are coming. Leaders of some countries are more skeptical than others, and some have already made greater adjustments than others. One of the most significant consequences of global climate change is that the icecap atop the Arctic Ocean is melting faster than even recent models predicted, with environmental and geopolitical consequences. **Box 1-3** introduces the notion of the **Anthropocene [22]**, a new epoch in the evolution of the Earth in which humans seem to have become, for better or worse, the predominant species affecting the natural environment. Climate change figures prominently in the Anthropocene and we will be returning to this issue in various chapters that follow.

Climate Regions We have just learned how variable climate can be, but in a human lifetime most of us see relatively little evidence of this variability—although that may change as a result of global warming. We talk about the weather (the immediate state of the atmosphere) in a certain place at a particular time, but as a technical term *climate* defines the aggregate, total record of weather conditions at a place, or in a region, over the entire period during which records have been kept. **Figure 1-6** may appear rather complicated, but even a glance at this map shows its utility. Devised long ago by Wladimir Köppen and later modified by Rudolf Geiger, it represents climatic regions through a combination of colors and letter symbols. In the legend, note that the **A** climates (rose, gold, and peach) are equatorial and tropical; the **B** climates (tan, yellow) are dry; the **C** climates (shades of green) are temperate, that is, moderate and neither hot nor cold; the **D** climates (shades of purple) are cold; the **E** climates (blue) are frigid; and the **H** climates (gray) prevail in highlands like the Himalayas/Tibetan Plateau and the Andes.

A good way to get a sense of how this map works is to start with a climate with which you are familiar. If you live in the southeastern United States (the large green **C** area extending from near the Great Lakes to Florida), you would feel at home—climatically—in the green area in southeastern South America, in southeastern Australia, and in southeastern China. If you live in the **B** zones of the U.S. Southwest, either in the

---

Box 1-3 Global Climate Change: The Anthropocene

About 20 years ago, the Dutch geochemist Paul Crutzen coined the term *Anthropocene* that has come to connote the arrival of a new epoch in the evolution of the Earth system. It indicates that the presence of humans and their influence on the planet has reached "geologic magnitude" in terms of climate, land usage, biodiversity shrinkage, and so on. Think about urbanization, deforestation, air pollution, water consumption, threats to the ozone layer, and the diversion or drying up of rivers. The Anthropocene follows the geologic epoch of the **Holocene [23]** which started about 12,000 years ago and, before that, the Pleistocene, which lasted about 2.6 million years. Considering that the Earth is about 4.5 billion years old, geologic time periods unfold at a huge time scale that is hard to reconcile with the identification of a new epoch that is supposed to have arrived merely a few decades ago. But that is precisely why the term *Anthropocene* has resonated: it is not just about science but also a call for action to control what are in some respects devastating human impacts on our planet, including global warming. Indeed, the term is now used widely beyond the discipline of geology, including in geography and other social sciences. Debates concerning the Anthropocene are mostly framed outside geology, particularly in Earth-system science, which focuses on the interactions of human activity and nature along a variety of dimensions. The total Earth system is hugely complex, and it is this complexity—notwithstanding a wide range of consistently disturbing evidence—that is at the root of (sometimes politically motivated) continued contestation of the notion of the Anthropocene.

Source: https://showyourstripes.info/

'Climate stripes' graphic designed by Ed Hawkins, a professor of climate science at Reading University in the U.K. The graphic shows annual mean global temperatures from 1850 (left) to 2018 (right). Lowest temperatures are indicated in dark blue, highest in dark red. Simple but accurate and science-based visualizations have become important in public debates about climate change.

WORLD CLIMATES
After Köppen–Geiger

A HUMID EQUATORIAL CLIMATE

Af	No dry season
Am	Short dry season
Aw	Dry winter

B DRY CLIMATE

BS	Semiarid	h=hot
BW	Arid	k=cold

C HUMID TEMPERATE CLIMATE

Cf	No dry season
Cw	Dry winter
Cs	Dry summer

a=hot summer
b=cool summer
c=short, cool summer
d=very cold winter

D HUMID COLD CLIMATE

Df	No dry season
Dw	Dry winter

E COLD POLAR CLIMATE

E	Tundra and ice

H HIGHLAND CLIMATE

H	Unclassified highlands

FIGURE 1-6 Interactive

tan (Texas) or the yellow (Arizona) area, conditions in much of Australia, southwestern Africa, or the Middle East would be familiar to you. This can even be taken to the city scale. If you enjoy the climate of San Francisco, you will rediscover it in Santiago (Chile), Cape Town (South Africa), Athens (Greece), and Perth (Australia). Details will differ, of course—higher wind incidence or maybe lower overall humidity—but, in general, the similarities will be greater than such differences. And if you live in Canada, you would be prepared for much of Russia and northeastern China.

Even at this relatively small geographic scale, Figure 1-6 can tell us a great deal about entire climate regions. For instance, among the equatorial/tropical (*Af*) climates, the areas with the darkest (rose) shade get the most rain, and it falls year-round, so that this is where tropical rainforests still stand. But in the peach-colored (*Aw*) areas, the rainfall regime is subject to dry seasons, and the vegetation reflects this by its very name: savanna. Much more open space and more widely scattered stands of trees prevail in these tropical grasslands. And a third kind of regime is signified by the *m* after the *A*: for *monsoon*, the annual copious rainy season on which the lives of hundreds of millions of people still depend.

Let us look at the world's climatic regions in greater detail.

Humid Equatorial (A) Climates

The humid equatorial, or tropical, climates are characterized by high temperatures all year and by heavy precipitation. In the *Af* subtype, the rainfall arrives in substantial amounts every month; but in the *Am* areas, the arrival of the annual wet *monsoon* (the Arabic word for "season" [see Chapter 9]) marks a sudden enormous increase in precipitation. The *Af* subtype is named after the vegetation that develops there—the **tropical rainforest**. The *Am* subtype, prevailing in part of peninsular India, in a coastal area of West Africa, and in sections of Southeast Asia, is appropriately referred to as the monsoon climate. A third tropical climate, the **savanna** (*Aw*), has a wider daily and annual temperature range and a more sharply seasonal distribution of rainfall.

Savanna rainfall totals tend to be lower than those in the rainforest zone, and savanna seasonality is often expressed in a "double maximum." Each year produces two periods of increased rainfall separated by pronounced dry spells. In many savanna zones, inhabitants refer to the "long rains" and the "short rains" to identify those seasons; a persistent problem is the unpredictability of the rain's arrival. Savanna soils are not among the most fertile, and when the rains fail hunger looms. Savanna regions are far more densely peopled than rainforest areas, and millions of residents of the savanna subsist on what they cultivate. Rainfall variability is their principal environmental challenge.

Dry (B) Climates

Dry climates are found in both lower and higher latitudes. The difference between the **BW** (true **desert**) and the moister **BS** (semiarid **steppe**) varies but may be taken to lie at about 25 centimeters (10 in) of annual precipitation. Parts of the central Sahara in North Africa receive less than 10 centimeters (4 in) of rainfall. Most of the world's arid areas have an enormous daily temperature range, especially in subtropical deserts (whose soils tend to be thin and poorly developed). In the Sahara, there are recorded instances of a maximum daytime shade temperature of more than 50°C (122°F) followed by a nighttime low of less than 10°C (50°F). But the highest temperature ever recorded on the Earth's surface is not in the Sahara: in 2013, the 56.7°C (134°F) measured in California's Death Valley a century earlier was officially recognized as the hottest.

Humid Temperate (C) Climates

As the map shows, almost all these midlatitude climate zones lie just beyond the Tropics of Cancer and Capricorn (23.5° North and South latitude, respectively). This is the prevailing climate in the southeastern United States from Kentucky to central Florida, on North America's west coast, in most of Europe and the Mediterranean, in southern Brazil and northern Argentina, in coastal South Africa, in eastern Australia, and in eastern China

Contrasting natural landscapes produced by humid-equatorial (**A**) and dry (**B**) climates in the Americas: the tropical-forest-swathed Guiana Highlands of southern Venezuela (left), and the vegetation-free cliffs lining southeastern California's arid Death Valley (right).

Jacques Jangoux/Alamy Stock Photo

Francesco Riccardo Iacomino/Moment/Getty Images

and southern Japan. None of these areas suffers climatic extremes or severity, but the winters can be cold, especially away from water bodies that moderate temperatures. These areas lie midway between the winter-free equatorial climates and the summer-free polar zones. Fertile and productive soils have developed under this regime, as we will note in our discussion of the North American and European realms.

The humid temperate climates range from moist, as along the densely forested coasts of Oregon, Washington, and British Columbia, to relatively dry, as in the so-called *Mediterranean* (dry-summer) areas that include not only coastal southern Europe and northwestern Africa but also the southwestern tips of Australia and Africa, central Chile, and Southern California. In these Mediterranean environments, the scrubby, moisture-preserving vegetation creates a natural landscape different from that of richly green western Europe.

Humid Cold (D) Climates

The humid cold (or "snow") climates may be called the continental climates, for they seem to develop in the interior of large landmasses, as in the heart of Eurasia or North America. No equivalent land areas at similar latitudes are found in the Southern Hemisphere; consequently, no *D* climates occur there. Great annual temperature ranges mark these humid continental climates, and cold winters and relatively cool summers are the rule. In a *Dfa* climate, for instance, the warmest summer month (July) may average as high as 21°C (70°F), but the coldest month (January) might average only 11°C (12°F). Total precipitation, much of it snow, is not high, ranging from about 75 centimeters (30 in) to a steppe-like 25 centimeters (10 in). Compensating for this paucity of precipitation are cool temperatures that inhibit the loss of moisture from evaporation and evapotranspiration (moisture loss to the atmosphere from soils and plants).

Some of the world's most productive soils lie in areas under humid cold climates, including the U.S. Midwest, parts of southwestern Russia and Ukraine, and northeastern China. The winter dormancy (when all water is frozen) and the accumulation of plant debris during the fall balance the soil-forming and enriching processes. The soil differentiates into well-defined, nutrient-rich layers, and substantial organic humus accumulates. Even where the annual precipitation is light, this environment sustains extensive coniferous forests.

Cold Polar (E) and Highland (H) Climates

Cold polar (*E*) climates are differentiated into true icecap conditions, where permanent ice and snow keep vegetation from gaining a foothold, and the tundra, which may have average temperatures above freezing up to four months of the year. Like rainforest, savanna, and steppe, the term *tundra* is vegetative as well as climatic, and the boundary between the *D* and *E* climates in Figure 1-6 corresponds closely to that between the northern coniferous forests and the tundra.

Finally, the climates labeled *H*—the unclassified highlands mapped in gray—resemble the *E* climates. High elevations and the complex topography of major mountain systems often produce near-Arctic climates above the tree line, even in the lowest latitudes such as the equatorial section of the high Andes of South America.

Here is an important qualification to keep in mind when studying Figure 1-6: this map is also a still-picture of a changing scene, a single frame from an unspooling film. Global climate never stops changing, and less than a century from now those changes may compel revisions to this map. If current predictions about rising sea levels turn out to be accurate, we may even have to start redrawing some familiar coastlines.

You will find larger-scale maps of climate in several of the regional chapters that follow, but it is always useful to refer back to this Köppen–Geiger map whenever the historical or economic geography of a region or country is under discussion. This world climatic map reflects agricultural opportunities and limitations as well as climatic regimes, and as such helps explain some enduring patterns of human distribution on our planet. We turn next to this crucial topic.

Realms of Population

Earlier we noted that population numbers by themselves do not define geographic realms or regions. Population distributions, and the functioning society that gives them common ground, are more significant criteria. That is why we can identify one geographic realm (the Austral) with less than 30 million people and another (South Asia) with more than 1.8 billion inhabitants. Neither population numbers nor territorial size alone can delimit a geographic realm. Nevertheless, the map of global population distribution shows some major clusters that are part of certain realms (**Fig. 1-7**).

Before we examine these clusters in some detail, remember that the world's human population now rounds off at 7.8 billion (and growing), confined to the landmasses that constitute less than 30 percent of our planet's surface, much of which is arid desert, inhospitable mountain terrain, or frigid tundra. After thousands of years of slow growth, world population during the nineteenth and twentieth centuries grew at an increasing rate. That rate has recently been slowing down, even imploding in some parts of the world. But consider this: it took about 17 centuries following the birth of Christ for the world to add 250 million people to its numbers; now we are adding 250 million about every three years. While the *rate* of population growth has declined in some parts of the world, in absolute terms the global population continues to grow apace and is expected to reach just under 10 billion by 2050.

This raises the important question as to whether there are limits to the Earth's carrying capacity—will there be enough food to go around? That question has become more and more pressing over the past decade due to rapidly rising food prices resulting from increased demand in China and India, a general dietary shift from grains to meat and vegetables, and the use of agricultural resources for the production of biofuels. The actual increase of population is only part of the problem; our

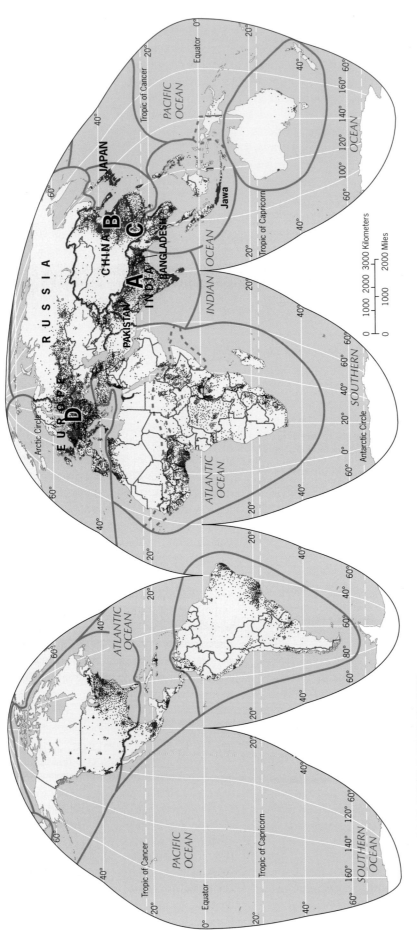

WORLD POPULATION DISTRIBUTION
One dot represents 100,000 people

FIGURE 1-7 Interactive

© J. Nijman, P.O. Muller and John Wiley & Sons, Inc

growing appetite for certain products is another. And we are not just talking about food—think drinking water, fossil fuels, and minerals as well. Therefore, it seems inconceivable that almost 10 billion people by mid-century could be consuming at the rate we do today in the developed world.

Major Population Clusters

One way to present an overview of the location of people on the planet is to create a map of **population distribution [24]** (Fig. 1-7). As you can see in the map's legend, each dot represents 100,000 people, and the clustering of large numbers of them in certain areas, as well as the near-emptiness of others, is readily evident. By the way, there is a technical difference between population distribution and population density, which is another way of showing where people live. Density maps reveal the number of persons per unit area.

- *South Asia*   The *South Asia* population cluster is centered on India and includes its populous neighbors, Pakistan and Bangladesh. This huge agglomeration of humanity focuses on the wide plain of the Ganges River (**A** in Fig. 1-7). South Asia recently became the world's largest population cluster, overtaking East Asia in the early 2010s.

- *East Asia*   The *East Asia* population cluster is centered on eastern China and includes the Pacific-facing Asian coastal zone from the Korean Peninsula to Vietnam. Not long ago, we would have reported this as a dominantly rural, farming population, but rapid economic growth and burgeoning urbanization have altered the picture. In China's interior river basins of the Huang (Yellow) and Chang/Yangzi (**B** and **C** on the map), and in the Sichuan Basin between these two letters, most of the people remain farmers. But the booming cities of coastal, and increasingly interior, China are attracting millions of new inhabitants, and in 2011 the Chinese urban population surpassed the 50 percent milestone.

- *Europe*   The third-ranking population cluster, *Europe*, also lies on the Eurasian landmass but at the opposite end from China. The European cluster, including western Russia, counts more than 700 million inhabitants, which puts it in a class with the two larger Eurasian concentrations—but there the similarity ends. In Europe, the key to the linear, east–west orientation of the axis of population (**D** in Fig. 1-7) is not a fertile river basin but a zone of raw materials for industry. Europe is among the world's most highly urbanized and industrialized realms, its human agglomeration sustained by factories and offices rather than paddies and pastures.

The three world population concentrations just discussed (South Asia, East Asia, and Europe) account for about 4 billion of the world's 7.8 billion people. No other cluster comes close to these numbers. The next-ranking cluster, eastern North America, is only about one-quarter the size of the smallest Eurasian concentrations. As in Europe, the population in this zone is concentrated in major metropolitan complexes; the rural areas are now relatively sparsely settled.

Toward an Urban World

Geographic realms and regions display varying levels of **urbanization [25]**, the proportion of the total population residing in cities and towns (which globally is estimated around 55 percent today). Some regions are urbanizing far more rapidly than others, a phenomenon we will investigate as we examine each realm. For now, it is important to be aware of the distinction between urbanization level and urban growth rates. The *urbanization level* refers to the proportion of a country's or region's total population living in cities (however defined by national governments). The *urban growth rate,* on the other hand, concerns the growth of the urban population over time, usually on an annual basis. So, for example, the U.S. urbanization level is 82 percent, whereas its yearly urban growth rate is around 1 percent; for Tanzania, the urbanization level is far less, 36 percent, but the urban growth rate is a much higher 5.1 percent. It makes sense that urbanization levels and growth rates are often inversely related: once the percentage of the population already living in cities reaches a certain level, urban growth rates will subside, as has been the case for the United States. Alternatively, when urbanization levels are low, there are ample opportunities for rapid urban growth.

Thus the fastest urban growth rates tend to occur in realms and regions in which a large proportion of the population resides in rural areas, as in Subsaharan Africa as well as South and East Asia. It is also worth noting—in these realms where urbanization levels remain low and urban growth rates are high—that a considerable number of **megacities [26]** have emerged over the past three decades (see **Box 1-4**). These huge urban agglomerations are defined as containing populations of greater than 10 million, and in 2018, 33 of them could be found on the world map. Among the latter, only two are in North America (New York and Los Angeles), whereas eight are located in South Asia and no less than six in China alone.

Realms of Culture

Try to imagine yourself in northeast Africa, aboard a riverboat on the White Nile, heading upstream (south) from Khartoum, Sudan. The desert sky is blue; the heat is searing. You pass by villages that look much the same: low, square, or rectangular dwellings, some recently whitewashed, others gray, with flat roofs, wooden doors, and small windows. The minaret of a modest mosque may rise above the houses, and there may be a small central square. There is very little vegetation; here and there a hardy palm tree stands in a courtyard. People on the paths wear

Box 1-4 From The Field Notes ...

"After a chat with the manager of a high-rise hotel in Mumbai, we are allowed access to the roof in early evening to take shots of the city's mushrooming skyline. In the foreground, waterfront-lining Marine Drive, also known as 'The Queen's Necklace,' faces the Arabian Sea as the setting sun shines faintly on some of the low-rise Art Deco apartment blocks that date back to the 1930s. We are at the southern tip of the Mumbai Peninsula, from where the British ruled their South Asian empire during colonial times. Today, the megacity has expanded far beyond this slender peninsula onto the surrounding mainland, and the metropolitan-area population exceeds 23 million. Looking northward, Mumbai's unrelenting densification presents itself as the proverbial concrete jungle."

© J. Nijman

long white or colored robes and headgear. A few goats lie in the shade, and children are playing on the dusty ground.

All of this is part of Sudan's rural **cultural landscape** [27], the distinctive attributes of a society imprinted on its portion of the world's physical stage. Continue your journey southward on the Nile, and you will soon witness a remarkable transition. Quite suddenly, the square, solid-walled, flat-roofed houses of Sudan give way to the round, wattle-and-thatch, conical-roofed dwellings of South Sudan. You may note that clouds have appeared in the sky: it rains more here, and flat roofs will not do. The desert has yielded to green. Vegetation,

natural as well as planted, grows between houses, flanking even the narrow paths. The villages seem less orderly, more varied. You'll see women, often in colorful dresses, carrying water in jars on their heads from a nearby well or the river to their huts. You have traveled from one cultural landscape into another, from Arabized, Islamic Africa to animist/Christian Africa. More broadly, you have crossed the boundary between two geographic realms (see the photo pair below).

Although no geographic realm is marked by only a single cultural landscape, these landscapes can help define both realms and regions. The cultural landscape of the high-rise

PATRICK BAZ/AFP Creative/Getty Images

Cultura Travel/Philip Lee Harvey/Getty Images

Contrasting settlement landscapes in Sudan. To the left, the town of Umm Dawban just north of the capital, Khartoum, which stands in sharp contrast with those in South Sudan (right, the village of Nyaro just south of the Sudan-South Sudan border).

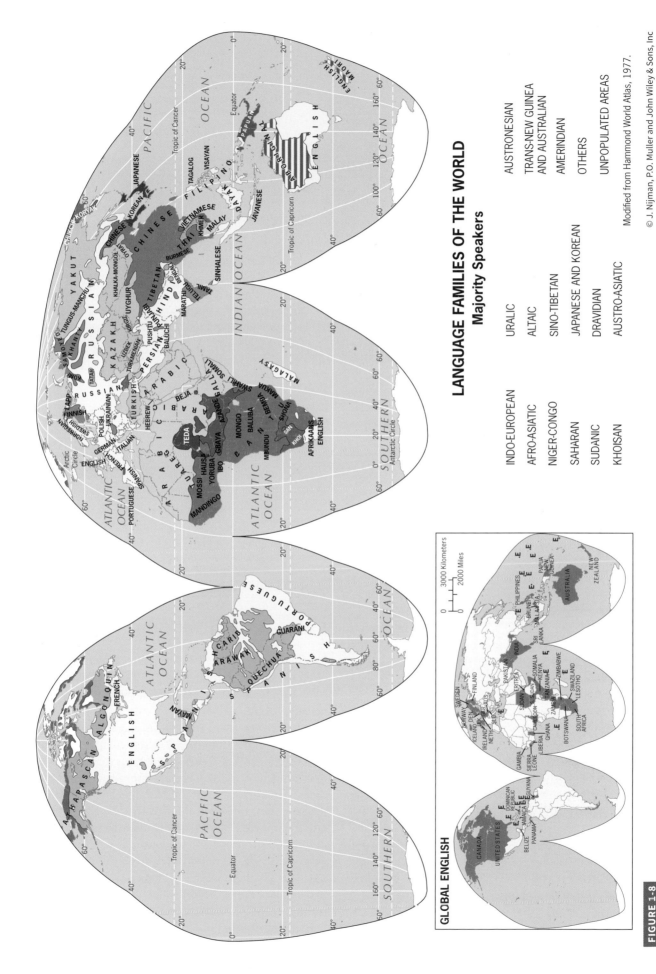

LANGUAGE FAMILIES OF THE WORLD
Majority Speakers

INDO-EUROPEAN

AFRO-ASIATIC

NIGER-CONGO

SAHARAN

SUDANIC

KHOISAN

URALIC

ALTAIC

SINO-TIBETAN

JAPANESE AND KOREAN

DRAVIDIAN

AUSTRO-ASIATIC

AUSTRONESIAN

TRANS-NEW GUINEA
AND AUSTRALIAN

AMERINDIAN

OTHERS

UNPOPULATED AREAS

Modified from Hammond World Atlas, 1977.

© J. Nijman, P.O. Muller and John Wiley & Sons, Inc

GLOBAL ENGLISH

FIGURE 1-8

North American city with its far-reaching suburbs differs from that of Europe; the meticulously organized, terraced paddies of Southeast Asia are unlike anything to be found in the rural cultural landscape of neighboring Australia. Important variations of cultural landscapes also occur within geographic realms, such as between highly urbanized and dominantly rural (and more traditional) areas, and they can be quite useful in the process of delineating several of the world's regions as well.

The Geography of Language

Language is the essence of culture. Most people tend to feel passionately about their mother tongue, especially when they believe it is threatened in some way. In the United States today, the English Only movement reflects many people's fears that the primacy of English as the national language is under threat as a result of immigration. As we will see in later chapters, some governments try to suppress the languages (and thereby the cultures) of minorities in mistaken attempts to enforce national unity, provoking violent reactions.

In fact, many languages emerge, thrive, and die out over time, and linguists estimate that the number of lost languages is in the tens of thousands—a process that continues. One year from the day you read this, about 25 more languages will have become extinct, leaving no trace. Just in North America, more than 100 native languages have been lost over the past half-century. Some major languages of the past, such as Sumerian and Etruscan, have left fragments in later languages. Others, like Sanskrit and Latin, live on in their modern successors. At present, about 6900 languages remain, half of them classified by linguists as endangered ("hot spots" include the Amazon Basin, the Andes to its west, Siberia, and northern Australia). By the end of this century, the bulk of the world's population will be speaking just a few hundred languages, which means that many millions will no longer be able to speak their ancestral mother tongues.

Scholars have tried for many years to unravel the historic roots and branches of the global "language tree," and their debates continue. Geographers trying to map the outcome of this research keep having to modify the pattern, so you should take **Figure 1-8** as a work in progress, not the final product. At minimum, there are some 15 so-called **language families**, groups of languages with a shared but usually distant origin. The most widely distributed language family, the Indo-European (shown in yellow on the map), includes English, French, Spanish, Russian, Persian, and Hindi. This encompasses the languages of European colonizers that were carried and implanted worldwide, English most of all.

Today, English serves as the national or official language of many countries and outposts, and it remains the **lingua franca**

Kumbh Mela (Allahabad, India, February/March 2019) is the world's largest religious festival, hosting 120 million people in a six-week period. It is held every three years, between four alternating holy places, of which Allahabad (now officially renamed Prayagraj), at the confluence of the Ganges and Yamuna rivers, is perhaps the most sacred of all. Hinduism is the world's third-biggest religion, with the preponderance of believers living in India.

(common second language) of government, commerce, and higher education in many multicultural societies (see Fig. 1-8 inset map). In the postcolonial era, English became the chief medium of still another wave of ascendancy now in progress: globalization. But even English may eventually go the way of Latin, morphing into versions you already can hear as you travel to different parts of the world. For example, in Hong Kong, Chinese and English are producing a local "Chinglish" you may hear in the first taxi you enter. In Lagos, Nigeria, where most of the people are culturally and ethnically Yoruba, a language called "Yorlish" is emerging. It is hard to keep up with the constant evolution of language.

Landscapes of Religion

Religion played a crucial role in the emergence of ancient civilizations and has shaped the course of world history. Hinduism, for instance, was one of the earliest religions that helped shape an entire realm (South Asia). Later, Buddhism, Christianity, and Islam emerged as major belief systems, often splitting up into various branches stretching across realms and regions. **Figure 1-9** shows the current distribution of world religions. Our world has become more complicated in recent times, and its patterns of religion are increasingly diffuse and dynamic. But today, we still find that geographic realms are often dominated by a single religion or family of religions: Christianity in Europe and the Americas, Islam in North Africa/Southwest Asia, Hinduism in South Asia, and Buddhism in mainland Southeast Asia. But the boundaries are usually fuzzy and frequently take the form of transition zones (e.g., between North and Middle America, and especially between North and Subsaharan Africa).

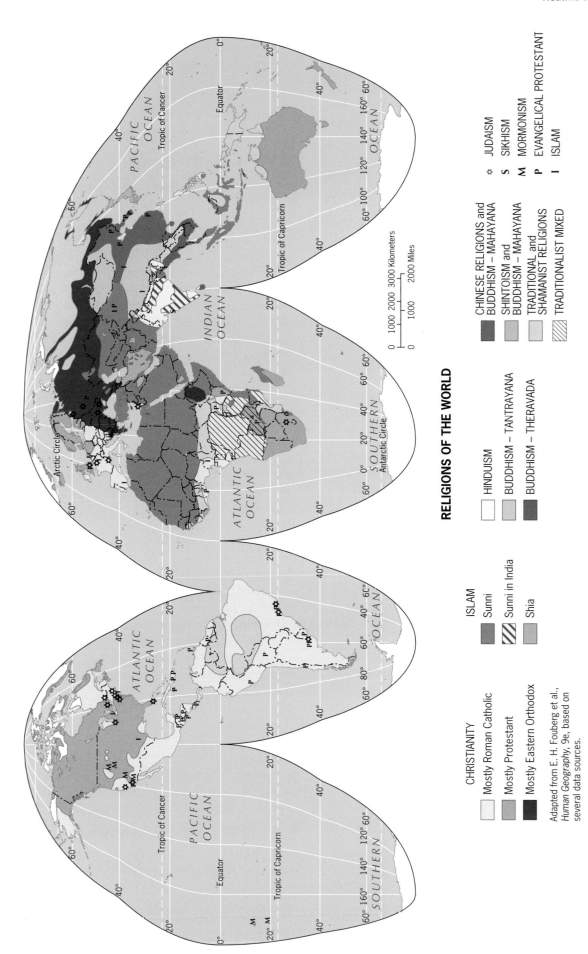

RELIGIONS OF THE WORLD

CHRISTIANITY

- Mostly Roman Catholic
- Mostly Protestant
- Mostly Eastern Orthodox

ISLAM

- Sunni
- Sunni in India
- Shia

- HINDUISM
- BUDDHISM – TANTRAYANA
- BUDDHISM – THERAVADA

- CHINESE RELIGIONS and BUDDHISM – MAHAYANA
- SHINTOISM and BUDDHISM – MAHAYANA
- TRADITIONAL and SHAMANIST RELIGIONS
- TRADITIONALIST MIXED

- ✿ JUDAISM
- S SIKHISM
- M MORMONISM
- P EVANGELICAL PROTESTANT
- I ISLAM

0   1000 2000 3000 Kilometers
0   1000   2000 Miles

Adapted from E. H. Fouberg et al., *Human Geography*, 9e, based on several data sources.

© J. Nijman, P.O. Muller and John Wiley & Sons, Inc

FIGURE 1-9

A World of States

Ours is a world of about 200 countries or **states [28]**. As Figure 1-4 and **Figure 1-10** suggest, geographic realms and regions are mostly assemblages of states, and the borders between them frequently coincide with the boundaries between countries— for example, between North America and Middle America along the U.S.–Mexico border. It is also possible for a realm boundary to cut across a state (Indonesia is one), so that state and realm boundaries do not always concur (see the transition zone between northern and Subsaharan Africa in Fig. 1-4).

The political territorial organization of the world within a system of states is based on the notion of **sovereignty [29]**. It is a concept from international law, which means that the government of a state reigns supreme within its borders. Normally, states recognize each other's sovereignty, but this becomes a matter of contention in times of conflict and war. Think, for example, of the ongoing conflict in which Russia is challenging Ukraine's sovereignty in that country's eastern provinces next to the Russian border. Or think of China's refusal to even consider the possible sovereignty of Taiwan.

The European Origins of the Modern State

In the long course of human history, the modern state is a relatively recent invention, and so is the international system of which it forms the cornerstone. The modern state emerged from other varieties of politico-territorial organization that date back to the earliest complex civilizations. In their studies of ancient history, scholars sometimes use the term *polity* or *proto-state* to indicate the difference. Ever since agricultural surpluses enabled the growth of large and prosperous towns, this was accompanied by the more sophisticated and centralized exercise of power and political organization. From these origins, the earliest states took shape.

It was not until the seventeenth century that European rulers and governments began to negotiate treaties that defined the state in international law. That is why the modern state is quite often described as based on the **European state model [30]**, with definitions of nationality and sovereignty. Often, the model assumed that the state and the nation were ideally conterminous, so that a **nation-state** would enclose an ethnically and culturally homogeneous people within a national boundary. That was never truly the case (even France, the "model of models," had its minorities), and today the ideal state is defined as a clearly and legally defined territory inhabited by a citizenry governed from a capital city by a representative government. As we shall discover in Chapter 5, not even in Europe itself are all governments truly representative. Nonetheless, the European state model has, for better or worse, been adopted throughout the world.

Although the modern state is a longstanding phenomenon, many states experience external and internal pressures and challenges. For instance, European states are challenged "from above" by the European Union (EU) as member-states transfer some of their power to Brussels (the EU's headquarters city), mainly because they believe it will be to their economic advantage. States can also be challenged "from below" by ethnic minorities, such as the Kurds in southeastern Turkey, or by regional secessionist movements such as the Catalonians in northeastern Spain.

Power and Geopolitics

As the term **geopolitics [31]** implies, geography plays a major role in the politics of the state, in the international relations among states within realms and regions, and even in warfare and terrorism. Geography does not, of course, explain everything, but it is certainly an essential dimension in the understanding of world politics. A state's power and influence are related to its size, access to natural resources, location vis-à-vis other states, trading routes, access to waterways, and the like. More than ever today, news reports about conflicts in any part of the world— whether on television, on the websites and hard-copy pages of leading newspapers, or in newsmagazines—are accompanied by maps crucial to the explanation of these clashes.

Even though the role of geography in political affairs can be regarded as fundamental, it certainly does not remain static. Relative geopolitical location itself is subject to change, especially in response to new technologies. For example, states and regions once deemed impregnable because they could not easily be reached by enemies (such as the United States prior to the age of aviation) can swiftly see their invulnerability disappear. It is not hard to imagine how the arrival of long-range missiles, state-of-the-art spy-satellite technology, advances in geospatial and mapping systems, and other high-tech military innovations such as drones can hugely impact a country's strategic planning and positioning.

The territorial functioning and integrity of many states in this century have further been challenged by the emergence, proliferation, and growing sophistication of terrorist groups. Today's most heavily contested zones of conflict are found in the North Africa/Southwest Asian realm (e.g., Syria, Iraq, Yemen) where longstanding sectarian disputes have been complicated by the rise of violent, radical organizations such as the Taliban, Al Qaeda, and ISIS. Increasingly, these threats are also being felt far beyond that realm, in places such as western Europe, Russia, and West Africa. You will learn about the geographic dimensions of these and other key conflicts in the chapters ahead. For now, it is important to realize that the world map of states is far from stable in certain realms and regions, and that states are not the only major geopolitical actors on the global scene.

Geographies of Development

Finally, as we prepare for our study of world regional geography, it is all too clear that realms, regions, and states do not enjoy the same level of prosperity. The field of **economic geography** focuses on spatial aspects of the ways people

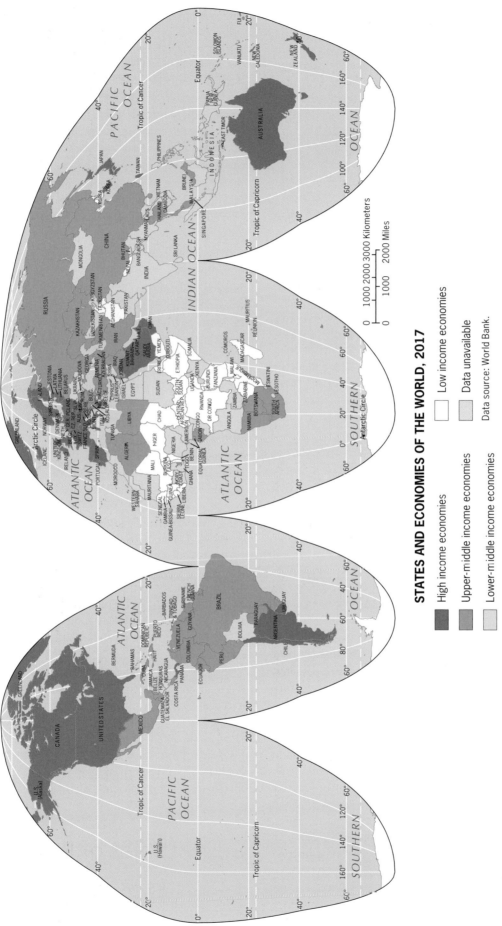

STATES AND ECONOMIES OF THE WORLD, 2017

High income economies
Upper-middle income economies
Lower-middle income economies
Low income economies
Data unavailable

Data source: World Bank.

FIGURE 1-10

make their living, and deals with patterns of production, distribution, and consumption of goods and services. As with all else in this world, these patterns reveal significant variation. Individual states report the nature and value of their imports and exports, farm and factory output, and many other economic data to the United Nations and other international agencies. From such information, economic geographers can measure the comparative well-being of the world's countries. The concept of **development** [32] is used to gauge a population's economic, social, and institutional growth and overall well-being.

As this definition indicates, the meaning of development is not confined to economics but can be expressed, for example, in access to schools or medical care, democratic institutions, environmental stewardship, social inclusiveness, political freedom, and the quality of governance. Development, therefore, is a broad theme that attracts attention not only from economic geographers but also from geographers who take a more comprehensive approach. In recent years, the dual notions of **sustainable development** [33] and **inclusive development** [34] have emerged as particularly important themes. The first focuses on the long-term viability of economic patterns in relation to the physical environment and natural resources—think of pollution, or finite resources like fossil fuels. The second emphasizes the extent of equal economic (and social) opportunities for different population groups, especially minorities and the poor.

Development in Spatial Perspective

Various schemes to group the world's states into economic-geographic categories have come and gone, and others will probably arise in the future. For our purposes, the classification scheme used by the World Bank (one of the agencies that monitor economic conditions across the globe) is the most effective. It sorts countries into four categories based on the success of their economies: (1) high-income, (2) upper-middle income, (3) lower-middle income, and (4) low-income. These categories, when mapped, display important regional patterns (see Fig. 1-10). Compare this map to our global framework (Fig. 1-4), and you can see the role of economic geography in the layout of the world's geographic realms. Also evident are regional contrasts within realms—for instance, between Brazil and its western neighbors, between South Africa and most of the rest of Subsaharan Africa, and between west and east in Europe. Thus the geographic scale at which we consider development is of critical importance.

A Complex, Core–Periphery World

The economic success of human societies on the Earth's surface tends to be concentrated in certain areas while bypassing others. This was true in ancient and more recent historical times, and it is true today. The contemporary world economy—an integrated, international spatial system—is structured within the framework of a dominant **global core** [35] and subordinate **global periphery** [36]. During the nineteenth century, the core more or less coincided with western Europe, controlling as it did vast areas of the world through its empires. In the twentieth century, the core expanded first to North America and then grew to include Japan, Australia, and New Zealand. By the 1980s, Hong Kong, Singapore, Taiwan, and South Korea had become part of the core as well. And as the twenty-first century opened, the newest entrant stepped forward: Pacific-fronting China. The regions constituting this still-evolving global core are mapped in tan in **Figure 1-11**.

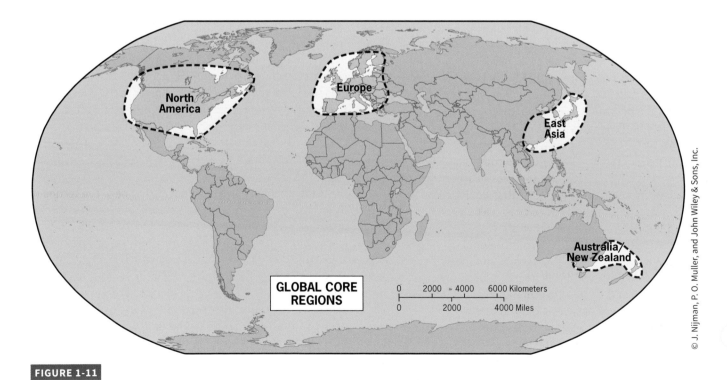

© J. Nijman, P. O. Muller, and John Wiley & Sons, Inc.

FIGURE 1-11

Note that the global core is not a contiguous area but rather consists of four dominant regions: North America, Europe, eastern East Asia, and Australia/New Zealand. Moreover, Singapore (not shown in Fig. 1-11) is often considered to be part of the Asian core region as well. Sometimes, the world outside the core is divided into a *periphery* and *semi-periphery*, where the latter comprises countries that occupy somewhat of a middle position between affluence and poverty, between dominance and dependency. Examples of these semi-periphery countries today would be Brazil, South Africa, and Saudi Arabia.

It is also important to keep in mind that **core–periphery relationships [37]** are not limited to the international scale. Countries as well can and do exhibit such patterns at the national scale. China is a particularly good example: its coastal provinces form the core, while the interior and westernmost reaches of the country comprise China's periphery. Except for a few special cases, all countries contain **core areas [38]**. These national cores are most often anchored by the country's capital and/or largest city: Paris (France), Tokyo (Japan), Buenos Aires (Argentina), and Bangkok (Thailand) are only a few notable examples. Larger countries may have more than one core area, such as Australia with its eastern and western coastal cores and intervening periphery. And uneven development also exists at the local scale, especially marking the complex socioeconomic landscapes of urban areas.

The world continues to exhibit major differences in productivity and well-being. One of the most intriguing economic-spatial outcomes of globalization is that a growing number of countries have accelerated their development—but this growth is often confined to specific city-regions, whereas the rest of the country remains quite poor. Figure 1-11, therefore, is provisional: it conveys a broad impression of the global core, but we realize that it is not really a continuous geographic space (and neither is the global periphery). The core-periphery concept is most useful in thinking about the organization of space as well as the geographies of power and dependence. But it is difficult to map precisely, not least because it operates simultaneously at multiple scales. For instance, the United States constitutes part of the global core; its northeastern seaboard anchors the national core area; and New York's Midtown and Lower Manhattan form the core of its largest city.

We also need to add a cautionary note about spatially aggregated statistics. The concept of development, as measured by data that reflect totals and averages for entire national populations, entails certain pitfalls that we should be aware of. When a state's economy is growing as a whole, and even when it is "booming" relative to other states, this does not automatically mean that every citizen is better off and the income of every worker is rising. Averages have a way of concealing regional variability and local stagnation. In very large states such as India and China, it is useful to assess regional, provincial, and even local economic data to discover to what extent the whole country is sharing in development. In the case of India, we should know that the State of Maharashtra (containing the megacity of Mumbai) is far ahead of most other States when it comes to sharing in the national economy. In China, the

coastal provinces of the Pacific Rim still overshadow those of the west. Thus national-level statistics can conceal as much as they reveal, and this is why *inclusive development* is becoming a leading policy goal around the world.

Uneven Development and Spatial Inequalities

Core–periphery relations imply uneven development across space: economic activity tends to cluster around certain locations or in certain regions, while others lag behind. The way the economy works results in **uneven development [39]** at the global scale (for instance, the North American realm is richer and more economically advanced than the South American realm) but also at finer scales. Thus in the United States there is a growing urban–rural divide, where larger metropolitan areas (though not all) gain most employment, production, and consumption, while rural regions lag behind. And a State like Connecticut is substantially more prosperous than a State like West Virginia. At the same time, inequalities are profound (and have been growing) within most metropolitan areas in the United States. We will return to this issue in the chapter on North America and elsewhere in the book.

Globalization

Globalization [40] is fundamentally a geographical process in which spatial relations—economic, cultural, political—shift to the macro-scale. What this means is that what happens in one place has repercussions in places ever more distant—thereby integrating the entire world into a "global village," linked ever more tightly by recent advances in communications and transport technologies. Globalization now comes into our homes via television, computers, and smartphones; news today has never traveled faster, and sometimes even government leaders turn to the Internet on their personal electronic devices to get the latest reports on international events. Note that some places are (much) more globalized and connected than others. Global cities, such as New York, Paris, or Singapore, are major nodes in worldwide networks of production, consumption, trade, and travel (see **Box 1-5**).

Globalization, however, is not all that new. The second half of the nineteenth century, for instance, also witnessed major advances in the intensification of global interdependence. It was particularly affected by such new technologies as the steamship, the railway, and the telegraph, which subsequently were followed by first-generation motor vehicles and aircraft. But with today's newest technologies, the global village is achieving an unprecedented level of long-distance interconnectivity. Accordingly, geography and our knowledge of the world's realms and regions have become more important than ever before in understanding and navigating this transforming macro-spatial reality.

Box 1-5 From the Field Notes …

"Would you like to float in an 'infinity pool' atop a 57-story skyscraper that provides amazing views? You can in Singapore. I spotted this futuristic-looking tower across the water from downtown and thought it would be worth a visit. It turned out to be the massive Marina Bay Sands Hotel (developed by the Sands Corporation of Las Vegas) that opened for business in 2010. Singapore is one of the best examples of a city propelled into prominence by the forces of globalization. Located strategically on one of the world's busiest shipping lanes (the Strait of Malacca) and possessing a fine harbor, this minuscule city-state was bound to benefit from expanding world trade. It now boasts the busiest transshipment port on Earth, and the city has also become a shopping magnet for elites all across the Southeast Asian realm. Like most 'world-cities,' Singapore is a hub of both production and consumption."

Globalization and Environmental Challenges

Globalization plays out in various spheres, from the environmental to the cultural to the economic. One of today's most pressing environmental issues is global warming, which threatens every corner of our planet. It is clear that the problem must be addressed internationally, yet it is far from easy to agree on mitigation strategies. Some countries are bigger polluters than others, some have more resources than others, and some are more developed than others. How to divide the burdens? Countries from around the world have been meeting for many years at United Nations Climate Change Conferences, the most recent one held in Madrid in late 2019. The main purpose is to reach agreement on collective reductions in the emission of greenhouse gases so as to limit global warming—but progress has been slow. **Figure 1-12** provides some insight into the geography of greenhouse gases, which emanate locally, yet have global consequences for which we all pay (see **Box 1-6**).

Globalization and Population Movements

Culturally, too, the world is coming closer together. This is apparent in the global spreading of new fashions, music, foods, and other innovations. In a more problematic way, it is also expressed in global migration flows (see **Box 1-7** and **Fig. 1-13**). In the past, **migration [41]** was relatively uncommon because most people were rooted in their home environment, where they lived out their entire lives. When residential relocation did occur, it used to be one-way, with people migrating from one place to another and then staying put. But over the past quarter-century, migration flows have intensified, in part

because people increasingly possess better knowledge about opportunities elsewhere. Moreover, it is now easier to travel back and forth, which allows migrants to maintain close ties with their countries of origin. Not surprisingly, as the number of highly mobile **transnational migrants** has risen, these transients have become major players in the spreading of cultures around the world. Examples include Algerians in Paris, Haitians in Montreal, Cubans in Miami, Mexicans in Los Angeles, Indians in Singapore, and Indonesians in Sydney.

Yet it is also important to keep in mind that people's mobility is often constrained because certain parts of this highly uneven world are so much better off than others. High-income countries exert the strongest pull on migrants—but all too often they cannot gain access. Millions of workers aspire to leave the poverty

Children refugees wash themselves in "Camp Moria" on the Greek island of Lesbos, September 25, 2018. Most of the people here have fled wars and violence in Syria, Afghanistan, and Iraq. Their intention is to move elsewhere in Europe, but many have been held in Moria for months under harsh circumstances that include shortages of food, water, and medicines and a lack of proper sanitation.

Box 1-6   Geographic Information Analysis

The Geography of CO_2 Emissions

Visualizations can reveal patterns that are difficult to see when using only tables or graphs. Of particular interest is when something like a bird species, or phenomena such as income or CO_2 emissions per capita are geographically clustered, or concentrated, in nearby locations. The corollary to such local geographic clustering is regional geographic variation across a map. What does the map of per capita CO_2 emissions display – more clusters or more variation? What do you think is related to, or possibly creating, the patterns of clustering and variations that you see on this map? Can you think of another phenomenon that would create similar patterns on a map?

Access your WileyPLUS course to interact with a dynamic version of this map and to engage with online exercises and questions.

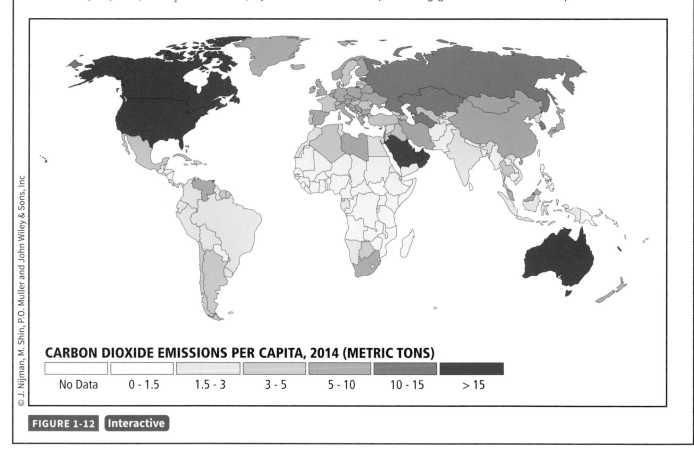

CARBON DIOXIDE EMISSIONS PER CAPITA, 2014 (METRIC TONS)

| No Data | 0 - 1.5 | 1.5 - 3 | 3 - 5 | 5 - 10 | 10 - 15 | > 15 |

© J. Nijman, M. Shin, P.O. Muller and John Wiley & Sons, Inc

FIGURE 1-12   Interactive

of the global periphery and seek a better life somewhere in the core. Trying to get there, every year many of them perish in the waters of the Mediterranean, the Caribbean, and the Atlantic. Others risk their lives at the barriers that encircle the global core as if it were a gated community—from the "security fence" between Mexico and the United States to the razor wire that encircles Spain's tiny outposts on North Africa's shore.

Migration has become a dominant political issue in much of the world, with the governments and populations of more developed countries becoming increasingly opposed to immigration, particularly illegal immigration or the granting of asylum to refugees. Populist politicians across Europe and in the United States have increasingly taken anti-immigrant positions, particularly where it concerns poor, less educated, and ethnically or religiously different migrants or refugees. Interestingly, many of these same countries actually face the growing demographic problems of low birth rates and graying populations, and from a longer-term economic perspective it can be argued that they actually need immigrants to stabilize their populations. Unquestionably, these flows of migrants and refugees underscore the enormous differences in prosperity and well-being across the world's realms and regions.

Regional and Thematic Geographies

At the beginning of this chapter, we introduced a map of the great geographic realms of the world (Fig. 1-4). We then addressed the task of dividing these realms into regions, and we used criteria ranging from physical geography to economic geography. The result is **Figure 1-14**. On this map, note that we display not only the world geographic realms but also the regions into which

Box 1-7 Geographic Information Analysis

Global Migration Flows

The realms and regions of the world are both interconnected and constantly changing. Such linkages and change are clearly apparent when we consider the movement of people from one place to another. To facilitate viewing global migration—or the flow of people among realms—consider the 'chord diagram' below. Each realm is assigned a color (e.g., South America is pink), and is identified on the outside of the circle. The thickness of the 'chords' or lines that connect the realms together correspond to the number of migrants moving to another realm. For the case of South

America, most migration occurs within the realm itself, and is followed by migrants moving to North America and Europe. Which realms are the destinations of global migration? Which realms are the origins of global migration? Can you think of particular source and destination countries through careful analysis of the chord diagram?

Access your WileyPLUS course to interact with a dynamic version of this figure and to engage with online exercises and questions.

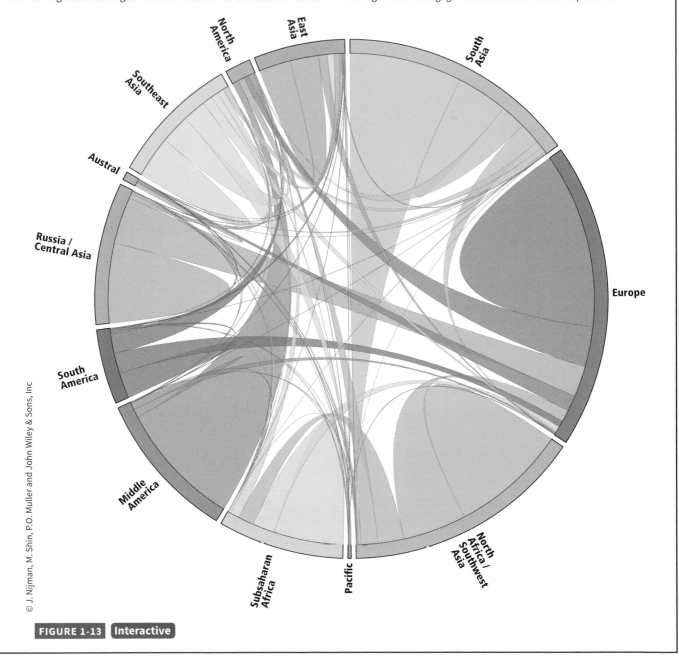

© J. Nijman, M. Shin, P.O. Muller and John Wiley & Sons, Inc

FIGURE 1-13 Interactive

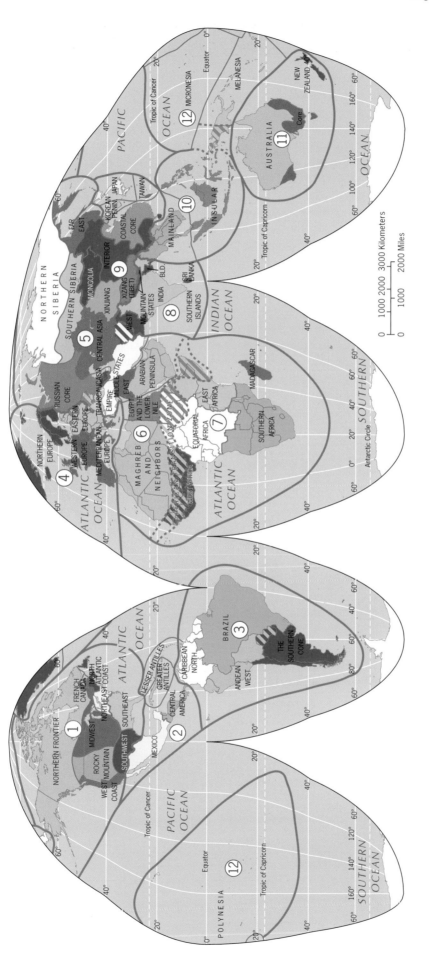

© J. Nijman, P. O. Muller, and John Wiley & Sons, Inc.

**WORLD GEOGRAPHIC REALMS
AND THEIR CONSTITUENT
REGIONS**

① NORTH AMERICA
② MIDDLE AMERICA
③ SOUTH AMERICA

④ EUROPE
⑤ RUSSIA/CENTRAL ASIA
⑥ NORTH AFRICA/
   SOUTHWEST ASIA

⑦ SUBSAHARAN AFRICA
⑧ SOUTH ASIA
⑨ EAST ASIA

⑩ SOUTHEAST ASIA
⑪ AUSTRAL REALM
⑫ PACIFIC REALM

FIGURE 1-14

they subdivide. The numbers in the legend reveal the order in which the realms and regions are discussed, starting with North America (1) and ending with the Pacific Realm (12).

As this introductory chapter has demonstrated, this book does not merely offer a series of descriptions of places and areas. We will combine the study of realms and regions with an overview of geography's ideas and concepts—the notions, generalizations, and basic theories that make the discipline what it is. We continue this approach in the chapters ahead so that you will become better acquainted with the world and with geography as a science. By now you are aware that geography is a wide-ranging, multifaceted discipline. It is often described as a social science, but that is only half the story: geography, in fact, straddles the divide between the social and the physical sciences. Many of the ideas and concepts you will encounter involve the multiple interactions between human societies and natural environments.

Regional geography allows us to view the world in an all-encompassing way. As we have seen, regional geography integrates information from many sources to create an overall image of our divided world. Those sources are not random: they represent topical or ***thematic geography***. Research in the systematic fields of geography makes our world-scale generalizations possible. As **Figure 1-15** shows, these thematic fields connect closely to those of other disciplines. Cultural geography, for example, is allied with anthropology; it

is the spatial perspective that distinguishes cultural geography. Economic geography focuses on the spatial dimensions of economic activity; political geography concentrates on the spatial imprints of political behavior. Other thematic fields include historical, medical, behavioral, environmental, and urban geography. We will also draw on information from biogeography, marine geography, population geography, geomorphology, and climatology (as we did earlier in this chapter).

These thematic fields of geography are so named because their approach focuses primarily on themes, not regions. Take the geographic study of cities, urban geography. Urbanization is a worldwide process, and urban geographers can identify certain human activities that all cities in the world exhibit in one form or another. But cities also display regional properties. The typical American city is quite distinct from, say, the European or the Chinese city. Regional geography, therefore, borrows from the thematic field of urban geography, but it injects this regional perspective. In the following chapters, we call upon these thematic fields to provide you with a better understanding of the world's realms and regions. As a result, you will gain insights into the discipline of geography as well as the regions we investigate.

We will begin our investigation of the world's realms and regions in North America. This is, after all, where most of you were born and raised—and it is from this geographic realm that your expanding worldview has evolved.

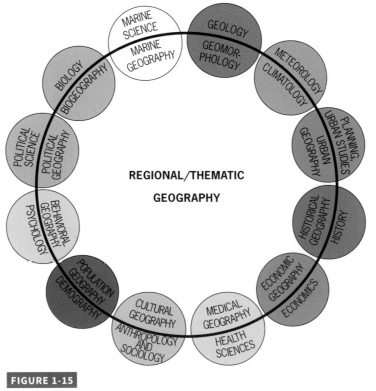

© J. Nijman, P.O. Muller and John Wiley & Sons, Inc

FIGURE 1-15

Points to Ponder

- According to some measures, China in 2015 surpassed the United States and became the world's biggest national economy.
- The global human population at the outset of the 2020s is 7.8 billion and is predicted to reach just under 10 billion by 2050. Will there be enough food and water to sustain everyone?
- Global warming is expected to cause a global rise in sea level somewhere between one and two feet by the end of this century. Which cities are most vulnerable?
- Fast forwarding the network society: the number of smartphones in the world surpassed 2.5 billion by 2020. Most of these devices have map navigation capabilities.

FIGURE 2-1

The Rio Grande Gorge Bridge in northern New Mexico's high desert, about 10 miles from the famous Native American town of Taos. The spring snow still caps the distant peaks of this lower eastern range of the Rockies. The bridge sits about 600 feet above the river which is barely visible towards the left of the image. From here the Rio Grande flows south to El Paso, Texas, and then continues southeastward as the natural boundary between Mexico and the U.S. until it empties into the Gulf of Mexico.

© J. Nijman

The North American Realm

IN THIS CHAPTER

- The U.S.–Mexico borderland
- Global climate change affecting North America
- The geography of U.S. presidential elections
- North America's high-tech regions
- The urban–rural divide
- Massive new oil reserves discovered in Texas
- Immigration and shifting demographics
- China's impact on the realm's Pacific coast

CONCEPTS, IDEAS, AND TERMS

Borderland **[1]**

Transition zone **[2]**

Physiographic region **[3]**

Continentality **[4]**

Rain shadow effect **[5]**

Federation **[6]**

Aquifer **[7]**

Fossil fuel **[8]**

Urban system **[9]**

American Manufacturing Belt **[10]**

Distribution center **[11]**

Intermodal connections **[12]**

Suburbanization **[13]**

Deindustrialization **[14]**

Central business district (CBD) **[15]**

Information economy **[16]**

Sharing economy **[17]**

Gig economy **[18]**

Gentrification **[19]**

GPS (Global Positioning System) **[20]**

Polycentric urban region **[21]**

Urban-rural divide **[22]**

Electoral geography **[23]**

Neighborhood effect **[24]**

Residential geography **[25]**

Sunbelt **[26]**

Migration **[27]**

Melting pot **[28]**

First Nations **[29]**

World-city **[30]**

Technopole **[31]**

Pacific Rim **[32]**

Great Recession **[33]**

Tar sands **[34]**

Boreal forest **[35]**

In our survey of the world's geography, we first turn to the Western Hemisphere. It contains the two great interconnected continents that separate the Atlantic and Pacific oceans and extend, very nearly, from pole to pole, flanked by numerous islands large and small, indented by gulfs and bays of historic and economic import, and endowed with an enormous range of natural resources. Two continents—North and South—form the Americas, but three geographic realms blanket them (North, Middle, and South). In the context of physical (natural) geography, North America—from Canada's Ellesmere Island

in the far north to the Isthmus of Panama in the south—is a continent. In terms of modern human geography, the northern continent is subdivided into the North American and Middle American realms along a transition zone marked by a political as well as a physical boundary between the United States and Mexico (**Fig. 2-1**; **Box 2-1**). In Texas, from the Gulf of Mexico upstream to El Paso/Ciudad Juárez, the Rio Grande forms this border. From El Paso westward to the Pacific Ocean, straight-line boundaries, hardened by fences and walls, separate the North from the Middle. We begin in North America.

Box 2-1 The U.S–Mexico Borderland: Boundary or Transition Zone?

Most of what is now the southwestern United States used to belong to Mexico. In the so-called Treaty of Guadalupe Hidalgo (1848), Mexico ceded the territory to the United States. Today, that history is still reflected here in the presence of substantial numbers of Mexicans, Mexican Americans, and Americans of Mexican descent. Therefore, in terms of cultural geography, the U.S.–Mexico **borderland** [1] should be viewed as a broad **transition zone** [2] marked by a decidedly Mexican imprint on

the landscape on both sides of the actual boundary. This duality extends into other spheres as well. The infusion of Mexican ethnicity profoundly shapes the demography of this quadrant of the country (see **Fig. 2-2**). And in the twenty-first century, the economies of the United States and Mexico are becoming ever more closely linked as their longstanding trade partnership intensifies—fully 80 percent of Mexico's exports were being shipped to its northern neighbor during 2019.

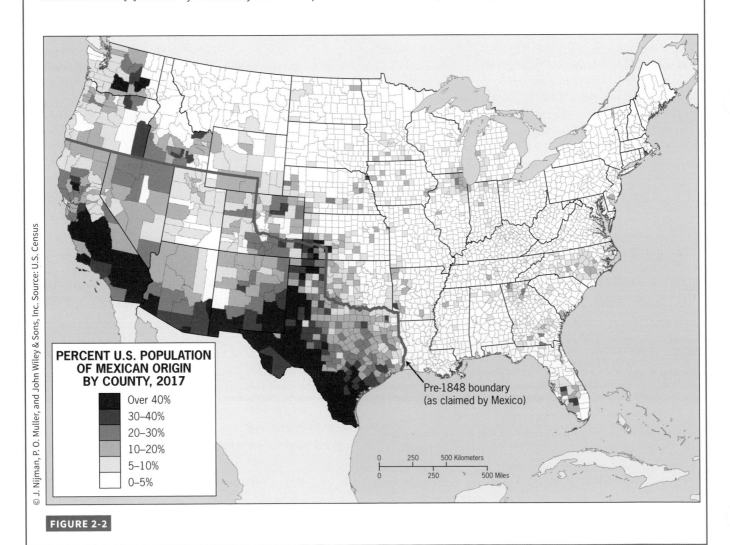

© J. Nijman, P. O. Muller, and John Wiley & Sons, Inc. Source: U.S. Census

PERCENT U.S. POPULATION OF MEXICAN ORIGIN BY COUNTY, 2017

- Over 40%
- 30–40%
- 20–30%
- 10–20%
- 5–10%
- 0–5%

Pre-1848 boundary (as claimed by Mexico)

FIGURE 2-2

Defining the Realm

North America is constituted by two of the world's most highly advanced countries in virtually every measure of social and economic development. Blessed by an almost unlimited range of natural resources and bonded by trade as well as culture, Canada and the United States are locked in a mutually productive embrace that is reflected by economic statistics: in an average recent year, about two-thirds of Canadian exports go to the United States, and just over half of Canada's imports come from its southern neighbor. For the United States, Canada is not quite as dominant, but it is still its leading export market and the second-biggest origin of imports (after China).

This realm is also defined on the basis of broad cultural traits, ranging from urban landscapes and a penchant for mobility to religious beliefs, language, and political persuasions (also see **Box 2-2**). Both countries rank among the world's most highly urbanized: nothing symbolizes the North American city as strongly as the skyscrapered panoramas of New York, Chicago, and Toronto—or the vast, beltway-connected suburban expanses of Los Angeles, Washington-Baltimore, and Houston. North Americans are also the most mobile people on Earth. Each year about one out of every nine individuals in the United States changes residence, a proportion that has leveled off lately but still leads the world.

Despite noteworthy multilingualism, in most areas English is the *lingua franca* of both countries. And notwithstanding all the religions followed by minorities, the great majority of both Canadians and Americans are Christians (whose U.S. total has declined sharply since 2007). Moreover, both states are stable democracies, and both have federal systems of government.

Driving from Michigan across Canada's Ontario to upstate New York does not present a sharp contrast in cultural landscapes. Americans visiting Canadian cities (and vice versa) find themselves in mostly familiar settings. Canadian cities have fewer impoverished neighborhoods, no ethnic ghettos (although low-income ethnic districts do exist), lower crime rates, and, in general, better public transportation; but rush hour in Toronto very much resembles rush hour in Chicago. The boundary between the two states remains porous, even after the events of 9/11 and the recurring alerts regarding terrorism. Americans living near the border shop for lower-priced medicines in Canada; Canadians who can afford it seek medical treatment in the United States. It is telling that the U.S. land border with Canada is nearly twice as long as the one with Mexico and is a lot easier to cross illegally in many places—yet it goes virtually unmentioned in U.S. debates about border security.

None of this is to suggest that Canadians and Americans don't have their differences—they do. Most of these differences may seem subtle to people from outside the realm, but they are important nonetheless, particularly for Canadians who at times feel dominated by Americans whose political or cultural values they do not always share. You can even argue over names: if this entire hemisphere is called the Americas, aren't Canadians, Mexicans, and Brazilians as well as others Americans too?

Box 2-2 Major Geographic Features of North America

1. North America encompasses two of the world's biggest states territorially (Canada is the second-largest in size; the United States is third).

2. Both Canada and the United States have federal administrative structures, and both are plural societies. Canada's pluralism is most strongly expressed in regional bilingualism. In the United States, divisions occur largely along geographic, ethnic, racial, and income lines.

3. North America's population is one of the world's most highly urbanized and mobile. Its slowing natural growth is compensated by continuing influxes of immigration.

4. By world standards, this is a rich realm with a diversified resource base, high average incomes, and high rates of consumption.

5. North America has long been home to one of the world's great manufacturing complexes that generated a highly developed urban system. In more recent times, North America has come to rely increasingly on a highly advanced information economy and high-technology industries.

6. The two countries heavily depend on each other for supplies of critical raw materials (e.g., Canada is a leading source of U.S. energy imports) and have long been each other's primary trading partners.

7. The U.S.–Mexico border separates the North American realm from Middle America; it is a sharp boundary in some respects but in other ways a broadening transition zone with substantial interaction and movement. The debate in the U.S. on the border with Mexico is also a reminder that boundaries are politically constructed to a considerable degree.

8. Continued immigration and high transnational mobility make for an exceptionally diverse multicultural realm—but increasingly, immigration is politically contested.

9. The North American realm is characterized by a growing urban–rural divide: economic growth is concentrated in major metropolitan complexes while many rural areas are struggling, with social and political consequences.

A stretch of the fortified fence along the Mexican border, suggesting a sharp boundary between the United States and Mexico. The wall slashes westward, between Tijuana, Mexico, and the southernmost suburbs of San Diego, California, as it makes its way toward the Pacific shore on the horizon. The cityscape of this part of Tijuana extends almost to the rusty barrier itself, while development on the U.S. side has kept its distance from this well-patrolled segment of the border.

Population Clusters

Although Canada and the United States share many historical, cultural, and economic qualities, they also differ in significant ways and thereby diversify this realm. The United States, somewhat smaller territorially than Canada, occupies the heart of the North American continent and, as a result, encompasses a greater environmental range. The U.S. population is dispersed across most of the country, forming major concentrations along both the (north–south trending) Atlantic and Pacific coasts (**Fig. 2-3**). The overwhelming majority of Canadians, on the other hand, live in an interrupted east–west corridor that extends across southern Canada, mostly within 300 kilometers (200 mi) of the U.S. border. The United States, again unlike Canada, is a fragmented country in that the broad peninsula of Alaska is part of it (offshore Hawai'i, however, belongs to the Pacific Realm).

Figure 2-3 reveals that the great majority of both the U.S. and Canadian population reside to the east of a north–south line drawn down the center of the realm, reflecting historic core-area development in the east as well as the later and still-continuing shift to the west, and, in the United States, also to the south. Certainly this map shows the metropolitan agglomerations that dominate the distribution of North America's population: you can easily identify cities such as Toronto, Chicago, Denver, Dallas–Fort Worth, San Francisco, and Vancouver. Overall, nearly 85 percent of the realm's population is concentrated in urban areas, a higher proportion even than Europe's.

POPULATION DISTRIBUTION: NORTH AMERICA

Population per Square Kilometer

- 0 to 5
- 5 to 10
- 10 to 50
- 50 to 100
- 100 to 500
- Over 500

© J. Nijman, M. Shin, P.O. Muller and John Wiley & Sons, Inc

FIGURE 2-3 Interactive

The population of the United States, which reached 330 million in early 2020, has been growing considerably faster than that of Canada, and the 400 million milestone is forecast to be surpassed in 2051. But this will increasingly depend on immigration because the birth rate in the U.S. (the number of births per 1000 people) has been dropping sharply in recent years. In other words, since natural population growth is stalling, this has to be compensated through immigration—not just to grow the population but also because the presence of generally young immigrants will counteract the aging of the population.

Although Canada's overall growth rate is significantly lower than that of the United States, immigration contributes proportionately even more to this increase than in the United States. With 37.5 million residents, Canada, like the United States, has been relatively open to legal immigration, and as a result both societies exhibit a high degree of cultural diversity in ancestral and traditional backgrounds. Indeed, Canada recognizes two official languages, English and French (the United States does not even designate English as such), and by virtue of its membership in the British Commonwealth, East and South Asians form a larger sector of Canada's population than Asians and Pacific Islanders do in the United States. In contrast, cultural diversity in the United States reflects large Hispanic (18.3 percent), African American (13.4 percent), and Asian (5.9 percent) minorities as well as a comprehensive range of other ethnic backgrounds (U.S. Census Bureau estimates for 2018).

Robust urbanization, substantial immigration, and cultural diversity are defining properties of the human geography of the North American realm. But before going further, we must also become familiar with the physical stage on which the human drama is unfolding.

North America's Physiography

Physiographic Regions

One of the distinguishing properties of the North American landmass, all of which lies on the North American tectonic plate, is the remarkable variation of regional landscapes. So well defined are many of these landscape regions that we use their names in everyday parlance—for example, when we say that we flew over the Rocky Mountains or drove across the Great Plains or hiked in the Appalachians. These landscapes are called **physiographic regions** [3], and nowhere else in the hemisphere is their diversity greater than to the north of the Rio Grande Valley (**Fig. 2-4**).

In the Far West, the Pacific Mountains extend all the way from Southern California through coastal Canada to Alaska. In the western interior, the Rocky Mountains form a continental backbone from central Alaska to New Mexico. Around the Great Lakes, the low-relief landscapes of the Interior Lowlands and the Great Plains to the west are shared by Canada and the United States, and the international boundary even divides the Great Lakes. In the east, the Appalachian Mountains (which, as

the cross-sectional inset shows, are no match for the Rockies) form a corridor of ridges, valleys, and plateaus that represent a familiar topography from Alabama and Georgia to Nova Scotia and Newfoundland. If there is a major physiographic region that belongs to only one of the realm's two countries, it is the Canadian Shield, scoured bare by the Pleistocene glaciers that deposited their pulverized rocks as fertile soil in the U.S. Midwest—that is, in the Interior Lowlands.

Climate

This diversity of landscapes is matched by a variety of climates. Take another look at Figure 1-6, and you can see the lineaments of landscape mirrored in the contrasts of climate. North America may not have it all—there are no areas of true tropical environment in the North American realm to speak of except at the southern tip of Florida—but it does exhibit a great deal of variation. North America contains moist coastal zones and arid interiors, well-watered plains, and even bone-dry deserts. On the world map, **Cf** and **Df** climates are especially good for commercial farming; note how large North America's share of these environments is.

Temperature-wise, this map leaves no doubt: the farther north you go, the colder it gets. And even though coastal areas derive moderation from warmer offshore waters, the rigors of **continentality** [4] (inland climate environment remote from ameliorating maritime influences) set in not far from the coast. Hot summers, frigid winters, and limited precipitation make these higher-latitude continental interiors difficult places to live. Figure 1-7 has much to do with the southern concentration of the Canadian people as well as the lower population densities that mark the realm's interior.

In the west, especially in the U.S., you can see the effects of the Pacific Mountains on inland areas. Moisture-laden air arrives from the ocean on prevailing west-to-east winds, and the mountain wall forces the air upward, cools it, condenses its moisture, and produces rain—the rain for which Seattle and other cities of the Pacific Northwest are (in)famous. By the time the air crosses the mountains and descends on the landward side, most of its moisture has been extracted, and the forests of the ocean-facing side quickly yield to scrub and brush. This **rain shadow effect** [5] extends all the way across the Great Plains; North America does not turn moist again until the Gulf of Mexico sends humid tropical air northward into the eastern interior via the Mississippi Basin.

In a very general way, therefore, and not including the coastal strips along the Pacific, nature divides North America into an arid west and a humid east. Again, the population map reveals more than a hint of this: draw a line approximately from Lake Winnipeg to the mouth of the Rio Grande, and look at the contrast between the (comparatively) humid east and drier west when it comes to population density. Water is a critical part of this story.

Between the Rocky Mountains and the Appalachians, North America lies open to air masses from the frigid north

FIGURE 2-4

and tropical south. In winter, southward-plunging polar fronts send frosty, bone-dry air masses deep into the heart of the realm, turning even cities like Memphis and Atlanta into iceboxes; in summer, hot and humid tropical air surges northward from the Gulf of Mexico, giving Chicago and Toronto a taste of the tropics. Such air masses clash in low-pressure systems along weather fronts loaded with lightning, thunder, and, frequently, dangerously destructive tornadoes. And the summer heat brings an additional threat to the Gulf-Atlantic Coastal Plain (Fig. 2-4): hurricanes capable of inflicting catastrophic devastation on low-lying

areas. These tropical cyclones also prune natural vegetation, replenish groundwater reservoirs, fill natural lakes, and flush coastal channels.

The climate across North America (and elsewhere) is subject to the effects of global warming. Some of this pertains to gradual change, such as sea-level rise or the slowly northward migration of certain species, both flora and fauna, and even agricultural zones (the Corn Belt, for example). Climate change can also make weather patterns more unpredictable and volatile, particularly in terms of rainfall (see **Box 2-3**).

Box 2-3 Global Climate Change in **North America**

Global climate change affects all realms, and North America is no exception. From west to east, its consequences are becoming ever more apparent and problematic. Due to changing weather patterns associated with climate change, California now experiences increasingly dry summers and wetter winters. The former contributes to the frequency and intensity of wildfires while the latter increases the risks of floods and mudslides. Northern and southern parts of the State experienced both in 2018. The photo on the left shows the conflagration unleashed by a massive mid-summer blaze near Yosemite National Park (2018 was the

deadliest year ever for wildfires in the State). The right photo shows a flooded street and sidewalk caused by exceptionally high tides on Florida's Miami Beach in the fall of 2015. Low-lying Miami is one of the U.S. cities most at risk of flooding from rising sea level triggered by global warming. The City of Miami Beach is implementing a continuous, proactive program of adaptation: raising roadways, installing enormous water pumps, bolstering building codes. But far too many other oceanfront municipalities facing similar risks have not even begun to confront this challenge.

USFS Photo/Alamy Stock Photo

Joe Raedle/Getty Images News/Getty Images

Great Lakes and Great Rivers

Two great drainage systems lie between the Rockies and the Appalachians: (1) the five Great Lakes that drain into the St. Lawrence River and the Atlantic Ocean, and (2) the mighty Mississippi–Missouri river system that carries water from a vast interior watershed to the Gulf of Mexico, where the Mississippi forms one of the world's major *deltas*. Both natural systems have been modified by human engineering. In the case of the St. Lawrence Seaway, a series of locks and canals has created a direct shipping route, via the Great Lakes and their outlet, from the Midwest to the Atlantic. The Mississippi has been fortified by (increasingly challenged) artificial levees that, while failing to contain the worst of flooding, have enabled farmers to cultivate some of the most fertile American soils.

Native Americans and European Settlement

When the first Europeans set foot on North American soil, the continent was inhabited by millions of people whose ancestors had reached the Americas from Asia via Alaska, and possibly also across the Pacific, more than 14,000 years earlier (and

perhaps as long as 30,000 years ago). In search of Asia, the Europeans misnamed them Indians. These Native Americans or *First Nations*—as they are now called, respectively, in the United States and Canada—were organized into hundreds of nations with a rich mosaic of languages and a great diversity of cultures.

The eastern nations were the first to bear the brunt of the European invasion. By the end of the eighteenth century, ruthless and land-hungry settlers had driven most of the native peoples living along the Atlantic and Gulf coasts from their homes and lands, initiating a westward push that was to devastate indigenous society. Today, less than 7 million Native Americans remain in the United States, and about one-fourth that number in Canada. In the United States, they are left with only about 3 percent of the national territory in the form of mostly impoverished reservations. Aboriginal peoples today are better off territorially in Canada where they hold titles to large tracts of land, especially in the northern sectors of British Columbia and Quebec (but keep in mind that northern Canada is nearly empty to begin with and that environmental conditions there can be quite harsh).

The realm's current population geography is rooted in the colonial era of the seventeenth and eighteenth centuries, which was dominated by Britain and France. The French sought mainly to organize a lucrative fur-trading network, while the British established settlements along the coast of what is now the eastern seaboard of the United States (the oldest, James-town, was founded in tidewater Virginia more than four centu-ries ago).

Cultural Foundations of the New Americans: Reality and Myth

The modern American creed, if one can be identified, is some-times idealistically characterized as exhibiting an adventurous drive, a liking for things new, a willingness to move, a sense of individualism, an aggressive pursuit of goals and ambition, a need for societal acceptance, and a firm sense of destiny. These qualities are not unique to modern U.S. culture, of course, but in combination they seem to have created a particular and per-vasive mindset that is reflected in many ways in this geographic realm. Generally speaking, these tendencies tend to translate into an intense pursuit of educational and other goals to fulfill high expectations concerning upward socioeconomic mobil-ity (at least among the aspiring classes). Thus, the *American Dream*, whether a genuine prospect or an ideological fabrica-tion, is very much an expression of this set of cultural values, and it is closely entwined with the character of North America as a land of immigrants.

Facilitating these aspirations is *language*. None of these goals would be within reach for so many were it not for the use of English throughout most of the realm. In Europe, by contrast, language inhibits the mobility so routinely practiced by Amer-icans: workers moving from Poland to Ireland, for example, found themselves at a disadvantage in competition with im-migrants from Ghana or Sri Lanka. In North America, a worker from Arkansas would not even consider language to be an issue when applying for a job in Toronto.

Also reflecting the cultural values cited above is the role of *religion*, which sets American (more so than Canadian) culture apart from much of the rest of the Western world. The over-whelming majority of North Americans express a belief in God, and a (declining) majority regularly attend church, in contrast to much of Europe. Two out of three people in the United States say they adhere to a religion, and about half say they attend a religious service once a week (but down sharply since 2000). Religious observance is a virtual litmus test for political lead-ers; no other developed country prints "In God We Trust" on its currency.

Figure 2-5 shows the mosaic of Christian faiths that blan-kets the realm, yet a map at this scale can only suggest the broadest outlines of what is a much more intricate pattern. Protestant denominations are estimated to number in the tens

FIGURE 2-5

of thousands in North America, and no map can capture them all. Southern (and other) Baptists form the majority across the U.S. Southeast from Texas to Virginia, Lutherans in the Upper Midwest and northern Great Plains, Methodists in a belt across the Lower Midwest, and Mormons in the interior West centered on Utah. Roman Catholicism prevails in most of Canada as well as the U.S. Northeast and Southwest, where ethnic Irish and Italian adherents form majorities in the Northeast and Hispanics in the Southwest. Behind these patterns lie histories of proselytism, migration, conflict, and competition, but tolerance of diverse religious (even nonreligious) views and practices remains a hallmark of this realm.

The Federal Map of North America

The two states that constitute North America may have arrived at their administrative frameworks with different objectives and at different rates of speed, but the result is unmistakably similar: their internal political geographies are dominated by straight-line boundaries of administrative convenience (**Fig. 2-6**). In comparatively few places, physical features such as rivers or the crests of mountain ranges mark internal boundaries, but by far the greater length of boundaries is ruler-straight. Even the international boundary between Canada and the United States west of the Great Lakes mostly coincides with a parallel of latitude, 49°N.

The reasons are all too obvious: the boundary framework was laid out before, in some areas long before, significant white settlement occurred. There was something arbitrary about these delineations, but by delimiting the internal administrative units early and clearly, governments precluded later disputes over territory and resources. In any case, neither Canada nor the United States was to become a *unitary (centralized) state*: both countries evolved into **federations [6].** In a *federal state*, power is shared between a country's central government and its highest-ranking political subdivisions. The Canadians call these first-order subdivisions *provinces*, whereas in the United States they are, as the country's name implies, *States*.

It is no trivial matter to take a careful look at some of the provinces and States that compartmentalize this realm because some are of greater consequence than others. We have already noted that Canada's pair of official languages includes French as well as English. Although French-speakers are in the minority, they are heavily concentrated in and around the province of Quebec, which occupies a unique position in federal Canada. In many ways, the pivotal Canadian province is Ontario, containing the country's largest and most globalized city (Toronto) as well as the capital (Ottawa) on its eastern river border with Quebec. In the United States, key States in the political and economic geography of the federation (indeed, the four most populated, in descending order) are California facing the Pacific, Texas bordering Mexico, Florida pointing toward the Caribbean and South America beyond, and New York with its window on the northern Atlantic Ocean.

The rectangular layout of the realm seems to symbolize North America's modernity and its stability, whereas the federal structure itself (with considerable autonomy at the State or provincial level) reflects the culture of freedom and independence. If federalism works within these highly advanced economies, why shouldn't the rest of the world emulate it? Federalism does indeed have its assets, but it also carries some significant liabilities.

The Distribution of Natural Resources

One of these liabilities has to do with the variable allotment of natural resources among the States and provinces. In Canada, Alberta is favored with massive energy reserves (and even more in the offing), but its provincial government is not always eager to see the federal administration take some of that wealth to assist less affluent provinces. In the United States, when oil prices are high, Texas benefits while California's budget suffers. So it is important to compare the map of natural resources (**Fig. 2-7**) with the map of States and provinces. Keep in mind, though, that the world market price of raw materials, including oil, is notoriously volatile: in early 2016, oil prices were almost one-third of what they were only two years earlier, with a massive impact on producing and consuming regions. In late 2019, the price for a barrel of crude oil was around $59, still far below the highs of 2012–2014.

Water certainly is a natural resource, and North America as a realm is comparatively well supplied with it despite concerns over long-term prospects in the American Southwest and the Far West, where some States depend on sources in other States. Another concern focuses on lowering the water tables in some of North America's most crucial **aquifers [7]** (underground water reserves) in which a combination of overuse and decreasing replenishment foretells the rise of supply problems.

North America is endowed with abundant reserves of *minerals* that are mainly found in three zones: the Canadian Shield north of the Great Lakes, the Appalachian Mountains in the east, and the mountain ranges of the west. The Canadian Shield contains substantial iron ore, nickel, copper, gold, uranium, and diamonds. The Appalachians yield lead, zinc, and iron ore. And the western mountain zone has significant deposits of copper, lead, zinc, molybdenum, uranium, silver, and gold.

In terms of **fossil fuel [8]** energy resources (oil, natural gas, and coal), North America is also quite well endowed. Nonetheless, the voracious demand of the United States, the highest in the world, cannot be met by domestic supplies alone and generates sizeable imports. Figure 2-7 displays the distribution of oil, natural gas, and coal deposits.

The leading *oil*-production areas lie (1) along and offshore from the Gulf Coast, where the floor of the Gulf of Mexico is yielding a growing share of the output; (2) in the Midcontinent

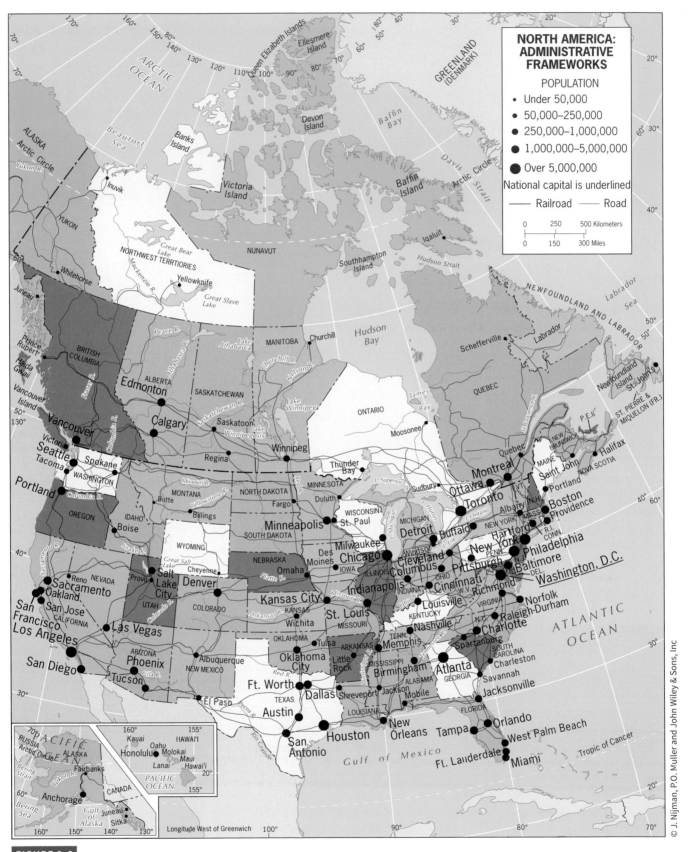

NORTH AMERICA: ADMINISTRATIVE FRAMEWORKS

POPULATION
- Under 50,000
- 50,000–250,000
- 250,000–1,000,000
- 1,000,000–5,000,000
- Over 5,000,000

National capital is underlined

— Railroad — Road

0 250 500 Kilometers
0 150 300 Miles

FIGURE 2-6

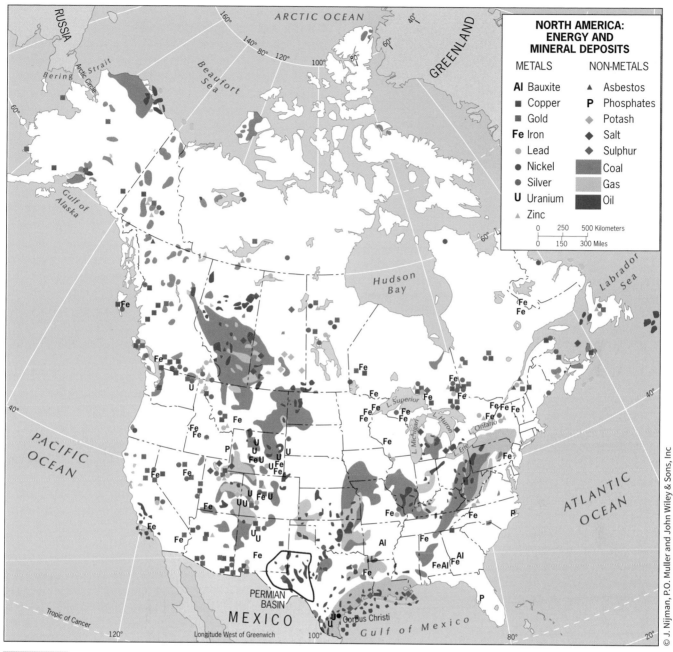

NORTH AMERICA: ENERGY AND MINERAL DEPOSITS

METALS

Al Bauxite
■ Copper
■ Gold
Fe Iron
● Lead
● Nickel
● Silver
U Uranium
▲ Zinc

NON-METALS

▲ Asbestos
P Phosphates
◆ Potash
◆ Salt
◆ Sulphur
Coal
Gas
Oil

0   250   500 Kilometers
0   150   300 Miles

FIGURE 2-7

District, from western Texas to eastern Kansas; and (3) along Alaska's North Slope facing the Arctic Ocean. An important development is taking place in Canada's northeastern Alberta, where oil is being drawn from deposits of *tar sands* in the vicinity of Fort McMurray.

In 2018, the U.S. Geological Survey announced the discovery of huge additional reserves in the Permian Basin, the oil- and gas-rich zone of western Texas and eastern New Mexico that is part of the Midcontinent District (Fig. 2-7). The Permian Basin covers an area about 250 miles in diameter that contains more than 7000 fields. The U.S.G.S. proclaimed that the new discoveries effectively doubled total U.S. oil reserves and increased gas reserves by 63 percent. In recent years, production in the Permian Basin has fueled

oil exports, yet it was only in 2015 that the federal government lifted restrictions on these exports for national security reasons. Now, however, the Permian Basin's output has risen so rapidly that there are not enough refineries to process the profusion of crude oil; moreover, Texas's Gulf Coast ports are increasingly hard-pressed to handle this export boom. Not surprisingly, crash programs are under way to expand both refining and port facilities.

The distribution of *natural gas* reserves resembles that of oilfields because petroleum and natural gas tend to form in similar geological formations. Gas output, too, has risen significantly in recent years, with the discovery of new reserves and as natural gas has become the fuel of choice for electricity generation in North America.

Wind farms have been spreading to many locales in North America as the clean-energy campaign to generate electricity from sources other than fossil fuels accelerates. This wind farm is located in the San Gorgonio Pass (near Palm Springs), one of the windiest places in Southern California. The tallest of the 2700 turbines here reach 53 meters (160 ft). This facility generates enough electricity to power about 300,000 homes.

The *coal* reserves of North America, perhaps the largest on the planet, are found in Appalachia, beneath the Great Plains of the U.S. and Canada, and in the southern Midwest, among other places. These reserves guarantee an adequate supply for centuries to come, although coal has become a less desirable fuel owing to its polluting emissions (especially carbon dioxide) that contribute to global warming.

Given the environmental challenges accompanying oil production and the geopolitical intricacies involved in the global oil trade (see Chapter 7) you might expect the use of nuclear energy to be more widespread, especially because it is relatively cheap to produce. The United States has just under 60 nuclear power plants that provide roughly one-fifth of all electrical energy (Canada has five plants that produce about 15 percent of its electricity). But ever since the near-calamitous radiation leak at Pennsylvania's Three Mile Island in 1979, fear of nuclear accidents has been pervasive. In fact, the federal government did not approve new nuclear plants again until 2012, and the first such facility opened in Tennessee in 2016.

Urbanization and the Spatial Economy

Industrial Cities

When the Industrial Revolution (which originated in England around 1800) crossed the Atlantic to America in the 1870s, it took hold so successfully and advanced so robustly that only 50 years later North America was surpassing Europe as the world's mightiest industrial stronghold. Industrialization

progressed in tandem with urbanization as manufacturing plants in need of large labor supplies were built in and near cities. This, in turn, propelled rural-urban migration (including the migration of African Americans from the South to northern cities during the first half of the twentieth century) and the growth of individual cities. At a broader scale, a *system* of new cities emerged that specialized in the collection and processing of raw materials and the distribution of manufactured products. This **urban system [9]** was interconnected through a steadily expanding network of railroads, themselves a product of the Industrial Revolution.

As new technologies and innovations emerged, and as specializations such as Detroit's automobile industry strengthened, the **American Manufacturing Belt [10]** (which included southern Ontario) swiftly evolved into the cornerstone of the North American core (**Fig. 2-8**). It contained the majority of the realm's industrial activity and leading cities, including New York, Chicago, Toronto, and Pittsburgh (the "steel city"), and in the first half of the twentieth century it became the world's most productive and important industrial region.

In the course of interlinked urbanization and industrialization, the spatial economy of North America underwent profound changes. The *primary* sector, involving the extraction of raw materials from nature (agriculture, mining, fishing), was rapidly mechanized, and the workforce in this sector shrank considerably. Employment in the *secondary* sector, using the input of raw materials and manufacturing them into finished products, grew rapidly. The *tertiary* sector, entailing all kinds of services to support production and consumption (banking, retail, transport), expanded as well. Both the secondary and tertiary sectors were overwhelmingly concentrated in urban areas.

Realm of Railroads

The United States first became a single, integrated, continental-scale economy when long-distance railroads were built during the middle decades of the nineteenth century. This rail network was launched in the Northeast and soon expanded into and across the Midwest, the realm's industrial heartland (**Fig. 2-9**). North America's flourishing network grew denser in the first half of the twentieth century, but then rapidly declined between 1960 and 1980. Railroading subsequently entered a new phase of development in the 1980s, marked by government deregulation and declining shipping costs. In the last few decades, those costs dropped by more than 50 percent, making the movement of goods by rail in the United States and Canada among the cheapest in the world, half of what it costs in Europe or in Japan. Presently, rail freight in North America accounts for more than one-third of the entire freight market (that also includes trucking, shipping, and air transport), which ranks it among the highest in the world's wealthy countries.

This far-flung rail infrastructure requires constant maintenance and frequent updating (which rail companies mostly pay for). One example is the new Alameda Corridor, a major

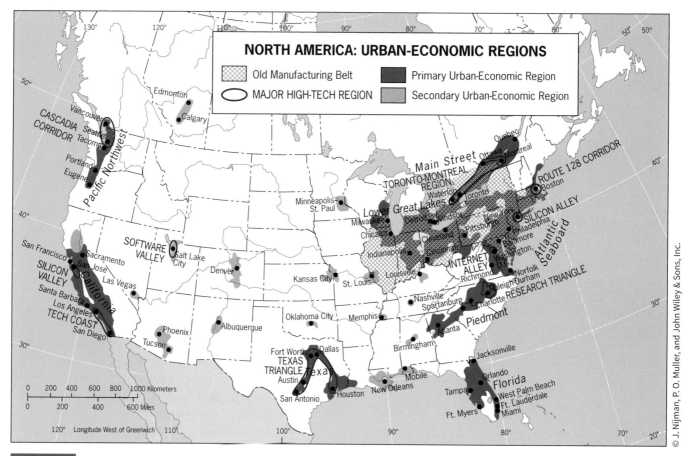

NORTH AMERICA: URBAN-ECONOMIC REGIONS

Old Manufacturing Belt   Primary Urban-Economic Region
MAJOR HIGH-TECH REGION   Secondary Urban-Economic Region

FIGURE 2-8

project involving a 20-mile-long rail cargo express line that links the ports of Long Beach and Los Angeles to the Southern California mainlines near downtown Los Angeles (which from there connect to the national rail system). It consists of a series of bridges and underpasses that allow unimpeded transport, separated from all other traffic. The opening of the enlarged Panama Canal in 2016, which translates into a major increase of trans-Pacific cargo headed for U.S. ports on the Gulf of Mexico and the Atlantic seaboard, requiring a commensurate increase in overland transfer by rail to get these goods to their destinations further inland (see **Box 2-4** and **Fig. 2-10**).

It should be noted that the North American rail system differs in two key respects from those in Europe and East Asia. First, it functions mostly as a freight-rail system, with passenger-rail transport at a minimum. And second, North America lags far behind China, Japan, South Korea, and Western Europe when it comes to high-speed rail, which plays an important role in the regional integration of major urban areas.

Deindustrialization and Suburbanization

Cities continued to evolve, along with the structure of the economy and the development of new transportation and communication technologies. The mass introduction of automobiles and the accompanying construction of a large-scale highway system, especially after World War II, had two major effects on cities. First, it resulted in much higher interconnectedness and mobility among cities. Second, it propelled a process of **suburbanization [13]**—the transformation from compact city to widely dispersed metropolis through the evolution of residential suburbs, often with their own clusters of office businesses and industries, sports and entertainment complexes, and myriad other amenities. As the newly urbanized suburbs increasingly captured major economic activities, many central cities saw their comparative status diminish.

Importantly, this transformation coincided with the **deindustrialization [14]** that devastated American cities in the 1970s, manifested in the loss of manufacturing jobs as a result of automation and the relocation of production to countries with lower wages. The suburban outer cities now became the primary destination of intrametropolitan migrants and of economic activity and employment in the service sector. Meanwhile, the **central business district (CBD) [15]** of the central city lost its role as the dominant metropolitan core, with many of its functions now reduced to serving the less affluent populations of the central city's close-in neighborhoods. The fate of Detroit (see photo page 50) is often cited as exemplifying this sequence of events, but to a certain degree all major U.S. cities suffered.

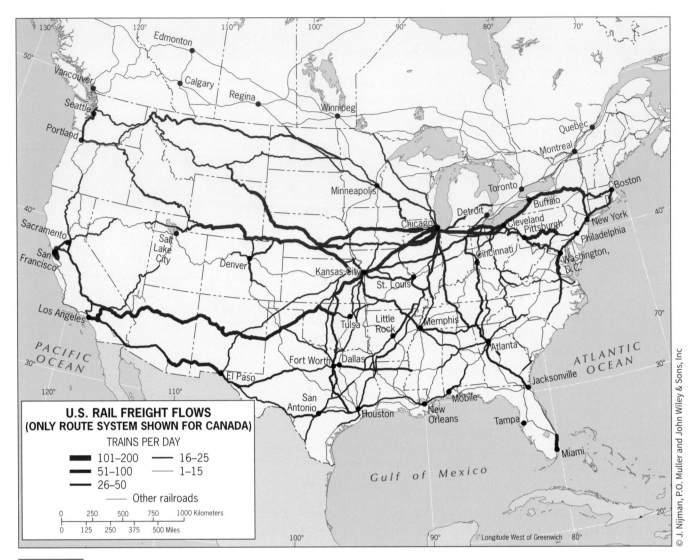

FIGURE 2-9

The Information Economy and City Regions

Since the 1980s, a number of North American cities have rebounded from the destructive effects of deindustrialization and unemployment in conjunction with the arrival of the **information economy [16]**. Many northern cities had lost population and experienced "hollowing out" for nearly two decades, but for a good number of them the tide had turned by the end of the twentieth century. Employment in the tertiary sector started to grow rapidly, especially in high-technology, producer services (such as consulting, advertising, and accounting), finance, research and development, and the like. The geography of the information economy is quite different from that of the old manufacturing industries and is still not fully understood. Some of it concentrates in established CBDs such as Midtown Manhattan; other economic activities center around important hubs of the Internet infrastructure to benefit from maximum bandwidth (e.g., North Dallas); and still other activities cluster in suburban areas containing large pools of highly skilled workers, notably suburban San Francisco's Silicon Valley, metropolitan Seattle, and North Carolina's Research Triangle. The largest and best known high-technology clusters in North America are mapped in Figure 2-8.

This new economy is heavily reliant on networking, both virtual and in real space, and therefore gravitates to high-density urban centers. Think **sharing economy [17]**, or **gig economy [18]**, both hallmarks of an emergent economy in which technology is as fundamental as proximity, in which coffee shops (as meeting places) are as vital as the next-job-finding app. This is producing what some have called the *great inversion*, the return of people to older urban centers. The revival of such communities has become a trend, often led by younger population groups. Yet, at the same time, the revitalization and upgrading—or **gentrification [19]**—of crumbling downtown-area neighborhoods drives up real estate values and property taxes, which forces lower-income, longtime residents from their homes.

Box 2-4 The Port of Savannah as a Global Gateway

The 2016 opening of the enlarged Panama Canal (see Chapter 3) has triggered major investments in port improvement up and down the U.S. Atlantic seaboard. These include New York–New Jersey, Charleston, Jacksonville, and Miami as well as Savannah, because a sizeable proportion of the shipping to and from these ports moves through the Panama Canal. Typical upgrades entail the raising of existing bridges, dredging to deepen ship channels, and state-of-the-art terminal modernization to accommodate the bigger, new-generation ("Neo-Panamax") container vessels that now transit the enlarged Canal.

Savannah, home to about 150,000 people, is mostly known as a small city (Georgia's oldest) with a rich colonial history, but that image overlooks the significance of its port. In fact, it is now the fourth busiest U.S. port in terms of container traffic, and for the past decade its growth has been second only to that of New York–New Jersey. Much of the port's prominence is based on Savannah's accessibility, particularly its proximity to such major regional expressways as Interstates 16, 20, and 95 as well as to metropolitan Atlanta—the highest-ranking **distribution center [11]** for the entire southeastern United States. Most of the seaborne cargo arriving at Savannah comes from Asia via the Canal. China alone accounts for nearly 30 percent of this trade (see **Figure 2-10**).

The ongoing Savannah Harbor Expansion Project, which is targeted for completion in 2021, is costing an estimated $700 million. Five feet (1.5 m) of sand and mud are being dredged along the 30-mile (50-km) stretch of the Savannah River between the port and its Atlantic outlet. The project also involves new and better **intermodal connections [12]** (mainly, ship to truck or train), port terminal upgrades, and new regional warehouse distribution facilities. It includes construction of a new rail depot that will allow double-decker cargo trains to be loaded right at the port.

The Expansion Project's overall goals are to accelerate the local growth of container traffic and to solidify Savannah's position as one of the realm's leading ports. But first, two challenges must be confronted. One is that the country's Pacific-coast ports can already accommodate larger container ships and may therefore be able to absorb even more of the U.S. oceanic trade with Asia. The other is that transshipment activities (transfer of goods to smaller ships that service ports with shallower harbors) are expected to grow at favored locations in the Caribbean Basin, which could divert cargo that would otherwise flow through America's East Coast ports. Time will tell if the port upgrading and infrastructural investments prove to be successful. Today, however, it is clear that Savannah's immediate economic future is heavily staked on its becoming a winner in the intense competition to capture more of the Asian container trade in the dawning neo-Panamax era.

A container ship glides up the Savannah River towards the port, which is located just upstream from the city's famous historic district. When the ongoing expansion project is completed, the port will be able to accommodate even bigger ships.

Michael Ventura/Alamy Stock Photo

PORT OF SAVANNAH
— Selected Interstate Highway
← Major Shipping Connection & Rank
▨ Regional Market, Port of Savannah

0   200   400   600 Kilometers
0   100   200   300   400 Miles

© J. Nijman, P.O. Muller and John Wiley & Sons, Inc

FIGURE 2-10

A counterexample to gentrification: an abandoned house in Detroit's Brush Park neighborhood, barely 2 miles (3 km) north of downtown, sits by itself during the summer of 2014. The poster child of a city ravaged by deindustrialization, the Motor City has followed a downward spiral in population, workforce, and tax base since the 1960s. In 2013, Detroit was forced to file for bankruptcy, by far the biggest American city ever to do so. Some say that Brush Park now possesses "regrowth potential" thanks to its low real estate values, but to date few developers have taken notice.

Northern California's *Silicon Valley*—the world's leading hub of computer research and development as well as the headquarters of the U.S. microprocessor industry—exquisitely illustrates the locational dynamics of this sector of the spatial economy. Proximity to Stanford, a world-class research university; adjacent to cosmopolitan San Francisco, the major local concentration of highly educated and skilled workers; a strong, thriving business culture; ample local investment capital; high-quality housing; and a scenic environment with good weather—all combine to make Silicon Valley the prototype for similar developments now springing up around the globe. Under different names (e.g., *technopolis* is used in Brazil, France, and Japan; *science park* in China, Taiwan, and South Korea), these ultramodern, campus-like complexes symbolize the digital economy just as the smoke-belching factory embodied the industrial age of the past (see **Box 2-5**).

The Polycentric Metropolis

Fly into any large American metropolitan region, and you will see the high-rises of *suburban downtowns* encircling the old central-city CBD, some of them boasting their own impressive skylines. The traditional CBD still tends to be situated at the center, much of its former cross-traffic diverted by beltways; but the outer city's CBD-scale nodes can be ultramodern and thriving. Large **polycentric urban regions [21]** are also the result of the amalgamation of previously separate cities and towns. For example, Miami and Fort Lauderdale used to be individual cities but are now part of a polycentric urban region of 6 million whose northern extension includes West Palm

Beach; similarly, San Francisco and San Jose have become integral parts of the Bay Area (see **Fig. 2-11**). Sometimes there is still one central CBD, but at other times there are multiple high-density urban centers of more or less equal standing. Figure 2-3, showing North America's population distribution, also reflects this development of large, often interconnected, polycentric urban regions. And if you go back to Chapter 1 and zoom into Figure 1-2 (night lights) and move across the realm, you will see the same.

Geographies of Inequality

If North America stands out as a realm of economic opportunities, inequality has always been a prevalent feature as well, more so in the United States than in Canada. One could argue that differences in income and well-being are part and parcel of all free (market) societies, where rewards are based on initiative, risk, talent, and hard work, all of which tend to vary within the population. But the particular organization of the economy and political system play a very important part—explaining, for instance, why Canada is generally less unequal. Think, for example, of the role of progressive taxes, college tuition fees, minimum wages, rent subsidies, or healthcare—all matters of government intervention, even in countries that typically embrace the free market. Of course, there are plenty

FIGURE 2-11

Box 2-5 Technology & Geography

GPS, Sensors, and Self-Driving Cars

No single geospatial technology is more ubiquitous than the Global Positioning System, or GPS. **GPS [20]** refers to the global network of satellites, sensors, and devices that are used to define absolute location, to determine elevation, and to calculate transportation routes. GPS satellites orbiting the Earth continuously transmit signals that are received by devices like our mobile phones to calculate precise locations. The signals are often enhanced by ground-based instruments such as cell phone towers with well known locations. Coupled with algorithms, digital maps, and other sensors, GPS is revolutionizing commerce, industry, and transportation.

Originally developed by the U.S. Department of Defense for military purposes, GPS is now one of the most widely used and applied geospatial technologies. In addition to providing directions on how to get from point A to point B, GPS technology serves as the foundation for autonomous or *self-driving vehicles*. Google's well-known self-driving car project is built on digital maps created with GPS, which are then enhanced by real-time data collected from sensors mounted on the cars themselves. According to Google, self-driving cars have already logged 1.5 million miles (2.4 million km) on the roads of California, Texas, and Washington State. Trials are expanding to other States in order to test performance under different conditions.

Once considered the stuff of science fiction, it is now envisioned that self-driving vehicles will become commonplace within the next two decades. The ramifications for urban planning, social behavior, transportation systems, and regional geography could be enormous.

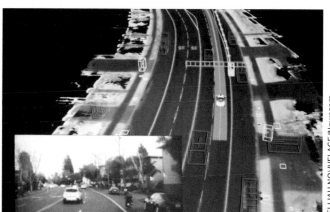

Left, a self-driving car traverses a parking lot at Google's headquarters in Mountain View, California, in 2016. On the right, video captured by a Google self-driving car in 2015: the vehicle's real-time sensing of its surroundings is visualized and coupled with the same street scene on film (lower-left corner).

of other countries that exhibit significant inequalities (e.g., South Africa, Brazil, and Russia), yet among so-called Western countries, the United States is increasingly regarded as being highly unequal.

Over the past two decades across North America, the most lucrative employment opportunities have shifted from manufacturing, construction, and other eroding middle-class job sectors into the burgeoning healthcare, technology, informatics, and energy sectors that require advanced levels of education and training. As a result, the pay-scale spectrum is beginning to resemble the bimodal pattern of the income-based residential fabric—expanding lower and upper extremes at the expense of a contracting middle. Not surprisingly, the States with greatest inequality are those with the highest levels of high-tech and information-based employment such as California, Texas, New York, Connecticut, and Massachusetts.

Inequality is often not only a matter of economics or social class. It is also influenced by gender, ethnicity, race, age, marital status, and immigration status. For example, an elderly, single, immigrant woman is much more likely to be poor than a married, middle-aged, white, U.S.-born man. In addition, and this is where geography comes in, a person's fortunes are strongly influenced by the area or neighborhood in which she or he was born and raised and/or is trying to earn a living. And inequality does not only refer to income: it is also reflected in health conditions, access to transportation, educational achievement, and even longevity. Much of this good or bad fortune for any individual or family depends on *where* you reside. In the United States today, the number of families living in middle-income neighborhoods is more than one-third lower than it was in 1970; during that same half-century, residential areas housing the wealthier and poorer extremes have doubled in size.

Although there are important regional differences across the realm (we will discuss each of its distinct regions later in this chapter), widening and deepening inequalities as well as uneven development are especially pronounced between urban and rural areas and within urban areas.

The Urban–Rural Divide

Much has been made in recent years about the importance of urban-rural dimensions in elections and that political identities are somehow forged, at least partially, in geographical context. **Figure 2-12** shows patterns of population growth and of people living in poverty within the State of Georgia. Greater Atlanta today accounts for nearly two-thirds of Georgia's economy, and has been growing steadily in terms of economy and population. However, aside from a few other urban centers such as Savannah or Columbus, the rest of the State is not doing nearly as well. Many rural areas experience shrinking job markets, dwindling populations, reduced services (e.g., schools), and growing poverty. In this State and many others, the **urban–rural divide [22]** is both reinforcing and widening major regional imbalances.

It can be argued that this divide plays an important part in the political polarization that has torn the country apart in recent years. **Box 2-6** and **Figure 2-13** discuss the geography of the presidential election of 2016, underscoring both substantial regional differences and the urban-rural divide.

Intraurban Inequalities

North American cities, particularly in the U.S., have always been marked by neighborhoods distinct in terms of social composition, income level, and racial/ethnic complexion. One's neighborhood is a key factor in shaping one's social identity, contacts, attitudes, aspirations, and life chances. This is the so-called **neighborhood effect [24]**. The economic well-being of a neighborhood is clearly reflected in its housing prices: the more expensive a residential district, the greater the earnings of its households.

Spotlighting high-end or lower-end residential areas in U.S. cities can often reveal correlations with race and/or ethnicity. As **Box 2-7** and **Figure 2-14** show, this is clearly the case in Los Angeles. Returning to the example of Greater Atlanta shown below, this metropolis may be Georgia's leading regional growth engine, yet its internal inequalities are deep and widening. In fact, measures of income distribution show that Atlanta is roughly twice as unequal as the United States overall, and in 2018 was ranked as the most unequal city in the nation.

Neighborhoods can be subject to change, and occasionally they can move "up" as with certain central cities discussed above. Often, older residents are displaced, reinvestment and redevelopment take off, and gentrification gathers momentum. For the displaced residents, such "improvement" is obviously unwelcome, but their concerns mostly carry little weight. In other cases, growing unemployment and declining income can result in a neighborhood losing its vitality. Many neighborhoods across U.S. metropolitan areas have in recent

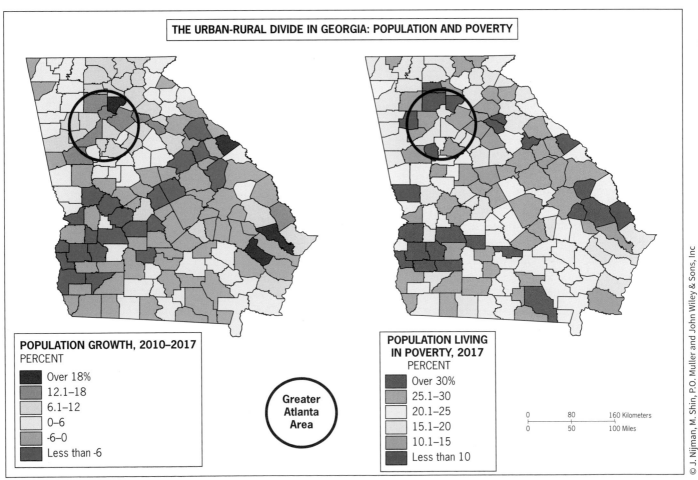

THE URBAN-RURAL DIVIDE IN GEORGIA: POPULATION AND POVERTY

POPULATION GROWTH, 2010–2017
PERCENT
- Over 18%
- 12.1–18
- 6.1–12
- 0–6
- -6–0
- Less than -6

Greater Atlanta Area

POPULATION LIVING IN POVERTY, 2017
PERCENT
- Over 30%
- 25.1–30
- 20.1–25
- 15.1–20
- 10.1–15
- Less than 10

0   80   160 Kilometers
0   50   100 Miles

FIGURE 2-12

Box 2-6 Geographic Information Analysis

Geography and Elections in the United States

Geography plays an important role in elections and democracy. Important insights into politics, geography, and society can be gain by visualizing election results. In the United States, presidential elections are frequently described as a competition for votes between Democratic or "blue" states versus Republican or "red" states. Figure 2-13 maps the 2016 U.S. presidential election using with circles that are: (1) sized according to county-level population; and (2) colored according to the level of support that

Donald Trump (red) and Hillary Clinton (blue) received at the polls. This visualization reveals notable differences in support between rural and urban areas, as well as significant variations within individual States. Which States are the most homogeneous or most diverse politically? What are some possible explanations for the observed differences in voting behavior between urban and rural areas?

Access your WileyPLUS course to interact with a dynamic version of this map and to engage with online exercises and questions.

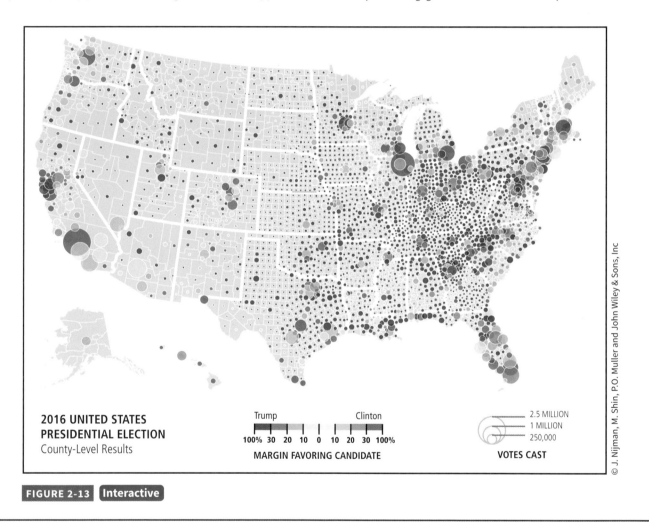

2016 UNITED STATES PRESIDENTIAL ELECTION
County-Level Results

Trump   Clinton
100% 30 20 10 0 10 20 30 100%
MARGIN FAVORING CANDIDATE

2.5 MILLION
1 MILLION
250,000
VOTES CAST

© J. Nijman, M. Shin, P.O. Muller and John Wiley & Sons, Inc

FIGURE 2-13   Interactive

---

times shown divergent paths, with some becoming increasingly expensive and doing well (for some people) while others have lost their spark and seem mired in stagnation. And what is true for neighborhoods, is true for individual people. Housing prices are again a good indicator of such disparities. The *average* price of an apartment in Midtown or Lower Manhattan in 2016 was $1.9 million, while 21 percent of New York City's population was living below the poverty line (less than $25,000 per year for a family of four). Such extreme contrasts are found in many other cities, with San Francisco always near the top of that list (**Box 2-8**).

The Making of a Multicultural Realm

Historical Patterns of Mobility and Immigration

Given the mobility of North America's population, the immigrant infusions that continue to diversify it, the economic changes reshaping it, and the forces acting upon it, the population

Box 2-7 Geographic Information Analysis

Race and Housing Values in Los Angeles

North America is sometimes described as a melting pot, but an analysis of the internal spatial distribution of metropolitan populations in terms of income, race, or ethnicity suggests otherwise. The visualizations in Figure 2-14 allow you to compare the **residential geography [25]** of African Americans (right map) with the distribution of home values (left map) across Greater Los Angeles. How would you describe the distribution of African Americans, and how does it relate to patterns of home values? What does it reveal about the nature of inequality in U.S. urban regions?

Access your WileyPLUS course to interact with a dynamic version of this map and to engage with online exercises and questions.

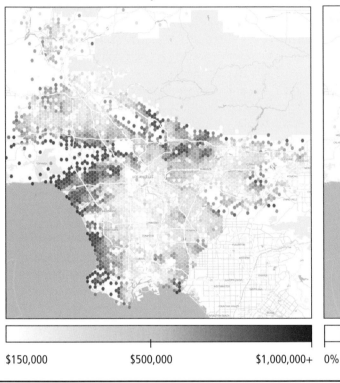

MEDIAN HOME VALUE, DOLLARS

$150,000 $500,000 $1,000,000+

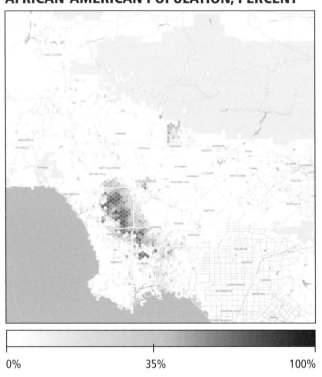

AFRICAN-AMERICAN POPULATION, PERCENT

0%   35%   100%

FIGURE 2-14 **Interactive**   © J. Nijman, M. Shin, P.O. Muller and John Wiley & Sons, Inc

Box 2-8 From the Field Notes ...

"Driving along San Francisco's 5th Street on a late afternoon in March 2016, I pass by northern California's largest homeless shelter, the St. Vincent de Paul Center. It is nearly 4 pm, and there is a line of about 60 or 70 people waiting for the doors to open (see photo). San Francisco ranks among the most expensive U.S. cities to live in, with a 2018 median home price of ca. $1.6 million. In the past few years, the city has witnessed an estimated 2500 home evictions per year, often because people living in much more modest housing cannot afford to pay the higher rent or because landlords sell the entire rental building to a developer who turns it into private-owner-occupied housing units. San Francisco now has an estimated 8000 homeless people, and this shelter serves 400 to 500 persons daily, with separate accommodations for men and women. They can have a bed for the night, a meal, shower, counseling, and basic medical help. The shelter closes from 9 am to 4 pm."

© J. Nijman

distribution map (Fig. 2-3) should be viewed as the latest still in a motion picture that has been unreeling for four centuries. Slowly at first, then with accelerating speed after 1800, North Americans pushed their settlement frontier westward to the Pacific. Even today, such shifts continue. Not only does the center of gravity of population continue to move westward, but within the United States it is also shifting southward—the latter a drift that has gained momentum since the 1960s as the advent of universal air conditioning made the so-called **Sunbelt [26]** States of the U.S. southern tier ever more attractive to internal migrants.

The current population map is the still-changing product of numerous forces. For centuries, North America has attracted a pulsating influx of immigrants who, in the faster-growing United States, were rapidly assimilated into the societal mainstream. Throughout the realm, people have arranged themselves geographically to maximize their proximity to evolving economic opportunities, and they have shown little resistance to relocating as these opportunities successively favored new locales.

During the past century, such transforming forces have generated a number of major **migrations [27]**, of which the still-continuing shift to the west and south is only the latest. Five others were (1) the persistent growth of metropolitan areas, first triggered by the impact of the late-nineteenth-century Industrial Revolution in North America; (2) the large-scale movement of African Americans from the rural South to the urban North during the latter stages of the industrial era; (3) the shift of tens of millions of urban residents from central cities to suburbs, and subsequently to exurbs even farther away from the urban core; (4) the return migration of millions of African Americans from the deindustrializing North back to the growing opportunities in the South (in metropolises such as Atlanta and Charlotte); and (5) the strong and steady inflow of immigrants from outside North America, including, in recent times, Mexicans, Cubans, and other Latinos; South Asians from India and Pakistan; and East and Southeast Asians from Hong Kong, Vietnam, and the Philippines.

In some respects, the need for immigration today is greater than ever because the North American population's rate of natural increase is at an historic low (with overall population growth the lowest in more than 80 years). The share of the foreign-born population in the U.S. is just under 15 percent, the highest since 1910. Moreover, their places of origin are not the same as a century ago: 41 percent of the immigrants who arrived since 2010 came from Asian countries, and another 39 percent from those of Middle and South America. Nearly half of these migrants are college-educated, and sizeable numbers have dispersed across States that traditionally have not attracted many immigrants. New York and California, States containing large immigrant populations, both saw increases of less than 6 percent since 2010. But the foreign-born population increased over the same period by 20 percent in both Kentucky and Tennessee, 13 percent in Ohio, and 12 percent in South Carolina. In Canada, the need for immigrant infusions to maintain a demographic balance is even greater because its rate of natural increase is even lower than in the United States. In 2020, the share of Canada's foreign-born population exceeded 22 percent, up from just over 20 percent a decade earlier.

The Challenge of Multiculturalism

At the same time, growing diversity, especially at a time of globalization and *transnationalism*, comes with challenges. The sheer dimensions of immigration into the United States, from virtually all corners of the world, creates an increasingly complex ethnic and cultural mosaic that tests **melting pot [28]** assertions at every geographic scale. Today there are more people of African descent living in the United States than in Kenya. The number of Hispanic residents in America is nearing half the population of Mexico. Miami is the second-biggest Cuban city after Havana, and Montreal is the largest French-speaking city in the world after Paris. In short, there are sufficient immigrant numbers for them to create durable societies within the overall national fabric. The challenge is to ensure that the great majority of the immigrants become full participants in that larger society.

Such are the numbers, and so diverse are the cultural traditions, that the melting pot that America once was is today morphing into something else. Spatially, the United States is now completing its transformation into a *mosaic culture*, an increasingly heterogeneous complex of separate, more or less uniform "tiles" whose residents spend less time than ever interacting and "melting." This new pattern applies not only to new immigrant groups but also to existing communities, underscored by the proliferation of walled-off, gated housing complexes in metropolitan areas across the realm. It has been called "the big sort," a wide-ranging process of segregation of people according to income, race, ethnicity, political identity, age, and so forth, where similar folks end up in the same city, county, town, or neighborhood, segregated from others.

In Canada, the melting-pot notion is put to the test with large numbers of (mostly affluent) Chinese immigrants arriving in Vancouver and elsewhere who purchase and renovate (or frequently replace) traditional homes in long-stable Vancouver neighborhoods, arousing the ire of numerous locals. The economic hardship in many smaller towns and rural areas appears to have invoked a political response to immigration as well. It is interesting to see that the strongest anti-immigrant stance, in North America and elsewhere, tends to be in geographic areas where cultural diversity is less pronounced.

The politically divisive, post-2016 debate about a "wall" along the U.S.-Mexico border reflected the different sentiments regarding immigration into the United States and the different ways people regard "others." Debates about immigration are often complicated by the fact that there are different kinds of immigrants, and by economic versus humanitarian perspectives. Most are legal immigrants and have obtained official permission for residency, but others are illegal (and there are various categories of illegal migrants); most are economic migrants who come to North America to advance their careers or to improve the well-being of their families, while others are refugees. The latter seek safe haven in North America and have escaped war, violence, and/or persecution in their country of origin. This debate is also complicated by the fact that migrants differ in their cultural and religious backgrounds, and that most refugees come from countries that are considered different. In the United States, the number of refugees admitted in 2018 (only 22,500) was the lowest since 1980; and among admitted refugees, the number of Muslims had dropped almost by half.

Regionalism in Canada: Divisive Forces

Canada's Spatial Structure

Territorially the second-largest state in the world after Russia, Canada is administratively divided into only 13 subnational entities—10 provinces and 3 territories (the United States has 50). The provinces—where 99.7 percent of all Canadians live—range in dimensions from tiny, Delaware-sized Prince Edward Island to Quebec, nearly twice the size of Texas (**Fig. 2-15**).

As in the United States, the smallest provinces lie in the northeast. Canadians call these four the Atlantic Provinces (or simply Atlantic Canada): Prince Edward Island, Nova Scotia, New Brunswick, and the mainland–island province named Newfoundland and Labrador. Gigantic, mainly *Francophone* (French-speaking) Quebec and populous, heavily urbanized Ontario, both flanking Hudson Bay, form the heart of the country. Most of western Canada (which is what Canadians call everything west of Ontario) is organized into the three Prairie Provinces: Manitoba, Saskatchewan, and Alberta. In the Far West, beyond the Canadian Rockies and extending to and along the Pacific Ocean, lies the tenth province, British Columbia.

Of the three territories in Canada's Arctic North, the Yukon is the smallest—but nothing is small in Arctic Canada. With the Northwest Territories and recently created Nunavut, the three territories cover almost 40 percent of Canada's total area. The population, however, is extremely small: only about 115,000 people inhabit this vast, frigid frontier zone.

Canada's population is clustered in a discontinuous ribbon, roughly 300 kilometers (200 mi) wide, nestled against the U.S. border and along ocean shores, whereas most of the country's energy reserves lie far to the north of this corridor. As such, the maps of Canada's human settlement (Figs. 2-3 and 2-15) bear out this generalization and emphasize the significance of environment. People

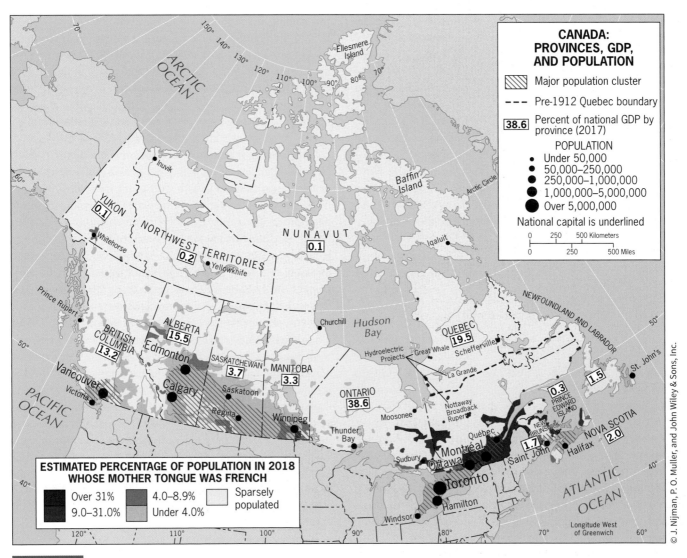

FIGURE 2-15

choose to live in the more agreeable areas to the south, but territorial control of abundant northern resources is highly important. This population pattern creates cross-border affinities with major American cities in several places: Toronto–Buffalo (see **Fig. 2-8**), Windsor–Detroit, and Vancouver–Seattle. Very little of this occurs in the three Prairie Provinces of the Canadian West, which for the most part adjoin sparsely settled North Dakota and Montana.

Cultural Contrasts

Canada's capital, Ottawa, is symbolically located astride the Ottawa River, which marks the boundary between English-speaking Ontario and French-speaking Quebec. Though comparatively small, Canada's population is markedly divided by culture and tradition, and this division has a pronounced regional expression. Linguistically, three-quarters of Canada's citizens speak English as their mother tongue, and nearly one-quarter speak French (almost one in five claim to be bilingual). The spatial concentration of more than 90 percent of the country's Francophones in Quebec accentuates this division (Fig. 2-15), and the province overwhelmingly continues to function as the historic, traditional, and emotional focus of French culture in Canada.

Over the past half-century, a strong nationalist movement has emerged in Quebec, and at times it has demanded outright separation from the rest of Canada. This ethnolinguistic division continues to be the litmus test for Canada's federal system, and the issue continues to pose a latent threat to the country's future national unity. The historical roots of the matter are related to French–British competition since the seventeenth century in which Britain gained the upper hand. To this day, the British monarch remains the official head of state (as Canada remains part of the British Commonwealth), represented by the governor general. It is one of the issues that riles Quebec nationalists.

There does seem to be an overall decline in support for Quebec's independence, however. This is partly due to a number of laws that have been enacted to reassure the primacy of Québécois culture in the province (a 2006 federal parliamentary resolution effectively recognizes the Québécois as a *distinct nation*). In addition, there appears to be a growing realization among the French-speaking people that continued integration within Canada combined with a high degree of autonomy is the best possible option. But the struggle for a maximum degree of autonomy continues, along with accompanying political and ethnic tensions.

Canada's Indigenous Peoples

On the wider Canadian scene, the culture-based Quebec struggle has stirred ethnic consciousness among the country's 1.7 million native peoples—dominated by **First Nations [29]**, but also including Inuit and Métis. They, too, have received a sympathetic hearing in Ottawa, and in recent years breakthroughs have been achieved with the creation of Nunavut as well as local treaties for limited self-government in northern British Columbia.

A foremost concern among these peoples—who constitute 5 percent of Canada's population—is that their aboriginal rights be protected by the federal government against the provinces of which they are a part. This is especially true for the largest First Nation of Quebec, the Cree, whose historic domain covers the northern half of the province. As Figure 2-15 shows, the territory of the Cree (north of the red dashed line) is no unproductive wilderness: it contains the James Bay Hydroelectric Project, a massive scheme of dikes and dams that has transformed much of northwestern Quebec and generates hydroelectric power for a huge market within and outside the province.

Regionalism and Ethnicity in the United States

In Canada, indigenous peoples were able to hold on to sparsely populated areas in the north. By contrast, in the United States the relentless push westward by European settlers and the U.S. government in effect dispossessed—and decimated—the original inhabitants. In the end, they were confined to some 300-plus reservations located mainly in the western half of the country (**Fig. 2-16**, lower-right map). As noted earlier, there are fewer than 7 million Native Americans left today, with the Navajo and Cherokee being the largest surviving nations. Whereas in Canada First Nations hold significant territorial power, in the United States Native Americans are in a far weaker position to challenge the federal government.

Within U.S. society, more so than in Canada, the social fabric and ethnic relations are determined by past and present waves of immigration. In certain cases, immigration has helped forge a particular regional geography. This was especially true in the case of the hundreds of thousands of Africans who were forcibly brought to the United States in bondage from the seventeenth to the mid-nineteenth centuries and put to work as slaves on plantations in the American South.

Slavery was abolished in 1865, however racial segregation persisted in a variety of legal forms until the civil rights movement crested in the 1960s. In violation of the Constitution, a wide range of State and local Jim Crow laws were enacted to codify the repression of African Americans. This legacy is still visible in the present-day geography of race in the United States, with blacks regionally concentrated in the Southeast, from eastern Texas to Maryland (Fig. 2-16, upper-left map). We noted earlier that substantial numbers of blacks migrated north to work in the industrial centers of the American Manufacturing Belt during the second quarter of the twentieth century; major African American population clusters persist in those northern cities, but they are not clearly apparent at the scale of this national map.

Race continues to be a sensitive issue in the United States, more so than in Canada, because of this terrible history, enduring rifts between blacks and whites, continuing deprivation among African Americans, and the recurrent racial injustices impeding the nation's evolution into a truly post-racial society.

Another highly visible regional expression of ethnic migration is the large concentration of Hispanics in the southwestern

AFRICAN-AMERICAN

ASIAN-AMERICAN

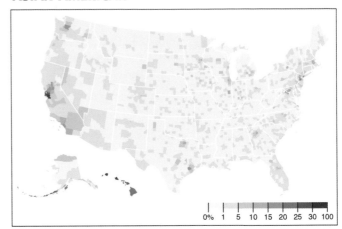

HISPANIC/LATINO (all races)

NATIVE AMERICAN

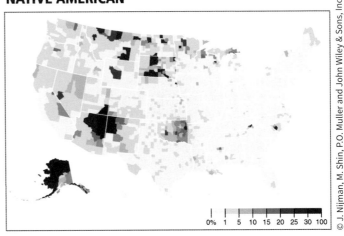

© J. Nijman, M. Shin, P.O. Muller and John Wiley & Sons, Inc

FIGURE 2-16 Interactive

United States (Fig. 2-16, lower-left map; see also Fig. 2-2). Since 1980, the number of Hispanics in the United States has more than tripled to surpass 60 million, forming the nation's biggest ethnic minority—and more than 18 percent of the country's total population. For some Anglophone Americans who fear the erosion of the status of English in their communities, language remains an emotional issue. Indeed, Hispanics are now dispersing and increasing their presence in many more parts of the United States, ranging from the rural areas of the Southeast and Midwest to thousands of urban communities from Massachusetts to Washington State to the small towns that still remain on the Great Plains.

Finally, the Asian immigrant population, whose numbers have risen over the past few decades, is geographically the most agglomerated of the leading ethnic minorities (Fig. 2-16, upper-right map). Ethnic Chinese remain the most numerous among the more than 20 million Americans who identify themselves as Asian, but this diverse minority also includes Japanese, Filipinos, Koreans, and Vietnamese as well as a sizeable cohort of South Asian Indians. Whatever their origins, the great majority have remained in California, where many have achieved economic success and upward mobility.

Regions of North America

Mountain ranges, coastal zones, deserts, low-lying plains, and a range of climatic environments all demand different kinds of human adaptation. Successive waves of immigrants brought with them a variety of cultures; sometimes the differences were subtle, but sometimes they were quite stark. As **Figure 2-17** shows, a majority of this realm's regions overlap both countries regardless

of the international boundary. We can regionalize North America's geography by combining functional and formal principles. The resulting map of nine regions is by no means the only solution, but for our general purposes these boundaries are the most pertinent, and the regional distinctions they facilitate will be helpful in further understanding the overall fabric of the realm.

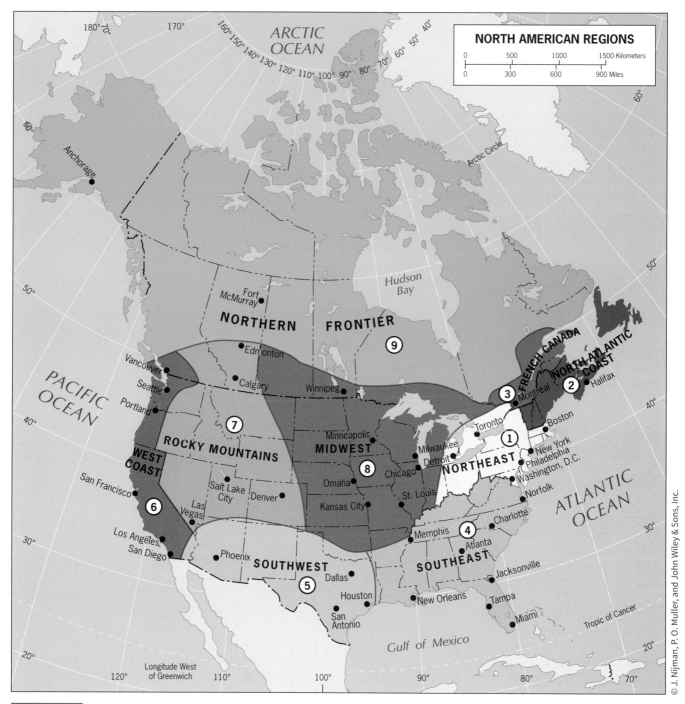

NORTH AMERICAN REGIONS

0   500   1000   1500 Kilometers

0   300   600   900 Miles

FIGURE 2-17

© J. Nijman, P. O. Muller, and John Wiley & Sons, Inc.

Region The Northeast (1)

The historic core region of the North American realm integrates what have long been the most prominent parts of the United States and Canada (Fig. 2-17). The region contains the largest cities and the federal capitals of both countries, the leading financial markets, the largest number of corporate headquarters, dominant media centers, prestigious universities, cutting-edge research complexes, and the busiest airports and intercity expressways. Moreover, both the U.S. and Canadian portions of the North American core still contain roughly one-third of their respective national populations. Political and business decisions, investments, and many other commitments made here affect not just North America but the world at large.

Nonetheless, the realmwide dominance of the Northeast has been declining. As was true of the American Manufacturing Belt—whose eastern anchor it forms—it once commanded unmatched supremacy. But deindustrialization and the subsequent emergence of the information economy have eroded that dominance as competitors to the south and west continue to siphon away some of its key functions. An example is eastern Michigan, or parts of Ohio, which have experienced long-term economic decline as jobs in the automobile industry have steadily disappeared over the past four decades.

Region North Atlantic Coast (2)

When you travel north from Boston into New Hampshire, Vermont, and Maine, you move across one of North America's historic culture hearths whose identity has remained strong for four centuries. Even though Massachusetts, Connecticut, and Rhode Island—traditional New England States—share these qualities, they have become part of the Northeast and became oriented to the New York urban region, so that on a functional as well as a formal basis, the North Atlantic Coast extends from the northern border of Massachusetts to Newfoundland, thereby incorporating all four Canadian Atlantic Provinces (Fig. 2-17).

Difficult environments, a maritime orientation, limited resources, and a persistent rural character have combined to slow economic development here. Primary industries—fishing, logging, and some farming—are mainstays, although recreation and tourism have boosted the regional economy in recent decades. Upper New England experiences the spillover effect of metropolitan Boston's prosperity to some degree, but not enough to ensure steady and sustained economic growth. Atlantic Canada has endured economic hard times as well in the recent past. For example, overfishing has seriously depleted fish stocks. Alternate opportunities focus on this subregion's spectacular scenery and the tourism it attracts. Economic prospects have also been boosted in Canada's poorest province, Newfoundland and Labrador, by the discovery of significant offshore oil reserves. Production is now well underway, and in the 2010s additional energy deposits were discovered both on and around Newfoundland Island.

Region French Canada (3)

Francophone Canada constitutes the inhabited southern portion of Quebec, focused on the St. Lawrence Valley from near Montreal northeastward to the river's mouth (Fig. 2-17). It also includes the French-speaking Acadians who reside in neighboring New Brunswick (Fig. 2-15). The French cultural imprint on the region's cities and towns is matched in farming zones by the narrow, rectangular *long lots* perpendicular to the river, also of French origin.

The economy of French Canada was historically less industrial and more rural than the adjacent Northeast region (and Toronto far outranks Montreal here). Nevertheless, Montreal's economic role has advanced since 2000, with information technology, telecommunications, and biopharmaceuticals all growing in importance. In some ways, French Canada is a richly endowed part of North America. Northern Quebec produces so much hydroelectricity that it exports the surplus to Ontario and to the northeastern United States. Most of this power is generated via the James Bay Project, a series of dams and power stations on the eastern side of Hudson Bay along the La Grande River (Fig. 2-15). Quebec is faring so well in the energy-production sphere that the recent discovery of natural gas in the St. Lawrence Basin was met with tempered enthusiasm. Many Québécois feel they have no need for gas: unlike hydropower, it is a nonrenewable resource that would boost carbon emissions and the likelihood of widespread water pollution.

Region The Southeast (4)

For more than a century following the U.S. Civil War (1861–1865), the Southeast (Fig. 2-17) remained in economic stagnation and cultural isolation from the rest of the country. During the 1970s, however, things changed so fast that a New South was born almost overnight. The Sunbelt migration stream drove people and enterprises into the long-dormant cities. Northeast-based companies looking for headquarters or branch offices found Atlanta, Charlotte, Tampa, Miami–Fort Lauderdale, and other urban areas both economical and attractive, swiftly turning them into boomtowns. Racial segregation had been dismantled in the wake of the civil rights movement. A new social order was matched by new facilities ranging from airports (Atlanta's quickly became one of the world's busiest) to theme parks open to all. Soon the nation was watching Atlanta-based CNN on television, vacationing at Orlando's Walt Disney World, and monitoring space flights originating at Cape Canaveral.

Lobster boats and wharf at Port Clyde, Maine. The cold, clean waters off Maine's rocky coast are ideal for lobster harvesting, and the industry is interwoven with the culture of the area. The annual catch exceeds 100 million pounds, and thanks to successful regulation it has been remarkably sustainable. Catching lobster eggs or egg-bearing females is prohibited; the only legal way to fish is by using traps (no nets); there is a limit of 800 traps per fisher, and they must be designed to allow smaller lobsters to escape. Maine's share of the North Atlantic region's lobster catch alone totals about $500 million annually.

Montreal's Rue Bonsecours. The old center of the city is marked by an unmistakably French cultural landscape. This restaurant and hotel (the brick building to the right) have been here since the early eighteenth century, predating the British rule of Canada.

The geography of development in the South, however, is rather uneven. Although many cities and certain agricultural areas have benefited, others have not; the South contains some of the realm's poorest rural areas, and the widening gap between rich and poor here is greater than in any other region of the realm. The "Silicon Alley" corridor of Northern Virginia's Washington suburbs, central North Carolina's Research Triangle, and metropolitan Atlanta's corporate campuses are locales that typify the New South; but Appalachia and rural Mississippi still represent the Old South, where depressed farming areas and stagnant small industries inhibit change.

Another, relatively small part of this region warrants special attention. South Florida used to be a particularly remote and nearly unpopulated corner of the United States. Before the twentieth century, it was too far from the northeastern core of the country (and even from closer urban centers such as Atlanta) to warrant development; the heat, humidity, and mosquitoes posed a formidable challenge to prospective settlers; and there were no natural resources to exploit. It was not until the 1960s that this subregion of the Southeast became fully integrated into the national economy and urban system—thanks to the widespread introduction of air conditioning and affordable air travel, not to mention the mass immigration of Cubans fleeing Fidel Castro's communist regime. Today, Miami has evolved into a **world-city [30]**, a globally linked, southward-oriented conduit between North America and the rest of the hemisphere (**Fig. 2-18**).

Region The Southwest (5)

As Figure 2-17 shows, by our definition the Southwest begins in east Texas and extends all the way to the eastern edge of Southern California. New Mexico, in the middle of the Southwest region, is the least developed economically and ranks low on many U.S. social indices, but its environmental and cultural attractions benefit Albuquerque and Santa Fe, with Los Alamos a famous name in research-and-development circles. To the west, Arizona's technologically transformed desert now harbors two large, coalescing

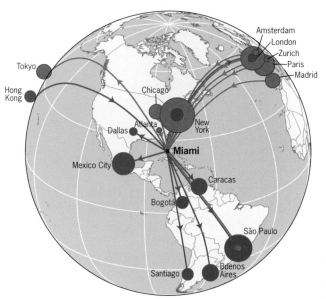

WORLD–CITY CONNECTIONS: THE CASE OF MIAMI

● Cities with largest number of headquarters of corporations conducting business in Miami

● Cities with largest number of branches of Miami-based corporations

© J. Nijman, P. O. Muller, and John Wiley & Sons, Inc.

FIGURE 2-18

metropolises—Phoenix and Tucson. Growth continues, but rising concerns about climate change add to the uncertainties of maintaining water supplies already stretched to the limit.

Texas leads this region in most respects. Its economy, once all but fully dependent on oil and natural gas (see the Natural Resources section in this chapter, which covers massive new discoveries of yet more oil and gas reserves in west Texas), has been restructured and diversified so that today the Dallas–Fort Worth–Houston–Austin–San Antonio triangle has become one of the world's most productive **technopoles [31]**—state-of-the-art, high-technology industrial complexes—especially Austin at the heart of it. This subregion is also a hub of international trade and the northern anchor of a North American Free Trade Agreement (NAFTA)–generated transnational growth corridor that extends into Mexico as far south as Monterrey.

The huge State of Texas, larger than France, is surpassed in areal size only by Alaska. Population-wise, with 30 million, it is second only to California (40 million). Besides its leading role in the production of oil and gas, the Lone Star State has also become a pacesetter, along with California, in the generation of wind energy, with wind turbines increasingly dotting the sparsely settled landscape of immense western Texas.

Region The West Coast (6)

The Southwest meets the Far West near metropolitan San Diego on California's border with Mexico, and from there the region we call the West Coast extends all the way north into Canada

The Austin City Limits Music Festival, 2018. Austin, Texas—highly regarded for its burgeoning technology-led economy—was one of the fastest-growing American cities during the 2010s. The city also occupies a prominent position on the U.S. popular culture scene. Austin is a product of the postindustrial Sunbelt boom of the past half-century, an anchor of the new high-tech industrial corridor that stretches from the Dallas–Fort Worth Metroplex in the north to Greater San Antonio in South Texas (see Fig. 2-8). In addition to its function as a far-reaching employment magnet for young people, Austin is considered to be one of the most livable urban areas in the nation.

where Vancouver forms the northern anchor in southwestern British Columbia (Fig. 2-17). We include almost all of California in this region but (for environmental as well as economic reasons) only the western portions of Oregon and Washington State. This is a particularly important part of North America and increasingly serves as a counterweight to the historic core area in the U.S. Northeast.

The West Coast is a major economic region in both this realm and in the global economy. It includes major cities such as San Francisco, San Diego, Portland, Seattle, and Los Angeles. And it contains America's most populous State, California, whose economy ranks among the world's ten largest *by country*. This region also encompasses one of the realm's most productive agricultural areas in California's Central Valley, magnificent scenery, and a culturally diverse population drawn by its long-term economic growth and pleasant living conditions.

Our regional definition is predominantly based on economic considerations and intensifying trade connections across the Pacific Ocean. This part of North America is deeply involved in the development of countries in eastern Asia. In

the postwar era, Japan's success had a salutary impact here, but what has since taken place in China, South Korea, Taiwan, Singapore, and other Asia–Pacific economies has created unprecedented opportunities. The term **Pacific Rim [32]** has come into use to describe the discontinuous regions surrounding the great Pacific Ocean that have experienced spectacular economic growth and progress over the past four decades: not only coastal China and various parts of East and Southeast Asia, but also Australia, South America's Chile, and the western shores of Canada and the conterminous United States.

Thus the Pacific Rim is a classic example of a *functional region*, with economic activity in the form of capital flows, raw-material movements, and trading linkages that generate urbanization, industrialization, and labor migration. As part of this process, human landscapes from Sydney to Santiago are being transformed within a 32,000-kilometer (20,000-mi) corridor that rings the globe's largest body of water. The functional economic integration of the various regions of the Pacific Rim has been accompanied by substantial international migration and, particularly in the major cities of the U.S. West Coast, has produced intricate multicultural landscapes (**Box 2-9**).

Box 2-9 From the Field Notes …

"On this street corner in central Los Angeles, four different cultures converge. Los Angeles is one of the most ethnically diverse metropolitan areas in the United States and in the world. More than a third of its population was born abroad, and more than half speak a language other than (or in addition to) English. The largest non-native ethnic group are Hispanics (mainly Mexicans and Middle Americans), but there also are many Asians, especially Chinese and Koreans. Cesar E. Chavez, after whom this avenue is named, was a charismatic Mexican-American farm worker turned union organizer who became nationally known as a leading civil rights activist. His birthday, March 31, is a State holiday in both California and Texas—and an optional holiday in Arizona(!). Chavez Avenue is a major east–west thoroughfare just north of downtown Los Angeles, and even a short stroll along it offers a sampling of global cultures."

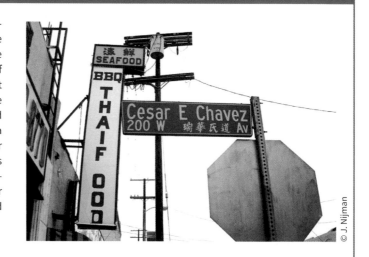

Region The Rocky Mountains Region (7)

Where the forests of the West Coast yield to the scrub of the rain shadow on the inland slopes of the mountain wall that parallels the West Coast, and the terrain turns into intermontane ("between the mountains") basins and plateaus, lies a region stretching eastward from the Sierra Nevada and Cascades to the Rockies, encompassing segments of southern Alberta and British Columbia; eastern Washington State and Oregon; all of Nevada, Utah, and Idaho; plus most of Montana, Wyoming, and Colorado. As Figure 2-17 shows, Edmonton, Calgary, and Denver are situated along the eastern edge of the Rocky Mountains region, and Salt Lake City, anchoring the Software Valley technopole and symbolizing the new high-tech era, lies at the heart of it.

Remoteness, dryness, and sparse population typified this region for a very long time. It became the redoubt of the Mormon faith and a place known for boom-and-bust cycles of mining, logging, and, where possible, livestock raising. However, in recent decades advances in communications and transportation technologies, sunny climates, wide-open spaces, lower costs of living, and growing job opportunities have combined to form effective pull factors for myriad outsiders. Every time an earthquake struck in California, eastward migration got a boost. And during the most recent ten-year period, millions of Californians emigrated from the overcrowded and overpriced Pacific coast to the inland high desert.

Development in the Rocky Mountains region centers on its widely dispersed urban areas, which are making this one of the realm's fastest-growing regions (albeit from a low base).

The sprawling Las Vegas suburb of Summerlin. In recent decades, Las Vegas has been one of the fastest-growing metro areas in the United States, and it also became the epicenter of the mortgage meltdown that accompanied the **Great Recession [33]**, which lasted from 2008 until early 2011. By 2012, nearly 70 percent of all mortgaged homes were "underwater," meaning the market value of the house was less than the mortgage balance (that number had declined to about 25 percent by 2016). The housing crisis was concentrated, here and elsewhere, in relatively new suburbs, where homes had come onto the market just before the bubble burst in 2008. For many, that crisis left a painful legacy that lingers to this day.

trekandshoot/Alamy Stock Photo

Here, thousands of new high-technology manufacturing and specialized service jobs have propelled a decade-long influx, slowed but not reversed by the Great Recession (2008–2010). Metropolitan Las Vegas typifies this accelerated growth trajectory. Far more than a gambling and entertainment magnet that draws some 40 million visitors annually, Las Vegas attracts in-migrants because of its history of job creation, relatively cheap land and low-cost housing, moderate taxes, and sunny if seasonally hot weather. Although the pace of development sharply leveled off during the late 2000s, by 2015 things were back on track (see photo below).

Region The Midwest (8)

This vast North American heartland extends from interior Canada to the borders of the Southeast and Southwest (Fig. 2-17). The region contains several noteworthy cities, including Chicago, Kansas City, Omaha, Minneapolis, and Winnipeg; nonetheless, agriculture is the dominant story here. As the map of U.S. farmlands (**Fig. 2-19**) shows, this is the nation's breadbasket. This is the land of the beef- and pork-producing Corn Belt (*Meat Belt* would be a more accurate label); of the mighty soybean, America's and the world's most rapidly expanding crop of the past half-century; and of spring wheat in the Dakotas and Canada's Prairie Provinces as well as winter wheat in Kansas. Indeed, the cities and towns of this region share histories of food processing, packing, and marketing.

This is also the scene of the struggle for survival of the family farm as the unrelenting incursion of large-scale corporate farming threatens a longstanding way of life. Even a cursory look at demographic statistics makes it clear why many States in the Midwest are losing population or gaining far fewer than the national average. The Great Plains, the western component of this region, is especially hard hit: younger people and those better off are leaving, whereas older and less affluent residents stay behind. Villages and small towns continue to die, and the notion of abandoning certain areas to return them to their natural state is being seriously discussed.

But things are not the same everywhere. The Midwest may not have as prominent a presence on the national scene as New York or California, and the region is not as wealthy, but with abundant coal and oil deposits (in Wyoming and North Dakota, respectively) as well as South Dakota's booming financial sector, unemployment rates here rank among the lowest in the nation. Buoyed by an economy largely driven by agriculture and mining, and with fossil-fuel production a leading growth industry, the effects of the Great Recession were minimal and short-lived in the Midwest.

Region The Northern Frontier (9)

Figure 2-17 leaves no doubt as to the dimensions of this final North American region: it is by far the largest of the realm, covering just about 90 percent of Canada and all of the

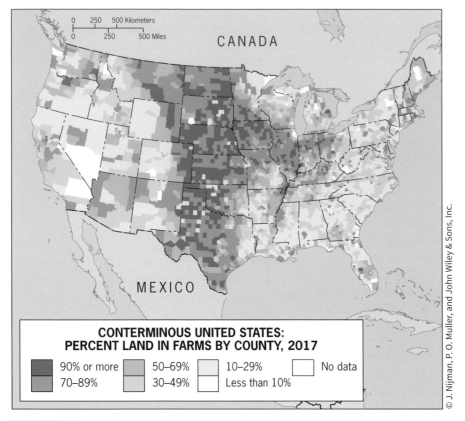

CONTERMINOUS UNITED STATES:
PERCENT LAND IN FARMS BY COUNTY, 2017

■ 90% or more	■ 50–69%	□ 10–29%	□ No data
■ 70–89%	■ 30–49%	□ Less than 10%	

FIGURE 2-19

© J. Nijman, P. O. Muller, and John Wiley & Sons, Inc.

Productive locations in the Northern Frontier constitute a far-flung network of mines, oil and gas installations, pulp mills, and hydropower stations that have spawned hundreds of small settlements as well as thousands of kilometers of interconnecting transportation and communications lines. Inevitably, these activities have infringed on the lands of indigenous peoples without the preparation of treaties or agreements, which has led to recent negotiations between the government and leaders of First Nations over resource development in the Northern Frontier. But essentially this gigantic region remains a frontier in the truest sense of that term.

Alaska's regional geography differs from the rest of the Northern Frontier in that the State contains several urban settlements and an incipient core area in the coastal zone around Anchorage. The population of Alaska (ca. 750,000) accounts for more than one-third of the region's total. Here also lies the only metropolis of any size, Anchorage (population: 300,000), a key node on the aviation network that links the realm to Pacific Asia. The State's internal communications (air and surface) are also better developed.

Finally, Nunavut merits special mention. Created in 1999, this newest Territory of Canada is the outcome of a major aboriginal land claim agreement between the *Inuit* people (formerly called Eskimos) and the federal government, and encompasses all of Canada's eastern Arctic as far north as Ellesmere Island (Figs. 2-6, 2-15), an area far larger than any other province or territory. With around one-fifth of Canada's total area, Nunavut—which means "our land"—contains only about 37,000 residents, of whom 85 percent are Inuit.

Climate change is likely to affect the Northern Frontier in as-yet-unforeseeable ways. As the map shows, this region extends all the way north to the Arctic Ocean; it therefore includes waters that, with continued global warming and associated contraction of the Arctic icecap, will open (possibly year-round) nearshore waterways that would dramatically alter global shipping routes and intercontinental distances—a topic discussed in more detail in Chapter 13. The question of ownership of these waters—and the submerged seafloor beneath them, coming within reach of exploitation—will be a matter of significant international contention in the decades ahead.

In 2010, Canada solidified its reputation as an environmentally conscious nation through the signing of an unprecedented agreement. The Canadian Boreal Forest Agreement involves 21 logging firms and nine environmental organizations as well as the Canadian government. As shown in Figure 2-20,

biggest U.S. State, Alaska. Not only does this region include the northern parts of seven of Canada's provinces, but it also comprises the Yukon and Northwest Territories as well as recently established Nunavut. The very sparsely populated Northern Frontier remains a land of isolated settlement based on the exploitation of newly discovered resources. The Canadian Shield, underlying most of the eastern half of the region, is a rich storehouse of mineral resources, including metallic ores such as nickel, uranium, copper, gold, silver, lead, and zinc. The Yukon and Northwest Territories have proven especially bountiful, with the mining of gold and especially diamonds (Canada currently ranks third in the world); at the opposite, eastern edge of the Shield, near Voisey's Bay on the central coast of Labrador, the largest body of high-grade nickel ore ever discovered is now open for extraction.

The province of Alberta experienced one of the most spectacular economic booms in the realm's history. Recent estimates indicate that Alberta's oil reserves, dominated by the Athabasca **tar sands [34]** (see **Fig. 2-20**), rank third in the world behind only those of Venezuela and Saudi Arabia. In the early twenty-first century, Alberta became Canada's fastest-growing province, and Calgary its fastest-expanding major city. The boom makes Calgary a magnet for business and highly skilled workers; the city's skyline has been spectacularly transformed, comparable to what has recently occurred on the Arabian Peninsula. But much depends on the price of oil, which has dropped significantly since the mid-2010s.

FIGURE 2-20

it imposes a moratorium on logging and any other development within a huge zone of 290,000 square kilometers (112,000 sq mi). Greenpeace, which originated in Canada, has called it the biggest agreement to date that preserves nature. The word "boreal" is derived from the Latin word for north;

boreal forests [35]—identical to the Russian *taiga* (snowforest) discussed in Chapter 6—are dominated by dense stands of coniferous (cone-bearing) needleleaf trees such as spruce, fir, and pine. The boreal forests of the Northern Frontier also constitute the habitat of such endangered wildlife species as the caribou, lynx, American black bears, and wolverines.

The Inuit hamlet of Pangnirtung (pop. 1425) on Nunavut's Baffin Island, located just south of the Arctic Circle (see Fig. 2-20). Baffin Island is larger in area than California, and has rugged, mountainous terrain that is snow-covered most of the year. These fjords were shaped by inland glaciers reaching the coast, just as in Norway.

Points to Ponder

- Whereas more than 70 percent of Canada's population resides within 200 kilometers (125 mi) of the northern U.S. border, only about 12 percent of the Mexican population lives that close to the southern U.S. border.

- The United States does not have an official language. Should it?

- North America has a very extensive freight-rail network but, if compared with other highly developed countries, passenger transport by rail is minuscule, and there is virtually no high-speed rail service.

- A Yale University poll during 2019 indicated that a record 73 percent of people in the United States said they believed climate change was real and important to them; 62 percent believed human activities were the leading cause of climate change.

FIGURE 3-1

Guadalupe Basilica church and the Mexico City skyline, as seen from Tepeyac Hill, which lies north of the city center. Mexico City is one of the largest metropolitan areas in the world and by far the biggest in Middle America. It also lies at the center of one of the hemisphere's oldest civilizations.

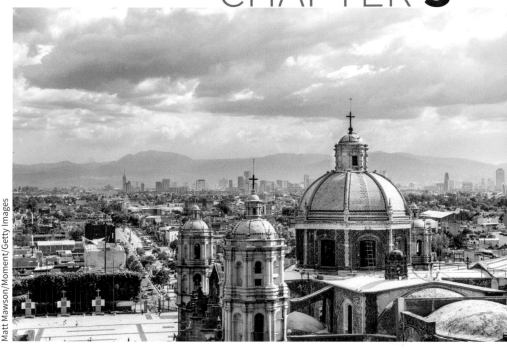

Matt Mawson/Moment/Getty Images

The Middle American Realm

IN THIS CHAPTER

- The Panama Canal: Doorway to U.S.–Asia trade
- Organized crime and violence in Middle America
- Puerto Rico still reeling from Hurricane Maria
- The paradox of U.S. involvement in Middle America
- Is small beautiful? The economic predicament of tiny island-nations

CONCEPTS, IDEAS, AND TERMS

Primate city	[1]	Tropical deforestation	[10]	Acculturation	[19]
Indigenous	[2]	Culture hearth	[11]	Transculturation	[20]
NAFTA	[3]	Mestizo	[12]	*Ejidos*	[21]
Borderland	[4]	Hacienda	[13]	Biodiversity hot spot	[22]
Maquiladoras	[5]	Plantation	[14]	Offshore banking	[23]
Land bridge	[6]	Small-island developing economies	[15]	Remittances	[24]
Archipelago	[7]	Economies of scale	[16]	Intermodal transport system	[25]
Hurricane Alley	[8]	Economic integration	[17]	Social stratification	[26]
Altitudinal zonation	[9]	Connectivity	[18]		

Look at a world or hemispheric map, and it is obvious that the Americas comprise two continents or landmasses: North America extending from Alaska to Panama and South America from Colombia to Argentina. But here we are reminded that continents and geographic realms do not necessarily coincide.

Between North America and South America lies the small but important geographic realm known as Middle America. Consisting of an impediment-filled mainland corridor and myriad Caribbean islands, Middle America is a particularly fragmented realm.

Defining the Realm

From **Figure 3-1** it is clear that Middle America is much wider than it is long. The distance from Baja California to Barbados is about 6000 kilometers (3800 mi), but from the latitude of Tijuana to Panama City it is only half that distance. In terms of total area, Middle America is the second smallest of the world's geographic realms. As the map shows, the dominant state of this realm is Mexico, larger than all its other countries and territories combined (see also **Box 3-1**).

A Realm of High Densities and Cultural Diversity

Middle America may be a rather small geographic realm by global standards, but comparatively it is densely peopled. Its population passed the 200-million milestone in 2011, more than half of it residing in Mexico alone, and the rate of natural increase (at 1.3 percent) remains above the world average. **Figure 3-2** shows Mexico's populous interior core area quite clearly, but note that the population in Guatemala and Nicaragua tends to cluster toward the Pacific rather than the Caribbean coast. Among the islands, this map reveals how crowded Hispaniola (containing Haiti and the Dominican Republic) is. In order to discern the pattern on the smaller islands of the eastern Caribbean and the Bahamas (some of which are also densely populated), you can zoom in on the interactive digital version of the map. In many Middle American countries, both on the mainland and in the islands, large **primate cities [1]** dominate the landscape, from Mexico City (nearly 21 million)

to Managua (1 million), which both contain about one-sixth of their national populations.

What Middle America lacks in size it makes up in physiographic and cultural diversity. This is a realm of soaring volcanoes and spectacular shorelines, of lush tropical forests and barren deserts, of windswept plateaus and scenic islands. It holds the architectural and technological legacies of ancient **indigenous [2]** civilizations. Today it is a mosaic of immigrant cultures from Africa, Europe, and elsewhere, richly reflected in music and the visual arts. Material poverty, however, is endemic: Haiti is often cited as the poorest country in the Americas; Nicaragua, on the mainland, is almost as badly off. As we will discover, a combination of factors has produced a distinctive but challenged realm between North and South America.

The "Latin" America Misnomer

Sometimes you will see Middle and South America referred to in combination as "Latin" America, alluding to their prevailing Spanish–Portuguese heritage. At the very least, this is an imprecise regional designation, much as Anglo America, a term once commonly used for North America, was also inappropriate. Such culturally based terminologies reflect historic power and dominance, and they tend to make outsiders of those people they do not represent. Thus you will not find the cultural landscape particularly "Latin" in the Bahamas, Jamaica, Trinidad, Belize, or lengthy stretches of Guatemala and Mexico.

But is Middle America sufficiently different from either North or South America to merit distinction as a separate

---

Box 3-1 Major Geographic Features of Middle America

1. Middle America is a relatively small realm consisting of eight mainland countries from Mexico to Panama and all the islands of the Caribbean Basin to the east.

2. Middle America's mainland constitutes a crucial barrier between Atlantic and Pacific waters. In physiographic terms, this is a land bridge connecting the continental landmasses of North and South America.

3. Middle America's cultural geography is complex. Various African and European influences dominate the Caribbean, whereas Spanish and indigenous traditions survive on the mainland.

4. This is a realm of intense cultural and political fragmentation. The presence of many small, insular, and remote countries poses major challenges to economic development.

5. Middle America's northern political boundary, despite current U.S. talk about building a wall, in many ways forms a dynamic borderland and transition zone.

6. In terms of area, population, and economic potential, Mexico leads the realm.

7. Many of the countries in this realm find themselves relatively isolated from the rest of the hemisphere.

POPULATION DISTRIBUTION: MIDDLE AMERICA

Population per Square Kilometer

0 to 5
5 to 10
10 to 50
50 to 100
100 to 500
Over 500

© J. Nijman, M. Shin, P.O. Muller and John Wiley & Sons, Inc.

FIGURE 3-2   **Interactive**

realm? Consider this: North America encompasses just a pair of states, and the entire continent of South America has only 12 states (plus France's dependency on the northeast coast). But far smaller Middle America, as we define it, incorporates more than three dozen political entities, including several dependencies (or quasi-dependencies) of the United Kingdom, the Netherlands, and France as well as a few constituent territories of the United States. Therefore, unlike South America, Middle America is a multilingual patchwork of independent states, territories in political transition, and residual colonial dependencies, with strong continuing ties to the United States and non-Iberian Europe. Middle America is defined in large measure by its vivid multicultural geography.

The Realm's Northern Boundary

The 3169-kilometer (1969-mi) land border between North America and Middle America is the longest in the world separating an affluent realm from a relatively poor one. The U.S.–Mexican boundary crosses half the continent from the Pacific to the Gulf of Mexico, yet Mexican cultural influences penetrate deeply into the southwestern States, and American impacts reach far inside Mexico. To some Mexicans, the border is a constant reminder of territory lost to the United States in historic conflicts; to many Americans, it is a symbol of economic contrasts and illegal immigration. Along the Mexican side, the effects of **NAFTA [3]**—the

North American Free Trade Agreement between Canada, the United States, and Mexico that went into effect in 1994—have transformed Mexico's economic geography.

Although the two realms are separated by a sharp political (and economic) boundary, this is a good example of a dynamic **borderland [4]** marked by considerable interaction, wherein the boundary can be more broadly viewed as the axis of a transition zone between North and Middle America. The implementation of NAFTA sparked an economic boom as Mexico became part of a free-trade zone and market now encompassing some 490 million people. A major beneficiary was the strip of Mexican land that ran along the boundary with the United States. The resulting boom changed urban landscapes all along the emerging border zone—but it could not, of course, close the massive economic gap between the two sides (see photo on next page).

Under NAFTA, factories based in Mexico can assemble imported, duty-free raw materials and components into finished products, which are then exported back into the U.S. market. Logically, most of these factories, called *maquiladoras* [5], are located as close to the U.S. border as possible. Thereby, manufacturing employment in the cities and towns along that border, from Tijuana on the Pacific to Matamoros at the mouth of the Rio Grande, expanded rapidly. Mexico now provides roughly 15 percent of all U.S. imports annually, the third-largest source of U.S. imports after Canada and China. China in particular has been increasing its share more rapidly, even though it faces tariffs that NAFTA-member Mexico (and Canada) does not. This puts pressure on Mexico to improve its competitiveness in order to build an even closer economic relationship with the United States and Canada.

David R. Frazier Photolibrary, Inc./Alamy Stock Photo

Automobiles wait at the border to enter the United States from Mexico at the Tijuana–San Ysidro port of entry just south of San Diego—the busiest land border crossing in the world. Here, as many as 70,000 northbound vehicles and 20,000 pedestrians enter the United States every day. There are currently 27 Mexico-bound inspection lanes and, after a new expansion, 34 U.S.-bound lanes.

A Troubled Border Zone

For several years, this has also been a troubled border zone in view of illegal border crossings by Mexicans and Central Americans looking for a better life in the United States. In 2018, there were nearly 400,000 illegal crossing attempts, but this number is part of a declining trend since 2000 because of tighter border controls and the increasing preference of potential migrants to remain in Middle America rather than head north (the number was about four times higher in 2000).

Another concern in the border zone is drug trafficking and associated organized crime and violence. Most cocaine is produced in northwestern South America, and the lion's share of what is destined for the U.S. market passes through Mexico (see Fig. 3-13). Much of the marijuana consumed within the United States is grown in western Mexico and similarly finds its way across the border. Contraband narcotics are mainly carried in trucks and cars, often imaginatively disguised as regular cargo (e.g., stuffed in teddy bears, inside hollowed-out furniture, or simply in boxes labeled as anything from candy to engine parts). It is the sheer volume of this illicit cross-border flow that makes it virtually impossible for authorities to conduct sufficient inspections. The infrastructure of cities and industrial areas on the Mexican side of the border further facilitates the operation of the drug trade.

The city of Ciudad Juárez seems ideally situated to handle this clandestine activity. It is located directly on the international boundary, and together with El Paso, Texas, on the U.S. side, forms a single metropolis—even though the two cities are separated by a continuous artificial barrier running across this entire urban area (Fig. 3-3). This map also shows the four ports of entry, which account for about 25 million border crossings every year. Juárez has grown quite rapidly and exhibits numerous manufacturing districts as well as a sprawling mosaic of neighborhoods,

ranging from clusters of upper-income *colonias* to the slums that house the poorest of the new migrants from the south. With a plentiful supply of poor people willing to take the risk of driving or carrying drugs across the border, Juárez also has dozens of *maquiladoras* that can be bribed to store, conceal, and transport the contraband north.

In 2006, the U.S. government, responding both to homeland security concerns (potential terrorists may be entering the country via Mexico) and the tidal wave of cocaine smuggling, initiated the construction of a fortified fence along the entire length of the Mexican border. The very idea is at odds with the ideals of closer economic and political cooperation and the economic integration of this border region, and many security experts questioned its efficacy.

In 2017, the U.S. administration embarked on a contentious new effort to extend a fortified barrier, or wall, along the entire length of the border—even though the precise meaning of the word *wall* (a wall, fence, or other kind of physical barrier, or a virtual barrier) remained unclear, as did the precise geographic locations where this new barrier should be built (see **Fig. 3-4**). The initiative was contested by many, including officials in various U.S. agencies such as Customs and Border Protection. First, data showed that illegal crossings were already on a downward trend since 2000. Second, data showed that most illegal migrants in the United States were those who had entered legally but stayed after their visas expired. Third, the bulk of illegal drugs smuggled into the United States come through legal ports of entry, as noted above, not across the wide swaths of the border that were not fortified (because any large-scale movement of drugs through such deserted areas would be highly conspicuous). Finally, the U.S.–Mexico land border is of very little significance in terms of terrorism threats. Known terrorist suspects typically arrive in the United States at airports and do not hail from Mexico or Middle America.

© J. Nijman, P.O. Muller and John Wiley & Sons, Inc.

FIGURE 3-3

THE U.S.-MEXICO BORDER:
NATURAL BOUNDARIES, CONSTRUCTED
BARRIERS, AND CROSSING POINTS

■ Official crossing point
▬ Current fencing along border
▬ Natural barrier along border
— Road

FIGURE 3-4

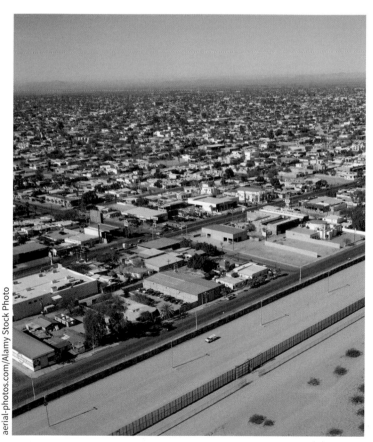

The boundary between Mexico and the United States displays some stunning cross-border contrasts. In this scene, the crowded Mexican town of San Luis Rio Colorado ends at a tall metal fence, beyond which the Arizona desert takes over. This photo, therefore, looks southwestward (south is toward the upper left corner). Like other Mexican towns pressed against the border, San Luis (the shorthand name used by locals) grew rapidly and chaotically in response to the opening of several *maquiladoras,* assembly plants fabricating products for export to the United States.

In 2019, the expansion of fortifications along the U.S.–Mexico border continued to be a highly politicized issue in the United States. Its symbolism mattered more to domestic politics and the presidential election of 2020 than to daily realities at the border.

Physical Geography of Middle America

A Land Bridge

The funnel-shaped mainland, a 4800-kilometer (3000-mi) connection between North and South America, is wide enough in the north to contain two major mountain chains and a vast interior plateau, but it narrows to a slim 65-kilometer (40-mi) ribbon of land in Panama. Here this strip of land—or *isthmus*—bends eastward so that Panama's orientation is east–west. Accordingly, mainland Middle America exemplifies what physical geographers call a **land bridge [6]**, an isthmian link between continents.

If you examine a globe, you can see some other present and former land bridges: Egypt's Sinai Peninsula between Asia and Africa, the (now-broken) Bering land bridge between northeasternmost Asia and Alaska, and the shallow waters between New Guinea and Australia. Such land bridges, though temporary features in geologic time, have played crucial roles in the dispersal of animals and humans across the planet. But even though mainland Middle America forms a land bridge, its internal fragmentation has always inhibited movement. Mountain ranges, swampy coastlands, and dense rainforests make contact and interaction difficult.

Island Chains

As shown in Figure 3-1, the approximately 7000 islands of the Caribbean Sea stretch in a lengthy arc from Cuba and the Bahamas eastward and then southward to Trinidad, with numerous outliers outside (such as Barbados) and inside (e.g., the Cayman Islands) the main chain. The four large islands in the Caribbean's east–west segment—Cuba, Hispaniola (containing Haiti and the Dominican Republic), Puerto Rico, and Jamaica—are called the *Greater Antilles*, and all the remaining smaller islands constitute the *Lesser Antilles*. The entire Antillean **archipelago [7]** (island chain) consists of the crests and peaks of mountain chains that rise from the floor of the Caribbean, the result of collisions between the Caribbean Plate and its neighbors (Fig. 1-5). Some of these crests are relatively stable, but elsewhere they contain active volcanoes, and almost everywhere in this realm earthquakes are an ever-present danger—in the islands as well as on the mainland. Add to this constant hazard the realm's seasonal exposure to Atlantic/Caribbean hurricanes, and it amounts to some of the highest-risk real estate on Earth.

Challenging Terrain and Atmosphere

The danger from below is dramatically illustrated in the Haitian, western half of Hispaniola, which is laced with geologic fault lines associated with the nearby boundary that separates the North American and Caribbean plates (see Fig. 3-19, inset map). On January 12, 2010, Haiti was devastated by a massive earthquake; the epicenter of that 7.0 temblor was located just outside Port-au-Prince and virtually destroyed this teeming, impoverished capital city of nearly 3 million. At least 300,000 people died, and within a week more than a million had fled to the countryside. It was the worst natural disaster in Haitian history but certainly not the first.

The environmental hazard from above comes in the form of hurricanes, powerful tropical cyclones that annually threaten the Caribbean Basin and its surrounding coastlines. The eastern half of Middle America is one of the most hurricane-prone areas in the world. One of the key conditions for the formation of hurricanes is very warm ocean water, which further heats the hot moist air rising above it to "fuel" the evolving storm. The prolonged Atlantic/Caribbean hurricane season extends from June 1 to December 1, with the greatest number of tropical cyclones occurring in August and September when seawater reaches its highest temperatures. Most storms travel in a westerly direction from their low-latitude spawning ground off the coast of West Africa, steered across the Atlantic by the trade winds to reach the Caribbean Sea. Once there, many of these cyclones follow similar routes within **Hurricane Alley [8]**, whose wide axis lies along all of the Greater Antilles and then broadens to include southern Florida, Mexico's Yucatán Peninsula, and all of the Gulf of Mexico (see photo below). On average, every season sees the development of four to eight major hurricanes, and rarely does a year go by without a destructive landfall on at least one of these densely populated areas.

Altitudinal Zonation of Environments

Continental Middle America and the western margin of South America are areas of high relief and strong environmental contrasts. Even though settlers have always favored temperate intermontane basins and valleys, people also cluster in hot tropical lowlands as well as high plateaus just below the snow line in South America's Andes Mountains. In each of these zones, distinct local climates, soils, vegetation, crops, domestic animals, and modes of life prevail. Such **altitudinal zones [9]** (diagrammed in **Fig. 3-5**) are known by specific names as if they were regions with distinguishing properties—as in reality they are.

The lowest of these vertical zones, from sea level to 750 meters (2500 ft), is known as the *tierra caliente*, the "hot land" of the coastal plains and low-lying interior basins where tropical agriculture predominates. Above this zone lie the tropical highlands containing Middle and South America's largest population clusters, the *tierra templada* of temperate land reaching up to about 1800 meters (6000 ft). Temperatures here are cooler; prominent among the commercial crops is coffee, while corn (maize) and wheat are the staple grains. Still higher, from approximately 1800 to 3600 meters (6000 to nearly

A satellite image of Hurricane Maria making landfall in Puerto Rico on September 20, 2017. It was the worst natural disaster on record for the island. It killed an estimated 3000 people, leveled around 160,000 homes, destroyed the power grid, and devasted agricultural areas. As the 2010s ended, recovery was still ongoing.

Claudia Weinmann/Alamy Stock Photo

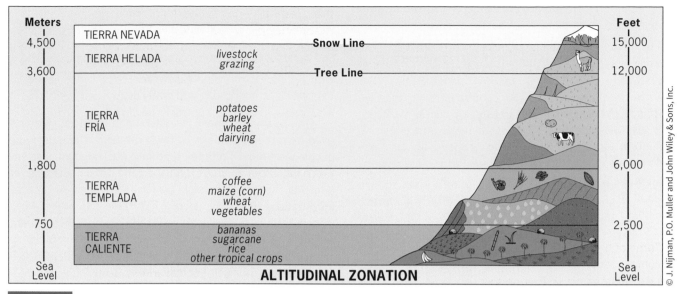

© J. Nijman, P.O. Muller and John Wiley & Sons, Inc.

FIGURE 3-5

12,000 ft), is the *tierra fría*, the cold country of the higher Andes where hardy crops such as potatoes and barley are mainstays. Above the tree line, which marks the upper limit of the *tierra fría*, lies the *tierra helada*; this fourth altitudinal zone, extending from about 3600 to 4500 meters (12,000 to 15,000 ft), is so cold and barren that it can support only the grazing of sheep and other hardy livestock. The highest zone of all is the uninhabited *tierra nevada*, a zone of permanent snow and ice associated with the loftiest Andean peaks. As we will see, the varied human geography of mainland Middle America and western South America closely reflects these diverse environments.

Tropical Deforestation

Before the Europeans arrived, two-thirds of continental Middle America (at lower altitudes) was covered by tropical rainforests. It is estimated that at present only about 10 percent of this vegetation remains. Between 2000 and 2010 alone, Central America lost almost 12 percent of its woodlands.

The causes of **tropical deforestation [10]** are related to the persistent economic and demographic problems of disadvantaged countries. In Central America, the leading cause has been the need to clear rural lands for cattle pasture as many countries, especially Costa Rica, became meat producers and exporters. Because tropical soils are so nutrient-poor, newly deforested areas are able to function as pastures for only a few years at most. These fields are then abandoned for other freshly cut lands and quickly become a ravaged landscape

(see photo below). Without the protection of binding tree roots, local soil erosion and flooding immediately become problems, affecting still-productive areas nearby. A second cause of deforestation is the rapid logging of tropical woodlands as the timber industry increasingly turns from the exhausted forests of the midlatitudes to harvest the rich tree resources of the equatorial zones in response to accelerating global demands for housing, paper, and furniture. A third major contributing factor is related to the region's population explosion: as more

Humberto Olarte Cupas/Alamy Stock Photo

Despite scattered attempts to reverse this landscape scourge, deforestation continues to afflict Central America. Its worst effects plague the steeper slopes of interior highlands, as here in western Panama. In the wake of recent deforestation, the land near the top of this hill has already begun to erode because in the absence of binding tree roots the copious tropical rains are making short work of the unprotected topsoil.

and more peasants are required to extract a subsistence from inferior lands, they have no choice but to cut down the remaining forest for both firewood and additional crop-raising space, and their intrusion prevents the trees from regenerating.

Cultural Geography

Mesoamerican Legacy

Mainland Middle America was the scene of the emergence of major ancient civilizations. Here lay one of the world's true **culture hearths** [11], a source area from which new ideas radiated outward and whose population could expand and make significant material as well as intellectual progress. Agricultural specialization, urbanization, trade, and transportation networks developed, and writing, science, art, and other spheres of achievement saw major advances. Anthropologists refer to the Middle American culture hearth as *Mesoamerica*, which extended southeast from the vicinity of what is now Mexico City to central Nicaragua. Its development is particularly remarkable because it occurred in highly different geographic environments, each presenting obstacles that had to be overcome in order to unify and integrate large territories. First, in the low-lying tropical plains of what is now northern Guatemala, Belize, and Mexico's Yucatán Peninsula, and perhaps simultaneously in Guatemala's highlands to the south, the Maya civilization arose more than 3000 years ago. Later, in the fourteenth century AD, far to the northwest on the high plateau in central Mexico, the Aztecs founded a major civilization centered on Tenochtitlán (present-day Mexico City) in the Valley of Mexico. Some of the greatest contributions of Mesoamerica's indigenous peoples came from the agricultural sphere and included the domestication of corn (maize), the sweet potato, cacao beans (the raw material of chocolate), and tobacco.

Spanish Conquest

Spain's defeat of the Aztecs in the early sixteenth century opened the door to Spanish penetration and supremacy. The Spaniards were ruthless colonizers, but not more so than other European powers that subjugated other cultures. The Spaniards enslaved the native population and were determined to destroy the strength of their society. Biology accomplished what ruthlessness could not have achieved in so short a time: diseases introduced by the Spaniards and the slaves they imported from Africa killed millions of indigenous people.

Middle America's cultural landscape was drastically modified. Unlike the indigenous peoples, who had utilized stone as their main building material, the Spaniards employed great quantities of wood and used charcoal for heating, cooking, and smelting metal. The onslaught on the forests was immediate, and rings of deforestation swiftly expanded around the colonizers' towns. The Spaniards introduced large numbers of cattle and sheep, and brought over their own crops (notably wheat) and farming equipment. Soon large wheat fields began to encroach upon the small plots of corn that the native people cultivated.

The Spaniards' most far-reaching cultural changes derived from their traditions as town dwellers. The indigenous people were moved off their land into nucleated villages and towns that the Spaniards established and laid out. In these settlements, the Spaniards could exercise the kind of rule and administration to which they were accustomed. The internal focus of each Spanish town was the central *plaza* or market square, around which both the local church and government buildings were located. The surrounding street pattern was deliberately laid out in gridiron form. Each town was located near what was thought to be good agricultural land (which often was not so good), so that the indigenous people could venture out each day and work in the fields. Packed tightly into these towns and villages, they came face to face with Spanish culture. Here they (forcibly) learned the Europeans' Roman Catholic religion and Spanish language, and they paid their taxes and tribute to a new master. Many of Middle America's leading cities still bear this Spanish imprint.

Collision of Cultures

But Middle America is not Spain, and its fragmented cultural mosaic reflects the collision of indigenous, Spanish, and other European influences. Indeed, in more remote areas in southeastern Mexico and interior Guatemala, indigenous societies survived, and to this day native languages prevail over Spanish (see **Fig. 3-6**).

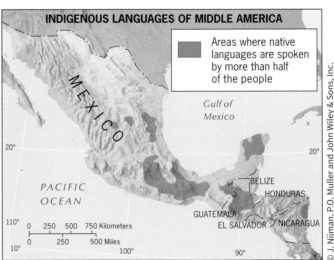

FIGURE 3-6

In Middle America outside Mexico, only Panama, with its twin attractions of interoceanic transit and gold deposits, became an early focus of Spanish activity. From there, following the Pacific-fronting side of the isthmus, Spanish influence radiated northwestward through Central America and into Mexico. The major arena of international competition in Middle America, however, lay not on the Pacific side but on the islands and coasts of the Caribbean Sea. Here the British gained a foothold on the mainland, controlling a narrow coastal strip that extended southeast from Yucatán to what is now Costa Rica. As the colonial-era map (**Fig. 3-7**) shows, in the Caribbean the Spaniards faced not only the British but also the French and the Dutch, all interested in the lucrative sugar trade, all searching for instant wealth, and all seeking to expand their empires.

Much later, after centuries of European colonial rivalry in the Caribbean Basin, the United States entered the picture and made its influence felt in the coastal areas of the mainland, not through colonial conquest but through the introduction of widespread, large-scale banana-plantation agriculture. The effects of these plantations were as far-reaching as the

impact of colonialism on the Caribbean islands. Because the diseases the Europeans had introduced were most rampant in these hot, humid lowlands (as well as the Caribbean islands to the east), the indigenous population that survived was too small to provide a sufficient workforce. This labor shortage was quickly remedied through the trans-Atlantic slave trade from Africa that transformed the population composition of the Caribbean Basin.

The cultural variegation of the Caribbean Basin is especially striking, and it is hardly an arena of exclusive Hispanic cultural heritage. For example, Cuba's southeastern neighbor, Jamaica (population 2.9 million, mostly of African ancestry), has a legacy of British involvement, while to the east in Haiti (11.3 million, overwhelmingly of African ancestry) the strongest imprints have been African and French. The Lesser Antilles also exhibit great cultural diversity. There are the (once Danish) U.S. Virgin Islands; French Guadeloupe and Martinique; a group of British-influenced islands, including Barbados, St. Lucia, and Trinidad and Tobago; and the Dutch St. Maarten (shared with the French) as well as the now-autonomous A-B-C islands of the former Netherlands

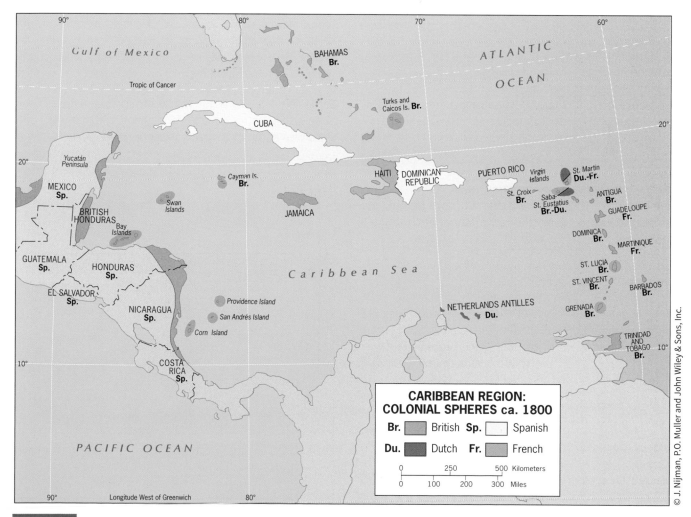

FIGURE 3-7

Antilles—Aruba, Bonaire, and Curaçao—off the northwestern Venezuelan coast.

Political and Economic Fragmentation

Independence

Independence movements stirred Middle America at an early stage. On the mainland, insurrections against Spanish authority (beginning in 1810) achieved independence for Mexico by 1821 and for the Central American republics by the end of the 1820s, resulting in the creation of eight different countries. The United States, concerned over European designs in the realm, proclaimed the Monroe Doctrine in 1823 to deter any European power from reasserting its authority in the newly independent republics or from further expanding its existing domains.

By the end of the nineteenth century, the United States itself had become a major force in Middle America. The Spanish-American War of 1898 made Cuba independent and placed Puerto Rico under the U.S. flag; soon thereafter, the Americans were in Panama constructing the Panama Canal across their newly acquired Canal Zone. Meanwhile, with U.S. corporations propelling a boom based on massive banana plantations, the Central American republics had become colonies of the United States in all but name.

Independence came to the Caribbean Basin in fits and starts. African-Caribbean Jamaica as well as Trinidad and Tobago, where the British had brought in a large South Asian population, attained full sovereignty from the United Kingdom in 1962; other British islands (among them Barbados, St. Vincent, and Dominica) became independent later on. France, however, retains Martinique and Guadeloupe as *overseas départements* of the French Republic, and the Dutch islands (see **Box 3-2**) are at various stages of autonomy. No less than 33 states are found on the political map of the Caribbean Basin today.

Regional Contrasts

There are some important contrasts, socially and economically, between the Middle American mainland on the one hand, and the Caribbean coasts and islands on the other. The mainland is dominated by Spanish as well as indigenous influences and also includes **mestizo [12]** sectors where the two ancestries mixed. Agriculture on the mainland is often organized around the **hacienda [13]**, a Spanish institution through which landowners possessed a domain whose productivity they might never push to its limits; the very possession of such a vast estate brought with it social prestige and a comfortable lifestyle. Native workers lived on the land—which may once have been their land—and had plots where they could grow their own subsistence crops. All this

---

Box 3-2 From the Field Notes . . .

"Driving around the small Caribbean island of Bonaire and coming face to face with this surreal landscape, I first thought I had experienced a mirage. These glistening white, perfectly cone-shaped hills are actually salt piles. This small (formerly Dutch) island off the coast of Venezuela possesses the perfect geographic conditions for salt production: it has a series of saltwater inlets, it is very hot and dry, and it lies in the zone of persistent trade winds. Remember that salt, nowadays taken for granted in every household, has long been one of the world's most precious spices and was widely used for the preservation of meat and fish. The Dutch began large-scale production on salt 'plantations' here in the 1620s; today this local industry is in the hands of the Antilles International Salt Company, and continues to be an important source of Bonaire's foreign revenues."

© J. Nijman

is written as though it is mostly in the past, but the legacy of the hacienda system, with its inefficient use of land and labor, still exists throughout mainland Middle America.

The Caribbean coasts (from Belize southward) and the Caribbean islands possess a very different cultural heritage, based on a fusion of European and African influences. Here, **plantations [14]** have long dominated the economies. The plantation, in contrast to the hacienda, is all about efficiency and profit. Foreign ownership and investment is the norm, as is production for export. Most plantations grow only a single crop, be it sugar, bananas, or coffee. Much of the labor is seasonal, being needed in large numbers mainly during the harvest period; such labor has been imported because of the scarcity of indigenous workers. With its "factory-in-the-field" operations, the plantation enables far more efficient use of land and labor than the hacienda. Profit and wealth, rather than social prestige, are the dominant motives for the plantation's establishment, operation, and perpetuation.

Small-Island Developing Economies

As noted, the Middle American realm is exceptional in the modest size of its territorial extent and population. But the number of countries is considerable, and they tend to be quite small. Caribbean islands often invoke images of paradise: beautiful scenery, tropical libations, and shiny blue waters. But the economic realities are almost invariably harsh. The limited land area and small population of most islands (those of the Lesser Antilles average less than half a million people), together with their insularity, relative inaccessibility, and low connectivity, pose formidable challenges that are common to **small-island developing economies [15]**.

What are the consequences of these geographical properties for economic development? First, natural resources are frequently limited, which requires a heavy reliance on imports made more expensive by added transport costs. Second, the cost of government is relatively high per capita: even the smallest population will require services such as schools, hospitals, and waste disposal. Third, these specialized services must often be brought in from elsewhere. And fourth, local production cannot really benefit from **economies of scale [16]** (unit-cost savings resulting from large-scale production). Consequently, local producers can be put out of business by cheaper imports, thereby driving up unemployment.

Given the Caribbean Basin's limited economic options, does the tourist industry provide better opportunities? The resort complexes, scenic treasures, and historic locales of Caribbean America attract more than 50 million visitors annually, with about half of them traveling on Florida-based cruise ships. But Caribbean tourism also has serious drawbacks. The invasion of poor communities by affluent tourists contributes to rising local resentment, which is fueled by the glaring contrasts of shiny new hotels towering over substandard housing, or luxury liners gliding past poverty-stricken villages. At the same time, tourism can debase local culture, which often is adapted to suit the visitors' tastes at hotel-staged "culture" shows. In addition, the cruise industry tends to monopolize revenues (accommodations, meals, and entertainment), with relatively few dollars flowing into the local economy. Finally, even though tourism does generate considerable income in the Caribbean, the intervention of island governments and multinational corporations can often remove opportunities from local entrepreneurs in favor of large operators and exclusive resorts.

The Push for Regional Integration

Another challenge for the countries of the Middle American realm is to foster greater **economic integration [17]** (the benefits of forging new international partnerships). Many of the countries on the mainland as well as in the Caribbean are poorly connected within the realm and are heavily dependent on major outside countries, particularly the United States. For the large majority of these countries, the United States is their primary trading partner. Consider this: of all the trade that engages Middle American countries, less than 10 percent occurs within the realm. And less than 1 percent takes place between the Caribbean Basin and the Middle American mainland.

Over the years, efforts have been made to advance economic integration and convert this realm into more of a *functional region*. There are presently two trade organizations in Middle America (in addition to Mexico's membership in NAFTA). *CARICOM* (Caribbean Community), established in 1989, consists of 15 full members, including nearby Guyana and Suriname in South America. To a certain degree, CARICOM emulates the European Union (EU), and in 2009 it even introduced a common passport; but unlike the EU, this organization has had only a superficial economic impact. The more influential *CAFTA-DR* (Central American Free Trade Agreement), which began operating in 2006, includes the United States, five Central American countries, and the Caribbean's Dominican Republic (hence the "DR" suffix). But this organization seems to be a mixed blessing: it may increase access to U.S. markets and lead to cheaper imports, yet it also reinforces the United States' dominant position at the expense of greater intra-realm integration.

The Importance of Regional Connections

The unusual layout of the Middle American realm offers an opportunity to consider the spatial dimensions of development. Geographical fragmentation and territorial size play their parts, but so does *accessibility* in the form of **connectivity [18]** in both the Caribbean Basin and on the mainland. **Figure 3-8** in **Box 3-3** provides a closer look at the relationship between development and connectivity.

Box 3-3 Geographic Information Analysis

Connectivity and Economic Development in Middle America

This visualization shows annual GDP per capita as well as the number of international air passengers in 2018. Mapping these variables allows us to analyze them in spatial context and compare their locational patterns. What seems to be the relationship between development and connectivity? What are the key geographic factors that shape connectivity within Panama and Mexico? (Also see the sections on these two countries later in the chapter).

Access your WileyPLUS course to interact with a dynamic version of this map and to engage with online exercises and questions.

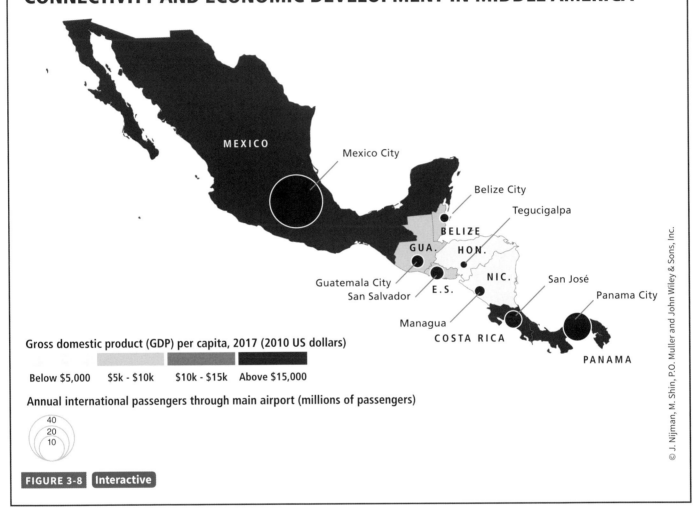

CONNECTIVITY AND ECONOMIC DEVELOPMENT IN MIDDLE AMERICA

© J. Nijman, M. Shin, P.O. Muller and John Wiley & Sons, Inc.

Gross domestic product (GDP) per capita, 2017 (2010 US dollars)

Below $5,000 $5k - $10k $10k - $15k   Above $15,000

Annual international passengers through main airport (millions of passengers)

40
20
10

FIGURE 3-8 Interactive

Regions of Middle America

The Middle American realm can be subdivided into four distinct geographic regions (the first two on the mainland; the other two in the Caribbean Basin): (1) *Mexico*, the giant of the realm in every respect; (2) *Central America*, the string of seven small republics occupying the land bridge from Mexico to South America; (3) the four islands of the *Greater Antilles*—Cuba, Jamaica, Hispaniola (containing Haiti and the Dominican Republic), and Puerto Rico; and (4) the numerous small islands of the *Lesser Antilles* (**Fig. 3-9**).

FIGURE 3-9

Region **Mexico**

Physiography

The physiography of Mexico is reminiscent of that of the conterminous western United States, although its environments are more tropical. **Figure 3-10** shows several prominent features: the elongated peninsula of Baja (Lower) California in the northwest, the far eastern Yucatán Peninsula, and the Isthmus of Tehuantepec in the southeast where the Mexican landmass tapers to its narrowest extent. Here in the southeast, Mexico's physiography most resembles that of Central America: a mountainous backbone forms the isthmus, curves southeastward into Guatemala, and extends northwestward toward Mexico City. Shortly before reaching the capital, this mountain range divides into two chains, the Sierra Madre Occidental to the west and the Sierra Madre Oriental to the east. These diverging ranges frame the funnel-shaped Mexican heartland, the center of which consists of the rugged, extensive Plateau of Mexico (the important Valley of Mexico lies near its southeastern end). Mexico's climates are characterized by dryness, particularly in the broad mountain-flanked north (also see Fig. 1-6). Most of the better-watered areas lie in the southern half of the country where a number of major population concentrations have developed (see Fig. 3-2).

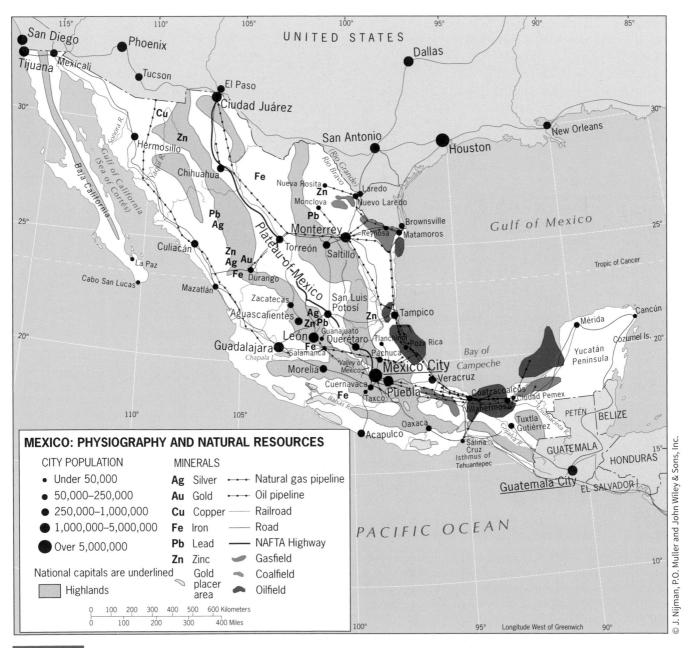

MEXICO: PHYSIOGRAPHY AND NATURAL RESOURCES

CITY POPULATION
- Under 50,000
- 50,000–250,000
- 250,000–1,000,000
- 1,000,000–5,000,000
- Over 5,000,000

National capitals are underlined
Highlands

MINERALS
Ag Silver
Au Gold
Cu Copper
Fe Iron
Pb Lead
Zn Zinc
Gold placer area

Natural gas pipeline
Oil pipeline
Railroad
Road
NAFTA Highway
Gasfield
Coalfield
Oilfield

0 100 200 300 400 500 600 Kilometers
0 100 200 300 400 Miles

© J. Nijman, P.O. Muller and John Wiley & Sons, Inc.

FIGURE 3-10

Regional Diversity

Physiographic, demographic, economic, historical, and cultural criteria combine to reveal a regionally diverse Mexico extending from the lengthy ridge of Baja California to the tropical lowlands of the Yucatán Peninsula, and from the economic dynamism of the U.S. borderland to the indigenous traditionalism of far southeastern Chiapas State. The country's core area, anchored by Mexico City and extending westward to Guadalajara, lies between the more Hispanic–mestizo north and the dominantly indigenous–mestizo south. East of the core area lies the Gulf Coast, once dominated by major irrigation projects and sprawling livestock-raising schemes but now the mainland center of Mexico's petroleum industry. The region south of the core is dominated by the rugged, Pacific-fronting Southern Highlands, where coastal Acapulco's luxurious resorts stand in stark contrast to the indigenous villages and communal-farm settlements of the interior. The dry, far-flung north stands in particularly sharp contrast to these southern regions. To go from comparatively well-off Monterrey in the far north to poverty-mired Chiapas adjacent to southern Guatemala is to observe the entire range of Mexico's regional geography.

Hundreds of thousands of North American retirees, and many affluent purchasers of second homes, have sought the sun and low-cost residential opportunities of Middle America, converting some areas into virtual exclaves of the North. From Mexico to Panama, waterfront real estate is among the attractions for these permanent and seasonal migrants, as here in Cabo San Lucas at the southern tip of Mexico's Baja California peninsula. This is yet another example of international connections that matter.

Population and Urbanization Patterns

Mexico's population expanded rapidly throughout the closing decades of the twentieth century, doubling in just 28 years; but demographers have recently noted a sharp drop in fertility, and they are predicting that Mexico's population (currently at 128 million) will likely cease growing altogether by about 2030. That will have important consequences for the country's economy and for the United States as well since it is likely to affect long-term migration patterns.

The distribution of Mexico's population relative to the country's 31 internal States is shown in Figure 3-2 and **Figure 3-11**. The largest concentration, containing the core area and more than half of the Mexican people, extends across the densely populated "waist" of the country from Veracruz State on the eastern Gulf Coast to Jalisco State on the Pacific. The center of this corridor is dominated by the most populous State, Mexico (**3** on the map), at whose heart lies the Federal District of Mexico City (**9**). In the dry and rugged terrain to the north of this central corridor lie Mexico's least-populated States. Southern Mexico also exhibits a sparsely peopled periphery in the hot and humid lowlands of the Yucatán, but to the southwest most of the highlands of the continental spine contain sizeable populations.

Another major feature of Mexico's population map is urbanization, driven by the pull of the cities (with their perceived opportunities for upward mobility) in tandem with the push of the economically stagnant countryside. Today, 83 percent of the Mexican people reside in towns and cities, a surprisingly high proportion for a less-developed country. Undoubtedly, these numbers are affected by the recent explosive growth of the region around Mexico City, which has surpassed 20 million and is home to one-sixth of the national population. Urbanization rates are at their lowest in the peripheral southern uplands where indigenous society has been least affected by modernization.

A Mix of Cultures

Nationally, the indigenous imprint on Mexican culture remains prevalent. Today, roughly 60 percent of all Mexicans are mestizos, about 30 percent are predominantly indigenous (of which 7 percent are full-blooded indigenous); almost all of the remaining 10 percent are Europeans. The Spanish influence in Mexico has been profound, but it has been met with an equally powerful thrust of indigenous culture. It has therefore not been a case of one-way European-dominated **acculturation [19]** but rather **transculturation [20]**—the two-way exchange of culture traits between societies in close contact. In the southeastern periphery (Fig. 3-6), several hundred thousand Mexicans still speak only an indigenous language, and millions more still utilize these languages daily even though they also speak Mexican Spanish. The latter has been strongly shaped by indigenous influences, as have Mexican modes of dress, cuisine, artistic and architectural styles, and folkways. This fusion of heritages, which makes Mexico unique, is the result of an upheaval that began to reshape the country just over a century ago.

Liane Cary/AGE Fotostock

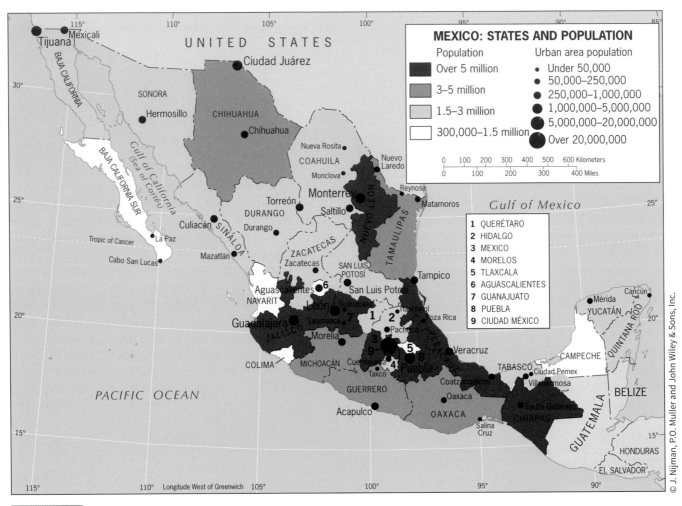

MEXICO: STATES AND POPULATION

Population
- Over 5 million
- 3–5 million
- 1.5–3 million
- 300,000–1.5 million

Urban area population
- Under 50,000
- 50,000–250,000
- 250,000–1,000,000
- 1,000,000–5,000,000
- 5,000,000–20,000,000
- Over 20,000,000

1 QUERÉTARO
2 HIDALGO
3 MEXICO
4 MORELOS
5 TLAXCALA
6 AGUASCALIENTES
7 GUANAJUATO
8 PUEBLA
9 CIUDAD MÉXICO

FIGURE 3-11

Agriculture: Fragmented Modernization

Following the Mexican Revolution of the 1910s and its ensuing land redistribution schemes, around half of the arable land was organized into so-called *ejidos* [21], communally-owned farms of 20 or more families. This meant an abrupt end for many large, single-family-owned haciendas. The overhauled system of land management was an indigenous legacy, and not surprisingly most *ejidos* lie in central and southern Mexico where nativist social and agricultural traditions are strongest. Today, about half of Mexico's land continues to be held in such social land-holdings. But some of this land is excessively fragmented and/or managed inefficiently with scarce capital, resulting in both inferior crop yields and widespread, persistent rural poverty. In the 1990s, the Mexican government attempted to privatize *ejidos*, hoping to promote consolidation and increase productivity, but that effort was unsuccessful.

At the same time, larger-scale commercial agriculture has diversified during the past three decades and made major gains in both domestic and export markets. The country's arid northern tier has led the way as major irrigation projects have been built on streams flowing down from the interior highlands. Along the booming northwestern coast of the mainland, which lies within a day's drive of Southern California, large-scale production of fruits, vegetables, cut flowers, and cotton has become the cornerstone of an increasingly profitable export trade to the United States and elsewhere.

Evolving Economic Geographies

The *maquiladoras* in the country's northern borderland—foreign-owned factories that assemble imported raw materials and components and then export finished manufactures back to the United States—account for about one-sixth of Mexico's industrial jobs and a major share of its total exports. *Maquiladoras* presently employ roughly two million Mexican workers. Many companies are headquartered in cities just north of the border, such as San Diego and El Paso, and this is where design, marketing, and other strategic work takes place. But most of the workforce is located south of the border at the assembly plants (which is why the cities on the Mexican side have grown so much faster than their U.S. counterparts). They produce electronics, machinery, clothing, construction materials, automobile parts, and much more. Although the impact of NAFTA has been widespread, it should be noted that Mexican employees work long hours for low wages with meager fringe benefits. Moreover, most reside in the overcrowded rudimentary settlements that encircle the burgeoning urban centers of the border zone. In 2018, the minimum daily wage in Mexico was raised to U.S. $5. It is estimated that 15 percent actually earn less, whereas more highly skilled workers in the north might make three times the minimum wage. The pay-scale disparity with the adjacent United States, however, remains enormous.

Mexico's government seeks to encourage the growth of higher-paying jobs in the more advanced economic sectors—particularly electronic goods—that are also less likely to be lost to other parts of the world. Expansion of education and training in high-technology and management fields is a critical first step. The northeastern city of Monterrey in the relatively high-income State of Nuevo Léon has become a successful model in this effort, nurturing both an international business community and an ultramodern industrial complex that has attracted leading multinational corporations. Here indeed lies hope for Mexico's future, and one frequently hears reports these days that rising labor costs in China may induce some manufacturers to return to northern Mexico, allowing them to more directly control flexible (so-called "on-time") production at a much closer distance.

States of Contrast

Countries with strong internal regional disparities face serious challenges that can be difficult to overcome and may intensify over time. Mexico's southernmost States—Chiapas, Oaxaca, and Guerrero, all bordering the Pacific Ocean—are by far the poorest. The States bordering the United States in the north, including Nuevo Léon, Chihuahua, and Baja California, have the highest incomes. Using rural poverty as a measure, we find that only about 10 percent of people residing in the countryside of the north rank in the poorest category—but in the south nearly 50 percent do.

Mexico's north–south divide is especially noticeable in the economic data mapped in **Figure 3-12**. In general terms, annual per capita income in the northern States exceeds U.S. $10,000; but in the southern States, it falls below $5000. Moreover, Mexico's infrastructure, already substandard, serves the south far less well than the north. Whatever the index—literacy, electricity use, water availability—the south lags by a wide margin. The population of the south not only trails far behind the rest of Mexico in overall development, but it is also the least well educated, the least productive agriculturally, and the most isolated in the country. Since the early 1990s, a radical group of Maya peasant farmers in Chiapas State, calling themselves the

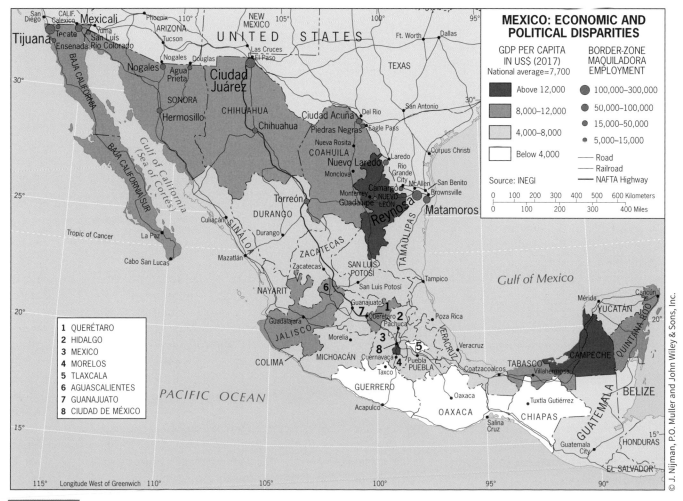

FIGURE 3-12

Zapatista National Liberation Army (ZNLA), has engaged in guerrilla warfare, demanding better treatment for Mexico's more than 30 million indigenous citizens. Despite widespread public support, their long struggle has not yielded noticeable results.

The Drug Wars

Over the past two decades, Mexico's efforts to achieve sustained economic progress have been hampered by the destructive presence of drug cartels and related organized crime. This problem has received worldwide attention as the cocaine-producing drug cartels centered in Colombia established new bases in U.S.-border cities in northern Mexico and launched a vicious war for supremacy.

Figure 3-13 provides an overview of the geography of this drug war. Cocaine is produced in Colombia, Bolivia, and Peru, and a large share of it enters the United States through Mexico. Bolivian and Peruvian cocaine, constituting approximately half of Mexico's "imports," is shipped by sea and illegally enters the country in Pacific Coast States such as Guerrero and Sinaloa. Most shipments originate in Colombia and find their way into Mexico along various routes: either overland via the Middle American

land bridge through Guatemala; by boat across the Caribbean and Gulf of Mexico to be smuggled into the eastern States of Yucatán, Quintana Róo, and Veracruz; or by boat on the Pacific side, along with the cocaine from Bolivia and Peru. The illicit powder is sometimes repackaged (and/or crystallized to form crack) in smaller quantities for "retailing" purposes prior to being smuggled into the United States, mainly through such border cities as Reynosa, Nuevo Laredo, Ciudad Juárez, and Tijuana (Fig. 3-13).

Interestingly, when Mexican criminal organizations were co-opted by the Colombian cartels, they demanded to be paid off in cocaine. This allowed the Mexicans to start up their own distribution networks; it also meant that some of the product "stuck around" and ended up on the streets of Mexican cities, resulting in heightened drug addiction and skyrocketing crime. Several major Mexican cartels sprang up and got involved in this drug-trafficking operation, and competition continues to literally be murderous. Territorial control is key in this business: each cartel dominates its own turf, with various components being heavily contested from time to time. The cartels themselves are not particularly stable, with countless mergers and splits occurring over time. The most powerful rivals include Los Zetas and the Sinaloa Cartel, which control huge swaths of eastern and

FIGURE 3-13

In 2012, students in Mexico City staged this protest against cocaine-trafficking violence. Although a large number of drug-related crimes go unreported, an official peak of 22,480 murders had been recorded for the year 2011. The year 2018 saw a new record, with estimates of 29,000 murders, most of them related to cartel rivalries.

western Mexico, respectively. According to U.S. government estimates, the Mexican drug cartels take in around $24 billion annually from drug sales in the United States.

Much of the worst violence occurs where control is contested, especially in transit areas, along primary transport routes, and near border zones. And it is not only about cocaine: the cartels also vie for control of Mexican-grown marijuana and heroin (poppy). By 2018, the murder total since the drug wars erupted was estimated to exceed 150,000. The intensity of the violence tends to fluctuate, depending mainly on the status of relations among the different cartels (Fig. 3-13).

Region The Central American Republics

A Land Bridge

Crowded onto the narrow segment of the Middle American land bridge between Mexico and the South American continent are the seven countries of Central America (**Fig. 3-14**). Territorially, they are all quite small; their population sizes range from Guatemala's 17.6 million down to Belize's 390,000. The land bridge

FIGURE 3-14

here consists of a highland belt flanked by coastal lowlands on both the Caribbean and Pacific sides. These uplands are studded with volcanoes, and local areas of fertile volcanic soils are scattered throughout them (**Fig. 3-15**).

The land bridge has a fascinating geologic and evolutionary history, and one famous study referred to it as the Monkey Bridge. For some 50 million years, North and South America were separated; the land bridge was formed only 3 million years ago, becoming a biologic highway of sorts for evolutionary exchange. The region contains only 1 percent of the Earth's land but 7 percent of all the world's natural species. The southern part (Costa Rica and Panama) is known as a global **biodiversity hot spot [22]**, even though deforestation has been a major problem. Human inhabitants have always been concentrated in the upland zone, where tropical temperatures are moderated by elevation and rainfall is sufficient to support a variety of crops.

Trapped in the Narcotics Trade

Unfortunately for Central America, during the 2010s the region was increasingly forced to accommodate the cocaine trade. Its intermediate location between northern South America and Mexico made the Central American land bridge the only possible route when narco-traffickers decided to expand their overland transit operations. Consequently, much of the cocaine that now reaches the United States through Mexico has passed through at least one Central American country (Fig. 3-13).

Given the region's poverty and severely limited economic opportunities, relatively large numbers of locals have been drawn into the potentially lucrative drug trade, sparking a steep rise in organized criminal activity and an explosion of violence, especially in the countries closest to Mexico—Guatemala, Honduras, and El Salvador (collectively known as the *Northern Triangle*) plus Belize. Over the last ten years, several Central American countries have had the highest murder rates in the world. In 2016, the most recent year for which reliable data are available, El Salvador and Honduras topped the list while Belize and Guatemala also ranked in the top ten. Central American governments, with few resources to call upon, were virtually powerless to respond to the criminal cartels and affiliated vicious youth gangs, while most businesspeople and civilians had learned long ago never to rely on the police or military. Today, in the Honduran capital of Tegucigalpa, private security forces (for those able to afford them) outnumber police personnel four to one.

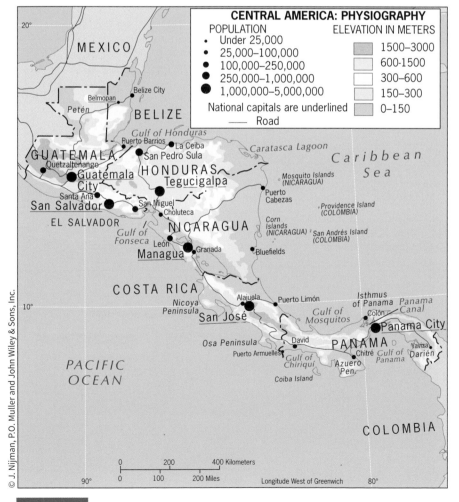

FIGURE 3-15

Guatemala

The westernmost of Central America's republics, Guatemala has more land neighbors than any other. Straight-line boundaries lying across the tropical forest mark much of the border with Mexico, creating the box-like Petén subregion between Chiapas State on the west and Belize on the east; also to the east lie Honduras and El Salvador (Fig. 3-15). This heart of the ancient Maya Empire, which remains strongly permeated by indigenous culture and traditions, has only a small window on the Caribbean but a longer Pacific coastline. Most populous of the seven republics with 17.6 million inhabitants (mestizos are in the majority at about 55 percent; indigenous number about 45 percent), Guatemala has seen a great deal of conflict, and military regimes have dominated political life. There is a deepening split between the poor indigenous populations and the better-off mestizos who continue to control the government, military, and land-tenure system.

Guatemala's economic geography demonstrates considerable potential but has long been shackled by unrelenting internal conflicts and the widening of one of the hemisphere's biggest gaps between rich and poor. In tandem, they keep the income of well over half the population below the poverty line. The country's mineral wealth includes nickel in the highlands and oil in the lower-lying north. Agriculturally, soils are fertile and moisture is ample over highland areas large enough to produce a wide range of crops, including superlative coffee. The 929-kilometer (577-mi) border with Mexico offers opportunities as well, but progress has been thwarted by out-of-control drug trafficking and violent crime.

Belize

Strictly speaking, Belize is not a Central American republic in the same tradition as the other six. Until 1981, this country, a wedge of land between northern Guatemala, Mexico's Yucatán Peninsula, and the Caribbean Sea (Fig. 3-15), was a dependency of the United Kingdom known as British Honduras. Slightly larger than Massachusetts and containing a minuscule population of just under 400,000 (many of African descent), Belize is much more reminiscent of a Caribbean island than of a continental Middle American state. Today, all that is changing as the demographic complexion of Belize is being reshaped. Thousands of residents of African descent have emigrated (many went to the United States) and were replaced by tens of thousands of Spanish-speaking immigrants. Most of the latter have been land seekers and refugees from strife in nearby Guatemala, El Salvador, and Honduras, and their proportion of the Belizean population has risen from one-third in 1980 to well over half today. With the newcomers now in the majority, Belize's cultural geography is becoming increasingly Hispanicized; in fact, since 2012 Spanish speakers outnumber those who use English or creole as a first language.

The Belizean transformation extends to the economic sphere as well. No longer just an exporter of sugar and bananas, Belize today is producing new commercial crops, and its seafood-processing and clothing industries have become major revenue earners. Also important is tourism, which annually lures about 400,000 vacationers to the country's Maya ruins, resorts, and newly legalized casinos; a booming specialty is *ecotourism*, based on the natural attractions of the country's near-pristine environment, including its magnificent offshore coral reefs. Belize is also known as a center for **offshore banking [23]**, a financial haven for foreign companies and individuals who endeavor to avoid paying taxes in their home countries.

Honduras

Honduras, in contrast with Guatemala, has a long Caribbean coastline and a relatively small window on the Pacific (Fig. 3-15). The country also occupies a critical place in the political geography of Central America, flanked as it is by Nicaragua, El Salvador, and Guatemala—all continuing to grapple with the aftermath of years of internal conflict, recent natural disasters, and the ongoing ravages of drug trafficking. Comparable in natural beauty and biodiversity to Costa Rica, Honduras is constrained by a defective infrastructure and inadequate funding (see **Box 3-4**).

Box 3-4 From the Field Notes . . .

"Invited by the Honduran government to survey the country's ecotourism potential, I toured the beautiful countryside with a National Geographic Society delegation. There is no doubt that Honduras is exceptionally endowed by nature and that there is considerable potential for a vibrant tourist industry. But this is a poor country, and the government has great difficulty making the required investments—in infrastructure, for example. We flew in these small propeller planes to get from the capital, Tegucigalpa, to the Maya ruins of Copán. There was no airport nearby, and overland travel is arduous and time consuming. Ecotourism by its very nature is small scale, and the revenues tend to be relatively modest, so it is a major challenge to raise the substantial funds needed for costly infrastructural improvements."

© J. Nijman

Agriculture, livestock, forestry, and limited mining form the mainstays of the economy, with the now-familiar Central American cluster of products—apparel, bananas, coffee, and shellfish—earning most of the external income. With just under 10 million inhabitants, about 90 percent mestizo, Honduras faces huge development challenges that are complicated by escalating crime and social dislocation.

In an effort to move forward, Honduras has tried to pursue a development strategy aimed at the creation of quasi-independent "special development regions." The idea is to attract foreign investment into these protected zones, which are to be independent in their economic, fiscal, and budgetary policies. This initiative follows the charter-cities model that has proven successful in Hong Kong, but critics fear that it may only clear the way for organized crime to take full control of these new economic enclaves—particularly in the besieged city of San Pedro Sula ("murder capital of the world" for much of the 2010s).

El Salvador

El Salvador is Central America's smallest country territorially, smaller even than Belize, but with a population 17 times as large (6.5 million) it is the region's most densely peopled. El Salvador adjoins the Pacific in a narrow coastal plain backed by a chain of volcanic mountains, behind which lies the country's heartland (Figs. 3-14, 3-15). Unlike neighboring Guatemala, El Salvador has a far more homogeneous population (86 percent mestizo and just 1 percent indigenous). Yet ethnic uniformity has not translated into social or economic equality or even opportunity. Whereas other Central American countries were sometimes derisively called "banana republics," El Salvador was a "coffee republic," and the coffee was produced on the huge landholdings of a few landowners and on the backs of a subjugated peasant labor force.

More affluent El Salvadorans who left the country in past decades, and who have succeeded in the United States and elsewhere, send substantial funds back home; these **remittances [24]** now provide the largest single source of foreign revenues (also see **Box 3-7**; Fig. 3-17). This has helped stimulate such industries as apparel and footwear manufacturing as well as food processing. But a major stumbling block to revitalization of the agricultural sector has again been land reform, and in this regard El Salvador's future remains uncertain.

Panama

Panama reflects some of the usual geographic features of the Central American republics. Its population of 4.3 million is about two-thirds mestizo but also contains substantial indigenous, European, and black minorities. Spanish is the official language, but English is also widely used. Ribbon-like and oriented east–west, much of Panama's topography is mountainous and hilly. Eastern Panama, especially Darien Province adjoining Colombia, is densely covered by tropical rainforest and contains the only remaining gap in the intercontinental Pan American Highway (Fig. 3-14). Most of the rural population lives in the uplands west of the Panama Canal; there, Panama produces bananas, coffee, sugarcane, rice, and, along its narrow coastal lowlands, shrimps and other seafood. Much of the urban population is concentrated in the vicinity of the famous canal, anchored by the cities at each end of this pivotal waterway.

Panama owes its existence to the idea of an artificial waterway connecting the Atlantic and Pacific oceans to avoid the lengthy circumnavigation of South America. The Panama Canal (see the inset map in Fig. 3-14) was opened in 1914, a symbol of U.S. power and influence in Middle America. It was both a massive feat of engineering and an especially dangerous undertaking in a hostile tropical environment. It is estimated that 25,000 workers died during the ten years of excavation and construction. The canal itself was contained within the cross-isthmian *Canal Zone*, an elongated strip of land 16 kilometers (10 mi) wide, with the waterway forming its centerline. This corridor was ceded to the United States in 1904 by newly independent Panama under a treaty granting the United States "all the rights, powers, and authority" in the Zone "as if it were the sovereign of the territory." In the 1970s, Panama sought to terminate U.S. control in the Canal Zone, and delicate negotiations were opened. In 1977, an agreement was reached on a staged withdrawal by the United States from the territory, first from the Canal Zone and then from the Panama Canal itself (a process completed on December 31, 1999).

Near the northern, Caribbean end of the Panama Canal lies the city of Colón, site of the Colón Free Zone, a huge trading hub designed to transfer and distribute goods on their way to South America. Near the southern, Pacific end lies the capital, Panama City, often likened to Miami because of its waterfront location and high-rise-dominated skyline. The capital is the financial center that handles the revenues generated by the Canal, but its skyscrapered profile also reflects the proximity of Colombia's illicit drug industry and associated money-laundering and corruption.

Until 2016, ships sailing through the Panama Canal could not be wider than 32 meters (106 ft). The Canal's much wider third lane, which opened in mid-2016, accommodates ("Neo-Panamax") vessels up to 55 meters (180 ft) in width that can carry two-and-a-half times as many shipping containers (up to 13,000) (see **Box 3-5** and **Box 3-6**). This will boost interoceanic traffic and increase business opportunities in Panama. U.S. ports on the Gulf of Mexico and the Atlantic seaboard will also get busier as they handle this heightened volume of cargoes. The ports of Miami and the New York–New Jersey metropolitan region are expanding their capacities, and leading retailers such as Walmart (a massive importer of Chinese goods) are building additional storage facilities at the Port of Houston and other key coastal distribution points (see also Box 2-4, the Regional Planning Case in Chapter 2).

Box 3-5 Panama Canal Expansion

Its 5000 containers piled high, a Panamax-sized cargo carrier squeezes through a lock in one of the Panama Canal's original lanes two years before the 2016 enlargement (left photo). The right photo was taken at the climax of opening day (June 26, 2016) as the first Neo-Panamax ship, carrying 9472 containers, approached the final new lock to complete its inter-oceanic transit at the Panama City outlet to the Pacific.

NEW FEATURES

1 Retaining basins fill lock chambers and retain 60 percent of the water for re-use.

2 Rolling lock gates slide in recesses, saving space and making maintenance work much easier.

3 Tugboats maneuver vessels into position, replacing costly towing locomotives.

New lane ▶

Locks transfer vessels from one elevation to another by filling closed chambers with water, or emptying them. When the level is even with that of the next chamber, the gates open.

NEO-PANAMAX SHIP

160 ft
60 ft
180 ft

PANAMAX SHIP
Pre-2016
maximum size

106 ft   42 ft
110 ft

Adapting to larger ships

The most recent container vessels and tankers are known as Neo-Panamax. They can hold up to 13,000 containers, 2.5 times the cargo capacity of ships able to fit in the existing locks.

Diagram showing the expansion and upgrading of the Panama Canal's lock system in order to accommodate ships that exceed the waterway's original ("Panamax") capacity. This massive construction project to build a third lane to handle "Neo-Panamax" vessels took 9 years to complete. It opened nearly two years late in mid-2016.

Mika Grondahl/National Geographic Image Collection

Box 3-6 Technology & Geography

The Shipping Container

Of all the innovations in transportation technology during the past century, the lowly shipping container continues to be one of the most significant to the global economy and trade between (and within) world regions. It is a technological breakthrough that is deceptively simple: how can a metal box measuring 20 (or now 40) feet (6/12 m) long, 8 feet (2.4 m) wide, and 8 feet high—or "twenty-foot equivalent units (TEUs)"—matter so much?

The ingenuity of the shipping container is fourfold. First, its uniformity allows universal, high-volume stacking and storage (increasingly enabling the use of supersized ships); it also facilitates mechanized or automated loading and unloading. Second, containers have driven the growth of highly efficient **intermodal transport systems [25]** involving ships, trains, and trucks. Because all these forms of transport are compatible with containers, they are often referred to as *intermodal freight containers.* Third, transport companies or clients can track individual containers with GPS, allowing for precise logistical planning. Finally, the durable steel frame of the container allows reuse, with the lifespan of the average container exceeding ten years.

An estimated 90 percent of the goods we consume travels by sea. The clothes we wear, the cars we drive, the television sets we watch, and the foods we eat—all mostly travel inside a container on a ship, then onto a railcar or the bed of a semi-trailer before reaching their final destination. Given these advances in shipping technology, logistics, improved port management, and even fuel efficiency, it is frequently more economical to put goods like food into a container to process overseas and then ship it back than it is to process locally.

Trade between the United States and East Asia (especially China and Japan) comprises a huge share of global container

The Mexican intermodal container port of Veracruz on the Gulf of Mexico. This port serves much of eastern and central Mexico via rail and road, and is dominated by trade with the United States and to a lesser extent Brazil, Venezuela, and Colombia.

shipping, and in 2018 nearly 40 percent of it (equal to about 14,000 ships per year) passed through Middle America's Panama Canal (**Fig. 3-16**). This share has increased following the widening of the Canal in 2016 (see Box 3-5), and is producing a decline in the flow of intermodal trade between East Asia and the U.S. West Coast. And if the plans for a second, Nicaragua Canal should ever come to fruition, the regional reorganization of shipping will be even more significant. It is all part of the dynamic reshaping of global transportation networks propelled by the advent of the shipping container.

FIGURE 3-16

Box 3-7 Geographic Information Analysis

Remittances Between the United States and Middle America

Remittances refer to money or other assets that migrants send back home. At the global scale, the World Bank estimates that (received) remittances exceeded $500 billion (USD) in 2018, with the Middle American realm contributing over $75 billion to this figure. **Figure 3-17** visualizes remittances sent from the United States to each country of the Middle American realm as a rectangle proportional to the amount it received in 2018. It is no surprise that Mexico received the most remittances from the U.S. in the realm, and today it ranks third in the world as a remittance recipient. Which factors do you think contribute to the notable variations in remittances across the Middle American realm—for instance, between Costa Rica and Guatemala despite their geographic proximity?

Access your WileyPLUS course to interact with a dynamic version of this figure and to engage with online exercises and questions.

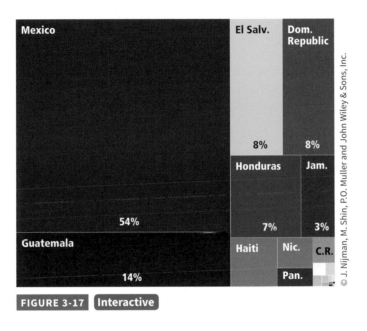

FIGURE 3-17 Interactive

© J. Nijman, M. Shin, P.O. Muller and John Wiley & Sons, Inc.

Nicaragua

Figure 3-14 underscores Nicaragua's regional position, tucked away in the heart of Central America. The Pacific coast follows a southeasterly direction here, but the Caribbean coast is oriented north–south so that Nicaragua forms a triangle of land, with its lakeside capital, Managua, situated on the mountainous, earthquake-prone, Pacific side (the country's core area has always been located here). The Caribbean side, where the uplands yield to a wide coastal plain of tropical rainforest, savanna, and swampland, has for centuries been home to indigenous peoples such as the Miskito, who have lived remote from the focus of national life. Options for this country of 6.6 million are limited: agriculture dominates the economy whereas manufacturing is weak. Like El Salvador, there is a heavy reliance on remittances from those who have emigrated.

In April 2018, people took to the streets to protest the downward spiraling living conditions and the inefficacy of their government. Clashes between anti-government protestors, pro-government groups, and security forces descended into virtual civil war. This caused an estimated 30,000 people from southern Nicaragua to cross the border into Costa Rica in search of safety. Costa Rica, while considerably more stable and economically advanced, has had difficulty accommodating the refugees, and their arrival has fueled anti-Nicaraguan sentiments among some of the population.

For decades, there has been talk of an interoceanic canal to rival the Pacific–Atlantic waterway across nearby Panama, even though the land bridge here is three times wider (see Fig. 3-14). What long seemed like a distant dream took a turn toward reality in 2013 when the Nicaraguan government officially declared that an international consortium of largely Chinese developers would forge ahead with the project. Planning and some preliminary construction began in 2014 even though many observers continued to question its feasibility. The project has drawn much international attention, but in 2018 reports surfaced that it had quietly ground to a halt because of the retreat of the Chinese investors.

Costa Rica

If there is one country that underscores Middle America's variety and diversity it is Costa Rica—because it differs significantly from its neighbors and from the norms of Central America as well. Bordered by two volatile countries (Nicaragua to the north and Panama to the east), Costa Rica is a nation with an old democratic tradition and, in this boiling cauldron, no standing army since 1949! Although the country's Hispanic imprint is similar to that found elsewhere on the Middle American mainland, its early independence, its good fortune to lie remote from regional strife, and its leisurely pace of settlement allowed Costa Rica the luxury of concentrating on its economic development. Perhaps most important, internal political stability has predominated over much of the 200 years since its independence from Spain in 1821.

Like its neighbors, Costa Rica (with a population of 5.1 million) is divided into environmental zones that parallel its coastlines. The most densely settled is the central highland zone, lying in the cooler *tierra templada*, whose heartland is the Valle Central (Central Valley), a fertile basin that contains Costa Rica's main coffee-growing area and the leading population cluster focused on the cosmopolitan capital of San José (Fig. 3-14).

The long-term development of Costa Rica's economy has given this country the region's highest standard of living, literacy rate, and life expectancy (though even here, one-fifth of the population is trapped in poverty). Agriculture continues to dominate (with bananas, coffee, tropical fruits, and seafood among the leading exports), and tourism—especially ecotourism—has expanded steadily. Despite the deforestation of more than 80 percent of its original woodland cover, Costa Rica is widely known for its superb scenery and for its (belated) efforts to protect what is left of its diverse tropical flora and fauna.

The Caribbean Basin

Fragmentation, Insularity, and Vulnerability

As Figure 3-9 reveals, the Caribbean Basin, Middle America's island arena, is constituted by a broad arc of numerous islands extending from the western tip of Cuba to the southern coast of Trinidad. The four larger islands or *Greater Antilles* (Cuba, Hispaniola, Jamaica, and Puerto Rico) are clustered in the east–west segment of this arc. The smaller islands or *Lesser Antilles* of the eastern segment extend southward within a crescent-shaped chain from the Virgin Islands

to Trinidad and Tobago. (Breaking this tectonic-plate-related regularity are the Bahamas as well as the Turks and Caicos, located north of the Greater Antilles, and many other islands too small to appear on a map at the scale of Fig. 3-9.)

On these myriad islands, whose combined land area accounts for only 9 percent of the Middle American realm, lie 33 states and numerous other political entities. (Europe's colonial flags have not totally disappeared from this archipelago, and the U.S. flag flies over both Puerto Rico and the Virgin Islands.) The inhabitants of these states and territories, however, constitute one-fifth of Middle America's total population, making this the most densely peopled component of the Western Hemisphere.

The island-nations of the Caribbean Basin are generally very small, and their territories are separated, often at considerable distances from other islands. As noted earlier in this chapter, these geographic conditions create a set of circumstances that have proven to be quite challenging. Economic opportunities are minimal, most goods and services are relatively expensive, and due to limited interaction with the outside world, the island societies tend to be static.

It is this vulnerability of the small Caribbean islands, perhaps even more so than is the case for the Central American countries, that has triggered the recent sharp intensification of cocaine trafficking across this far-flung island chain. From the main cocaine distribution source in Colombia, a growing quantity of the product is now being routed eastward across the South American continent—via Venezuela, northernmost Brazil, and the three coastal Guianas—thereby entering the Antillean archipelago from the southeast. From there it proceeds to the United States, either directly or, more likely, indirectly through northern Central America and Mexico. The stressful impacts of the expanding drug trade on the fragmented, poverty-mired societies of the Lesser and Greater Antilles are identical to those that plague the central segment of the continental land bridge: widening cocaine use and addiction among local populations; surges of organized criminal activity; and an alarming rise in the overall level of violence.

Social Cleavages

Social stratification [26] in most Caribbean islands is rigid, and upward mobility is decidedly limited. Class structures tend to be closely associated with ethnicity, and island societies still carry imprints of colonial times. The historical geography of Cuba, the Dominican Republic, and Puerto Rico is suffused with Hispanic culture; Haiti and Jamaica carry stronger African legacies. But the reality of this ethnic diversity is that European lineages still hold the advantage. Hispanics tend to be in the best positions in the Greater Antilles; people who have mixed European–African ancestries rank next. The great bulk of this social pyramid is constituted by the least privileged: the Afro-Caribbean majority. In virtually all Caribbean societies, the advantaged minorities hold disproportionate power and exert overriding influence. In Haiti, the mixed European–African minority accounts for less than 5 percent of the population but has long held most of the power. In the neighboring Dominican Republic, the pyramid of power puts Hispanics (16 percent) at the top, the mixed sector (73 percent) in the middle, and the Afro-Caribbean minority (11 percent) at the bottom. In the Caribbean social mosaic, historical advantage has a way of perpetuating itself.

The composition of the population of the islands is further complicated by the presence of Asians from both China and India. During the nineteenth century, the emancipation of slaves and subsequent local labor shortages brought some far-reaching solutions. More than 100,000 Chinese emigrated to Cuba as indentured laborers; and Jamaica, Guadeloupe, and particularly Trinidad saw nearly 250,000 South Asians arrive for similar purposes. To the African-modified forms of English and French heard in the Caribbean Basin, therefore, can be added several Asian languages. The ethnic and cultural diversity of the societies of Caribbean Middle America indeed seems endless.

Region The Greater Antilles

The four islands of the Greater Antilles (whose populations constitute 90 percent of the Caribbean Basin's total) contain five political entities: Cuba, Haiti, the Dominican Republic, Jamaica, and Puerto Rico (Fig. 3-9). Haiti and the Dominican Republic share the island of Hispaniola.

Cuba

The largest Caribbean island state in terms of both territory and population (11.3 million), Cuba lies only 145 kilometers (90 mi) from the southernmost island of the Florida Keys (**Fig. 3-18**). Havana, the capital, lies almost directly south of outermost Key West on the northwestern coast of the elongated island.

Cuba was a Spanish possession until 1898 when, with U.S. help in the ten-week-long Spanish-American War, it achieved independence. Fifty years later, an American-backed dictator was fully in control, and by the 1950s Havana had become an American playground. The island was ripe for revolution, and in 1959 Fidel Castro's insurgents gained control, thereby converting Cuba into a communist dictatorship and a client state of the Soviet Union. Nearly a million Cubans fled the island for the United States, and Miami quickly became the second-largest Cuban city after Havana.

In the fall of 1962, the world was on the brink of nuclear war when the United States called on the Soviet Union to remove its newly installed nuclear missiles from Cuba (aimed at the United States) or face reprisals. In the end, the Soviets conceded, and in return Washington promised not to attack Cuba. The United States also continued to retain the Guantánamo Bay Naval Base located near the southeastern tip of the island (Fig. 3-18), which was perpetually leased to the United States by the Cuban government in 1903. Most recently, Guantánamo for some time became notorious as a detention camp for 9/11-related terrorists as well as prisoners of war from both the Iraq and Afghanistan conflicts.

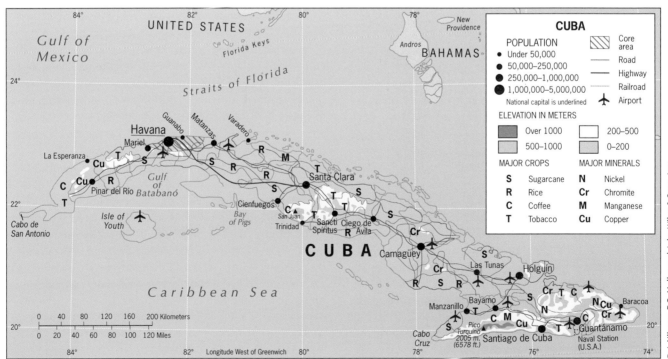

© J. Nijman, P.O. Muller and John Wiley & Sons, Inc.

FIGURE 3-18

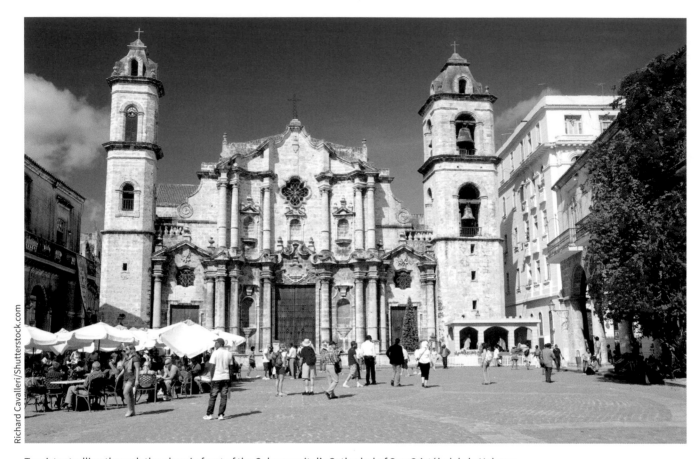

Richard Cavalleri/Shutterstock.com

Tourists strolling through the plaza in front of the Cuban capital's Cathedral of San Cristóbal de la Habana, a Spanish Baroque architectural gem completed in 1777. With the recent lifting of the U.S. embargo, the ranks of these tourists are sure to swell as new waves of American visitors are expected to descend on Cuba in the 2020s.

Castro's rule survived the 1991 demise of the Soviet Empire despite the loss of its benefactor's subsidies and the sugar markets on which the country had long relied. As the map indicates, sugar was Cuba's economic mainstay for many years; the plantations, once the property of rich landowners, extend all across the island. But sugarcane is losing its position as the top-ranked Cuban foreign exchange earner. Mills have been closing down, and cane fields are being cleared for other crops as well as for pastures. However, Cuba has additional economic opportunities, especially in its highlands.

Cuba has three mountainous zones, of which the southeastern Sierra Maestra is the highest and most extensive. These highlands create considerable environmental diversity as reflected by sizeable timber-producing tropical forests and varied soils on which crops ranging from tobacco to coffee to rice to subtropical and tropical fruits are grown. Rice and beans are the staples, but Cuba is unable to meet its dietary needs and so must import food. The central and western savannas support livestock-raising. Even though Cuba has only limited mineral reserves, its nickel deposits are extensive and have been mined for more than a century.

In 2006, Fidel Castro became too ill to lead the country and turned over the reins to his brother Raúl. Since then, a series of modest liberalization measures have taken effect. These changes were undoubtedly necessitated by the persistently deplorable condition of the Cuban economy. In 2014, encouraged by signs of reform in Cuba and a growing consensus that the American embargo of the island was now hurting the people more than the regime by depriving the country of the means to develop, the U.S. government announced the lifting of its embargo in a series of stages. In 2015, diplomatic relations between the two countries were fully restored, and in 2016 Barack Obama became the first U.S. president to visit the island since 1928. This also brought to an end the highly preferential treatment of Cuban refugees (e.g., compared to Haitians) in the United States. The so-called "wet foot/dry foot" policy, which implied that Cubans who made it onto land in the U.S. were given asylum while those intercepted on the seas were sent back, was rescinded in 2017. In 2018, Raúl Castro was succeeded by Miguel Diaz-Canel—for the first time in nearly 60 years, Cuba was no longer ruled by the Castros. Although the country is still officially communist, economic liberalization measures continue apace.

Jamaica

Across the Caribbean's deep Cayman Trench from southern Cuba lies Jamaica, and a cultural gulf separates these two countries as well. A former British dependency, Jamaica has an almost entirely Afro-Caribbean population. As a member of the British Commonwealth, it still recognizes the British monarch as the head of state, represented by a governor-general. The effective head of government in this democratic country, however, is the prime minister. English remains the official language here, and British traditions still linger.

Smaller in size than Connecticut and with just under 3 million people, Jamaica has experienced a steadily declining national income over the past few decades despite its relatively slow population growth. Tourism has become the largest source of income, but the markets for bauxite (aluminum ore), of which this island was once a leading exporter, have dwindled. And like other Caribbean countries, Jamaica has difficulty making money from its sugar exports. Jamaican farmers also produce crops ranging from bananas to tobacco, but the country faces the disadvantages on world markets common to those in the global periphery. Meanwhile, Jamaica must import all of its oil and much of its food because the densely populated coastal flatlands suffer from overuse and shrinking harvests.

The capital, Kingston, lies on the southeastern coast and reflects Jamaica's economic struggle. Almost none of the hundreds of thousands of tourists who visit the country's beaches, explore its Cockpit Country of awesome limestone towers and caverns, or populate the many cruise ships calling at Montego Bay and other points along the north coast, get even a glimpse of what life is like for the ordinary Jamaican.

Jamaica has also become an investment target of Chinese companies, mainly in infrastructure. A major new expressway running north–south across the island was completed in 2016 at a cost of $600 million, all paid for by the Chinese. In return, the Jamaican government turned over 1200 acres of land along that highway corridor to the company, which will build and manage its own tourist resorts. Among Jamaicans, the overall benefits for the island are disputed.

Haiti

Already the longtime poorest state in the Western Hemisphere, Haiti in the autumn of 2008 was staggered by no less than four tropical cyclones that struck within a few weeks of each other. The storms and the floods they unleashed killed more than 800 people and uprooted some 800,000. Then, barely 15 months later, came the monstrous earthquake of January 12, 2010, which produced devastation on a truly unimaginable scale. The capital city, Port-au-Prince, lay in ruins, most of the country's infrastructure collapsed, and the Haitian government became effectively invisible. Schools, hospitals, and railroads as well as seaports and airports all came to a standstill. Millions became unemployed overnight. But all of that paled in comparison to the scale of human suffering. In the months following the quake, the estimated death toll reached and then surpassed 300,000; at least another 300,000 victims were injured, many permanently disabled. About 1.5 million homeless people initially fled the capital for the countryside, and nearly a decade after the disaster Haiti's homeless still numbered in the tens of thousands.

It was a tragedy of colossal proportions, its horrendous images broadcast around the world on television and the Internet. Large-scale international emergency aid arrived, but it became clear that, in the long run, much of the affected area would have to be totally rebuilt. In the catastrophe's aftermath, millions of Haitians continue to subsist in the most challenging circumstances.

Haiti lies along a particularly dangerous tectonic fault zone where the North American and Caribbean plates meet (**Fig. 3-19**, inset map). The quake of 2010, geologists say, had been in the making for centuries, and others could follow in the future. The country also lies astride the axis of Hurricane Alley, and even a single year without a major storm is something to be grateful for.

Few countries in the world have a more checkered history than Haiti, where political instability, repression, and material deprivation have been constants since the early nineteenth century. The GNI per capita of Haiti's 11.3 million residents is barely one-sixth of Jamaica's, a level below even that of many impoverished African countries. Compared to other countries in Middle and South America, its infant mortality rate is nearly three times as high. A shocking one-third of elementary-school-age girls have never even attended school. The negative consequences of Haiti's irreversible environmental degradation (**Box 3-8**) are staggering. And because of its scant natural resources and almost nothing to offer in the international trading arena, most of Haiti's meager public expenditures depend on direct foreign aid. As for the limited private consumption that takes place, it relies heavily on remittances from the Haitian diaspora in such cities as Miami, Paris, New York, and Montreal.

Dominican Republic

The mountainous Dominican Republic has a larger share of Hispaniola than Haiti (Fig. 3-19) in terms of territory (64 percent of the island), but at 10.8 million the Dominican Republic's population ranks just behind that of Haiti. It has a far more robust resource base than its neighbor to the west. Nickel, gold, and silver have long been exported along with sugar, tobacco, coffee, and cocoa, and tourism is the leading industry. Yet the Dominican Republic's economic performance since 2000 has been erratic and continues to remain below its potential.

Puerto Rico

The largest U.S. domain in Middle America, this easternmost and smallest island of the Greater Antilles region covers 9000 square kilometers (3500 sq mi) and has a population of 2.9 million. Puerto Rico is larger than Delaware and more

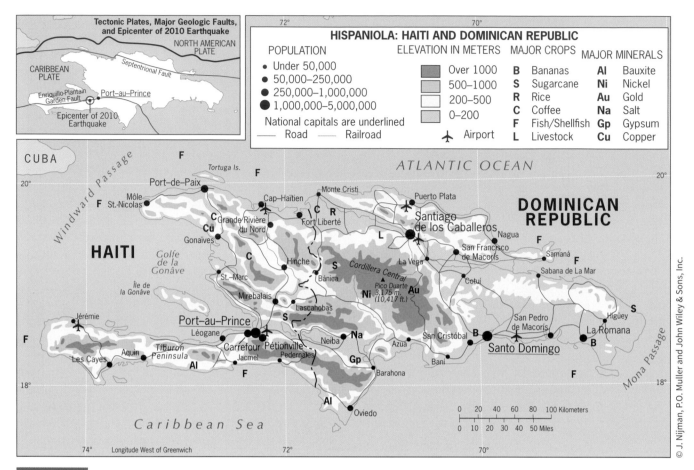

FIGURE 3-19

populous than Iowa. Most Puerto Ricans are concentrated in the urbanized northeastern sector of this rectangular island (**Fig. 3-20**). The island's population has been declining for well over a decade, mostly due to net out-migration to the mainland United States.

Puerto Rico fell to the United States over a century ago during the Spanish-American War of 1898. Since the Puerto Ricans had been struggling for some time to free themselves from Spanish imperial rule, this transfer of power was, in their view, only a change from one colonial master to another.

Box 3-8 From the Field Notes . . .

"Fly along the political boundary between Haiti and the Dominican Republic, and you see long stretches of the border marked by the starkest possible contrast in vegetation: denudation prevails to the west in Haiti while the forest prospers on the Dominican (eastern) side. On the Haitian side, overpopulation, widespread grinding poverty, and chronic government mismanagement—all exacerbated by the snail's pace of earthquake recovery—combine to create one of the Caribbean's most jarring cultural-landscape divides."

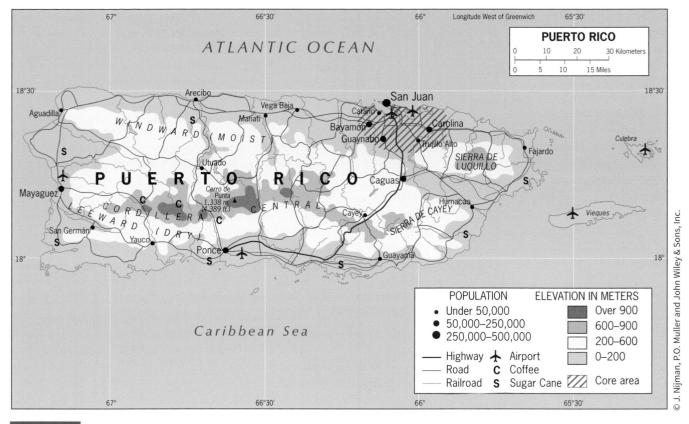

© J. Nijman, P.O. Muller and John Wiley & Sons, Inc.

FIGURE 3-20

As a result, the first half-century of U.S. administration was difficult, and it was not until 1948 that Puerto Ricans were permitted to elect their own governor. When the island's voters approved the creation of a commonwealth in a 1952 referendum, Washington, D.C. and San Juan, the seats of government involved, entered into a complicated arrangement. For example, Puerto Ricans are U.S. citizens but pay no federal taxes on their local incomes.

Despite these apparent advantages, Puerto Rico has not thrived under U.S. administration. Long dependent on a single-crop economy (sugar), the island based its industrial development during the 1950s and 1960s on its comparatively cheap labor, tax breaks for corporations, political stability, and special access to the U.S. market. Consequently, pharmaceuticals, electronic equipment, and apparel top the current list of exports, not sugar or bananas. But this industrialization failed to stem a tide of emigration that carried more than one million Puerto Ricans to the New York City area alone. The same wage advantages that favored corporations kept many Puerto Ricans poor or unemployed.

Most Puerto Ricans are fed up with their lack of economic progress. In recent referendums in 2012 and 2017, a majority voted in favor of becoming the 51st State of the U.S. Discussions about implementation (which has political implications for the deeply divided U.S. Congress) as well as the process and conditions of accession are likely to continue for many years.

In 2017, the island suffered a direct hit from catastrophic, Category-5, Hurricane Maria (see photo p. 72), and recovery efforts were still ongoing two years later. The destruction of the island's infrastructure (including electrical power and water provision) was so severe that it had to be virtually rebuilt from scratch.

Region The Lesser Antilles

As Figure 3-9 shows, the Greater Antilles are flanked by two clusters of islands: the extensive Bahamas/Turks and Caicos archipelago to the north and the Lesser Antilles to the southeast. **The Bahamas**, a former British colony that is now the closest Caribbean neighbor to the United States, alone consists of nearly 3000 coral islands—most of them rocky, barren, and uninhabited—but about 700 carry vegetation, of which roughly 30 are inhabited. Centrally positioned New Providence Island houses most of the country's 400,000 inhabitants and also contains the capital, Nassau, a leading tourist attraction.

The Lesser Antilles are grouped geographically into the Leeward Islands and the Windward Islands, a (climatologically incorrect) reference to the prevailing airflows in this tropical zone. The Leeward Islands extend southward from the U.S. Virgin Islands to the French dependencies of Guadeloupe and

Martinique, and the Windward Islands from St. Lucia south and then west to the Dutch-affiliated islands of Aruba and Curaçao off the northwestern Venezuelan coast (Fig. 3-9).

It would be impractical to detail the individual geographic characteristics of each island of the Lesser Antilles, but we should note that nearly all share the insularity and environmental risks of the Caribbean Basin—major earthquakes, volcanic eruptions, and hurricanes; that they confront, to varying degrees, similar social and economic problems in the form of limited domestic resources, overpopulation, soil exhaustion, land fragmentation, and rising levels of crime and violence; and that tourism has become the leading industry for many.

Under these kinds of circumstances one looks for any hopeful sign, and such an indication comes from **Trinidad and Tobago**, the two-island republic at the southern end of the Lesser Antilles (Fig. 3-9). This country (population: 1.4 million) has embarked on a natural-gas-driven industrialization boom that could turn it into a future economic tiger. Trinidad has long been an oil producer and has also become the largest supplier of liquefied natural gas for the United States. Many of the new industrial facilities have agglomerated at the state-of-the-art Point Lisas Industrial Estate outside the capital, Port of Spain, and they have propelled Trinidad to become the world's leading exporter of ammonia and methanol. Natural gas also is an efficient fuel for the manufacturing of metals, and

steelmakers as well as aluminum refiners have been attracted to locate here. With Trinidad lying just off the Venezuelan coast of South America, it is also a sea-lane crossroads that is strongly connected to the vast, near-coastal supplies of iron ore and bauxite that are mined in nearby countries, most importantly the Brazilian Amazon.

But Trinidad is an exception. It is a reflection of the predicament of small-island developing economies that, after the initial wave of decolonization in the 1950s and 1960s, decided they were probably better off affiliating with the European country that had ruled them. Thus, for example, Guadeloupe remained with France, and the Cayman Islands are still British. The former Netherlands Antilles passed up full independence to continue its affiliation with the Dutch, but did so individually and with considerable autonomy: **Curaçao**, **Aruba**, and **St. Maarten** are now "countries" within the Kingdom of the Netherlands, whereas the other Caribbean islands of the former Dutch Empire have acquired the status of overseas municipalities. Although economic logic seems to dictate continued association with the former colonial powers, local politics and pride demand some minimal degree of autonomy and avoidance of paternalistic European interference. The combined result is often a complicated legal framework that shapes the contemporary political status of the islands.

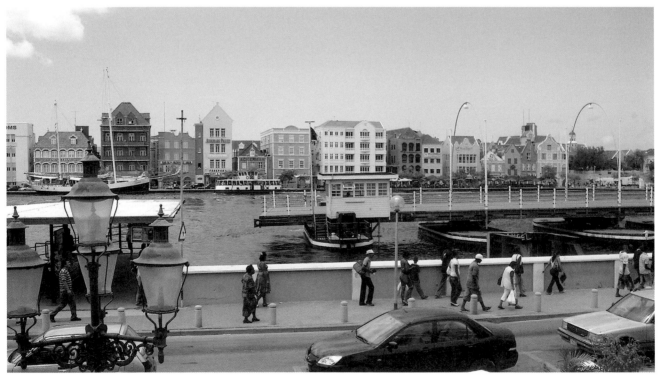

The city of Willemstad, Curaçao, provides a playful, pastel-colored version of traditional Dutch architecture that dates back to colonial times. Like many other Caribbean islands, Curaçao faces economic challenges that relate to its small size and relative isolation.

© J. Nijman

Points to Ponder

- The United States is the single most important trade partner of just about every country and territory in this realm—is that a plus or a minus for Middle America?

- Demand for illegal drugs in the United States is a basic driver of drug production and trafficking in South and Middle America, which in turn poses a huge problem for these smaller and less-developed countries.

- Most of the small Caribbean island-nations contain less than half a million people—is that enough to sustain a viable economy?

- In Mexico, there has been talk of the so-called China card: if the United States wants to "build a wall" and turn its back on Mexico and Central America, it could lead to greater involvement of China in this realm and a concomitant loss of U.S. influence.

- Could the immense devastation in Puerto Rico and The Bahamas, respectively caused by Hurricanes Maria (2017) and Dorian (2019), eventually result in a more advanced and resilient infrastructure for water, energy, and Internet services?

FIGURE 4-1

© J. Nijman, P.O. Muller and John Wiley & Sons, Inc.

The Patagonia region of Argentina, covering much of South America's far south, is known for its stunning natural beauty. The photo looks west, down the road to El Chalten, a small town in Santa Cruz province that has been named Argentina's *Capital Nacional del Trekking*. The southernmost extension of the Andes is visible in the background, with the peak of Mount Fitz Roy reaching into the clouds. Not far behind it lies the border with Chile and about 200 kilometers (125 miles) beyond that, the southern Pacific Ocean. At 50° South, this place is as far from the equator as Winnipeg, Canada, in the opposite hemisphere.

Adria Photography/Moment/Getty Images

The South American Realm

IN THIS CHAPTER

- Venezuela in free fall
- Global climate change and vanishing Lake Poopó
- Comparing race relations in Brazil and the U.S.
- China's economic footprint in South America
- Medellín: From cocaine capital to model city
- Challenges of regional integration

CONCEPTS, IDEAS, AND TERMS

Unity of place **[1]**

Indigenous **[2]**

Altiplano **[3]**

Land alienation **[4]**

Liberation theology **[5]**

Cultural pluralism **[6]**

Commercial agriculture **[7]**

Subsistence agriculture **[8]**

Remote sensing **[9]**

Uneven development **[10]**

Supranationalism **[11]**

Rural-to-urban migration **[12]**

Informal sector **[13]**

Barrio **[14]**

Favela **[15]**

Megacity **[16]**

Central business district (CBD) **[17]**

Gini Index **[18]**

Dependencia theory **[19]**

Insurgent state **[20]**

Failed state **[21]**

Neoliberal policies **[22]**

Landlocked country **[23]**

Human Development Index **[24]**

Triple Frontier **[25]**

Primate city **[26]**

Viticulture **[27]**

Elongation **[28]**

Buffer state **[29]**

Entrepôt **[30]**

Forward capital **[31]**

Cerrado **[32]**

Growth-pole concept **[33]**

Of all the continents, South America has the most familiar shape—a gigantic triangle connected by mainland Middle America's tenuous land bridge to its neighbor in the north. South America also lies not only south but mostly east of its northern counterpart. Lima, the capital of Peru—one of the continent's westernmost cities—lies farther east than Miami, Florida. Thus South America juts out much more prominently into the Atlantic Ocean toward southern Europe and Africa than does North America. But lying so far eastward means that South America's western flank faces a much wider Pacific Ocean, with the distance from Peru to Australia nearly twice that from California to Japan.

Defining the Realm

As if to reaffirm South America's northward and eastward orientation, the western margins of the continent are rimmed by one of the world's longest and highest mountain ranges, the Andes, a giant wall that extends unbroken from Tierra del Fuego near the continent's southern tip in Chile to northeastern Venezuela in the far north (**Fig. 4-1**). The other major physiographic feature of South America dominates its central north—the Amazon Basin; this vast humid-tropical amphitheater is drained by the mighty Amazon River, which is fed by several prominent tributaries (see **Box 4-1**). Much of the remainder of the continent can be classified as plateau, with the most important components being the Brazilian Highlands that cover most of Brazil southeast of the Amazon Basin, the Guiana Highlands located north of the lower Amazon Basin, and the cold Patagonian Plateau that blankets the southern third of Argentina. Figure 4-1 also reveals two other noteworthy river basins beyond Amazonia: the Paraná-Paraguay Basin of south-central South America, and the Orinoco Basin in the far north that drains interior Colombia and Venezuela.

Explorers' Continent

It was here in northern South America that the great German explorer and scientist Alexander von Humboldt, one of the founders of the modern discipline of geography, embarked on his legendary expeditions in the early nineteenth century. After landing on the coast of Venezuela and trekking across the continent's northern interior, the 30-year-old von Humboldt was struck by the area's biodiversity, its majestic natural beauty, and the adaptive abilities of the human populations. He discovered and named many species of flora and fauna, traversed tropical grasslands and jungles, met with indigenous peoples, crossed dangerous rivers, and reached the summit of the highest mountain in the Americas climbed by Europeans at the time (Ecuador's Chimborazo). He compiled large numbers of maps and was one of the first scientists to note how the coastlines of eastern South America and western Africa fitted together like pieces of a jigsaw puzzle, speculating about the continents' geologic movements. Von Humboldt was a key figure in the rise of the modern discipline of geography because of his many discoveries as well as his views on the **unity of place** [1]—that in a particular locality or region intricate connections exist among climate, geology, biology, and human cultures. As such, he laid the foundation for modern geography as an *integrative discipline* marked by a spatial perspective.

Almost three centuries earlier, the opposite end of South America was the scene of a crucial stage in the first circumnavigation of the globe by the expedition led by Ferdinand Magellan. Once it became clear that Columbus had stumbled upon America, not India, Spanish and Portuguese explorers continued their efforts to discover a westward passage from the Atlantic to the Pacific. This took them to the far south along what is now the Argentinean coast. Magellan's ships set sail for the treacherous waters of what he named the *Estrecho de Todos los Santos*—now called the Strait of Magellan (Fig. 4-1). Of the (remaining) four ships in Magellan's fleet, only three made it through the menacing, 600-kilometer (375-mi)-long passage; the fourth turned back and crashed on the rocks in the icy waters of the Southern Ocean.

Box 4-1 Major Geographic Features of South America

1. South America's physiography is dominated by the Andes Mountains in the west and the Amazon Basin in the central north. Much of the remainder is plateau country.

2. Almost half of the realm's area and just under half of its total population are concentrated in one country—Brazil.

3. South America's population remains concentrated along the continent's margins. Most of the interior is sparsely peopled, but portions of it are now undergoing significant development.

4. Interconnections among the states of the realm are improving. Economic integration has become a growing force but is still at an early stage.

5. Regional economic contrasts and disparities strongly persist, both in the realm as a whole and within individual countries.

6. Cultural pluralism prevails in almost all of the realm's countries and is often expressed regionally.

7. Headlong urban growth continues to mark much of the South American realm, and urbanization overall is today on a par with the levels of the United States and western Europe.

8. This realm contains abundant natural resources that, although important for exports, create considerable dependency on the ups and downs of world markets. The prevalence of natural resources can also carry the risk of highly unequal control over those resources.

Myriad Climates and Habitats

On the world map, if the Russia/Central Asia realm is the widest in east–west extent, South America is the longest measured from north to south. (Within South America, no other country is more emblematic of this elongated geography than Chile, averaging only about 150 kilometers [90 mi] in width but 4000 kilometers [2500 mi] in length.) As a consequence of this huge latitudinal span, from about 12°N to 56°S, this realm contains an enormous variety of climates and vegetation. Combine this with substantial variation in elevation from west to east, and it is clear why South America has such an impressive range of natural habitats.

Take another look at the map showing the global distribution of climates (Fig. 1-6) and note the variation in climatic types, particularly in the realm's northwest and southern half. Travel northeast from Lima, Peru, for about 600 kilometers (375 mi) and you encounter no less than four different climate zones: arid, highland, and two varieties of humid tropical. A transect of similar length from Santiago, Chile, eastward across the continent to Buenos Aires, Argentina, will take you through five climate zones: interior highland, arid, and semiarid environments bracketed by a different humid–temperate climate along each coast. Vegetation in South America varies accordingly, from lush tropical rainforests to rocky and barren snow-covered mountaintops to grasslands, fertile as well as parched. This natural diversity also makes for considerable cultural differences, as we shall soon discover.

States Ancient and Modern

Thousands of years before the first European invaders appeared on the shores of South America, aboriginal peoples now referred to as **indigenous [2]** had migrated into the continent via North and Middle America and founded societies in coastal valleys, in river basins, on plateaus, and in mountainous locales. These societies achieved different and remarkable adaptations to their diverse natural environments, and by about 1000 years ago, a number of regional cultures thrived in the elongated valleys between mountain ranges of the Andes from present-day Colombia southward to Bolivia and Chile. These high-altitude valleys, called *altiplanos* [3], provided fertile soils, reliable water supplies, building materials, and natural protection to their inhabitants.

The Inca State

One of these *altiplanos*, surrounding the city of Cuzco in what is now southeastern Peru, became the core area of South America's greatest indigenous empire, that of the *Inca*. The Inca were expert builders whose stone structures (among which Machu Picchu near Cuzco is the most famous [photo on next page]), roads, and bridges helped to unify their vast empire; they also proved themselves to be efficient administrators, successful farmers and herders, and skilled manufacturers; their scholars studied the heavens, and their physicians even experimented with brain surgery. Great military strategists, the Inca integrated the peoples they vanquished into a stable and well-functioning

state, an amazing accomplishment given their never-ending need to cope with treacherous, high-relief terrain (**Box 4-2**).

As a minority ruling elite in their far-flung empire, the Inca were at the pinnacle within their rigidly class-structured, highly centralized society. So centralized and authoritarian was their state that a takeover at the top was enough to gain immediate power over all of it—as a small army of Spanish invaders discovered in the 1530s. The European invasion brought a quick end to thousands of years of indigenous cultural development and changed the social order forever (**Box 4-3**).

The Iberian Invaders

The modern map of South America started to take shape when the Iberian (Spanish and Portuguese) colonizers began to understand the location and economies of the indigenous societies. The Inca, like Mexico's Maya and Aztec peoples, had accumulated gold and silver at their headquarters, possessed productive farmlands, and constituted a ready labor force. Not long after the defeat of the Aztecs in 1521, Francisco Pizarro sailed southward along the continent's northwestern coast, learned of the existence of the Inca Empire, and withdrew to Spain to organize its overthrow. When he returned to the Peruvian coast in 1531, the diseases brought by Spaniards in earlier years had already devastated part of the Inca population. Pizarro's army was quite small, but the military and psychological advantages of horses and guns proved decisive (together with a cunning strategy involving the treacherous murder of the Inca emperor, Atahualpa). Only two years later, his party rode victorious into Cuzco.

The new order that emerged in western South America placed the indigenous peoples in serfdom to the Spaniards. Great haciendas were formed by **land alienation [4]** (the takeover of indigenously held land by foreigners), taxes were instituted, and a forced-labor system was introduced to maximize the profits of exploitation. Lima, the west-coast headquarters of the Spanish conquerors, soon became one of the richest cities in the world, its wealth based on the exploitation of vast Andean silver deposits. Lima also became the seat of the Viceroyalty of Peru, the political administrative district (later subdivided into a triad of viceroyalties) whereby its South American territories were incorporated into the Spanish Empire.

Meanwhile, another vanguard of the Iberian invasion was penetrating the east-central part of the continent, the coastlands of present-day Brazil. This area had become a Portuguese sphere of influence, and Brazil's boundaries were bent far inland to include almost the entire Amazon Basin. Eventually, the country came to be only slightly smaller in territorial size than all the other South American countries combined. This westward thrust was the result of Portuguese and Brazilian penetration, particularly by the *Paulistas*, the settlers of São Paulo who needed indigenous slave labor to run their plantations.

Independence and Isolation

Despite their adjacent location on the same continent, their common language and cultural heritage, and their shared national

Box 4-2 From the Field Notes …

"Climbing up to Machu Picchu, South America's best-known pre-Columbian site, is a thrill. This fifteenth-century 'City of the Incas' is located on an Andean ridge (elevation: 2400 meters/7900 ft) about 80 kilometers (50 mi) northwest of Cuzco, the long-time capital of the Inca civilization and still a major city today. Machu Picchu is believed to have been built (ca. 1450) as an estate for the emperor Pachacuti but was abandoned less than a hundred years later at the time of the Spanish Conquest. It first became known to the outside world in 1911, and restoration (now nearing the halfway point) has been ongoing ever since. Archeologists have long been fascinated as well, and their intensive on-site fieldwork and research efforts continue. All these buildings were put in place using manual labor at an elevation where most visitors, myself included, have to carefully pace themselves to avoid shortness of breath."

© J. Nijman

problems, the countries that arose out of South America's Spanish viceroyalties (together with Brazil) until quite recently existed in a considerable degree of isolation from one another. Distance as well as physiographic barriers reinforced this separation, and most of the realm's population agglomerations still adjoin the coast, mainly the eastern and northwestern coasts (**Fig. 4-2**). The viceroyalties existed primarily to extract riches and fill Spanish coffers. In Iberia, there was little interest in developing the American lands for their own sake. Only after those who had a stake in Spanish and Portuguese America had made it their new home and had therefore rebelled against Iberian authority did things begin to change, and then very slowly. Thus South America was saddled with the values, economic outlook, and social attitudes of eighteenth-century Iberia—not the best tradition from which to begin the task of forging modern countries.

Certain isolating factors had their effect even during the wars for independence. Spanish military strength was always concentrated at Lima, and those territories that lay farthest from their center of power—Argentina as well as Chile—were the first to gain independence from Spain (in 1816 and 1818, respectively). In the Caribbean north, Simón Bolívar led the burgeoning independence movement, and in 1824 two decisive military defeats there spelled the end of Spanish control in South America.

This joint struggle, however, did not produce much unity because no fewer than nine countries emerged from the three former viceroyalties. It is not difficult to understand why this fragmentation took place. With the Andes intervening between Argentina and Chile and the Atacama Desert between Chile and Peru, overland distances seemed even greater than they really were, and these obstacles to contact proved quite effective. Hence, from their outset the new countries of South America began to grow apart amid friction that occasionally escalated into local wars. In fact, only within the past few decades have the countries of this realm finally begun to recognize the mutual advantages of increasing cooperation and to make lasting efforts to steer their relationships in this direction.

Box 4-3 From the Field Notes …

"Cuzco is the historic capital of the Inca Empire, located high in the Peruvian Andes at an elevation of 3400 meters (11,200 ft). The city center is a UNESCO-designated World Heritage Site and attracts around 2 million tourists each year. I drove out of the valley up the hillside to get a good view of the layout of the center, so typical of Spanish colonial cities. The focal Plaza de Armas with its parks and benches is surrounded by the Cathedral of Santo Domingo and other religious and municipal buildings, and adjacent blocks around them are used for commerce and retailing. Located farther away from the Plaza are newer residential areas. This historic Inca capital was sacked in the early sixteenth century by Pizarro's *conquistadores*; few Inca ruins remain in the city center, but just outside Cuzco lies the impressive Sacsayhuaman fortress, with most of its walls still intact. This is not a very large city—it is too remote for that—but Cuzco's population has tripled over the past three decades to nearly 500,000."

© J. Nijman

POPULATION DISTRIBUTION: SOUTH AMERICA

Population per Square Kilometer

0 to 5
5 to 10
10 to 50
50 to 100
100 to 500
Over 500

© J. Nijman, M. Shin, P.O. Muller and John Wiley & Sons, Inc

FIGURE 4-2 **Interactive**

The Population Map—Then and Now

If we were able to reconstruct a (pre-Columbian) map of South America's population before the arrival of the Europeans, it would differ sharply from the current one displayed in Figure 4-2. Indigenous societies inhabited not only the Andes and adjacent lowlands but also riverbanks in the Amazon Basin, where settlements numbering in the thousands subsisted on fishing and farming. They did not shy away from harsh environments such as those of the island of Tierra del Fuego in the far south, where the fires they kept going against the bitter cold led the Europeans to name the place "land of fire."

Today, the realm's population distribution has been transformed. Many of the indigenous societies succumbed to the European invaders, not just through warfare but also because of the diseases the Iberian conquerors brought with them. It is estimated that 90 percent of native Amazonians died within a few years of contact, and even the peoples of isolated Tierra del Fuego are no longer there to build their fires. From one end of South America to the other, the European intrusion spelled disaster.

Spanish and Portuguese colonists penetrated the interior of South America, but the great majority of the settlers stayed on or near the coast, a pattern that still predominates. Almost all of the realm's major cities have coastal or near-coastal locations, and the current population distribution map gives you the impression of a continent still awaiting penetration and settlement. But another look at Figure 4-2 will reveal a swath of population located well inland from the settlements along the northwestern coast, most clearly in Peru but also extending northward into Ecuador and southward into Bolivia. That is the legacy of the Inca Empire and its incorporated peoples, surviving in their highland redoubt and continuing to number in the tens of millions.

Cultural Geographies

Indigenous Reawakening

Since the turn of our new century, South America's long-down-trodden indigenous peoples have been staging a social, political, and economic reawakening. In several of the realm's countries, where their numbers are large enough to translate into political strength, they have begun to realize their potential. This is especially true in the Andean countries: indigenous peoples, descendants of the societies controlled by the Inca Empire, constitute no less than 45 percent of Peru's national population, and in Bolivia they are in the majority at 62 percent.

Today, newly empowered indigenous political leaders have emerged and are bringing the plight of the realm's aboriginal peoples to both local and international attention. South America's native peoples were conquered, decimated by foreign diseases, robbed of their best lands, subjected to involuntary labor, denied the right to grow their traditional crops, socially discriminated against, and swindled out of their fair share of the revenues from resources in their traditional domains. They may still rank among the realm's poor, but they are increasingly asserting themselves. For a number of South American states, the consequences of this movement are evident in the shifting political landscape.

The indigenous reawakening is in part related to changing religious practices in South America. Officially, just over 80 percent of the population is Roman Catholic, and traditionally South Americans tend to be viewed as devout followers of the Vatican. But recent surveys show that many do not attend church regularly and that more than half of all Catholics describe themselves as believers of the doctrine—rather than adherents of the Church. Since the late nineteenth century, the Catholic Church has often been criticized for its conservative position on social issues and for always siding with the establishment. During the 1960s, a powerful movement known as **liberation theology [5]** emerged in South America and subsequently gained followers around the world. This movement was a blend of Christian religion and humanist philosophy that interpreted the teachings of Christ as a quest to liberate the impoverished masses from oppression. Although the Church has tried to make some amends, it could not avoid losing popular support, especially among indigenous peoples.

African Descendants

When the Portuguese began to develop their territory on the coast of what is now Brazil, they turned to the same lucrative activity that their Spanish rivals had pursued in the Caribbean—the plantation cultivation of sugar for the European market. And they, too, found their workforce in the same source region: millions of Africans (nearly half of all who came to the Americas) were brought in bondage to the tropical Brazilian coast north of Rio de Janeiro. Not surprisingly, Brazil now has South America's largest black population, which is still heavily concentrated in the country's poverty-mired northeastern States. With Brazilians of direct or mixed African ancestry today accounting for 57 percent of the population of

211 million, the Africans decidedly constitute a major component in the immigration of foreign peoples into South America.

Ethnic Landscapes

The cultural landscape of South America, similar to that of Middle America, is a layered one. Indigenous inhabitants had cultivated and crafted diverse landscapes throughout the continent, some producing greater impacts than others. When the Europeans arrived, the cultural transformation that resulted from depopulation severely impacted the environment. Native peoples now became minorities in their own lands, and Europeans introduced crops, animals, and ideas about land ownership and land use that irreversibly changed South America. They also brought in African slaves from various parts of Subsaharan Africa. Europeans from non-Iberian Europe also began emigrating to South America, particularly during the first half of the twentieth century. Japanese settlers arrived in Brazil and Peru during the same era. All of these elements contributed to shaping the present-day ethnic complexion of this realm. **Figure 4-3** shows the distinct concentrations of indigenous and African cultural dominance, as well as areas where these groups are largely absent and people of European ancestry predominate.

Of course, ethnic origins are not always so straightforward, and patterns can change as a result of internal migrations and ethnic mixing. In recent years, research on individual DNA and genetics has taught us a lot about the regional and group origins

SOUTH AMERICA: DOMINANT ETHNIC GROUPS

- African
- Mestizo
- European
- Indigenous

© J. Nijman, P.O. Muller and John Wiley & Sons, Inc.

FIGURE 4-3

of people. On the map, Argentina is indicated as having a predominantly European ancestry. More specifically, recent research shows that, for the average Argentine, nearly 80 percent of his or her genetic structure is European, 18 percent indigenous, and 2 percent African. If nobody had mixed ancestors, these percentages would translate perfectly into European, indigenous, and African shares of the population—but of course that is not the case. Many people do not have precise knowledge of their ethnic ancestry, and although some may be descendants of those with a single origin, many others exhibit at least some degree of mixed ancestry. In the aggregate, however, there is no doubt that the population of cone-shaped southern South America is predominantly of European origin. Thus, overall, South America is a realm marked by **cultural pluralism [6]**, where peoples of various ethnicities and cultures cluster in adjacent areas but generally do not mix. The result is a cultural mosaic of almost endless variety.

Economic Geography

Agricultural Land Use and Deforestation

The internal divisions of this cultural kaleidoscope are further reflected in the realm's economic landscape. In South America, larger-scale **commercial**—or market (for-profit)—**agriculture [7]** and smaller-scale **subsistence agriculture [8]** (primarily for farmers' own use) exist side by side to a greater degree than anywhere else in the world (**Fig. 4-4**). The geography of plantations and other commercial agricultural systems was initially tied to the distribution of landholders of European background, whereas subsistence farming (such as highland mixed subsistence–market and agroforestry–shifting cultivation) is historically associated with the spatial patterns of indigenous peoples as well as populations of African and Asian descent.

At a broader level of generalization, agricultural systems and land use in South America also vary in close relationship to physiography, as one would expect. Forestry and agroforestry prevail in the Amazon Basin; ranching dominates in the grasslands of the Southern Cone; and different forms of agriculture, with or without irrigation, are found in an extensive zone from northeastern Brazil to northern Argentina, as well as scattered among pockets of moderate elevation in the Andean highlands. Land uses throughout South America are changing rapidly today, owing mainly to ongoing deforestation and the introduction or expansion of new crops. The fastest-growing crop is soybeans, the cultivation of which now dominates much of east-central Brazil and spills over into adjacent areas of Paraguay, Uruguay, and Argentina (Fig. 4-4).

Deforestation is a global problem, but it has been a particularly acute problem in Brazil. In the past, deforestation was attributed mainly to small-scale landholders, colonists who had made their way into the Amazon rainforest to eke out a new living. The deforestation crisis intensified as *agro-industrial* operations, engaged in large-scale production for the export market, increasingly penetrated Amazonia and transformed its land

SOUTH AMERICA: AGRICULTURAL SYSTEMS

- Oases
- Scattered Pastoralism
- Non-Soybean Grain Crops
- Dairy Products and Flowers for Export
- Mixed Dry Farming
- Soybeans
- Cattle
- Plantation
- Agroforestry & Shifting Cultivation
- Highland Mixed Subsistence-Market
- Mixed
- Nonagricultural areas

0 500 1000 1500 Kilometers
0 500 1000 Miles

© J. Nijman, P.O. Muller and John Wiley & Sons, Inc.

FIGURE 4-4

cover. The exact loss of rainforest is difficult to measure, but it has been estimated that Brazil has lost an area of rainforest almost the size of Venezuela since 1980 (see **Box 4-4**). Since then, the government has had little success in reducing the rate of deforestation while increasing the area of protected forests—but in absolute terms, deforestation today continues unimpeded.

As South American economies modernize, agriculture remains highly important. Almost one-fifth of this realm's workforce is still employed in the primary sector that includes farming, cattle-raising, and fishing—a much greater share than in North America. The South American contribution to global trade in grains, soybeans, coffee, orange juice, sugar, and many other crops is significant and still increasing. And in addition to all these successful agricultural endeavors, illegal farming thrives in the form of narcotics production, particularly cocaine.

Box 4-4 Technology & Geography

Remote Sensing and Deforestation

It is said that "a picture is worth a thousand words," but satellite imagery of the Earth can also leave us speechless. The indirect capture of images by satellites orbiting the Earth is referred to as **remote sensing** [9]. Remote sensing helps us to understand, define, and try to reshape the regions of the world and the complex interrelationships between the human and physical environments. In addition to capturing images illuminated by sunlight, sensors on satellites can also detect portions of the electromagnetic spectrum that are invisible to humans (e.g., infrared) and can also transmit signals like radar to measure the surface of the Earth. Coupled with advances in computer modeling and with the compelling bird's eye perspective of satellite images, new insights and greater understandings of processes and patterns behind phenomena such as urbanization, agricultural development, and climate change can be obtained.

Since 1972, NASA's Landsat program has captured remote sensing images for use in science, government, business, and industry. As the world's longest continuously operating platform for acquiring remotely sensed imagery, Landsat is instrumental in identifying and recording regional-scale changes across the surface of the Earth such as the deforestation of the Amazon rainforest (see photo). Without remote sensing imagery, it was extremely difficult to estimate rates of deforestation, and the information could be easily manipulated by governments or interest groups. In that sense, remote sensing technologies have contributed immensely to our objective knowledge of land-use change around the world.

National Remote Sensing Centre Ltd/Science Source

This is what the Amazon's equatorial rainforest looks like from an orbiting satellite after the human onslaught in preparation for settlement. The colors on this Landsat image emphasize the destruction of the trees, with the dark green of the natural forest contrasted against the pale green and pinks of the leveled forest. The linear branching pattern of deforestation is the preferred approach here in Rondônia State's Highway BR-364 corridor. But farming is not likely to succeed for very long, and much of the cleared land will be abandoned. Then the onslaught would resume to clear additional land—a pernicious cycle the Brazilian government has committed itself to ending by 2030.

The Geography of Cocaine

All of the cocaine that enters the United States comes from South America, mainly from Colombia, Peru, and Bolivia. Within these three countries, cocaine annually brings in billions of (U.S.) dollars and "employs" tens of thousands of workers, thereby constituting a powerful industry. The first of the three stages of cocaine production involves the extraction of coca paste from the coca plant, a raw-material-oriented activity that is located near the areas where the plant is grown. The main zone of coca cultivation is along the eastern slopes of the Andes and adjacent tropical lowlands in Bolivia, Peru, and Colombia (**Fig. 4-5**).

Coca leaves harvested in the source areas of the Andes and nearby interior lowlands make their way to local collection centers, located at the convergence of rivers and trails, where coca paste is extracted and prepared. The second stage of production involves the refining of that coca paste (about 40 percent pure cocaine) into cocaine hydrochloride (more than 90 percent pure), a lethal concentrate that is diluted with substances such as sugar or flour before being sold on the streets to consumers. Cocaine refining requires sophisticated chemicals, carefully controlled processes, and a labor force skilled in their supervision. Most of this processing stage is dominated

by Colombia and takes place in the lowland central-south and east, beyond the reach of the Bogotá government.

The final stage of production entails the distribution of cocaine to the marketplace, which depends on an efficient, clandestine transportation network that leads into the United States. Private planes operating out of remote airstrips were the preferred "exporting" method until about 1990, and in the 1970s and 1980s Miami was a major center of cocaine distribution in the United States. But aggressive U.S. measures along the coasts of Florida caused a shift toward Mexico (and the Northern Triangle of Central America). Today, most cocaine still reaches the United States through smuggling across the Mexican border (a topic elaborated in Chapter 3). Since around 2005, another route has opened: small planes fly the product eastward from Colombia to Venezuela and the Guianas; from there it is transshipped northward by boat or plane to various Caribbean islands; subsequently, it is routed westward to Central America or Mexico, sometimes even directly into the United States.

Industrial Development

The geography of industrial production varies markedly in this realm. Brazil and Chile have shown rapid growth; Argentina has been in long-term decline; and countries such as Peru,

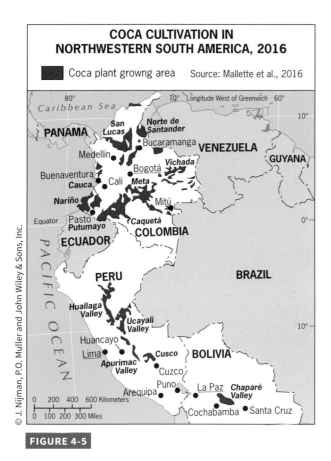

COCA CULTIVATION IN NORTHWESTERN SOUTH AMERICA, 2016

■ Coca plant growng area   Source: Mallette et al., 2016

FIGURE 4-5

Recognizing that free trade may well be able to solve many of the realm's economic–geographic problems, governments are now pursuing several avenues of heightened international cooperation. This is a good example of the concept of **supranationalism [11]**, which geographers define as a voluntary association in economic, political, or cultural spheres of three or more independent countries willing to yield some measure of sovereignty for their mutual benefit (a concept elaborated in Chapter 5).

In 2020, South America's republics were affiliating with a number of trading blocs, of which the following pair are the most important:

- **Mercosur/l** (**Mercosur** in Spanish; **Mercosul** in Portuguese): Launched in 1995, this Common Market established a free-trade zone and customs union that now links Brazil, Argentina, Uruguay, Paraguay, and Venezuela (seven additional South American countries are associate members). The organization aspires to be the dominant free-trade association for all of South America, but political agendas tend to prevail over economic needs. Thus the path toward full economic integration is proving to be both slow-paced and obstacle-ridden. In 2016, Venezuela was indefinitely suspended from the organization due to its highly criticized economic policies. In 2020, Bolivia was in the final stage of attaining full membership.

- **Pacific Alliance (PA):** Inaugurated in 2012, this bloc was founded by Mexico, Colombia, Peru, and Chile (the organization stretches across Middle and South America). They swiftly formed a free-trade area along the western edge of the continent and announced their commitment to further economic integration "with a clear orientation to Asia." Whereas Mercosur/l countries are often entangled in political debates, the PA has advanced steadily, exhibiting a collective economic growth rate almost twice as high. Candidates for full membership, led by Panama, Costa Rica, and Ecuador, are lining up; 57 other countries—including Canada, the U.S., Japan, China, and 25 European states—have acquired observer status. The rise of the Pacific Alliance over the past decade, and its articulation of a strong liberal economic stance in comparison with Mercosur/l, have spawned notions of an east–west "continental divide" regarding the realm's evolving economic geography.

Ecuador, and Bolivia have long struggled to modernize their economies and improve standards of living. Within countries, there is substantial **uneven development [10]** as well. Most manufacturing is concentrated in and around major urban centers, leaving vast empty spaces in the interior of the realm.

Brazil, in particular, increasingly drew attention over the past decade for its momentous growth and the rising sophistication of a number of its economic sectors. In the past, typical Brazilian exports were foods and footwear. Today, the leading exports include oil (thanks to the discovery of huge offshore reserves), steel, and state-of-the-art Embraer aircraft.

Despite recent impressive economic growth across much of the realm, it is clear that the longstanding isolation of its individual countries remains a serious impediment. Most governments agree that tighter integration, political as well as in terms of transportation infrastructure, would greatly enhance economic opportunities.

Economic Integration

The realm continues to be economically fragmented, and only about 25 percent of its exports remain in South America (a very low share compared to Europe, where intra-realm exports account for nearly 70 percent). Nonetheless, things are moving in a positive direction. With mutually advantageous trade the catalyst, a new continent-wide spirit of cooperation is blossoming at every level. Cross-border rail, road, and pipeline projects, stalled for years, are multiplying steadily. And importantly, investments now flow more freely than ever from one country to another, most notably in the agricultural sector.

Dependence on Raw Materials

During the first decade of this century, South America's economy in the aggregate grew robustly by about 5 percent annually. Much of this growth occurred in Brazil, Chile, and Peru, but other parts of the continent also fared well. A leading driver was the steadily rising demand for the realm's abundant raw materials in the global marketplace, in no small part resulting from the explosive growth (and accompanying raw-material appetite) of China. The Asian giant gobbled up Brazilian soybeans, Chilean copper, Peruvian silver, Venezuelan oil, and so much more. Commodity prices skyrocketed, and South America reaped the benefits.

The economic picture since the 2000s, however, is quite different. Commodity-driven growth leveled off markedly after

2011, much of it in response to slowing Asian demand and lowered prices on the global market. Although exports of raw materials are still substantial, the heyday hype of the *commodity boom* has faded. In its place has been a resurfacing of the pre-boom realization that heavy reliance on the production of raw materials (still around 50 percent of the realm's exports today) is a risky longer-term proposition—for a variety of reasons.

First, the increasing demand for commodities tends to drive up the value of a producing country's currency, which in turn limits sales of other exports (including processed products with higher added value) as they become more expensive for foreign trading partners. Second, higher revenues from the export of raw materials can undermine the development of local manufacturing and other processing industries. Some economists now employ the term *premature deindustrialization*, wherein domestic industries erode as government policies overly emphasize the exporting of commodities at a time of rising value in the local currency. Third, resource management, because of complex ownership issues and exploitation rights, can invite corruption and complacency (note, for instance, that major discoveries of oil reserves rarely result in a society's advancement). Finally, and most significantly, the great majority of resources (especially key minerals and energy deposits) are nonrenewable, making it essential that producing countries plan for an economic future without them. Thus, in the aftermath of South America's commodity boom in the early twenty-first century, its longer-term benefits may well depend on the ability of national governments to successfully refocus their energies on nurturing home-grown industries.

At any rate, by 2020 most South American economies had gone through nearly a decade of economic stagnation followed by a slow recovery. Brazil, having been celebrated for years as the new powerhouse among emerging economies, went through a particularly painful contraction, recording its worst economic performance in two decades. The situation was also particularly dire in Argentina and, most of all, in Venezuela (which is overwhelmingly dependent on world market prices for oil).

Urbanization and Its Regional Expressions

Rural-to-Urban Migration

As in most other realms, South Americans are leaving the land and migrating to towns and cities. This realm's modern urbanization process got an early start and has been intensifying since 1950 as the growth of urban settlements has averaged 5 percent a year. Meanwhile, rural areas have grown by less than 2 percent annually over those same seven decades. As a result, the realm's urban population has climbed to over 80 percent, ranking it with those

NASA image created by Jesse Allen, using EO-1 ALI data provided courtesy of the NASA EO-1 team

An aerial image of one of South America's most spectacular raw-material sites—northern Brazil's Carajás Iron Ore Mine, located where the southeastern rim of the Amazon Basin meets the Brazilian Highlands in the center of Pará State. This may be the world's richest as well as largest proven deposit of iron ore, whose open-pit mines steadily deepen and sprawl outward as minerals are extracted from the surface—one stripped-away, environmentally degrading layer at a time. Annually, more than 300 million metric tons are hauled away via the dedicated, 850-kilometer (535-mi) railroad to the Atlantic port of São Luis.

of Western Europe and the United States. These numbers underscore not only the dimensions but also the durability of the **rural-to-urban migration [12]** from the countryside to the cities.

In South America, as in Middle America, Africa, and Asia, people driven by the poverty of rural areas are attracted to the towns and cities. Both *push* and *pull* factors are at work. Rural land reform has been very slow in coming, and for this and other reasons every year tens of thousands of farmers simply give up and leave, seeing little or no possibility of economic advancement. The urban centers lure them because they are perceived to provide opportunity—the chance to earn a regular wage, visions of education for their children, better medical care, upward social mobility, and the excitement of life in a city.

But the move itself can be very challenging. Big cities in developing countries are often surrounded and invaded by slum-like neighborhoods, and this is where the urban immigrant most often finds a first—and sometimes permanent—abode in makeshift housing, often without basic amenities and sanitary facilities. But still the people come, hopeful for a better life. Many newcomers earn their first cash income by becoming part of the **informal sector [13]**, in which workers are undocumented and money transactions are beyond the control of government. The entry-level settlements consist mostly of self-help housing, vast shantytowns known as **barrios [14]** in Spanish-speaking South America and as **favelas [15]** in Brazil. Some of their entrepreneurial inhabitants succeed more than others, transforming parts of these shantytowns into beehives of activity that can propel resourceful workers toward a middle-class existence. The fastest growing city on the list is Brasília, the capital of Brazil that was built from scratch in 1960 and that continues to attract migrants to this day. Another city that stands out is Caracas, the Venezuelan capital, the only major city in South America to *lose* several hundred thousand people as a result of the deepening economic and humanitarian crisis of recent years.

Regional Patterns

The generalized spatial pattern of South America's urban reconfiguration is displayed in **Figure 4-6**, a *cartogram* (specially transformed map) of the continent's population. Here we see not only the realm's countries in population space relative to one another, but also the proportionate sizes of individual large cities within their total national populations.

Regionally, the Southern Cone (countries colored orange) is the most highly urbanized. Today in Argentina, Chile, and Uruguay, almost all of the population resides in cities. Ranking below them in urbanization is Brazil (in green). The next highest group of countries (yellow) borders the Caribbean Sea in the north. Not surprisingly, the Andean countries (gray) constitute the realm's least urbanized zone. Figure 4-6 tells us a great deal about the relative positions of the largest metropolises within their countries. Five of them—São Paulo, Rio de Janeiro, Buenos Aires, Lima, and, most recently, Bogotá—rank among the world's **megacities [16]**, in which the metropolitan population exceeds 10 million. The size

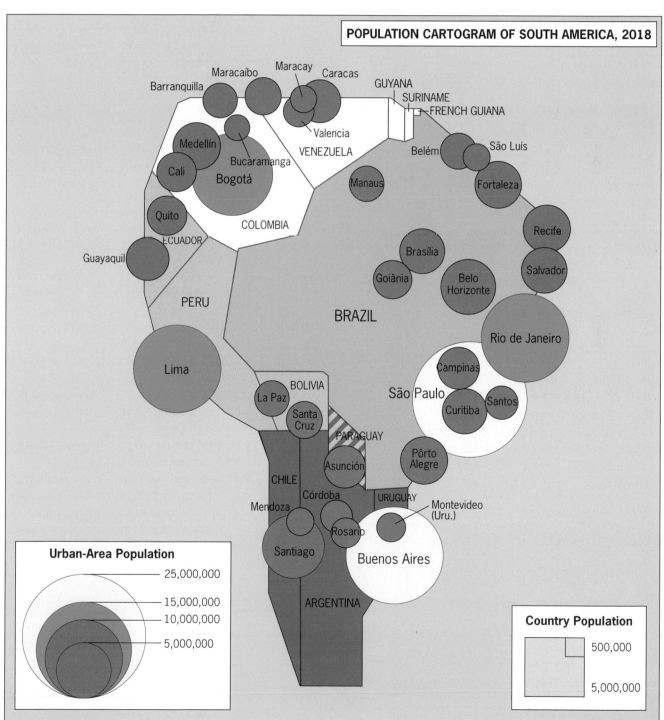

© J. Nijman, P.O. Muller and John Wiley & Sons, Inc.

FIGURE 4-6

and density of these megacities pose enormous challenges in terms of housing, mobility, pollution, water provision, and overall environmental sustainability (see photo of São Paulo).

Brazil's São Paulo is the eleventh-largest metropolis on Earth and, with more than 21 million people, by far the biggest in this realm. Most of its growth over the past half-century is the result of massive rural in-migration. Severely overcrowded São Paulo has grown so rapidly that certain neighborhoods exhibit some of the world's highest residential densities. Moreover, this sprawling megacity has become notorious for its gridlocked automobile traffic and choked transit systems, which produce commuting times that are about triple those in major U.S. cities. This photo was taken from a taxi-helicopter, an elitist new mode of transportation used routinely by high-level business executives to leapfrog the congestion below.

Problems of Inequality

An important indicator of South America's economic growth in this century has been the steady expansion of its middle class. The World Bank recently reported that this socioeconomic stratum had increased its share of the realm's income-earning population from 20 to 29 percent between 2000 and 2010. But that period coincided (as noted above) with strong economic growth that was in part due to the commodity boom. As stagnation set in over the past half decade, it appears that inequality stalled as well; that is, the improvements of the 2000s did not carry over. By most measures, the disparity between rich and poor is still wider in this realm than in any other, with wealth concentrated in the hands of a tiny minority by an especially lopsided margin (see **Box 4-5** and **Fig. 4-7**).

The United States and China in South America

The Shadow of the United States

The United States has long played a key role in this realm, beginning in 1823 with the Monroe Doctrine's assertion that European colonial powers no longer had rights to South America. During the Cold War (1945–1990), the United States became politically involved in a number of countries, mainly to keep Soviet influence out of the Western Hemisphere (above all in Chile during the early 1970s). This effort did not always enhance the local standing of the United States. In fact, anti-Americanism never seems far from the surface in South America, based heavily on the U.S. record of supporting right-wing dictatorships as well as continuing suspicions that the Americans still harbor imperialistic intentions.

Relations with the United States have never been smooth as they evolved in a consistently asymmetrical manner. Thus, South America today attracts only about 4 percent of all U.S. foreign trade. Yet the U.S. remains the biggest trading partner for the realm as a whole, accounting for a large share of the exports and imports of all South American countries. In 2018, the U.S. was the biggest export destination for goods from Colombia, Venezuela, and Ecuador, and the second biggest destination for goods from Peru, Brazil, Chile, Argentina, and Guyana. In the past, the economic dominance of the United States was considerably greater, because European or Asian involvement in this realm used to be quite minimal. It was hardly a coincidence that *dependencia* **theory [19]** originated in South America during the 1960s. It was a new way of thinking about economic development that interpreted the persistent poverty of certain countries as the outcome of their unequal relations with the world's more affluent countries. Whatever the current validity of *dependencia*, the asymmetry remains, and U.S. foreign policy, as viewed from South America, is not always credible. Ironically, America's special interests in this realm are now under increased pressure from another major power that would have been furthest from President Monroe's mind back in 1823: China.

China's Growing Economic Clout

China has in recent years displaced the United States as the leading export destination for goods from Brazil, Chile, and Peru. And China is a close second, after the U.S., for goods from Colombia and Venezuela. Other South American countries may soon be next. This remarkable economic rebalancing, and its possible political ramifications, has yet to fully register in the United States. Nearly invisible just 20 years ago, China has made its presence felt all across South America by establishing new embassies and consulates, buying up companies, partnering joint ventures, financing infrastructure projects and development assistance, and sending as well as inviting highest-level trade delegations. Their motives are clear: the Chinese still need large supplies of key raw materials such as oil, copper, and a plethora of minerals to reinforce the ongoing economic development of their country. Just as importantly, they are also seeking to expand their markets for Chinese exports, and the steadily expanding middle-income consumer classes of Brazil and other countries make potentially lucrative targets.

For a number of South American countries, China in some ways poses the same challenges as the United States did in the past—an asymmetry of relations that can lead to dependency and undercut the development of domestic economic sectors. The economic slowdown of China in recent years, accompanied by downward-trending demand for raw materials, has had immediate and deep repercussions for many South American economies, and underscores the risks of this asymmetry and precarious dependence.

Box 4-5   Geographic Information Analysis

The South American realm has long been considered to be the most unequal in the world. One way to measure inequality is with the **Gini Index [18]** or coefficient (named after the Italian statistician Corrado Gini). A Gini value of 0 indicates that income is equally distributed across society, whereas a value of 100 indicates that all income is held by a single recipient. The Gini score for Sweden is a modest 27.3, and the United States comes in at 41.1. Figure 4-7 shows the Gini Index for South America's countries and a number of major cities. Examine the

geographic patterns of national- and urban-scale income inequality across the realm. How does inequality in South America compare to that in the United States? In which cases is there a discrepancy between the Gini scores of cities and the countries in which these cities are located? How would you explain this variation?

Access your WileyPLUS course to interact with a dynamic version of this map and to engage with online exercises and questions.

INCOME INEQUALITY IN SOUTH AMERICAN STATES AND CITIES

City population (millions)

20
10
5

GINI index (World Bank estimate; most recent year available 2010 or later)

| No Data | Below 44 | 45 - 49 | 50 - 55 | Above 55 |

© J. Nijman, M. Shin, P.O. Muller, and John Wiley & Sons, Inc.

FIGURE 4-7 **Interactive**

Regions of South America

On the basis of physiographic, cultural, and political criteria, South America can be divided into four clearly defined regions (**Fig. 4-8**):

- **The Caribbean North**, located almost entirely north of the equator, consists of five entities that display a combination of Caribbean and South American features: Colombia, Venezuela, and those that represent three historic colonial footholds by Britain (Guyana), the Netherlands (Suriname), and France (French Guiana).

REGIONS OF SOUTH AMERICA

POPULATION
- 50,000–250,000
- 250,000–1,000,000
- 1,000,000–5,000,000
- Over 5,000,000

National capitals are underlined

REGIONS
- Caribbean North
- Andean West
- Southern Cone
- Brazil

© J. Nijman, P. O. Muller, and John Wiley & Sons, Inc.

FIGURE 4-8

- **The Andean West** is formed by four republics that share a strong indigenous cultural heritage as well as powerful influences resulting from their Andean physiography: Ecuador, Peru, Bolivia, and, transitionally, Paraguay.

- **The Southern Cone**, for the most part located south of the Tropic of Capricorn (latitude 23½°S), includes three countries: Argentina, Chile, and Uruguay (all with strong European imprints and little remaining indigenous influence) plus aspects of Paraguay.

- **Brazil** occupies an enormous part of interior and eastern South America. In the Amazon Basin of the north, its own interior overlaps with the continent's "green heart"—often referred to as "the world's lungs"—the largest tropical rainforest on Earth. Brazil is South America's giant, accounting for just about half the realm's territory as well as population. In Brazil the dominant Iberian influence is Portuguese, not Spanish, and here Africans, not indigenous peoples, form a significant component of the demography and culture.

Each of these four regions shares some important commonalities in terms of physical environment, ethnic origins, cultural milieu, and international outlook. But there also are important differences, especially involving economic performance, democratic functioning, and social stability.

Region The Caribbean North

The countries of South America's northern tier have something in common besides their coastal location: each has a coastal tropical-plantation zone based upon the Caribbean colonial model. Especially in the three Guianas, early European plantation development encompassed the forced immigration of African laborers and eventually the absorption of this element into the population matrix. Far fewer Africans were brought to South America's northern shores than to Brazil's Atlantic coasts, and tens of thousands of South Asians also arrived as contract laborers and stayed as settlers, so the overall situation here is not comparable to Brazil's. Moreover, it is also distinctly different from that of the rest of South America.

To this day, Guyana, Suriname, and French Guiana still display the coastal orientation and plantation dependency with which the colonial period endowed them, although the logging of their tropical forests is penetrating and ravaging the interior. In Colombia and Venezuela, however, farming, ranching, and mining drew the population inland, overtaking the coastal-plantation economy and establishing more diversified economies (**Fig. 4-9**).

Colombia

Imagine a country more than twice the size of France but with only three-quarters of the French population, with an

FIGURE 4-9

environmental geography so varied that it can produce crops ranging from the temperate to the tropical, and possessing world-class oil reserves as well as many other natural resources. This country is situated in the crucial northwestern corner of South America, with 3200 kilometers (2000 mi) of coastline on both Atlantic (Caribbean) and Pacific waters, closer than any of its neighbors to the markets of the north and sharing a border with giant Brazil to the southeast. This nation uses a single language and adheres to one dominant religion.

In many respects, then, Colombia seems have numerous favorable assets, but its development has been volatile. Colombia has a history of strife and violence; its political life is unstable, its population divided through extreme inequality, and its economy impaired. Its spectacular, scenic physical geography also divides its population of just over 50 million into clusters not sufficiently interconnected to foster integration; even today, this enormous country has an astonishingly low total mileage of four-lane highways. Its proximity to U.S. markets seems a mixed blessing: at the root of Colombia's latest surge of internal conflict lies its role as one of the world's leading producers of cocaine and other illicit narcotics.

As Figure 4-9 indicates, Colombia's physiography is mountainous in its Andean west and north and comparatively flat in its interior. Colombia's scattered population tends to cluster in the west and north (Fig. 4-2), where the resources and the agricultural opportunities (including the coffee for which Colombia is famous) lie.

Colombia's population clusters continue to be poorly interconnected. Several lie on the Caribbean coast, centered on Barranquilla, Cartagena, and Santa Marta, old colonial entry points. Others are anchored by major cities such as Medellín and Cali. What is especially interesting about Figure 4-9 in this context is how little development has taken place along the country's lengthy Pacific coast, where the port city of Buenaventura, across the mountains from Cali, is the only settlement of any size.

The map also shows that Colombia and neighboring Venezuela share the oil and gas reserves in and around the "Lake" Maracaibo area ("Lake" because this is really a gulf with a narrow opening to the sea), but Venezuela has the bigger portion. Recent discoveries, however, have boosted Colombia's production along the base of its easternmost Andean cordillera, and our map shows a growing network of pipelines from the interior to the coast. Meanwhile, the vast and remote southeastern interior proved fertile ground for that other big Colombian money-maker—illicit drugs.

Cocaine and Insurgencies: An Intractable Mix

The ascendance of the narcotics industry, fueled by outside (especially U.S. but also Brazilian and European) demand and coupled with its legacy of violence, crippled the state for decades and threatened its very survival. Drug cartels based in the cities controlled vast networks of producers and exporters; they infiltrated the political system, corrupted the military and police, and waged wars with each other that cost tens of thousands of lives and destroyed much of Colombia's social order. The drug cartels also organized their own armed forces to combat initiatives by the Colombian government to control the clandestine narcotics economy. Meanwhile, owners of large haciendas in the countryside hired private security guards to protect their properties, banding together to expand these units into what quickly became, in effect, private armies. Colombia was in chaos as narco-terrorists committed vicious acts of violence in the cities, rebel forces and drug-financed armies of the political "left" fought against paramilitaries of the political "right" in the countryside, and Colombia's legitimate economy, from coffee-growing to tourism, suffered grievously. To further weaken the national government, the rebels even took to blowing up oil pipelines.

Colombia's problems were compounded by the presence of various rebel groups (most notably, the Revolutionary Armed Forces of Colombia, or FARC) that contested government control of several provinces and that established mutually convenient partnerships with the cocaine cartels. FARC initially pursued a purely political or ideological agenda, but their widespread guerrilla activities soon brought them into contact with the cocaine cartels, which had their own reasons for wanting to exercise territorial control. In essence, this led to close, if murky, ties between the cartels and the rebel armies: the insurgents had found themselves a source of funding, while the cartels linked up with a convenient source of military support against government forces in times of need. In the course of the 1990s, FARC assumed the role of what political geographers call an **insurgent state [20]**, whereby rebels hold total control over part of the state's territory (**Fig. 4-10**). It appeared to be just a matter of time before the national government in Bogotá would lose control, the status quo would slide into disintegration, and Colombia would devolve into a **failed state [21]** (a country whose institutions have collapsed and in which anarchy prevails).

Colombia's Rebound

Since 2002, however, Colombia has mounted successful campaigns to defeat the rebels militarily and to persuade them through legal means to give up their arms. Long-running peace talks with FARC finally resulted in a "definitive" ceasefire in June 2016. A half-century of conflict had left at least 260,000 people dead and more than seven million displaced. By 2016, rebel influence had diminished as the government had restored control over most of its territory—effectively transforming Colombia from an incipient failed state into a more stable, secure, and potentially prosperous country. It has proven difficult, though, for the government to reach political agreements with the rebels, and details and implementation of the accords were still being worked out in 2020. And there is as yet no end in sight concerning the production of cocaine. As Figures 4-5 and 4-10 show, coca cultivation is still going strong in various parts of the country. In part, this is related to ongoing operations of the drug cartels, but it may also reflect the decision of local farmers (in areas where coca has been a mainstay) to grow what they consider the highest revenue crop as well as the continuing failure of the government to support them in raising legal crops.

Nonetheless, the government matched its domestic counteroffensive against the rebels with an international campaign to help revive the economy, promoting market-oriented, business-friendly policies. The country's strong economic performance over the past decade is based on revenues from fossil fuels, metallic ores, tourism, and farm products, particularly exports of coffee and cut flowers (Colombia is the leading U.S.

FIGURE 4-10

Venezuela is concentrated in the northern and western parts of the country, where the Venezuelan Highlands form the eastern spur of the northern end of the Andes system. Most of Venezuela's 28.4 million people are concentrated in these uplands, which include the capital of Caracas, its early rival Valencia, and the commercial/industrial centers of Barquisimeto and Maracay.

The Venezuelan Highlands are flanked by the Maracaibo Lowlands and "Lake" Maracaibo (which is actually a brackish tidal bay of the Caribbean Sea) to the northwest and by an immense plainland of savanna country, known as the *Llanos*, in the Orinoco Basin to the south and east (Fig. 4-9). The Maracaibo Lowlands, at one time a disease-infested, sparsely peopled coastland, today constitute one of the world's leading oil-producing areas; much of this fuel is drawn from reserves that lie beneath the shallow waters of the "lake" itself. The country's third-largest city, Maracaibo, is the focus of the petroleum industry that transformed the Venezuelan economy in the 1970s; Venezuela has the largest proven oil reserves in the world, in 2019 estimated at more than 300 billion barrels.

The *Llanos* on the southern side of the Venezuelan Highlands and the Guiana Highlands in the country's far southeast, like much of Brazil's interior, are in a nascent stage of development. The 300- to 650-kilometer- (180- to 400-mi)-long *Llanos* slope gently from the base of the northernmost Andean spur to the Orinoco's floodplain. Their mixture of savanna grasses and scrub woodland facilitates cattle grazing on higher ground, but widespread wet-season flooding of the more fertile lower-lying areas has thus far inhibited the plainland's commercial farming potential (much of the development of the *Llanos* to date has been limited to the exploitation of its massive oil reserves). Crop-raising conditions are more favorable in the *tierra templada* zones at moderate altitudes in the Guiana Highlands. Economic integration of this even more remote interior zone with the heartland of Venezuela was spearheaded by the discovery

supplier). Bogotá, the capital, centers the country's leading economic region and has become one of the realm's primary business complexes. The city of Medellín—former base of the legendary drug kingpin Pablo Escobar, and designated the "world's murder capital" in the late 1980s—has recently made a comeback and is now internationally regarded as a model for innovative urban design (photo below).

Venezuela

A long boundary separates Colombia from Venezuela, its neighbor to the northeast. Most of what is important in

In 2013, Medellín, Colombia, was voted the world's most innovative city by the U.S. Urban Land Institute. Striking innovations include cable cars and escalators on the city's steep hillsides that have drastically reduced local travel time. Through thriving public–private partnerships, the city has invested in new parks, schools, and libraries, especially in poor neighborhoods. This erstwhile capital of Colombia's cocaine industry may have been plagued by violence in the 1980s and 1990s, but Medellín's turnaround since 2010 has attracted attention and acclaim from urban planners around the world.

of rich iron ores during the 1950s on the northern side of the Guiana Highlands southwest of Ciudad Guayana near the mouth of the Orinoco River. Local railroads were built to connect the mines with the Orinoco, and from there ores have long been shipped directly by sea to foreign markets.

Venezuela in Crisis   Venezuela is one of those countries in which a rich natural resource base can actually contribute to deep political and economic problems. The country has long been in political turmoil because its governments had the bad habit of living off oil profits while neglecting long-term productive investments in the economy to better the lives of most Venezuelans. A potential problem with natural resources is that they can fairly easily be monopolized and controlled, with the benefits accruing to only a small number of people. Further, the over-reliance on oil exports is felt most painfully when prices on the world market fall.

By the mid-1990s, the economy descended into a severe recession and widespread social unrest. With more and more Venezuelans enraged at the way their oil-rich country was approaching bankruptcy without making progress toward a more equitable distribution of the national wealth, the way was cleared for the rise to power of the socialist radical Hugo Chávez in 1999 (with an apparent mandate from the urban poor and the increasingly penurious middle class). The newly elected president took a strong stand against the so-called **neoliberal policies [22]** of the past, policies that generally favor deregulation, free trade, and favorable conditions for capital and corporations.

Chávez's rule, however, quickly turned both autocratic and economically destructive. During his long presidency that ended with his death in 2013, relatively high oil prices brought in consistently high revenues totaling well over (U.S.) $800 billion. But there was very little to show for it: infrastructural improvements were hardly to be seen, bloated public-sector companies chronically underperformed, and the entrepreneurial sector of the population had fled the country soon after Chávez came to power. As for the poor, they intermittently benefited from new social programs, but life did not improve for them. Today, all Venezuelans are worse off, bedeviled by constant shortages of basic foods and vital retail products thanks to the staggering incompetence of the autocratic regime of Nicolás Maduro, who succeeded Chávez in 2013.

Since 2016, the Venezuelan economy has been in free fall, and by 2019 the country's GDP had shrunk by about a quarter. There were severe shortages of food, water, medicine, and even gasoline. Hyperinflation was rampant, and crime, violence, and protests flared. By the end of 2019, more than four million people had fled the country, often on foot, to Brazil, Colombia, Ecuador, and Peru (see photo). The United States (especially southern Florida) is also home to a sizeable portion of the Venezuelan diaspora, mostly those who left in the earlier years of Chávez's "Bolivarian Revolution."

Although an election in 2019 gave Maduro a new presidential term, it was universally viewed as rigged. The country then veered toward a major leadership crisis as another politician, Juan Guaido (the chairman of the Venezuelan National Assembly, similar to the position of Speaker of the House in the U.S.),

RAUL ARBOLEDA/AFP/Getty Images

Venezuelan migrants climb on a truck near Cúcuta, a Colombian city just across the border from Venezuela, to take them further into Colombia (see Fig. 4-9). This photo was taken in February, 2019, when Venezuela was in grave economic and political turmoil. About one-sixth of the population has fled the country since 2013.

proclaimed himself as the "acting President of Venezuela." Despite receiving declarations of support from 52 countries, within weeks Guaido's movement failed and soon evaporated.

The "Three Guianas"

Three small political entities form the eastern flank of the realm's northern region: Guyana, Suriname, and French Guiana. They are good reminders of why the name "Latin" America is so inappropriate: the first is a legacy of British colonialism and employs English as its official language; the second is a remnant of Dutch influence where Dutch is still official among its polyglot of tongues; and the third is still a dependency—of France. None has a population exceeding 800,000, and all three exhibit social characteristics and cultural landscapes far more representative of the Caribbean Basin than South America. Here British, Dutch, and French colonial powers acquired possessions and established plantations, brought in African and Asian workers, and created economies similar to those that mark a number of the nearby Caribbean islands.

Guyana, with just under 800,000 people, has nearly as many inhabitants as Suriname and French Guiana combined. When Guyana became independent in 1966, its British rulers left behind an ethnically and culturally divided population in which people of South Asian (Indian) ancestry now make up about 44 percent and those of African background (including Afro-European ancestry) 30 percent. This makes for contentious politics, given the religious mix that is approximately 57 percent Christian, 28 percent Hindu, and 7 percent Muslim. Guyana remains dominantly rural, and plantation crops continue to figure prominently among exports (gold from the interior is the most valuable single product).

Guyana ranks among the realm's poorest and least urbanized countries, and it is ever more strongly affected by the expanding drug industry of South America's north. Its thinly populated interior, beyond the reach of antidrug campaigns, has become a staging area for cocaine distribution to the United States via Middle America (and increasingly to Brazil and Europe). A recent study suggests that Guyana's take from the drug trade already amounts to at least one-fifth of the country's total income.

Oil may soon become a very important factor in the economy, however, because of recently discovered large offshore reserves (Fig. 4-9). According to some estimates, Guyana's GDP could increase fivefold by 2030, making it one of the richest countries in the realm—though the example of neighboring Venezuela confirms that the existence of great natural resources is no guarantee of economic prosperity.

Suriname, actually progressed more rapidly than Guyana did after it was granted independence in 1975, but unending political instability swiftly ensued. The Dutch colonists brought South Asians, Indonesians, Africans, and even some Chinese to their colony, making for a notably fractious state. More than 100,000 residents—about one-quarter of the entire population—emigrated to the Netherlands, and were it not for support from its former colonial ruler, Suriname's economy would have collapsed. Agriculturally, its rice farms provide Suriname with self-sufficiency, even allowing for some exports (which is dominated by plantation crops). Suriname's leading income producer continues to be its bauxite (aluminum ore) that is mined in a zone across the middle of the country. To this short list can also be added the promise of those offshore oil finds noted previously, which may soon be earning important revenues.

Suriname's cultural geography is seasoned by the many languages spoken by its 600,000 inhabitants. Dutch is the official tongue, but other than from officials and in schools it is not heard much. A mixture of Dutch and English, *Sranan Tongo*, serves as a common language, but one also hears indigenous languages, Hindi, Chinese, Indonesian, and even some French Creole on the streets.

French Guiana, the easternmost outpost of the Caribbean North region, is the last remaining dependency on the mainland of South America. This territory is an anomaly in other ways as well. Consider this: it is almost as large as South Korea (population: just above 50 million) with a population totaling only 290,000. Its status is that of an overseas *département* of France, and its official language is French. Nearly half the population resides in the immediate vicinity of the coastal capital, Cayenne.

In 2020, there still was no prospect of independence for this decidedly underdeveloped relict of the former French Empire. Gold remains the most valuable export, and a small fishing industry sends some exports to France. But what really matters in French Guiana is the European Space Agency's launch complex at Kourou on the Atlantic coast, which accounts for more than half of the territory's entire economic activity—from plantation farming to spaceport—but with minimal involvement, or benefits, for most ordinary people.

Region The Andean West

The second regional grouping of South American states—the Andean West (**Fig. 4-11**)—is dominated physiographically by the great Andes mountain chain and historically by indigenous peoples. This region encompasses Peru, Ecuador, Bolivia, and transitional Paraguay, the last with one foot in South America's West and the other in the South (Fig. 4-8). Bolivia and Paraguay also are the realm's only examples of a **landlocked country [23]**.

Spanish conquerors overpowered the indigenous nations, but they did not reduce them to small minorities as happened to so many native peoples in other parts of the world. Today, around a quarter of the residents in these four countries are indigenous, and well over half is of mixed ancestry—ethnic categories can be notoriously difficult to define, and data are somewhat unreliable (an issue elaborated in the section on Brazil). Among the indigenous population, there are many distinct communities, often with their own language or dialect (e.g., Peru contains over 50 different indigenous groups).

This region has long been South America's poorest economically, with lower incomes, higher numbers of subsistence farmers, and fewer opportunities for job-seekers than elsewhere in the realm. For a very long time, the urbane lives of the land-owning elite have been worlds apart from the hardscrabble existence of the landless peonage (the word *peón* is an old Spanish term for an indebted day laborer). But in recent years, the Andean West, like the realm as a whole, has been stirring, and oil and natural gas are leading the way.

Peru

Peru straddles the Andean spine for more than 1600 kilometers (1000 mi) and is the largest of the region's four republics in both territory and population (32.5 million). Physiographically and culturally, Peru divides into three parts (Fig. 4-11). Lima and its port, Callao, lie at the center of the *desert coast* subregion, and it is symptomatic of the cultural division prevailing in Peru that for nearly 500 years the capital city has been positioned on the western periphery, not in a central location in a basin of the Andes (moreover, Peru had only a single indigenous president even though aboriginal peoples constitute nearly half of the population). From an economic point of view, however, the Spaniards' choice of a headquarters on the Pacific coast proved to be sound, for the coastal strip has become commercially the most productive part of the country. A thriving fishing industry contributes significantly to the export trade; so do the products of irrigated agriculture in some 40 oases dispersed across the arid coastal plain, which include fruits such as citrus and olives, and vegetables such as asparagus (a big money-maker) and lettuce.

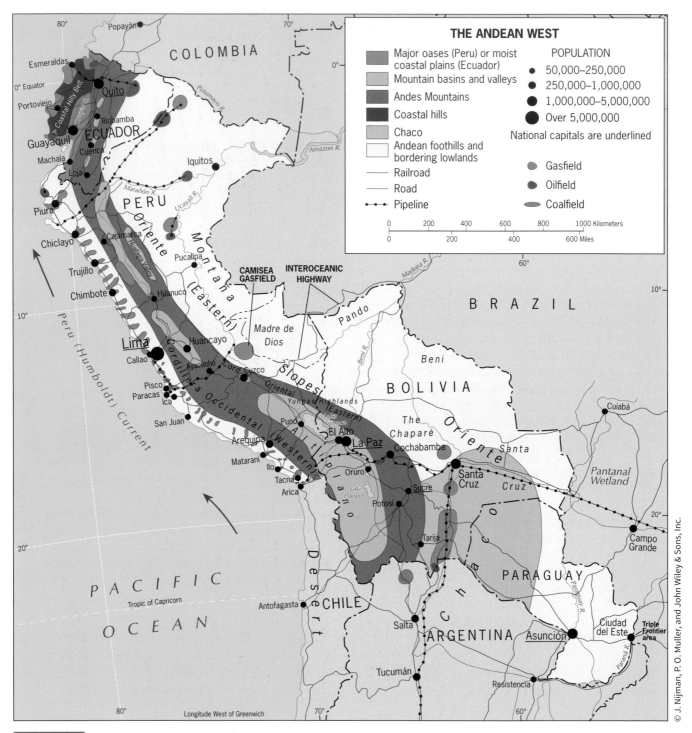

FIGURE 4-11

The Andean or *Sierra* subregion occupies just about one-third of the country and is the ancestral home of the largest component in the total population, the speakers of Quechua, the *lingua franca* that emerged during the Inca Empire. Their physical survival during the harsh Spanish colonial regime was made possible by their adaptation to the high-altitude environments they inhabited.* The Andean subregion is not nearly as important economically as the coast, but it does yield copper, zinc, and lead from mining centers, the largest of which is Cerro de Pasco. Peru is also the world's seventh-largest producer of gold, the price of which has been soaring in recent years,

contributing to this country's steadily growing economy; most of the mines are located in the "gold belt" of the northernmost Peruvian Andes, but illegal strip mining is widespread in the Amazon rainforest of the southeastern interior. In the high valleys and intermontane basins, the indigenous population is concentrated either in isolated villages, around which many

*This is a good place to become reacquainted with the concept of *altitudinal zonation* that is elaborated in Figure 3-5 and accompanying text. The subregions of the Andean West represent the best place on Earth to apply this fivefold classification of vertical environments. The five *tierras,* in ascending elevational order, are the *caliente, templada, fría, helada,* and *nevada.*

people engage in subsistence agriculture, or in the more favorably located and fertile areas where they are tenants, peons on white- or mestizo-owned haciendas.

Of Peru's three subregions, the East or *Oriente*—the inland-facing slopes of the Andean ranges that lead down to the Amazon-drained, rainforest-covered *montaña*—is the most isolated. The focus of the eastern subregion, in fact, is Iquitos in the far north, a city that looks east rather than west because it can be reached by oceangoing vessels sailing 3700 kilometers (2300 mi) up the Amazon River across northern Brazil. Iquitos grew rapidly during the Amazon's wild-rubber boom of the early twentieth century but swiftly declined thereafter; now it is finally growing again, reflecting Peruvian plans to open up the eastern interior. Today, this subregion, and perhaps Peru as a whole, appears to be on the threshold of a new era owing to major new discoveries of oil and gas reserves in the *Oriente*. Indeed, while still one of the poorer countries in this realm, since the mid-2000s Peru has experienced a higher economic growth rate than any other. A trans-Andean pipeline already transports natural gas from the Camisea reserve (north of Cuzco) to a conversion plant on the coast south of Lima (Fig. 4-11). From there, an underwater connector pipeline extends to an offshore loading platform for tankers taking it to the U.S. market. Additional interior gas reserves are now being tapped, and they only heighten Peruvian aspirations to become a major energy producer.

Peru's economic development, however, is not proceeding without adversity. Environmentalists and supporters of indigenous rights are raising issues throughout the *Oriente* (even as political activists argue against the terms of trade that the Peruvian government has accepted with the big oil companies). They complain that the oil- and gas-led exploitation of the unprotected East will harm both the environment and local communities, and that the new energy revenues will only further benefit the already-favored coastal residents of Peru—leaving the ever more disadvantaged inhabitants of the interior even farther behind.

Ecuador

On the map, Ecuador, the smallest of the three Andean West republics, appears to be just a northern corner of Peru. But that would be a misrepresentation because Ecuador also possesses a complete range of regional variations (Fig. 4-11). It has a coastal belt; an Andean zone that may be narrow (less than 250 kilometers [150 mi]) but by no means of lower elevation than elsewhere in the region; and an *Oriente*—an eastern subregion that is as sparsely settled and as economically marginalized as that of Peru. As with Peru, just about half of Ecuador's population of 17.4 million is concentrated in the Andean intermontane basins and valleys, and the most productive subregion is the coastal strip. There, however, the similarities end.

Ecuador's Pacific coastal zone consists of a belt of hills interrupted by lowlands, of which the most important lies in the south between the hills and the Andes, drained by the Guayas River. Guayaquil—the country's largest city, main port, and leading commercial center—forms the focus of this subregion (**Box 4-6**). Unlike Peru's coastal strip, Ecuador's is not a desert: it

---

Box 4-6 From the Field Notes . . .

© H. J. de Blij

"I can't remember being hotter anyplace on Earth, not in Kinshasa, not in Singapore … not only are you near the equator here in steamy Guayaquil, but the city lies in a swampy, riverine lowland too far from the Pacific to benefit from any cooling breezes and too far from the Andean foothills to enjoy the benefits of suburban elevation. Guayaquil is not the disease-ridden backwater it used to be (left photo): its port is modern, its city-center waterfront on the Guayas River has been renovated, its international airport is the hub of a commercial complex. This has grown into a metropolis of 2.9 million, made possible in part by oil revenues. Beating the heat and glare is an everyday priority: entire streets have been covered by makeshift tarpaulins and more permanent awnings to protect shoppers (right). Talk to the locals, though, and you find that there is another daily concern: the people 'up there in the mountains who rule this country always put us in second place.' Take the 45-minute flight from Guayaquil to cool and comfortable Quito, the capital, and you're in another world, and you quickly forget Guayaquil's problems. That's just what the locals here say the politicians do."

consists of fertile tropical plains not afflicted by excessive rainfall. Seafood (especially shrimp) is a leading product, and these lowlands support a thriving commercial agricultural economy built around bananas, cacao, cattle-raising, and coffee on the hillsides. Moreover, Ecuador's western subregion is also far less Europeanized because this white component of the national population is only about one-third the size of Peru's.

A greater proportion of whites are engaged in administration and hacienda ownership in the central Andean zone, where most of the Ecuadorians who are indigenous also reside—and, not surprisingly, where land-tenure reform is an explosive issue. The differing interests of the Guayaquil-dominated coastal lowland and the Andean-highland subregion focused on the capital (Quito) have long fostered a deep regional cleavage between the two. This schism has not lessened in recent years, and autonomy and other devolutionary remedies often are openly discussed in the coastlands.

In the rainforests of the *Oriente* subregion, oil production is expanding as a result of the discovery of additional reserves. Some analysts predict that eastern Ecuador as well as Peru will prove to contain energy resources comparable to those buried

FIGURE 4-12

along the inland side of the Andes in Colombia and Bolivia. These developments could well portend the dawning of a new "oil era" accompanied by the economic transformation of the Ecuadorian *Oriente*. Indeed, oil already tops the country's list of exports, and to prepare for the next stage, the energy industry's infrastructure is being modernized and upgraded.

Bolivia

Nowhere are the problems typical of the Andean West more acute than in the volatile, landlocked country of Bolivia. Before reading further, take a careful look at Bolivia's regional geography in Figure 4-11 and **Figure 4-12**. Bolivia is bounded by remote peripheries of both Brazil and Argentina, mountainous Andean highlands as well as intermontane *altiplanos* (high-elevation basins and valleys) of Peru, and coveted coastal zones of northern Chile. As the maps show, the Andes in this zone broaden into a vast mountainous complex some 700 kilometers (450 mi) wide.

A landlocked location does not always imply insurmountable challenges (think of Switzerland), but in the majority of cases it poses major limitations. There are 45 landlocked countries in the world, and almost all can be considered poor. Indeed, of the 15 lowest-ranking countries on the UN's **Human Development Index [24]**, eight have no coastline. They have no direct access to ports; transport is costly; they are dependent

on neighboring countries; the construction of cross-border infrastructures is costly and complicated; and overland trade is easily disrupted by possible conflicts in those adjoining countries. The economic impact is difficult to measure, but a recent research paper estimated that Bolivia's GDP would be 20 percent higher if it had a corridor (through northern Chile) with direct access to the Pacific Ocean.

On the boundary between Peru and Bolivia, freshwater Lake Titicaca lies at 3700 meters (12,500 ft) above sea level and helps make the surrounding *Altiplano* livable by mitigating the coldness in its vicinity (that includes the heart of modern Bolivia focused on its capital, La Paz). On the cultivable land that encircles the lake, potatoes and *tierra fría* grains have been raised for centuries dating back to pre-Inca times. However, the *Altiplano's* other large water body, Lake Poopó, is now experiencing a very different fate (see **Box 4-7**).

The Bolivian state is the product of the Hispanic impact, and the country's indigenous peoples (who comprise almost two-thirds of the national population of 11.5 million) no more escaped the loss of their land than did their Peruvian or Ecuadorian counterparts. What made the richest Europeans in Bolivia wealthy, however, was not land but minerals. The town of Potosí in the eastern cordillera became a legend for the immense deposits of silver in its vicinity; tin, zinc, copper, and lead were also discovered there, and together with oil and gas are still leading sources of foreign revenue. And more recent finds show Bolivia contains

Box 4-7 Global Climate Change on Bolivia's *Altiplano*

The Drying Up of Lake Poopó

This pair of NASA satellite images—2013 on the left; 2016 on the right—shows the dramatic pace of the dessication of Lake Poopó, which used to be Bolivia's second-largest lake (after Lake Titicaca that straddles the border with Peru). This lake used to be about twice the size of Greater Los Angeles. Figure 4-12 still shows the lake, and during the spring parts of it still fill, but for most of the year it now stands almost completely dry. Global environmental change is producing longer and more intense droughts, and may also be driving the retreat of Andean glaciers that supply the water in the first place. The winter of 2014 seemed to mark a distressing turning point, with first-time reports of rotting fish and dying birds on the dried-out lake bed. This is not only an ecological catastrophe, but also has unleashed dire economic consequences for local indigenous communities whose livelihoods have depended on the lake for centuries. And, as is so often the case, climate change works its effects in combination with unsustainable human practices: environmental scientists say that the excessive diversion of water to support nearby mining activity is also to blame. Thus what remains is a growing expanse of salt flats that steadily reduces agricultural opportunities even if sufficient water supplies were available.

AP Images/NASA Earth Observatory

a significant share of the world's lithium reserves—essential for batteries that power electric motor vehicles.

Landlocked, physiographically bisected, culturally bifurcated, and economically divided, Bolivia is a severely challenged state. Moreover, the country's prospects are worsened by its political geography: look again at Figure 4-12 and you can see that Bolivia's nine provinces are regionally divided between indigenous-majority *departments* (as subnational units are called here) in the upland west and departments with mestizo majorities in the much-lower-lying east. As noted above, La Paz, the national seat of government, is located on the indigenous-majority Andean *Altiplano*, but many mestizo Bolivians do not recognize it as such.

The Santa Cruz Department, like the others in the *Oriente*, stands in sharp contrast to those of the Andes in the west. Here the hacienda system (discussed in Chapter 2) persists almost unchanged from the colonial era, its profitable agriculture supporting a wealthy aristocracy. Recently, this eastern sector of Bolivia has proved to contain abundant energy resources as well, adding to its economic advantages. So, there is talk—abetted by frequent public demonstrations—in support of autonomy, even secession here (photos next page). But it must also be kept in mind that neither the haciendas nor the energy industry could operate without the indigenous labor force.

Paraguay

Paraguay, Bolivia's landlocked neighbor to the southeast, is one of those transitional countries lying between regions and exhibiting properties of each (see Fig. 4-8). Certainly, Paraguay (population: 7.1 million) is not an Andean country. It has no highlands of consequence. Its well-watered eastern plains give way to the dry scrub of the Gran Chaco in the northwest. Nor does it have clear, spatially entrenched ethnic divisions between indigenous, mestizo, and other peoples as do Bolivia and Peru. But aboriginal ancestry dominates the ethnic complexion of Paraguay, and continuing protests by landless peasants are similar to those taking place elsewhere in the Andean West. Moreover, the native language, Guaraní, is so widely spoken (by over 90 percent of the people) that this is one of the world's most thoroughly bilingual societies.

Today, Paraguay's longstanding isolation is eroding, most evident in the hot, semiarid Chaco of the northwest. Here, rapid deforestation has opened up a burgeoning new cattle-ranching frontier. For aboriginal peoples like the Ayoreo, however, the impact is disastrous because their traditional livelihoods are threatened with imminent destruction (the word for *bulldozer* in their language means "attacker of the world"). The contraction of the Chaco has also sounded alarms among biologists and conservationists because so much of it remains unexplored and is certain to contain myriad undiscovered plant and animal species.

Looking southward, there is little in Paraguay's economic geography to compare to Argentina, Uruguay, or Chile. Paraguay's low GDP resembles that of countries of the Andean West, not the far more advantaged Southern Cone region. This is also one of South America's least urbanized states, and poverty dominates the countryside as well as the slums girdling the capital, Asunción, and other towns (Fig. 4-11). Almost a quarter of the population lives at or below the official poverty level. Moreover, research suggests that 1.6 percent of the population owns about 80 percent of the land, which may be a record high for this measure of inequality in South America.

In the east, perhaps half a million Brazilians have crossed the border to settle in Paraguay, where they have created a

AIZAR RALDES/AFP/Getty Images.

RODRIGO BUENDIA/AFP/Getty Images

Bolivia's political geography is as divided as its physical geography, the power base of the indigenous population lying on the *Altiplano* and in the adjacent mountains, and that of the mestizo minority centered in the interior lowlands. These photos, taken a day apart in 2014 in Santa Cruz (now Bolivia's largest city by far) where mestizo strength is concentrated, show the crucial difference of opinion: a majority of mestizos want autonomy for their eastern provinces (left); virtually all indigenous peoples protest the prospect of such autonomy (right), seen as a prelude to secession and independence that would fatally fracture their country.

thriving commercial agricultural economy that produces soybeans, livestock, and other farm products exported to or through Brazil. This puts Paraguay in a geographically difficult position, and it often complains that Brazil is not living up to its regional trade obligations, thereby creating unacceptable difficulties for Paraguayan exporters. Meanwhile, politicians raise fears that Brazilian immigration is creating a steadily expanding foreign enclave within Paraguay, where people speak Brazilian Portuguese (including in the local schools), Brazil's rather than Paraguay's flag flies over public buildings, and a Brazilian cultural landscape is steadily emerging in the eastern border zone.

Another quintessentially geographical problem stemming from Paraguay's long-term weakness lies in the southeast, where the borders of Brazil, Argentina, and Paraguay converge in a chaotic tangle of smuggling, money-laundering, political intrigue, and even terrorist activity. This hub of lawlessness—centered on the town of Ciudad del Este, but with local spillovers into the territory of all three countries—is known as the *Triple Frontera* or **Triple Frontier [25]** (Fig. 4-11). It should also be noted that the Triple Frontier contains a sizeable, tristate Middle Eastern community, identified by the U.S. government as a source of funding for Islamic militant groups—an allegation repeatedly rejected by the governments of Argentina, Brazil, and Paraguay.

Region The Southern Cone

Argentina

The largest Southern Cone country by far is Argentina, whose territorial size ranks second only to Brazil in this geographic realm; its population of 45 million ranks third overall after Brazil and Colombia. Although Argentina exhibits a great deal of physical-environmental variety within its boundaries, the overwhelming majority of the Argentines are concentrated in the physiographic subregion known as the *Pampa* (a Spanish word meaning "plain"). **Figure 4-13** underscores the degree of clustering of Argentina's inhabitants on the land and in the cities of the Pampa. It also shows the relative emptiness of the remaining six subregions shown on this map: the scrub-forested *Chaco* in the northwest; the mountainous *Andes* in the west, along whose crestline lies the boundary with Chile; the arid plateaus of *Patagonia* south of the Rio Colorado; and the undulating transitional terrain of intermediate *Cuyo, Entre Rios* (also known as *Mesopotamia* because it lies between the Paraná and Uruguay rivers), and the *North*.

The Argentine Pampa is the product of the past 175 years. During the second half of the nineteenth century, when the great grasslands of the world were being opened up (including those of interior North America, Russia, and Australia), the economy of the long-dormant Pampa began to emerge. The food needs of industrializing Europe grew by leaps and bounds, and the advances of the Industrial Revolution—railroads, more efficient ocean transport, refrigerated ships, and agricultural machinery—helped to make large-scale commercial meat and grain production in the Pampa not only feasible but also highly profitable. Large haciendas were laid out and farmed by tenant workers; railroads were built and radiated ever farther outward from the booming capital of Buenos Aires, soon bringing the entire Pampa into production.

Argentina once was one of the richest countries in the world. Its historic affluence is still reflected in its architecturally splendid cities whose plazas and avenues are flanked by ornate public buildings and private mansions. This is true not only of the capital, Buenos Aires, at the head of the Rio de la Plata estuary—it also applies to interior cities such as Mendoza and Córdoba. The cultural imprint is dominantly Spanish, but it was diversified by a massive influx of Italians and smaller but

FIGURE 4-13

a few highly populated and urbanized provinces that hold almost all the political power and a large number of provinces with very small populations (**Fig. 4-14**). Buenos Aires Province has over 17 million people (out of a total 45 million), whereas Tierra del Fuego has barely more than 150,000. A dozen additional provinces contain less than 750,000 inhabitants, so that the larger ones besides dominant Buenos Aires are also disproportionately influential in domestic politics, especially Córdoba and Santa Fe. Corruption and mismanagement reached so high a level in 2012 that the authoritative international weekly, *The Economist*, decided it would no longer include statistics provided by the Argentine government; pointedly, the newsmagazine also announced that it did not want to be part of the government's "deliberate attempt to deceive voters and swindle investors"—a reflection of the country's sorry condition for the past decade.

If one compares Argentina to Paraguay, it is striking how much stronger Argentina's position is in terms of natural endowments and geographic opportunities. About a century ago, Argentina was regarded as highly developed and, in some quarters, even ahead of Western Europe and the U.S. Bottom line: things could be so much better here than they presently are.

Chile

For over 4000 kilometers (2500 mi) between the crestline of the Andes and the coastline of the Pacific lies the narrow strip of land that is the Republic of Chile (**Fig. 4-16**). On average just 150 kilometers (90 mi) wide (and only rarely over 250 kilometers [150 mi] wide), Chile is the world's quintessential example of what **elongation [28]** means to the functioning of a state. Accentuated by its north–south orientation, this severe territorial attenuation

influential numbers of British, French, German, and Lebanese immigrants. This has also long been one of the realm's most urbanized countries: 93 percent of its population is concentrated in cities and towns, a higher percentage even than western Europe or the United States. Fully one-third of all Argentinians live in metropolitan Buenos Aires, by far the leading industrial complex.

But political infighting and economic mismanagement have combined over the long term to shackle a vibrant and varied economy. Part of the problem, it appears, lies in the country's lopsided geography, with an enormous gap between

not only results in Chile extending across numerous environmental zones; it has also contributed to the country's external political, internal administrative, and general economic challenges. Nevertheless, throughout most of their modern history, the Chileans have made the best of this potentially disruptive centrifugal force: from the beginning, the sea has constituted an avenue of longitudinal communication; the Andes Mountains continue to form a barrier to prevent encroachment from the east; and when confrontations loomed at the far ends of the country, Chile proved to be quite capable of coping with aggressive neighbors.

The name Mendoza is synonymous with Argentinean wine. As our maps show, Mendoza is a province as well as a city in the Cuyo subregion of west-central Argentina, and both are massively involved in wine production at some of South America's biggest winemaking facilities. This has always been red-wine country, and its best-known varietals are Malbec and Cabernet Sauvignon. Even though Mendoza Province produces more than 80 percent of Argentina's wine, this still-expanding industry must contend with the area's semiarid environment and increasingly relies on irrigation networks to meet its growing water needs. This view is toward the west, whose horizon is filled entirely by the mighty wall of the Andes; beyond the mountain range lies central Chile, another South American producer of world-class wines (also see Box 4-8).

FIGURE 4-14

As Figure 4-16 indicates, Chile is a three-subregion country. About 90 percent of its 18.9 million people are concentrated in what is called Middle Chile, where Santiago, the capital and largest city, and Valparaíso, the chief port, are located. North of Middle Chile lies the Atacama Desert, wider, drier, and colder than the coastal desert of Peru. To the south of Middle Chile, the coast is punctuated by a plethora of fjords and islands, the topography is mountainous, and the climate—wet and cool near the Pacific—soon turns drier and much colder against the Andean interior.

Not surprisingly, the land in Middle Chile is the most fertile and valuable, and is marked by some important wine-growing regions (see **Box 4-8**). In contrast, there is hardly any agriculture to be found in either the North or South (Fig. 4-4). The three subregions are also apparent on the realm's cultural map, with Europeans dominating the Middle, mestizos prevailing in the north, and indigenous groups forming the ethnic majority in the South (Fig. 4-3).

Nevertheless, prior to the 1990s, the arid Atacama in the North accounted for more than half of Chile's foreign revenues. It contains the world's largest exploitable deposits of nitrates, which were the country's economic mainstay before the discovery of methods of synthetic nitrate production in the twentieth century. Subsequently, copper became the dominant export (Chile again possesses the world's largest reserves, which now account for more than half of all export revenues). It is mined in several places, but the main concentration lies on the eastern margin of the Atacama near Chuquicamata, not far from the port of Antofagasta. The Atacama Desert also contains lithium, a rare mineral rapidly gaining strategic value because it is a key component in today's electric batteries (see photo on p. 127).

Chile has in recent decades embarked on a highly successful program of free-market economic reform that brought stable growth, lowered inflation as well as unemployment, reduced poverty, and attracted foreign investment. The last is of particular significance because these new international connections enabled the export-led Chilean economy to diversify and develop in some badly needed new directions. Copper remains at the top of the export list, but many other mining ventures

Box 4-8 Geographic Information Analysis

Viticulture [27], or wine-growing, in Argentina, Chile, and everywhere else is very much a function of geography and climate (at a finer scale, soil and slopes matter a great deal as well). And climatic conditions, especially temperatures, are partially dependent on latitude. **Figure 4-15** shows the location of the world's major wine-producing regions. Note the importance of the upper and lower temperature limits as shown on the map. Which latitudinal

zones contain almost all of the world's wine regions? But wine production does not occur everywhere within these latitudinal zones. Why not? Consult Figure 1-6 in Chapter 1 (which displays the global distribution of climate types), and explain in greater detail.

Access your WileyPLUS course to interact with a dynamic version of this map and to engage with online exercises and questions.

WINE GROWING REGIONS OF THE WORLD

Annual Wine Production (million hectoliters)

50
30

Average Annual Temperature (degrees Celsius)

| Below -30 | -30 to -20 | -20 to -10 | -10 to 0 | 0 to 10 | 10 to 20 | 20 to 30 | 30 to 40 |

FIGURE 4-15 **Interactive**

© J. Nijman, M. Shin, P.O. Muller and John Wiley & Sons, Inc.

IVAN ALVARADO/Newscom

Lithium is a small but critical component of batteries—all batteries. It is a scarce commodity, and the world market price has been going up steadily along with demand. It is quickly turning into what some call a strategic natural resource, even though the overall value (due to the tiny quantities involved) of lithium trade is still quite modest. Bolivia, Chile, and Argentina together contain about 75 percent of all known world reserves. This photo was taken in the Atacama Desert of northern Chile, looking east toward the Andes. Lithium is made from the brine (water saturated with salt) obtained from pools in the surrounding salt flats. This brine is pumped into rectangular ponds that, as the water evaporates, turn into different shades of blue, yellow, and green, like a giant artist's palette. Once the brine is sufficiently concentrated, it is collected and transported to factories in which various potassium compounds and the prized lithium powder are extracted.

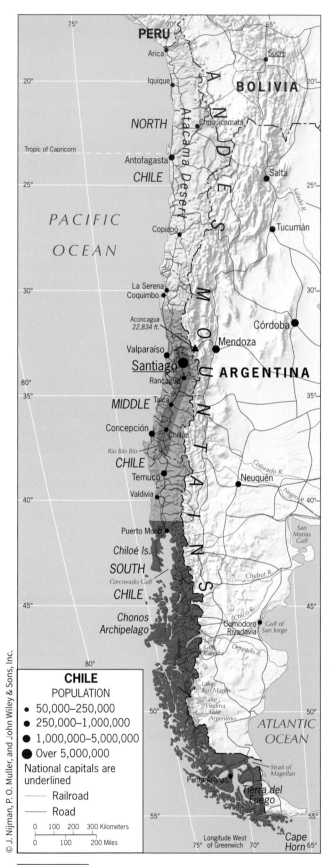

CHILE
POPULATION
- 50,000–250,000
- 250,000–1,000,000
- 1,000,000–5,000,000
- Over 5,000,000

National capitals are underlined

— Railroad
— Road

0   100   200   300 Kilometers
0   100   200 Miles

FIGURE 4-16

have been launched. In the agricultural sphere, fruit and vegetable production for export has soared. Industrial expansion is taking place as well, though at a more leisurely pace, and new

manufactures have included an array of goods that range from basic chemicals to computer software.

Chile's increasingly globalized economy has propelled the country into a prominent role on the international scene. Its regional commerce has also been growing steadily in conjunction with Chile's new role as a full member of the recently formed Pacific Alliance, which now seems to be outpacing Mercosur/l.

Uruguay

Uruguay, in comparison with Argentina or Chile, is compact, small, and rather densely populated. It is about the size of Florida but with one-sixth the population. This **buffer state [29]** of the early independence era, lying between (then) potentially hostile Argentina and Brazil, has evolved into a fairly prosperous agricultural country—in effect, a smaller-scale Pampa, though possessing less favorable soils and topography. Montevideo, the coastal capital, contains over a third of the country's population of 3.5 million; from here, railroads and roads radiate outward into the productive agricultural hinterland. In the immediate vicinity of Montevideo lies Uruguay's major farming zone, producing vegetables and fruits for the metropolis as well as wheat and fodder crops; most of the rest of the country is used for grazing cattle and sheep, with beef products, wool and textile manufactures, and hides dominating the export trade. Tourism is another major economic activity as Argentines, Brazilians, and other visitors flock to the Atlantic beaches at Punta del Este and other thriving seaside resorts. In terms of per capita income, Uruguay has long been one the realm's better performing countries.

Region Brazil: Giant of South America

By every measure, Brazil is South America's giant. It is so large that it has common boundaries with all the realm's other countries except Ecuador and Chile (Fig. 4-8). Its tropical and subtropical environments range from the equatorial rainforest of the Amazon Basin to the humid-temperate climate of the far south. Territorially, Brazil ranks fifth in the world, and its population size is sixth; and in both categories, it accounts for just about half of this entire realm.

The emergence of Brazil as the regional superpower of South America and an economic powerhouse in the world at large was one of the main stories of the early twenty-first century. Known largely for its coffee and other farm products in the past, by 2010 Brazil had become a leading manufacturer of machinery, chemicals, high-tech goods, and aircraft. For about ten years, Brazil's economic performance ranked it among the most productive emerging-market countries on Earth. In the process, it became the world's sixth-largest national economy (before slipping back to ninth, where it has remained since 2015).

Brazil's rapid post-2000 rise can be attributed to two main factors. First, after a long period of dictatorial rule by a minority elite that used the military to stay in power, Brazil embraced democratic government in 1989 and did not look back. The era of military coups and repeated crushing of civil liberties was over. Second, with its vast storehouse of natural resources, Brazil benefited enormously from increased demands for commodities in the world market, mainly from China and India. In some ways, this political and economic turnaround paralleled Chile's, yet because of its sheer size Brazil's matters more.

However, Brazil also epitomizes the excessive dependence of this realm on raw materials, and its susceptibility to the vagaries of the world market. When, around 2012, the demand for commodities began to slow and prices came down, the Brazilian economy felt the impact almost immediately. By 2013, economic growth had fizzled out, and in 2015 and 2016 GDP growth was the lowest in 25 years (negative 3.8 percent). Between 2017 and 2019, economic growth moved into positive territory again, but only modestly. Moreover, major political scandals surfaced and had a paralyzing effect on both the government and the recession-plagued economy. Thus the positive hype of the 2000s has now given way to talk of continuing crisis.

Population and Culture

Brazil's population grew rapidly during the world's twentieth-century population explosion. But over the past three decades, the rate of natural increase has plunged from nearly 3.0 percent to 0.7 percent today, and the average number of children born to a Brazilian woman has been more than halved from 4.5 in 1975 to 1.7 in 2019. It is a demographic trend consistent with Brazil's overall modernization since 1980.

Brazil's population of 211 million is as diverse as that of the United States. In a pattern quite familiar in the Americas, the indigenous inhabitants of the country were decimated following the European invasion (less than 700,000 now survive, about two-thirds of them deep within the Amazonian interior). Africans came in very large numbers as well, and Brazil's culture is infused with African themes, a quality that has marked it from the very beginning. So many Africans were brought in bondage to the city and hinterland of Salvador in Bahia State (**Fig. 4-17**) from what is today Benin (formerly named Dahomey) in West Africa that Bahia has become a veritable outpost of African culture. Indeed, with 57 percent (120 million) of its population claiming African heritage, Brazil can be said to have the second-largest black population in the world, after Nigeria.

Yet another significant, though small, minority began arriving in Brazil in 1908: the Japanese, whose descendants remain concentrated in the States of São Paulo and Paraná. These Japanese-Brazilians (numbering nearly 1.5 million) form the largest ethnic Japanese community outside Japan, and in their multicultural environment have risen to the top ranks of Brazilian society as business leaders, urban professionals, and commercial farmers. Committed to their Brazilian homeland as they are, the Japanese community also retains its linkages with Japan, resulting in many a trade connection.

Brazilian society has made some progress in dealing with its racial divisions. Blacks do remain the least advantaged among the country's major population groups, and community leaders continue to complain about discrimination. But ethnic mixing in Brazil is so pervasive that hardly any group is unaffected—so in the end, official census statistics concerning "blacks" and "Europeans" are hard to decipher (see **Box 4-9**).

Brazilians have a strong national culture, expressed in a traditional adherence to the Roman Catholic faith (though declining under Protestant-evangelical and secular pressures), universal use of a modified form of Portuguese as the common language ("Brazilian"); and a set of lifestyles in which soccer, "beach culture," distinctive music and dance, and an intensifying national pride are fundamental ingredients.

Inequality and Poverty

For all its accomplishments in multiculturalism, Brazil remains a country of stark social inequalities (**Fig. 4-18**). Although such inequality is hard to measure precisely, South America is frequently cited as the geographic realm exhibiting the world's widest gap between affluence and poverty (Box 4-5). And within South America, Brazil has one of the greatest of all.

At the start of this century, the Brazilian government implemented a set of policies aimed at bringing relief to the poor while maintaining robust economic growth. It enacted land reform and increased access to education for Brazil's masses. It has also launched a subsidy program that has had significant results in the poorest States. This *Bolsa Familia* (Family Fund) plan, instituted under former President Fernando Cardoso in the 1990s, provides families with small payments of cash to keep their children in school and ensure their vaccinations against diseases that especially afflict the poor. In just a few years this program proved to be so successful that it became a model for antipoverty campaigns the world over.

Regional inequalities were being targeted as well by government policies. Major new legislation was passed in 2012 to allow revenues from oil to be shared by all of Brazil's States (previously this income was mainly channeled to the three oil-producing States of Rio de Janeiro, Espirito Santo, and São Paulo). But, with the commodity boom quickly becoming a distant memory, such revenues dwindled and inequality in Brazil, as elsewhere in the realm, either stagnated or resumed widening.

Development Prospects

In the long run, Brazil's prospects are favorable. The country is richly endowed with mineral resources, including enormous iron and aluminum ore reserves, extensive tin and manganese deposits, and proven massive supplies of offshore oil and

FIGURE 4-17

natural gas (Fig. 4-17). Other significant energy developments involve state-of-the-art hydroelectric facilities and the successful integration of sugarcane-based anhydrous ethanol with gasoline, allowing Brazil's motor-vehicle owners to use this "gasohol" instead of costlier imported petroleum (all Brazilian cars are configured to use this biofuel, whose ethanol content must be at least 18 percent). Besides these endowments, Brazil has more arable land than any other country on Earth. Brazilian soils sustain a bountiful agricultural output that makes

the country the world's leading exporter of coffee, soybeans, orange juice, sugar, tobacco, ethanol, beef, and chicken.

Industrialization has driven Brazil's ascendancy as a global economic power. As a result, annual revenues from industrial exports surpassed those from agriculture more than a decade ago. The relatively low wages of its workers and the mechanized efficiency of its steelmakers enable Brazil to produce that commodity at half the cost of steel made in the United States (which caused American steel producers to demand

Box 4-9 Comparing the Meaning of Race in Brazil and the U.S.

Internationally, race relations in the U.S. are often compared to either South Africa or Brazil, because of their common histories as white-settler societies and the disadvantaged position of blacks or African Americans. The U.S. and Brazil are, respectively, the biggest countries in their realms, both were at the receiving end of a large-scale slave trade, and both came to be dominated by European settlers. Over time, about 400,000 Africans were brought to the United States in bondage, while about five million were transported to Brazil. In both countries, today, blacks are more likely to be poor and lead less prosperous lives. In modern times, both societies generated a powerful myth about the progressive nature of their society: the "American dream" is supposed to allow anybody, any family of any background, to rise to a good standard of living; the Brazilian equivalent is the myth of "racial democracy." The latter suggests that every individual, regardless of race, possesses equal rights and equal chances in life.

The actual understanding or definition of race differs substantially in the two countries, and they underscore that "race" very much involves a socially constructed set of categories. Brazil's census recognizes five official "racial" categories: White, Black, Brown, Yellow, and Indigenous. For the U.S., the main official categories are White, Black (African American), Native American, Asian, and Native Hawaiian or Other Pacific Islander. In both the Brazilian and the U.S.

census, race is self-identified by the individual in question. In the U.S., individuals may identify with more than one category. Interestingly, the share of the population that identify as Black is 13 percent in the U.S. and only 8 percent in Brazil; at the same time, in the U.S. only 3 percent of the population identified with more than one race, while in Brazil 43 percent identifies as Brown or mixed-race.

Why this vast difference in mixed racial identities between the two countries? In the U.S., race seems to be almost a binary: you are either black or white, a notion dating back to the so-called one-drop rule from the time of slavery: if you have even the slightest African American in you, you are black. This rule came about, most historians agree, because it would guarantee a steady supply of slave labor. The binary seems to hold to this day. Former President Obama had one white parent and one black parent, but he was typically categorized as black, not as white. In Brazil, the concept of race is more fluid and multifaceted, and the mixed-race (Brown) category is in itself highly diverse and constitutes a majority. In other words, the U.S. has a color line, whereas Brazil has a color continuum. As noted, these identities are self-reported and reflect how populations view themselves. These views can change over time. Next time when you fill out a census form, you may be one of a growing number of people who wonder about your individual identity and where you fit.

protectionist measures). Other industries have been growing as well, lessening reliance on world market prices for raw materials and allowing for a more stable, mature, national economy.

Brazil's Subregions

Brazil is a federal republic consisting of 26 States and the federal district of the inland capital, Brasília (Fig. 4-17). As in the United States, the smallest States lie in the northeast and the larger ones farther west; their populations range from about 600,000 in the northernmost, peripheral Amazon State of Roraima to more than 45 million in burgeoning São Paulo State. Although Brazil is about as large as the 48 contiguous United States, it does not exhibit a clear-cut physiographic regionalism. Even the Amazon Basin, which covers just about 60 percent of the country, is not entirely a plain: between the tributaries of the great river lie low but extensive tablelands. Given this physiographic ambiguity, the six Brazilian subregions that will now be discussed exhibit no absolute or even generally accepted boundaries. In Figure 4-17, those boundaries have been drawn to coincide with the borders of States, making identifications easier.

The **Northeast** was Brazil's source area, its culture hearth. The plantation economy was established here at an early date, attracting Portuguese planters, who soon began to import the country's

largest contingent of African slaves to work in the sugarcane fields. But the ample rainfall occurring along the coast soon gives way to drier and more variable patterns in the interior,

BRAZIL: INCOME DISPARITIES (GDP PER CAPITA IN US$, 2016)

- Over 12,000
- 8,000–12,000
- 5,000–8,000
- Under 5,000

Source: Brazil Institute of Geography and Statistics. Regional Account Reports.
3.48 Brazilian Reais = 1 US$ approx. (2016)

FIGURE 4-18

which is home to about half of the region's approximately 60 million people. This semiarid inland backcountry—called the *sertão*—is not only seriously overpopulated but also contains some of the worst poverty to be found anywhere in the Americas. The Northeast produces less than one-sixth of Brazil's gross domestic product, but its inhabitants constitute more than one-fourth of the national population. Moreover, this subregion contains half of the country's poor and exhibits a literacy rate 20 percent below Brazil's mean as well as an infant mortality rate twice the national average.

Much of the Northeast's problems are rooted in its unequal system of land tenure. Farms must be at least 100 hectares (250 acres) to be profitable in the hard-scrabble *sertão*, a size that only large landowners are able to afford. Moreover, the Northeast is tormented by a monumental environmental problem: the cyclical recurrence of devastating droughts at least partly attributable to *El Niño* (periodic events of sea-surface warming off the continent's northwestern coast that skew regional weather patterns).

The Northeast has in recent times received greater attention from the central government, largely in the form of federally funded infrastructure projects and incentive-driven investment (think oil money). In cities such as Recife and Salvador, hordes of peasants driven from the land constantly arrive to expand the surrounding shantytowns. As yet, few of the generalizations about emerging Brazil apply here, but there are some bright spots. A petrochemical complex has been built near Salvador, creating thousands of jobs and luring foreign investors. Irrigation projects have nurtured a number of productive new commercial agricultural ventures. Tourism is booming along the entire Northeast coast, whose thriving beachside resorts attract tens of thousands of vacationing Europeans (flight times are under 8 hours). Recife has now spawned a budding software industry and a major medical complex. And Fortaleza is the center of new clothing and shoe industries that have already put the city on the global economic map.

The **Southeast** has been modern Brazil's *core area*, with its major cities and leading population clusters. Gold first drew many thousands of settlers, and other mineral finds also contributed to the influx—with Rio de Janeiro serving as the terminus of the "Gold Trail" and then as the long-time capital of Brazil until 1960. "Rio" became the cultural capital as well, the country's most internationalized city, **entrepôt** [30], and tourist hub. The third quarter of the twentieth century brought another mineral age to the Southeast, based on the iron ores around Lafaiete carried to the nearby steelmaking complex at Volta Redonda (Fig. 4-17). The surrounding State of Minas Gerais (the name means "General Mines") formed the base from which industrial diversification in the Southeast has steadily mushroomed. The burgeoning metallurgical center of Belo Horizonte paved the way and is now the endpoint of a rapidly developing,

ultramodern manufacturing corridor that stretches 500 kilometers (300 mi) southwest to metropolitan São Paulo (Fig. 4-17, striped zone).

São Paulo State is both the leading industrial producer and the primary focus of ongoing Brazilian development. This economic and geographic powerhouse accounts for nearly half of the country's GDP, with a thriving economy that today matches Argentina's in overall size. Not surprisingly, this subregion is growing phenomenally (it contains more than 20 percent of Brazil's population) as a magnet for upwardly mobile migrants, especially from the Northeast.

The wealth of São Paulo State was built on its coffee plantations (known as *fazendas*), and Brazil is still the world's leading producer. But coffee today has been eclipsed by other farm commodities. One of them is orange juice concentrate (here, too, Brazil leads the world). São Paulo State now produces more than double the annual output of Florida, thanks to a climate all but devoid of winter freezes, ultramodern processing plants, and a fleet of specially equipped tankers that ship the concentrate to markets around the globe. Another major crop is soybeans, and in 2019 Brazil surpassed the U.S. to become the world's number-one producer.

Matching this agricultural prowess is the State's industrial strength. The revenues derived from the coffee plantations provided the necessary investment capital, ores from Minas Gerais supplied the vital raw materials, São Paulo City's outport of Santos facilitated access to the ocean, and immigration flows from Europe, Japan, and other parts of Brazil contributed to the increasingly skilled labor force. As the capacity of the expanding domestic market grew, the advantages of central location and large-scale agglomeration nailed down São Paulo's primacy. This also resulted in metropolitan São Paulo becoming the country's—and South America's—foremost industrial complex and megacity. But this rapid progress was accompanied by massive problems of overcrowding, pollution, and congestion, which now severely challenge the sustainability of this gigantic urban dynamo (photo, p. 112).

The **South** consists of three States, whose combined population totals around 30 million: Paraná, Santa Catarina, and Rio Grande do Sul (Fig. 4-17). Southernmost Brazil's exceptional agricultural potential has long attracted a steady flow of European immigrants. They introduced their advanced farming methods to several areas in this part of the country. Portuguese rice farmers clustered in the valleys of Rio Grande do Sul, where tobacco production has now propelled Brazil to become the world's leading exporter. The Germans, specialists in raising grain and cattle, occupied the somewhat higher areas to the north and Santa Catarina. The Italians selected the highest slopes, where they established thriving vineyards. All of these fertile lands proved highly productive, and with growing markets in the mushrooming urban areas to the north, this tri-State subregion has become Brazil's most affluent corner.

Economic development within the South is not limited to the agricultural sector. Coal from Santa Catarina and Rio Grande do Sul is shipped north to the steel plants of Minas Gerais. Local manufacturing is growing as well, especially in Pôrto Alegre and Tubarão. During the 1990s, a major center of the computer software industry was established in Florianópolis, the island-city as well as State capital just off Santa Catarina's coast. Known as *Tecnópolis*, this budding technopole continues to grow by capitalizing on its seaside amenities, skilled labor force, superior air-travel and global communications linkages, and government and private-sector initiatives to support new companies.

The **Interior** subregion—constituted by the sprawling States of Goiás, Mato Grosso, and Mato Grosso do Sul—is also known as the *Central-West*. This is the subregion that Brazil's developers long sought to make a part of the country's productive heartland, and in 1960 the new capital of Brasília was deliberately situated on its margins (Fig. 4-17).

By locating the new capital city in the wilderness 650 kilometers (400 mi) inland from its predecessor, Rio de Janeiro, the nation's leaders dramatically signaled the opening of Brazil's development thrust toward the west. Brasília is noteworthy in another regard because it represents what political geographers call a **forward capital [31]**. A state will sometimes relocate its capital to a sensitive area, perhaps near a peripheral zone under dispute with an unfriendly neighbor, in part to confirm its determination to sustain its position in that contested zone. Brasília does not lie close to a contested area, but Brazil's interior was an internal frontier to be conquered by a growing nation. Spearheading that drive, the newly built capital occupied a decidedly forward position.

Despite the subsequent growth of Brasília to 3.3 million inhabitants today (which includes a sizeable ring of peripheral squatter settlements), it was not until the 1990s that the Interior subregion began its economic integration with the rest of Brazil. The catalyst was the exploitation of the vast *cerrado* **[32]**—the fertile savannas that blanket the Central-West and make it one of the world's most promising agricultural frontiers (at least two-thirds of its arable land still awaits development). As with the U.S. Great Plains, the flat terrain of the *cerrado* is one of its main advantages because it facilitates the large-scale mechanization of farming with a minimal labor force. Another benefit is rainfall, more prevalent here than in the Great Plains or Argentine Pampa. The increasingly dominant local crop is soybeans (see pink zone in Fig. 4-4), and Brazil entered the 2020s as the top global producer.

The **Northwest** is Brazil's territorially largest and most rapidly developing subregion, which consists of the seven States of the Amazon Basin (Fig. 4-17). This was the scene of the great rubber boom of more than a century ago, when the wild rubber trees in the *selvas* (tropical rainforests) produced huge profits and the central Amazon city of Manaus enjoyed a brief period of wealth and splendor. But the rubber boom had ended by 1920, and for most of the six decades that followed, Amazonia was a stagnant hinterland lying remote from the centers of Brazilian settlement. All that changed quite dramatically during the 1980s as new development began to stir throughout this awakening subregion (**Fig. 4-19**), which continues to be the scene of the world's largest migration into virgin territory as more than 200,000 new settlers arrive each year.

Brazil's Northwest is also widely known for its high rates of deforestation. Between 1995 and 2005, an area of 20,000 square kilometers (7700 sq mi) was cleared away every year. Some believe that the worst might be over because the Brazilian government had pledged to achieve an 80 percent reduction in deforestation by 2020 and to terminate the practice by 2030. At least part of its thinking is that Brazil has enormous potential as a "green economy" with its abundance of land, water, and sunshine. Today, nearly half of the country's energy is already obtained from renewable sources, mostly biofuels (vs. around 13 percent in the United States), and it is becoming easier to envision Brazil as a pacesetting environmental power. Still, it is hard to imagine that the government's ambitious goals will be achieved anywhere near 2020 and 2030, respectively—not least because at the same time it has embarked on major infrastructural projects deep in the Amazon Basin (Fig. 4-19).

As Figure 4-15 shows, development projects abound in the Amazon Basin today. One of the most durable is the *Grande Carajás Project* in central Pará State, a huge, multifaceted scheme centered on the world's largest known deposit of iron ore in the hills around Carajás (photo, p. 110). Besides its vast iron-mining complex, other new construction in the vicinity includes the Tucuruí Dam on the nearby Tocantins River and an 850-kilometer (535-mi) railroad to the Atlantic port of São Luis. This ambitious development project also emphasizes the exploitation of other minerals, cattle-raising, crop farming, and forestry. What is taking place here is a manifestation of the **growth-pole concept [33]**. A growth pole is a location at which a set of activities, given a start, will expand and generate widening ripples of development in the surrounding area. According to this scenario, the stimulated hinterlands could one day cover one-sixth of all Amazonia.

Understandably, hordes of settlers have descended on this part of the Amazon Basin. Those seeking business opportunities have been in the vanguard, but they have been followed by masses of lower-income laborers and peasant farmers in search of jobs and land ownership. The initial stage of this colossal enterprise has boosted the fortunes of many urban centers, most of all Manaus, northwest of Carajás. Here, a thriving industrial complex specializing in the production of electronic goods has emerged within the free-trade zone adjoining the city thanks to the outstanding air-freight facilities and operations at Manaus's ultramodern airport. At the same time, in and around Manaus, a new era of urban expansion

THE AMAZON BASIN

- Amazon River basin
- Rainforest
- Indigenous groups
- —— Major road
- —— Railroad
- ==== Proposed railroad

HYDROELECTRIC DAMS
- □ Planned
- ◻ Under construction
- ◼ In operation

OIL PRODUCTION
- Leased areas
- Exploration proposed

© J. Nijman, P. O. Muller, and John Wiley & Sons, Inc.

FIGURE 4-19

Ian Trower/AWL Images/Getty Images

The Ponte Rio Negro, iconic symbol of the ongoing development thrust into South America's interior, with the Manaus riverfront in the foreground. This first major bridge in the Amazon Basin and Brazil's longest, at 3.5 kilometers or about 2 miles, was completed in 2011 and spans the mighty Rio Negro just upstream from its confluence with the even greater Amazon itself. Not surprisingly, this most impressive river junction on the South American continent spawned the central Amazon's largest city, Manaus, well over a century ago. Hemmed in until recently by dense rainforests and these enormous rivers, Manaus's new bridge connects the city to its satellite, Iranduba. The latter is now booming as a result of its much-upgraded accessibility and serves as the jumping-off point on the west side of the Rio Negro, its own suburbs spearheading the outward thrust of metropolitan Manaus into an entirely new development sector.

has been unleashed by the opening of the spectacular Rio Negro bridge (photo, p. 134) in the early 2010s.

But many problems have also surfaced as the tide of pioneers has rolled across central Amazonia. One of the most tragic involved the Yanomami people, whose homeland in Roraima State was overrun by thousands of claim-stakers (in search of newly discovered gold), who triggered violent confrontations that ravaged the fragile aboriginal way of life.

Points to Ponder

- The distance from Bogotá, Colombia, to Chile's southernmost Cape Horn (6600 km/4100 mi) is greater than the distance from Paris, France, to Delhi, India.

- Even though the Amazon rainforest lies almost entirely within Brazil, it is often referred to as "the lungs of the Earth." Should its preservation be a global responsibility?

- Is it a coincidence that both liberation theology and *dependencia* theory originated in this realm?

- Bolivia, Peru, and Chile together contain no less than three-quarters of the world's known reserves of lithium, an essential component of electrical batteries of every stripe.

FIGURE 5-1

On June 25th of 2019, Parisians escaped the heat in their homes (most do not have air conditioning) and took to public fountains to cool themselves off, such as this one at Trocadero esplanade. In June and July of 2019, various European countries experienced successive unprecedented heatwaves, breaking records that were set only a year earlier. For some it was a playful event but for many others, especially the elderly, the heat brought severe discomfort and worse. According to official government figures, it caused nearly 1,500 deaths in France alone. The recurring extreme temperatures were widely attributed to global climate change.

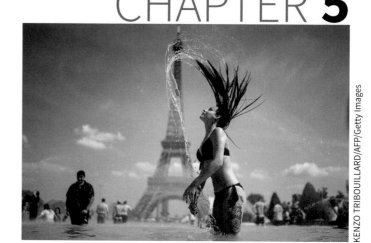

KENZO TRIBOUILLARD/AFP/Getty Images

The European Realm

IN THIS CHAPTER

- Immigration and national politics
- Europe's extreme weather events: The new normal?
- Geopolitics in the Baltics
- Islam in Europe
- Brexit: The continuing saga
- The making of North Macedonia
- The EU: Balancing between East and West

CONCEPTS, IDEAS, AND TERMS

Transition zone **[1]**	Transferability **[14]**	Primate city **[27]**
Extreme weather events **[2]**	Central business district (CBD) **[15]**	Site **[28]**
Land hemisphere **[3]**	Centrifugal forces **[16]**	Situation **[29]**
City-state **[4]**	Centripetal forces **[17]**	Estuary **[30]**
Geographic information system (GIS) **[5]**	Supranationalism **[18]**	Conurbation **[31]**
Digital elevation model **[6]**	Euro zone **[19]**	Landlocked location **[32]**
Local functional specialization **[7]**	Schengen Area **[20]**	World city **[33]**
Industrial Revolution **[8]**	Four Motors of Europe **[21]**	Break-of-bulk **[34]**
Sovereignty **[9]**	Devolution **[22]**	*Entrepôt* **[35]**
Nation-state **[10]**	Suwalki Gap **[23]**	Shatter belt **[36]**
Nation **[11]**	Asylum **[24]**	Balkanization **[37]**
Indo-European language family **[12]**	Microstate **[25]**	Irredentism **[38]**
Complementarity **[13]**	Urban system **[26]**	Exclave **[39]**

A case can be made that, over the past five centuries, Europeans have influenced and changed the rest of the world more than the people of any other realm. For good or bad, much of the world would look quite different today if it had not been for Europe. Europe's colonial empires spanned the globe and transformed societies far and near. Millions of Europeans migrated from their homelands to parts of the Eastern Hemisphere as well as the Americas, changing (and sometimes nearly obliterating) traditional communities and creating new societies from Australia to North America. Colonial power and economic incentive combined to propel the movement of millions of imperial subjects from their ancestral homes to distant lands: Africans to the Americas, Indians to Africa, Chinese to Southeast Asia. In agriculture, industry, politics, and other spheres, Europe generated revolutions—and then, through imperialism and colonialism, exported those revolutions across the world. The European realm itself was also changed through these global connections, through immigration, economic growth fueled with raw materials from distant shores, and cultural influences (for example, food).

During the seven decades since World War II (1939–1945), however, and particularly over the past four, Europe has decidedly become a more "ordinary" realm. The days of empire are now long gone. National economies face the same global challenges as their counterparts elsewhere. And Europe no longer plays a leading role in the world economy, even though it continues to rank among the most highly developed geographic realms. If the claim can still be made today that Europe is "special," it is with an eye on the remarkable process of economic and political integration that shaped the evolution of the *European Union (EU)* since the 1950s. Although the European realm is highly diverse in a cultural and geographical sense, most of it is also tightly integrated, with considerable economic powers devolved from national governments to Brussels, the EU headquarters city. This integration itself, it should be noted, varies across Europe as well—making for an especially complex geographic realm.

Defining the Realm

As **Figure 5-1** shows, Europe is a realm of peninsulas and islands on the western margin of the world's largest landmass, Eurasia. It is a realm containing 600 million people and 40 countries, but it is territorially quite small. The European realm is bounded on the west, north, and south by Atlantic, Arctic, and Mediterranean waters, respectively (**Box 5-1**). But where is Europe's eastern limit? Each episode in the historical geography of Eastern Europe has left its particular legacy in the cultural landscape.

Twenty centuries ago, the Roman Empire ruled much of Eastern Europe (Romania is a cartographic reminder of that period); for most of the second half of the twentieth century, the Soviet Union controlled nearly all of it. During the intervening two millennia, Christian Orthodox church doctrines spread from the southeast, and Roman Catholicism advanced toward the northwest. Turkish (Ottoman) Muslims invaded and created an empire that reached the vicinity of Vienna. By the time the Austro-Hungarian Empire ousted the Turks more than a century ago, millions of Eastern Europeans had been converted to Islam (Albania and Kosovo still remain predominantly Muslim countries). Meanwhile, it is often said that Western Europe's civilization had its cradle in ancient Greece, but that lies farther still to the southeast, beyond the former Yugoslavia.

As we shall see, the geographic extent of Europe has always been debatable. Currently, it is a highly contentious issue in terms of European Union expansion and relations with Russia—as underscored by the turmoil that continues

---

Box 5-1 Major Geographic Features of Europe

1. The European geographic realm lies on the western flank of the Eurasian landmass.

2. Though territorially small, Europe is heavily populated and fragmented into 40 states.

3. European natural environments are highly varied, and Europe's resource base is rich and diverse.

4. Europe's geographic diversity, cultural as well as physical, created strong local identities, specializations, and opportunities for trade and commerce.

5. The European Union (EU) is a historic and unique effort to achieve multinational economic integration and, to a lesser degree, political coordination.

6. Europe's relatively prosperous population is highly urbanized and aging rapidly.

7. Despite Europe's momentous unification efforts, east–west contrasts still mark the realm's regional geography; to a somewhat lesser extent, cultural differences persist between north and south.

8. The influx of immigrants and refugees, which reached crisis proportions around 2015, is still testing the stability of the EU and has caused a shift toward nationalism and populism in many European countries.

9. The Baltic region has become a critical geopolitical testing ground in the relations between Europe and Russia.

to plague eastern Ukraine. Europe's eastern boundary is a dynamic one, and it has changed with history. Some would say that there really is no clear boundary, that Europe's ambience, so to speak, just thins out toward the east. Our present definition places Europe's eastern boundary between Russia and its six European neighbors to the west. That definition is based on several geographic criteria that include European–Russian contrasts in territorial dimensions, geopolitical developments, cultural properties, and history.

Figure 5-1, as you may already have observed, places the Crimean Peninsula inside Russia as a consequence of its *de facto* annexation from Ukraine in 2014. This does not mean that we as geographers endorse that forcible takeover. But it does show that state boundaries frequently materialize through violence and the will of the victors. Thus the global map of states is in many ways the product of power struggles, territorial conflicts, and war— especially in Europe. In fact, it can be argued that much of Eastern Europe constitutes a **transition zone [1]** between the European and Russian-dominated realms (which is particularly notable in Belarus and eastern Ukraine). That argument could be made on the basis of language, religion, and/or functional spatial interaction—but it does not translate into simple claims of territorial control.

Physical Geography

From the balmy shores of the Mediterranean Sea to the icy peaks of the Alps, and from the moist woodlands and moors of the Atlantic fringe to the semiarid prairies north of the Black Sea, Europe presents an almost infinite range of natural environments (**Figs. 5-2** and **5-3**). Moreover, as the realm has experienced an increase in extreme weather events in recent years (**Box 5-2**), Europe's dynamically wide range of climate zones has become quite palpable.

Europe's peoples have benefited from a large and varied store of raw materials. Whenever the opportunity or need arose, the realm proved to contain what was required. Early on, these requirements included cultivable soils, rich fishing waters, and wild animals that could be domesticated; in addition, extensive forests provided wood for houses and boats. Later, coal and mineral ores propelled industrialization. More recently, Europe proved to contain sizeable deposits of oil and natural gas.

Europe's area may be small, but its physical landscapes are varied and complex. Regionally, we identify four broad units: the Central Uplands, the southern Alpine Mountains, the Western Uplands, and the North European Lowland (Fig. 5-3).

The **Central Uplands** are at the center of Europe. They are a region of hills and low plateaus loaded with raw materials

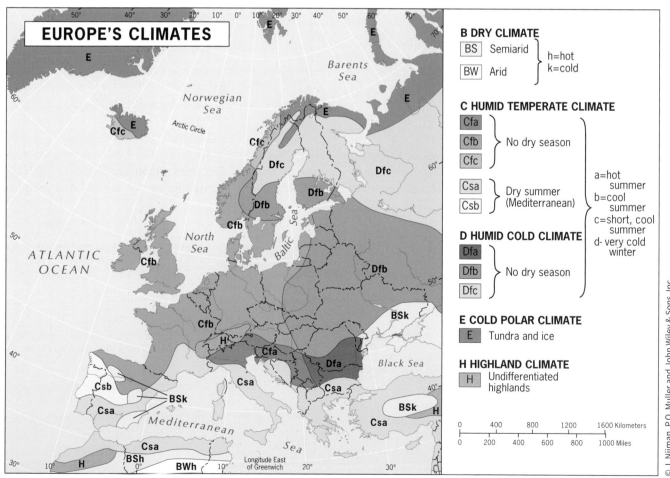

© J. Nijman, P.O. Muller and John Wiley & Sons, Inc.

FIGURE 5-2 Interactive

FIGURE 5-3

whose farm villages grew into towns and cities when the Industrial Revolution transformed this realm.

The ***Alpine Mountains***, a highland region named after the Alps, extends eastward from the Pyrenees on the French–Spanish border to the Balkan Mountains near the Black Sea, and include Italy's Appennines as well as the Carpathians of Eastern Europe (also see **Box 5-3** and **Figure 5-5**).

The ***Western Uplands***, geologically older, lower-lying, and more stable than the Alpine Mountains, extend from northern Scandinavia through western Britain and Ireland to the heart of the Iberian Peninsula in Spain.

The ***North European Lowland*** stretches in a lengthy arc from southeastern Britain and southwestern France across Germany and Denmark into Poland and Ukraine, from where it continues far into Russia. Also known as the Great European Plain, this has been an avenue for human migration time after time, so that complicated cultural and economic mosaics developed here and together produced a jigsaw-like political map. As Figure 5-3 shows, many of Europe's major rivers and connecting waterways serve this populous region, where a number of Europe's leading cities (London, Paris, Berlin, Amsterdam, Copenhagen, Warsaw) are located.

Box 5-2 Global Climate Change in Europe: Extreme Weather Events

Global climate change, or overall global warming, does not simply imply increasingly warm weather. *Climate* pertains to long-term weather characteristics, averaged over substantial periods of time; *weather* refers to a snapshot, the weather at a particular moment in time or over a short period of time. While temperatures have been rising around the world, slowly, on average, there may well be days or weeks or even months of unusually cold weather in different parts of the world. In fact, the strongest indication of global climate change is the increased frequency of **extreme weather events** [2], be they storms, droughts, heatwaves, or extreme cold spells. Scientists labeled the year 2018 "the most extreme year on record for weather events in Europe," with a total of 117 extreme events. The 2018 summer heatwaves broke records across the realm, and especially so in the northern regions. The European Arctic witnessed the highest temperatures ever since records were kept: 25 consecutive days of maximum temperatures above 25 degrees Celsius (77 degrees Fahrenheit). Wildfires swept through Sweden and Greece (around much the same time as the devastating California fires). Germany, the Netherlands, Denmark, Austria, and parts of southeastern Europe experienced severe heat and long dry spells. Across Europe, the heat and fires killed over a hundred people. The drought led to crop failures, and the Danube and Rhine rivers were at record low levels, inhibiting water transportation. By September and October, severe rainstorms caused flooding in France and Spain, again killing dozens of people. Adapting to, and preparing for, different kinds of extreme weather events is a difficult and costly challenge, but in many parts of the world, it is increasingly apparent that there is no alternative (besides major efforts needed to mitigate global warming). Scientists' warnings that extreme weather events are likely to become the new normal were borne out in 2019, when successive heatwaves in June and July across many European countries broke a number of records set only a year earlier.

AP Images/Michael Probst

It is October 24, 2018, and people are standing on a sandbank in the Rhine River near Bingen (west of Frankfurt), Germany. Heatwaves and droughts led to record low levels of the river, causing chaos for the inland shipping industry, environmental damage, and reportedly billions of euros of losses.

Locational Advantages

Europe's *relative location*, at the crossroads of the **land hemisphere** [3], creates considerable efficiency for contact with much of the rest of the world (**Fig. 5-4**). A "peninsula of peninsulas," Europe is nowhere far from the ocean and its avenues of seaborne trade and conquest. Hundreds of kilometers of navigable rivers, augmented by an elaborate system of canals, open the interior of Europe to its neighboring seas and to the shipping lanes of the world. The Mediterranean and Baltic seas, in particular, were critical in the development of trade and capitalism in early modern times and in the emergence of Europe's early **city-states** [4], such as Venice, Italy, in the south (photo below) and Lübeck, Germany, in the north.

© J. Nijman, P.O. Muller and John Wiley & Sons, Inc.

EUROPE IN THE LAND HEMISPHERE

3000 kilometers=ca. 1900 miles

FIGURE 5-4

Visions Of Our Land/The Image Bank/Getty Images

The Rialto Bridge, one of Venice's famous landmarks, dates from the time that Venice was one of Europe's most powerful and richest city-states. The first bridge over the Grand Canal at this spot dates back to the twelfth century; the present span was completed in 1591.

Box 5-3 Technology & Geography

What is GIS?

The ubiquity of maps and mapping is a relatively new phenomenon and can be attributed to **geographic information systems** or **GIS [5]**. GIS takes the digital data that we collect with satellites, sensors, or smartphones, and puts them on a map. What distinguishes GIS from other technologies is its ability to integrate diverse datasets using geography. This process is called overlay. Overlaying one or more data layers on top of each other based on location reveals spatial patterns and relationships, and supports analysis and decision-making. When satellite data are processed and integrated with other data, the power of GIS is revealed. For example, the maps below overlay a **digital elevation model** or **DEM [6]** showing a sea level rise scenario across a coastal portion of Malta. GIS permits us to see the impact of a six-meter rise in sea level on different areas and landscape features (e.g., roads, historic sites, tourist attractions). GIS and digital mapping are valuable tools for understanding and mitigating the effects of climate change, unexpected events, and natural disasters.

Access your WileyPLUS course to interact with a dynamic version of this map and to engage with online exercises and questions.

EXPLORING SEA LEVEL RISE IN MALTA WITH GIS

A) Base imagery only

B) Sea level change layer added

C) Roads layer added

D) Sites of historical significance highlighted

© J. Nijman, M. Shin, P.O. Muller and John Wiley & Sons, Inc.

FIGURE 5-5 **Interactive**

Do not overlook the scale of the maps in this chapter. Europe is a realm of moderate distances and close proximities. This propinquity, combined with substantial cultural differences, makes for intense interaction, the constant circulation of goods and ideas. That has been the hallmark of Europe's geography since early modern times.

Modern Historical Geography

The Industrial Revolution

From the late eighteenth century on, Europe experienced a steady process of intensifying **local functional specialization [7]**.

The **Industrial Revolution [8]**, the beginning of which is often attributed to the invention of the steam engine by Scotsman James Watt (1736–1819) in 1780, was essential in this process, and it would provide Europe with a critical edge in world affairs for a long time to come. The Industrial Revolution gathered momentum throughout the nineteenth century, and Britain was at its epicenter. The power loom revolutionized the weaving industry. Iron smelters, long dependent on Europe's dwindling forests for fuel, could now be concentrated near coalfields. Engines could move locomotives as well as power looms. Ocean shipping entered a new age. The British had an enormous advantage because they controlled the flow of raw materials and held a monopoly over products that were in demand globally. In Britain, manufacturing regions developed near coalfields in the English Midlands, at Newcastle to the northeast, in southern Wales, and along Scotland's Clyde River around Glasgow.

After 1850, the Industrial Revolution spread eastward from Britain into continental Europe, and by 1890 had reached as far as western Russia. In mainland Europe, the axis of industrialization followed a west-to-east belt of major coalfields extending roughly along the southern margins of the North European Lowlands; each gave rise to manufacturing complexes that can still be seen on the map today in northern France and Belgium, Germany's Ruhr Basin, the Czech Republic's Bohemia, southern Poland's Silesia, and eastern Ukraine's Donets Basin (Donbas). Abundant iron ore deposits lie in a similar west–east belt and together with coal provide the key raw materials for manufacturing steel.

As in Britain, this cornerstone heavy-metal industry spawned new concentrations of economic activity, which grew steadily as millions migrated in from the countryside to fill expanding employment opportunities. Densely populated and heavily urbanized, these industrial agglomerations became the backbone of Europe's world-scale population cluster (labeled **D** in Fig. 1-7) as well as an enduring major feature on the realm-scale map of Europe's population distribution (**Fig. 5-6**).

POPULATION DISTRIBUTION: EUROPE

Population per Square Kilometer

0 to 5
5 to 10
10 to 50
50 to 100
100 to 500
Over 500

FIGURE 5-6 Interactive

© J. Nijman, M. Shin, P.O. Muller and John Wiley & Sons, Inc.

It is important to remember that, especially in the second half of the nineteenth century and the early twentieth century, industrialization in Europe and European imperialism and colonialism in other parts of the world were in some ways closely related—the raw materials obtained overseas (e.g., cotton) served as important inputs for the growing European industries; overseas markets were also sometimes significant destinations for European products.

Political Transformations

The Industrial Revolution unfolded against the backdrop of the ongoing formation of Europe's national states, a process that had been underway since the 1600s. These political revolutions generally involved the centralization of royal power by monarchies and the usurpation of the former city-states in larger national territories—which frequently produced violent reconfigurations of the realm's evolving political geography. Most historians point to the Peace (Treaty) of Westphalia in 1648 as a key step in the emergence of Europe's state system, ending decades of war and recognizing territories, boundaries, and the **sovereignty [9]** of countries. But this was only the beginning, and these so-called *absolutist states*, in which monarchs held all the power and the people had few if any rights, would not last.

One of the most dramatic episodes to follow was the French Revolution (1789–1795), which soon ended the era of absolutist states. More gradual political transformations occurred in the Netherlands, Britain, and the Scandinavian countries. But a number of other components of Europe would remain much longer under the tight control of authoritarian (dictatorial) regimes headed by monarchs or despots. By the late nineteenth century, Europe became the arena for the competing ideologies of *liberalism, socialism*, and *nationalism* (national spirit, pride, patriotism). However, a generation later the rise of extreme nationalism (*fascism*) in the 1930s plunged the realm into one of the most violent wars the world has ever seen.

A Fractured Map Europe's political map has always been one of intriguing complexity. As a geographic realm, Europe occupies only about 5 percent of the Earth's land area, but that minuscule territory is fragmented into some 40 countries—more than one-fifth of all the states in the world today. Therefore, when you look at Europe's political map, the question that arises is how did so small a geographic realm come to be divided into so many political entities? Europe's map is a legacy of its feudal and monarchic periods, when powerful kings, barons, dukes, and other rulers, rich enough to fund armies and powerful enough to forcibly extract taxes and tribute from their domains, created bounded territories in which they reigned supreme. Royal marriages, alliances, and conquests actually simplified Europe's political map. In the early nineteenth century, there still were 39 separate German States; a unified Germany as we know it today would not emerge until the 1870s.

State and Nation Europe's political transmutations produced a prototype of politico-geographical organization known as the **nation-state [10]**, a territorial state embodied by its culturally distinctive population. The term **nation [11]** refers to a people with a single language, a common history, and a similar ethnic background. In the sense of *nationality*, it relates to citizenship (legal membership) in the state. Only a handful of states today are so culturally homogeneous that, spatially, their culture practically coincides with the state. Europe's prominent nation-states of a century ago—France, Spain, Great Britain, and Italy—are today multicultural societies, with *nation* defined more by an intangible spirit of nationalism and emotional commitment than by cultural or ethnic homogeneity.

Contemporary Europe: A Dynamic Realm

Cultural Diversity

The European realm is home to peoples of numerous cultural-linguistic stocks, including not only Latins, Germanics, and Slavs but also minorities such as Finns, Magyars (Hungarians), Basques, and Celts. This diversity of ancestries continues to be an asset as well as a liability. It has generated not only interaction and exchange, but also conflict and war.

It is worth remembering that Europe's territory is just over 60 percent the size of the United States, but that the population of Europe's 40 countries is almost twice as large as America's. This population of just under 600 million speaks numerous languages, almost all of which belong to the **Indo-European language family [12]** (see Fig. 1-9 and **Fig. 5-7**). Nonetheless, most of those languages are not mutually understandable; some, such as Finnish and Hungarian, are not even members of the Indo-European family. When Europe's unification efforts began after World War II, one major problem was to determine which languages to recognize as "official." That problem still prevails, although English has become the realm's unofficial *lingua franca* (common language). During a visit to Europe, though, you would find that English is more commonly usable in the big cities than in the countryside and more in Western Europe than farther east. Europe's multilingualism remains a rich cultural legacy, but it is also a barrier to further integration.

Another major divisive force confronting Europeans involves religion. Europe's cultural heritage is steeped in Christian traditions, but sectarian strife between Catholics and Protestants, which plunged parts of the realm into bitter and widespread conflict, still divides communities and, as until recently in Northern Ireland, can still provoke violence. Some political parties still carry the name "Christian"—for instance, Germany's Christian Democrats.

FIGURE 5-7

LANGUAGES OF EUROPE

0   200   400   600 Kilometers
0   100   200   300 Miles

MAJOR INDO-EUROPEAN BRANCHES

GERMANIC GROUP

WESTERN GERMANIC
1 Dutch
2 German
3 Frysian
4 English

NORTHERN GERMANIC
5 Danish
6 Swedish
7 Norwegian
8 Icelandic
9 Faeroese

ROMANCE GROUP

10 Portuguese   16 Rhaeto-Romansch
11 Spanish   17 Romanian
12 Catalan   18 Corsican-Italian
13 Provençal   19 Sardinian-Italian
14 French   20 Walloon
15 Italian

SLAVIC GROUP

WEST SLAVONIC   EAST SLAVONIC   SOUTH SLAVONIC
21 Polish   25 Russian   28 Slovene
22 Slovak   26 Ukrainian   29 Serbo-Croatian
23 Czech   27 Belarussian   30 Macedonian
24 Lusatian   31 Bulgarian

OTHER INDO-EUROPEAN BRANCHES

CELTIC GROUP

BRITANNIC   GAELISH
32 Breton   34 Irish Gaelic
33 Welsh   35 Scots Gaelic

BALTIC GROUP

36 Latvian   37 Lithuanian

HELLENIC

38 Greek

THRACIAN/ILLYRIAN GROUP

39 Albanian

INDO-IRANIAN GROUP

40 Romani (dispersed)

URALIC LANGUAGE FAMILY

FINNO-UGRIC GROUP

41 Finnish   44 Estonian
42 Karelian 45 Hungarian
43 Saami   46 Komi

SAMOYEDIC GROUP

47 Samoyedic

ALTAIC LANGUAGE FAMILY

TURKIC GROUP

48 Turkish

OTHER LANGUAGES

BASQUE

49 Basque

Areas with significant
concentrations of other
languages (usually
adjacent national languages)

Boundary between languages

After Murphy, 1998.

More generally, Christianity has gradually lost adherents since the 1960s as secularization gathered momentum, especially in Western Europe. The Roman Catholic Church, long powerful in much of Europe, has steadily been losing its grip on society, and many church institutions—schools, universities, unions, political parties, charities, and clubs—have been hollowed out. Moreover, many churches have closed down or have been converted into art galleries, public meeting halls, and even corporate offices.

Today, a new force roils the religious landscape: the rise of Islam. In southeastern Europe, this takes the form of new Islamic assertiveness in an old Muslim bastion: the (Turkish) Ottoman Empire left behind millions of converts from Bosnia to Bulgaria, many of whom are now demanding greater political representation and power. In the west, this Islamic resurgence results from the relatively recent infusion of millions of Muslim immigrants from former colonies in North Africa and other parts of the far-flung Islamic world. Here, while mosques overflow with the faithful, churches stand nearly empty as a witness to secularism among Europeans.

Spatial Interaction

If not a single culture, then what does unify Europe? The European realm is best understood as a large-scale *functional region*, a fully interdependent realm held together by highly developed spatial-economic and political networks. Contemporary Europe has seized on the realm's abundant geographic opportunities to create a huge, intensively used network of spatial interaction linking places, communities, and countries in countless ways. This spatial interaction operates on the basis of two key principles.

First, regional **complementarity [13]** means that one area produces a surplus of a commodity required by another area. The mere existence of a particular resource or product is no guarantee of trade: it must be needed elsewhere. When two areas each require the other's products, we speak of *double complementarity*. Europe exhibits countless examples of such complementarity, from local communities to entire countries. Industrial Italy needs coal from Western Europe; Western Europe needs Italy's farm products.

Second, the ease with which a commodity can be transported by producer to consumer defines its **transferability [14]**. Distance and physical obstacles can raise the cost of a product to the point of unprofitability. But Europe is small, distances are short, and the Europeans have built a most efficient transport system of roads, railroads, and canals linking navigable rivers. Taken together, Europe's enormously diverse economic regions, its ultra-efficient transportation infrastructure, and its far-reaching free-trade agreements within the European Union make for a supremely interdependent geographic realm.

A Highly Urbanized Realm

About three out of every four Europeans live in towns and cities, and in Western Europe urban population percentages range even higher. Large cities are production centers as well as marketplaces, and they also form the crucibles of their national cultures. Europe's major cities tend to have long histories. They are compact, and in general the European cityscape looks quite different from its North American counterpart.

Seemingly haphazard inner-city street systems impede traffic; central cities may be picturesque, but they are also cramped. European city centers tend to be more vibrant today than cities in the United States. They offer a mix of businesses, government functions, tourism, shopping facilities, educational and arts institutions, and entertainment as well as housing for upper-income residents.

Wide residential sectors radiate outward from the **central business district (CBD) [15]** across the rest of the central city, often inhabited by particular income groups. Beyond the central city lies a sizeable suburban ring, but even here residential densities are much higher than those in the United States because the European preference has been to live in apartments rather than detached single-family houses, and to set aside recreational spaces in "greenbelts." There also is a greater reliance on public transportation, which further concentrates the realm's urban and suburban development patterns. However, traditions may now be starting to change. One indication is the recent growth of cutting-edge suburban business centers that increasingly compete with the CBD in many parts of urban Europe (see **Box 5-4**).

European Unification

Realms, regions, and countries can all be subject to dividing and unifying forces that cause them to become more or less cohesive and stable political units. Where the region or country is home to diverse populations with different political goals, it may prove difficult to avoid divergence and territorial fragmentation. Political geographers use the term **centrifugal forces [16]** to identify and measure the strength of such division, which may result from religious, racial, linguistic, political, economic, or other regional factors.

Centrifugal forces are measured against **centripetal forces [17]**, the binding, unifying glue of the state or region. General satisfaction with the system of government and administration, shared identities, legal institutions, and other functions of the state (notably including its treatment of minorities) can ensure stability at the state level.

Since World War II ended in 1945, Europe has witnessed a steady process of integration and unification at a much broader geographic scale. A majority of European states and their leaders recognized that closer association and regional coordination formed the key to a more stable, prosperous, and secure future. Centripetal forces have thus far prevailed, but, as the EU has expanded to include ever more countries, recent years have witnessed growing dissent and criticism, particularly in the countries where it all began in the late 1950s.

Box 5-4   From the Field Notes ...

"If you were to be asked what city is shown here, would Paris spring to mind? Most images of the French capital show venerable landmarks such as the Eiffel Tower or Notre Dame Cathedral. I am taking this photo from the top of the famous Arc de Triomphe, looking toward another Paris, the burgeoning northwestern business district just beyond the city line named *La Défense*. There, Paris escapes the height restrictions and architectural limitations of the historic center and displays an ultramodern complexion. Glass-box skyscrapers reflect the vibrant global metropolis this is, and the central landmark is the 'Cube,' a huge open structure admired as well as reviled (as was the Eiffel Tower in its time). *La Défense* is ingeniously incorporated into Paris's urban design, with a straight-line, broad avenue connecting it to the Arc de Triomphe and then connecting, via the Champs Élysées, to the Place de la Concorde. In terms of a grand geographic layout, Paris is hard to beat."

© J. Nijman

Postwar Motivations

At the end of World War II, much of Europe lay shattered, its cities and towns devastated, its infrastructure wrecked, its economies ravaged. If this was one of the world's most developed realms at the beginning of the twentieth century, its economic prowess had been almost completely destroyed by 1945. One of the primary motives for integration and collaboration among Western European countries, therefore, was rapid economic recovery.

The integration process was also driven by political considerations. From the perspective of countries such as France, the Netherlands, and Belgium, one of the key issues was to control defeated Germany in the postwar years, and this could only be done through close political cooperation. Thus European integration was from the start both an economic and a political affair, and concerns about Germany played a major part.

From afar, the United States was a strong supporter of (Western) European integration. The end of World War II had witnessed accelerating tensions with the Soviet Union, which had taken control over the bulk of Eastern Europe. Furthermore, communist parties seemed poised to dominate the political life of major Western European countries. The United States was intent, therefore, to have a firm hand in that part of Europe and to keep communist influences at bay. Under the so-called *Marshall Plan*, massive U.S. financial support, exceeding $130 billion in today's money, poured in to 16 European countries, including the western half of Germany not controlled by the Soviets as well as Turkey.

As these economies recovered, Germany (West Germany, that is, until the Cold War ended in 1989) became firmly embedded in a pan-European structure, the United States assumed a more distant role, and the motivations driving European unification developed a new emphasis. Economically, the process has increasingly been propelled by European corporate interests—the need to facilitate an ever larger and more efficient open market that could compete globally with the U.S., China, and Japan. Politically, twenty-first-century goals are more about stabilizing a much larger and more diverse European community that now approaches the western edge of Russia. Thus, the focus has shifted from Germany to an enlarged and far more complicated geopolitical zone. The latter consists of numerous sovereign states that must maintain satisfactory relations with the Russian giant next door.

The Unification Process

European integration has been a complex, gradual, and step-by-step process aimed at the creation of a single territorial economy across Europe. It has involved (1) agreements on free trade across boundaries without tariffs or border controls; (2) the abolition of protectionist economic policies; (3) harmonization of national laws (e.g., on taxation) to establish a level playing field; (4) political cooperation and the creation of pan-European executive, judicial, and legislative bodies; (5) the introduction of a single currency, the *euro*; and (6) agreements to transfer specific powers from national governments to EU institutions.

European integration is a classic example of **supranationalism [18]**, the voluntary association in economic, political, or cultural spheres of independent states willing to yield some measure of sovereignty for their mutual benefit. **Box 5-5** provides a timeline of the key milestones and shows that European integration has in no small part been a matter of steady institution building. **Figure 5-8** maps European Union membership, membership of the **Euro zone [19]** (not all EU countries have given up their national currency), and membership in the so-called **Schengen Area [20]** (26 countries that have abolished internal border controls and that have a single visa policy for non-EU visitors). Note that these three membership areas do not overlap entirely: for instance, Poland is part of the EU but has not adopted the euro; Norway is not part of the EU but is inside the Schengen Area; and Ireland is in the EU and has adopted the euro, but is not part of the Schengen Area. It is a complex mosaic, indeed.

In some of the latest developments, Croatia became the 28th country to join the EU in 2013; Lithuania, in 2015, became the 19th member-country to adopt the euro; and in 2016, the United Kingdom voted to bail out—we will return to the "Brexit" saga later in the chapter.

The European Union is not just an organization on paper. It has a profound impact on national economies, on the role of individual states, and on the daily lives of its member-countries' nearly 450 million citizens.

Box 5-5 Timeline of Supranationalism in Europe

1948 Organization for European Economic Cooperation (OEEC) established.

1949 Council of Europe created.

1957 Treaty of Rome signed, establishing European Economic Community (EEC) (effective 1958); also known as the Common Market and "The Inner Six," the founding member-countries were France, (then) West Germany, Italy, Belgium, the Netherlands, and Luxembourg. European Atomic Energy Community (EURATOM) Treaty signed (effective 1958).

1967 European Community (EC) inaugurated.

1968 All customs duties removed for intra-EC trade; common external tariff established.

1973 United Kingdom, Denmark, and Ireland admitted as members of EC, creating "The Nine."

1979 First general elections for a European Parliament held; new 410-member legislature convenes in Strasbourg, France. European Monetary System established.

1981 Greece admitted as member of EC, creating "The Ten."

1985 The Schengen Agreement results in the abolition of border checks between the Benelux countries, Germany, and France. The Schengen Area is gradually expanded to include 26 EU countries by 2008.

1986 Spain and Portugal admitted as members of EC, creating "The Twelve." Single European Act ratified, targeting a functioning European Union in the 1990s.

1993 Single European Market goes into effect. Modified European Union Treaty ratified, transforming the EC into the European Union (EU).

1995 Austria, Finland, and Sweden admitted into EU, creating "The Fifteen."

1999 European Monetary Union (EMU) goes into effect.

2002 The euro is introduced in 12 countries; the Euro zone expands gradually over time (see Fig. 5-8), but not all EU members choose to give up their national currency.

2004 Historic expansion of EU from 15 to 25 countries with the admission of Cyprus, the Czech Republic, Estonia, Hungary, Latvia, Lithuania, Malta, Poland, Slovakia, and Slovenia.

2005 Proposed EU Constitution is rejected by voters in France and the Netherlands.

2007 Romania and Bulgaria are admitted, bringing total EU membership to 27 countries. Slovenia adopts the euro.

2010 Financial crisis strikes heavily-indebted Greece, requiring massive EU bailout. Later in the year, Ireland follows and raises fears of similar crises in Portugal, Spain, and Italy. The future of the EMU is clouded; the value of the euro declines after a long rise against the dollar.

2013 Croatia gains entry as the 28th EU member-state.

2014 Latvia adopts the euro.

2015 A popular referendum in Greece results in a majority rejecting another EU bailout and related austerity measures. A "Grexit" seems closer than ever before, but the government eventually agrees on a renegotiated package. Lithuania adopts the euro, increasing Euro zone membership to 19 countries.

2016 The intensifying refugee crisis across Europe stresses the EU and produces divisive impacts in member-countries. The United Kingdom holds its "Brexit" referendum, with a 52-percent majority voting to leave the EU.

2017 Ukraine and the EU sign an EU Association Agreement, giving preferential treatment to Ukrainian trade and increased free movement of Ukrainians in the Schengen Area. Preparations are made for Ukraine to apply for full EU membership in 2024.

2019 In October 2019, the U.K. Parliament voted, with a small majority, in favor of the Withdrawal Agreement Bill, which would set in motion a transition period during 2020 and withdrawal from the EU by the end of 2020.

FIGURE 5-8

Consequences of Unification

One Market

EU directives are aimed at the creation of a single market for producers, consumers, businesses, and workers. Corporations should be able to produce and sell anywhere within the Union without legal impediments, whereas workers should be able to move anywhere in the EU and find employment without legal restrictions. In order to make this happen and to keep things manageable, member-states have had to harmonize a wide range of national laws from taxation to the protection of the environment to educational standards.

One major step toward union came with the introduction of a single central bank (with considerable power over, for example, interest rates) and a single currency, the euro. The euro was also meant to symbolize Europe's strengthening unity and to establish a counterweight to the no-longer-almighty American dollar. In 2002, 12 of the (then) 15 EU member-countries withdrew their own currencies and began using the euro, with only the United Kingdom, Denmark, and Sweden opting to stay out (Fig. 5-8). More recently, Slovakia (2009), Estonia (2011), Latvia (2014), and Lithuania (2015) joined the Euro zone. The single currency, hailed as a major triumph at its inception, has evolved into a major bone of contention because it has significantly reduced the freedom of member-states to formulate fiscal policies. We will return to the ongoing "euro crisis" shortly.

A New Economic Geography

The establishment and continuing development of the EU have generated a new economic landscape that today not only transcends the old but is fundamentally reshaping the realm's regional geography. By investing heavily in new infrastructure and by smoothing the flows of money, labor, and products, European planners have dramatically reduced the divisive effects of their national boundaries. And by acknowledging demands for greater freedom of action by their provinces, States, departments, and other administrative subunits of their countries, European leaders unleashed a wave of economic energy that has transformed some of these subnational units into powerful engines of growth.

Four of these burgeoning hubs are especially noteworthy, to the point that many policy analysts refer to them as the **Four Motors of Europe** [21]: (1) France's southeastern **Rhône-Alpes Region**, centered on the country's second-largest city, Lyon; (2) **Lombardy** in north-central Italy, focused on the industrial metropolis of Milan; (3) **Catalonia** in northeastern Spain, anchored by the cultural and manufacturing nucleus of Barcelona; and (4) **Baden-Württemberg** in southwestern Germany, headquartered by the high-tech city of Stuttgart.

Despite economic integration and the harmonization of policies, **Figure 5-9** also shows that major inequalities persist among Europe's regions as well as within EU member-states. This is an important reminder that, as noted in Chapter 1, uneven development is a pervasive phenomenon that plays out at every geographic scale—in some ways, European integration has deepened regional inequalities within a number of countries because some regions are much better positioned than others to seize the opportunities of a single economic market (e.g., northern Italy versus southern

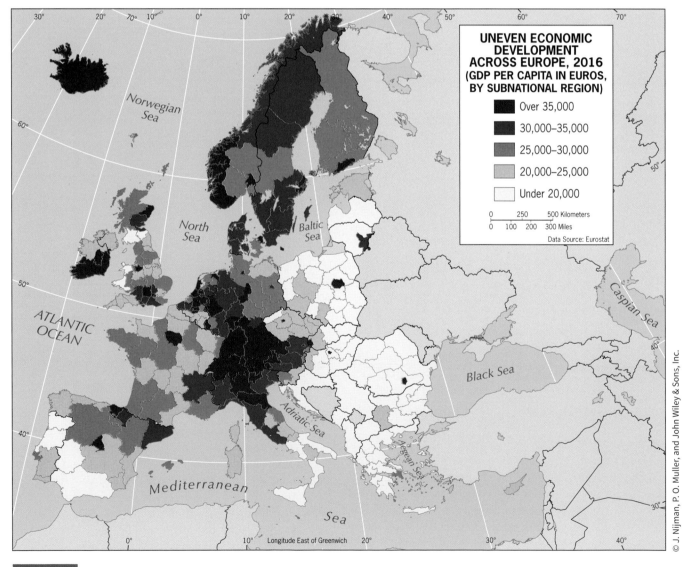

UNEVEN ECONOMIC DEVELOPMENT ACROSS EUROPE, 2016 (GDP PER CAPITA IN EUROS, BY SUBNATIONAL REGION)

- Over 35,000
- 30,000–35,000
- 25,000–30,000
- 20,000–25,000
- Under 20,000

Data Source: Eurostat

FIGURE 5-9

Italy). Figure 5-9 also shows that major urban areas tend to have higher per capita incomes than the rest of their country (e.g., Athens or Dublin).

Diminished State Power and New Regionalism

As states relinquish some of their power to Brussels and express agreement with the ideal of realm-wide integration, some of the provinces and other subdivisions *within* states have seized the opportunity to assert their own cultural identity as well as particular economic interests. Often, the local governments in these subregions simply bypass the central government in their national capital, dealing not only with each other but also directly with foreign governments as their business networks expand to the global scale.

Thus, even as Europe's states have been working to join forces in the EU, many of those same states are confronting severe centrifugal stresses. The term **devolution [22]** has come into use to describe the powerful centrifugal forces whereby regions or peoples within a state, through negotiation or active rebellion, demand and gain political strength and sometimes autonomy at the expense of the center. Most states exhibit some level of internal regionalism, but the process of devolution is set into motion when a key centripetal binding force—the nationally accepted idea of what a country stands for—erodes to the point that a regional drive for autonomy, or for outright secession, is launched. As **Figure 5-10** shows, more than a dozen European countries are now affected by devolutionary pressures to varying degrees.

States respond to such pressures in a number of ways, ranging from accommodation to suppression. One way to deal with these centrifugal forces is to give historic regions (such as Scotland or Catalonia) certain rights and privileges formerly held exclusively by the national government. Another answer is for EU member-states to create new administrative divisions that will allow the state to meet regional demands, often in consultation with, or under pressure from, the EU.

© J. Nijman, P. O. Muller, and John Wiley & Sons, Inc.

FIGURE 5-10

Recent Challenges of Integration

The Changing Risks of Expansion

Expansion was always an EU objective, and the subject never fails to arouse passionate debate. Will the incorporation of weaker economies undermine the strength of the overall organization? Should the ties and cooperation among existing members be deepened and solidified before other, less prepared countries are invited to join? Keep in mind that EU members must adhere to strict economic policies and harmonize their political systems and laws (see **Box 5-6**). This is much easier for prosperous countries with longstanding democratic traditions than for poorer nations with a volatile political past.

Numerous structural implications arise from this expansion, affecting all member-countries. First, some of the pre-2004 EU's less affluent countries, which used to be on the receiving end of EU subsidy programs, now had to pay up to support the much poorer new eastern member-states. Second, disputes also intensified over representation at EU's Brussels headquarters. And, most of all, the incorporation of weaker national economies tends to compromise the ability of the EU to move forward in unison because the weaker members are unable to keep their fiscal deficits within limits, and sometimes their industries are hurt by the more developed, competitive industries of the wealthier members. The inclusion of countries like Romania and Bulgaria, especially, has often been criticized in this regard.

The Challenges of a Single Currency

The kind of economic incompatibility among EU countries described in the previous section has come into sharper focus since the arrival of the single currency. Since the late 2000s, the "euro crisis" has been an almost standard news item throughout the realm as it laid bare the limitations that accompanied the shift to a single currency and a common monetary policy. The evolving problems are not just about the euro itself, but about the inability or unwillingness of EU members to comply with fiscal standards (e.g., avoiding government deficits) and apply them as solutions within their individual economies. Keep in mind that each EU country still has its own budget as well as its own revenues and expenditures. But even though the EU budget itself remains quite small (about 1 percent of the EU's GDP), the budget rules governing the Union are in crucial ways determined by Brussels.

The economic difficulties confronted by several European national governments during the 2010s intensified because they had only recently introduced the single European currency—which significantly constrained their policy options. Some of the EU's weaker member-states, most of all Greece, accumulated enormous debts that violated terms in the agreement that established the Euro zone. Moreover, having a single currency implied that countries could no longer independently pursue monetary policies such as devaluations (a common approach to promoting exports) or the lowering of their interest rates (used to stimulate the economy). In other words, as countries such as Greece faced deepening economic problems, their hands were tied by their dependence on the European Central Bank—whose primary mission is to maintain the purchasing power (price stability) of the euro.

Box 5-6 From the Field Notes ...

"The Grasshopper, one of Amsterdam's many 'coffee shops.' For the past four decades, the Netherlands has permitted legalized sales and consumption of moderate amounts of marijuana. The general consensus is that marijuana is not as harmful as hard drugs and that legalization helps to decriminalize its commerce and use. This policy is at odds with legislation in most of the rest of the EU, especially with bordering Germany and nearby France. It used to be that coffee shops were found only in Amsterdam and other major cities, but in recent years they have opened in towns along the Dutch border to attract a foreign clientele (another byproduct of heightened cross-border interaction). As a result, once quiet and conservative rural villages (where lifestyles differ enormously from those in Amsterdam) started to experience heavy traffic of buyers and users from abroad—not something the town folk were likely to appreciate. Largely for this reason, the national government has turned coffee shops in the border zone into membership-only establishments whereby only Dutch citizens and residents are eligible to join. Initially, there were plans to make this change across the entire country, including the big cities. As you might expect, reactions to this proposal varied geographically: it was supported in the rural border areas, but most people in Amsterdam and other cities viewed it as an infringement on traditional liberties and a bad business decision to boot. Said one Amsterdammer and steady customer of The Grasshopper with a tone of sarcasm: 'I never thought I'd experience this: some provincials trying to decide what happens in Amsterdam. Not in my lifetime!' In Amsterdam's coffee shops, it is very much business as usual."

© J. Nijman

The euro crisis began in earnest in 2010 when it came to light that Greece had accumulated so massive a government deficit that an emergency bailout from Brussels was essential to avoid immediate bankruptcy and economic chaos. A year later, the same happened to Ireland and Portugal, and Greece received its second bailout in 2012. By then, a "Grexit" loomed larger than ever before, but Greek membership survived. In 2019, the economy of these member-states had mostly stabilized, especially the main urban areas (Dublin, Lisbon, and Athens).

"Brexit": A Historic Referendum

Perhaps the most critical challenge to the EU came in 2016 from the United Kingdom (UK), often described as the "reluctant European." There had been growing opposition in the UK to continued EU membership, and the government decided to hold a popular referendum on June 23, 2016. Those favoring "Leave" argued that EU rules constrained the British economy, and that the UK was asked to contribute too much to the spectrum of funds that were distributed among other member-states. More generally, they opposed the idea of "ever closer union" as a growing threat to the UK's sovereignty. And most important, they also claimed that the UK must control its own borders in order to reduce the number of immigrant workers and refugees.

On the other hand, those favoring "Remain" argued that the benefits of membership outweighed the cost. Thanks to the EU, they claimed, the UK was able to send more than half of its exports to EU countries, London had become one of the most influential nodes in the global economy, and the country was able to play a prominent role in EU decision-making (especially concerning the free movement of people between the European continent and the UK). In their view, severing ties with the EU would not solve the refugee problem and could even make it worse by depriving the British of a voice in realm-wide policy-making on this critical issue. Moreover, the EU supporters maintained, the UK was committed to staying outside the Euro zone (and therefore remain unaffected by its fiscal crises), and that immigrant workers contributed significantly to the UK economy.

The well-known outcome of the 2016 referendum was that approximately 52 percent of the electorate voted for the UK to leave the EU. In every way, this was a hotly contested choice that deeply divided the country. The prime minister of the pro-EU government felt compelled to resign, and new elections were called. Throughout the EU, and in much of the rest of the world, the result was deplored. Financial markets everywhere were shaken by the news, and the UK economy, in particular, entered a dark period of uncertainty as the long, complicated process of EU severance got underway.

A new government was installed in 2016, but for more than three years it failed to convince a Parliamentary majority of a proper plan of transition, and it also failed to negotiate precise terms with the EU. The prime minister (Theresa May) was finally forced to resign in mid-2019, giving way to yet another government (led by Boris Johnson) that aimed to leave the EU, but it too failed to earn majority support in Parliament

before it handily won year-end re-election. Overall, it is fair to say that the Brexit controversy has been very much a political and politicized dispute in the absence of a clear plan that also laid out the consequences of EU withdrawal.

One the most difficult issues relates to Scotland and Northern Ireland: they are part of the United Kingdom, but their majorities voted to remain in the EU. Thus for the UK, Brexit also means confronting major internal political challenges. In the case of Scotland, where a strong independence movement has been building for decades, this could lead to a break-up of the United Kingdom. As for Northern Ireland, should it stay with the UK after an EU exit, a "hard border" would be created between the Republic of Ireland (an independent country that remains in the EU) and Northern Ireland. Considerable apprehension and criticism of this possibility already exists on both sides of that border, so it is not inconceivable that Brexit could result in the reunification of Ireland.

During 2019, divisions seemed to only deepen, with anti-Brexit demonstrators taking to the streets on various occasions, demanding a second referendum (photo below). But that was not to be. Somehow, Prime Minister Johnson was able to forge a slim majority of support in the parliament and on October 19 of 2019 it decided, by 329 to 299 votes, to approve the so-called Withdrawal Agreement Bill. Its intention was to set in motion a transition period during 2020 and a withdrawal by the end of that year.

The Remaining Outsiders

A number of other countries and territories still remain outside the EU (Fig. 5-8). The first group includes five of the seven states that emerged out of former Yugoslavia, plus Albania. Here in the western Balkan Peninsula, most are troubled politically as well as economically. Only Slovenia and Croatia have achieved

On October 19 of 2019, hundreds of thousands of anti-Brexit protesters took part in a march through central London to demand a second referendum on the outcome of Brexit – only the most recent of various demonstrations in the preceding year. The protest coincided with an emergency session of Parliament where MPs voted on Prime Minister Johnson's EU withdrawal deal. The demonstrators did not get their referendum and the Parliament voted to empower the prime minister to proceed with plans for the the UK's withdrawal from the EU.

Wiktor Szymanowicz/Barcroft Media/Getty Images

full EU membership. The rest—Serbia, Bosnia and Herzegovina, North Macedonia, Montenegro, and Kosovo—rank among Europe's poorest and most ethnically fractured states, yet they do contain almost 20 million people whose circumstances could worsen outside the "EU club." Better, EU leaders reasoned, to move them toward membership by demanding political, social, and economic reforms.

The second group of outsiders is composed of Ukraine and its neighbors. The government of Ukraine itself has expressed interest, but its electorate is strongly (and regionally) divided—as well as overshadowed by a civil conflict that has put everything on hold (see discussion later in this chapter). Moldova views the EU as a potential supporter in the struggle against devolution along its eastern border, and a country not even on the map—Georgia, across the Black Sea from Ukraine—has proclaimed its interest in future membership for a similar reason: to find an ally against Russian intervention in its internal affairs (including a brief 2008 military incursion).

Finally, we should take special note of another long-mentioned candidate for EU membership—Turkey. Some EU leaders have long wanted to include (NATO member) Turkey mainly for geostrategic reasons, but others argued that Turkey was just not "European" enough. The Turkish case for accession has been thrown into a different light today because of the country's pivotal role in the refugee crisis that has engulfed

Europe (to be discussed shortly). At the same time, the autocratic tendencies of the longtime Turkish president Erdogan will do little to persuade many in the EU to advance Turkey's candidacy.

European Geopolitics

NATO Expansion

The emphasis on EU developments tends to overshadow another important pan-European institution—the *North Atlantic Treaty Organization (NATO)*—a far-reaching security alliance that partially overlaps with the EU (**Fig. 5-11**). NATO, however, is a very different organization. First, it is primarily concerned with politics and security and not with economics: NATO's purpose is to provide military protection for its members. Second, NATO is led by the United States, which provides the bulk of the alliance's military muscle, and also includes Canada. And third, NATO includes non-EU member Turkey, a country that is hugely important because of its strategic location on the southeastern edge of Europe nearest to the Middle East.

NATO was created in 1949 to provide a security umbrella for Western Europe under U.S. leadership (in response, the Soviets and their satellites created the Warsaw Pact). When the

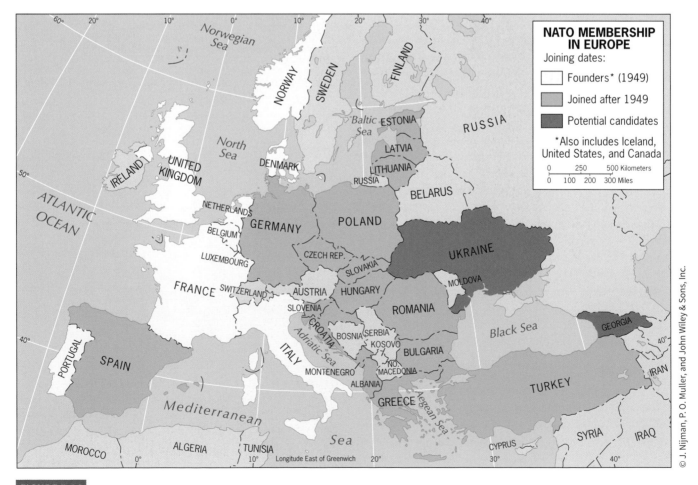

FIGURE 5-11

© J. Nijman, P. O. Muller, and John Wiley & Sons, Inc.

Cold War ended in 1991 with the disintegration of the Soviet Union, NATO became a more general security organization that would provide stability across Europe. It also became more concerned with emerging, twenty-first-century threats to its member-states: terrorism, attacks by "rogue" states with weapons of mass destruction, and cyber-warfare (a very real and growing danger with critical global implications). However, as NATO opened its doors to membership for Eastern European countries (Fig. 5-11), it initiated an eastward expansion—not unlike that of the EU—that produced a renewed standoff with Russia. Most recently, the organization was joined by Bulgaria, Romania, Estonia, Latvia, and Lithuania (all in 2004); Croatia and Albania (both in 2009); and Montenegro (2017).

Geopolitics in the Baltics

In recent years, the area around the Baltic Sea has become something of a testing ground of European–Russian relations. Four issues are in play, and they are all interconnected (**Fig. 5-12**). First, the former Soviet republics of Estonia, Latvia, and Lithuania are now members of NATO and are sitting on Russia's doorstep. They have always been considered critical to Russian security interests (which is why the Soviet Union in past times effectively annexed them, until it had to let them go when the USSR imploded in the late twentieth century). Moreover, there are sizeable ethnic Russian minorities in these countries (24 percent in Latvia, 28 percent in Estonia, and 9 percent in Lithuania), adding to Moscow's interest in this region.

Second, over land, the Baltic states are only connected to other NATO territory at what is called the **Suwalki Gap [23]**, a 65-kilometer (40-mile) stretch of border between Lithuania and Poland (Fig. 5-12). It is named after the nearby Polish town of Suwalki. This is a vulnerable land connection because it could be fairly easily blocked by Russian intervention from Kaliningrad (a highly militarized Russian exclave on the Baltic Sea) or from Belarus, a close Russian ally. If so, the Baltic states would be cut off from their allies. There is significant militarization of the Suwalki Gap on both the Russian and the Polish side.

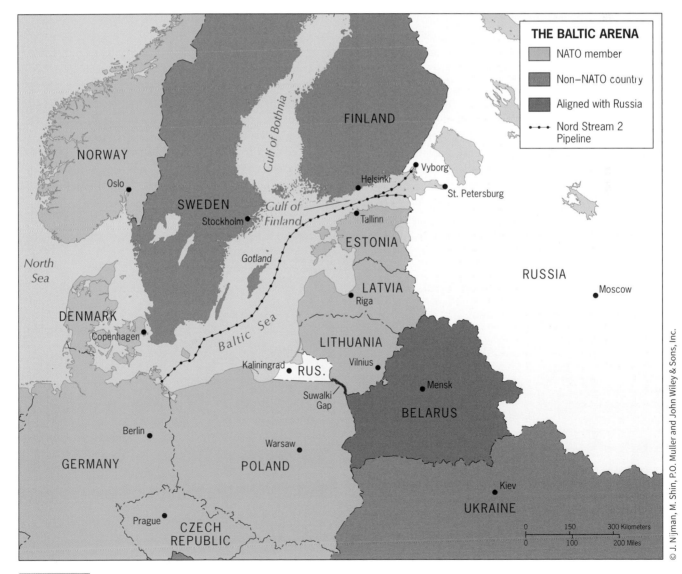

FIGURE 5-12

Third, it does not alleviate the sense of insecurity in the Baltic states that the two countries across the Baltic Sea, Sweden and Finland, are not members of NATO, allowing Russian naval forces to dominate this body of water, with major naval bases in St. Petersburg and Kaliningrad. In fact, Sweden, while officially neutral, appears itself increasingly concerned about threats from Russia. In 2018, the Swedish government embarked on a major preparedness campaign in case of war, including the construction of a series of underground bunkers. It has also been militarizing the small but strategically located island of Gotland (Fig. 5-12), and it has been seeking closer cooperation with NATO.

Fourth, Russia is constructing a new gas pipeline, "Nord Stream 2," from the western Russian city of Vyborg, along the bottom of the Baltic Sea, to northern Germany (Fig. 5-12). It is a mega-project, financed entirely by the Russians, to supply gas to Germany, and it could be finished by 2020. But there are already gas pipelines from Russia to Germany and other parts of Western Europe, so why is this new pipeline needed? The existing pipelines go through Poland and Ukraine, two countries with conflict-ridden relations with Russia (we will elaborate on the relationship between Russia and Ukraine later in this chapter). Both countries depend in large part on gas from Russia, and there have been times that Russia threatened to use the supply of gas as political leverage. But cutting off gas to either Poland or Ukraine inevitably meant also depriving customers further west and jeopardizing Russian business interests. The Nord Stream 2 will solve this problem for Russia, as it will be in a position to threaten to cut off Poland or Ukraine while continuing to supply Germany and other Western European countries. The Poles, clearly, are watching the construction of the new pipeline with concern (and with a sense of betrayal by Germany, which agreed to this new pipeline). For Russia, the new pipeline will be one more reason to maintain control over the Baltic Sea.

Altogether, this part of Europe is seeing increased military buildup and tension, with increased Russian assertiveness; the three Baltic states trying for a posture that does enough to deter Russia but not too much to provoke it, and they have to trust on a continued NATO (and U.S.) security commitment; Poland allowing for NATO to organize a growing military presence near its northern borders; Sweden increasing its military capabilities and seeking out closer relations with NATO; and Germany, as the realm's most central and economically powerful country, trying to maintain geopolitical stability in the middle of it all.

Europe's Refugee Crisis

Another challenge that has faced Europe in recent years, and that is sometimes framed in terms of a security threat, is the refugee crisis. The turmoil and conflict that have plagued parts of North Africa and Southwest Asia since 2011 have triggered an ever larger stream of refugees into (relatively nearby) Europe. The contrasts between these two realms are enormous: one is quite rich, politically stable, highly integrated, democratic, progressive, and largely Christian; the other is mostly poor, conflict-ridden, often war-torn, fragmented, authoritarian, conservative, and mostly Islamic. The story of this refugee crisis is as much about globalization and the "shrinking" of distance as it is about stark socioeconomic juxtapositions and cultural encounters.

The statistics are astounding: the total number of verified illegal border crossings into Europe rose from about 400,000 in 2014 to more than two million in 2015. This huge sudden increase coincided with a major shift of refugee flows from the central Mediterranean to the eastern Mediterranean and the Balkans. Those coming across the Mediterranean mostly originated in North Africa and (to a lesser extent) Subsaharan Africa, while those seeking entry into Europe through the Balkans came mostly from Syria and other sources in Southwest Asia (photos below).

The EU countries on the front line are Greece and Italy. From there, refugees or migrants (the distinction has become

Refugees trying to reach Europe across the central Mediterranean (left) and the Balkan routes (right). The left-hand photo was taken on April 17, 2016, off the coast of Libya, from where more than 130 migrants cast off to reach Italy. Within hours, their boat broke in two, and according to news reports, 108 lives were saved as 28 drowned. The right-hand photo, taken on March 6, 2016, shows a group of refugees in Greece approaching its northern border with (now North) Macedonia. By this time, however, the Macedonian government had erected a border fence, making it virtually impossible to cross.

politically contentious in Europe) typically try to travel to EU countries they believe offer them the best chance of acceptance and a good living. In practice, this means that many are destined for Germany and Sweden. Once they have reached their destination, many apply for **asylum [24]** or legally protected residency status. **Figure 5-13** shows the number of refugees by country in 2017, with Germany leading, followed by France, Sweden, the United Kingdom, and Italy. Most refugees are from Syria (29 percent), Afghanistan (14 percent), and Iraq (10 percent).

Once people have set foot inside EU borders (e.g., in Italy or Greece), they are free to move around to any of the 26 countries that are part of the *Schengen Area* (Fig. 5-8). It has been proposed, especially by Germany, that all EU member-states should accept their share of asylum seekers, but most countries have been less than forthcoming. Some, particularly those on the Balkan route, have seized the initiative and put up border fences to avoid being overwhelmed by refugees (e.g., Hungary, North Macedonia, Austria). Certain other EU countries have refused to take in any refugees at all, or only minimal numbers (e.g., Poland, Slovenia). As might be expected, most refugees would prefer to settle in Germany or Sweden than in Romania or Lithuania, and the problem of distributing refugees has become both intractable and politically divisive in the EU.

In the meantime, Italy, and especially Greece, became swamped by refugees at the same time they were experiencing difficult economic times. When North Macedonia closed its border, thousands of makeshift refugee camps sprang up across northern Greece. It was this increasingly untenable situation in Greece that drove the EU to an agreement with Turkey. More than half the illegal border-crossers into Europe in 2015 (amounting to 1.2 million people) came via Turkey. From Turkey, they either crossed the Aegean Sea to Greece or entered Greece or Bulgaria via overland routes. The deal implied that Turkey would close and control its land borders with Greece and Bulgaria, that it would stop Aegean Sea crossings, and that it would accept returned refugees from Greece. In exchange, the EU pledged U.S. $3.2 billion in financial support to Turkey in order to run its expanded refugee centers, visa-free travel for Turks to the EU, *and* promises to accelerate Turkey's bid for EU membership. This agreement did have the intended effect of reducing the influx of migrants and refugees into the EU: illegal border crossings into Europe dropped from 1.2 million in 2015 to just over 100,000 in 2018.

The refugee problem and its possible solutions have had a divisive effect on both the EU and individual countries. As noted previously, it surrounded the UK's 2016 Brexit referendum, and it has fueled support for populist right-wing political parties from France to Denmark and from the Netherlands

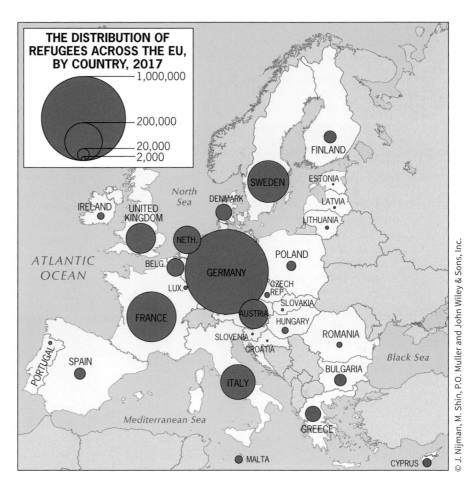

FIGURE 5-13

to Poland. The difficulty of distinguishing between "migrants" who seek a better economic future versus "refugees" who want to escape deadly conflict at home has complicated matters, as has the fact that certain aging European countries would clearly benefit from an influx of young immigrants (**Box 5-7**): of all asylum-seekers in the EU in 2015, 83 percent were under the age of 35. And then there is the challenge of rising Islamist terrorism in Europe over the past decade. It is generally not directly connected to the refugee crisis (most terrorism suspects are born in the EU or have entered the EU legally), but because myriad asylum seekers are Muslims, for many Europeans the politics of these issues have become intertwined.

The Specter of Islamic Terror in Europe

Since the beginning of this century, Europe has experienced a series of terrorist attacks that were claimed by Islamic militant organizations:

- May 2017: 22 people were killed and hundreds wounded in the Arena bombing in Manchester, UK.
- March 2016: 32 people were killed and hundreds wounded in bombings in Brussels.
- November 2015: 130 people died and hundreds were wounded in a series of attacks in Paris.

- January 2015: 12 people were killed in the Paris office of the satirical newspaper *Charlie Hebdo* in retaliation for its articles on Islamic fundamentalism.

- March 2012: Six people, including Jewish schoolchildren, were killed in Toulouse, France.

- July 2005: 52 London commuters died when suicide bombers blew themselves up on subway trains and a bus.

- March 2004: A series of bombs on commuter trains killed 191 people and injured nearly 2000 in metropolitan Madrid.

Box 5-7 Geographic Information Analysis

Demography is Destiny?

A popular saying proclaims that "Demography is Destiny"—because tomorrow's age-group distribution depends on today's birth rates and growth. Understanding population growth/decline—which also includes migration—is critical to understanding Europe's future. The geographic variation of demographic "destinies" is clearly visible on maps of the European realm. Some populations rank among the oldest and fastest-aging in the world, and some are the preferred destinations for migrants from outside Europe. Examine the patterns in **Figure 5-14**. Considering their shares of elderly population, which countries would benefit most from an infusion of immigrants?

Access your WileyPLUS course to interact with a dynamic version of this map and to engage with online exercises and questions.

EUROPE: Population aged 65 and over (percent of total population)

| No Data | < 12.5% | 12.5% - 15% | 15% - 17.5% | 17.5% - 20% | 20% - 22.5% | > 22.5% |

FIGURE 5-14 **Interactive**

© J. Nijman, M. Shin, P.O. Muller and John Wiley & Sons, Inc.

Most of these attacks were said to be the work of ISIS (Islamic State of Iraq and Syria), the fundamentalist, terrorist organization that controls parts of those countries and Libya (see Chapter 7). Islamist terrorism in Europe has been on the rise in the past few decades, and the reasons are not always clear. On one hand, they are grounded in the involvement of the EU and individual European countries (e.g., France) in the conflicts of the Middle East; on the other hand, numerous European cities are home to relatively large numbers of disaffected, radicalized, second-generation Islamic youth—who were born in Europe, but who somehow do not fit in. Although the attackers were linked to the Islamic State, it is hard to determine the extent to which their actions were centrally coordinated or locally initiated.

These events have helped fuel right-wing, anti-immigration, and anti-Muslim sentiments—even if the overwhelming majority of European Muslims are as critical or fearful of Islamist terror as everyone else. The challenge for European governments is to fight terrorism without alienating this Muslim majority, and to prevent radicalization of young Muslims by building a more inclusive society.

Migration, Multicultural Challenges, and Prospects of Integration

Immigration from Outside the EU

It seems ironic, amid all the problems revolving around refugees and Islamist terrorism, that most European countries are in dire need of immigrants—and that when these immigrants arrive, they largely come from non-Western societies and are often Muslim.

There was a time when Europe's population was (in the terminology of population geographers) exploding, sending millions to the New World and other colonies while still growing at home. But these days Europe's native population, unlike most of the rest of the world's, is shrinking. To keep a given population from declining, the (statistically) average woman must bear 2.1 children. For Europe as a whole, that indicator was 1.59 in 2017, and several countries recorded numbers well below that. Such negative population growth poses serious challenges for any nation. When the age-group profile becomes top-heavy, the number of workers whose taxes pay for the social services of the elderly goes down, leading to reduced pensions and dwindling funds for healthcare. It is estimated that, for the EU as a whole, the working-age population will shrink by 6 percent between 2015 and 2030. In some countries, that decline will be more severe: in Germany, for example, the working-age population is projected to drop another 13 percent by 2030.

The most obvious way to counter the adverse economic effects of negative population growth is to promote immigration, particularly of youthful working-age people and families with young children. This is why a number of European countries, which half a century ago experienced net out-migration, are now committed to encouraging in-migration (Box 5-7; Fig. 5-14).

The integration of Muslim immigrants is a demanding process. The majority are generally more religious than the realm's Christians. They arrived in a Europe where religious institutions are weakening and where certain cultural norms are incompatible with Islamic traditions. Integration of Muslim communities into the national fabric has been slow; their education and income levels are considerably below those of the host population; and unemployment among Muslims is usually higher. In Western European countries, Islamic immigrants are heavily concentrated in metropolitan areas. Therefore, in major cities such as Hamburg, London, Paris, and Brussels, the Muslim population percentage is considerably higher than the national average.

There are positive signs of integration as well, an example being the 2016 election of Sadiq Khan—a son of Pakistani immigrants and a Muslim—as the mayor of London. Undoubtedly, he would describe himself first as a Londoner. Across Europe, former immigrants are increasingly elected to office in local and national government. Importantly, there are signs of generational shifts: the first generation of immigrants, around the 1960s, often maintained their original culture and often could not even speak the language of their new country. The second generation was raised and educated in the new country of their birth, but their lives were sometimes torn between competing cultural identities. It is often suggested that second-generation (Islamic) immigrants are the more likely to be radicalized. The third generation is now emergent, and they may have more in common with other millennials than with the culture of their grandparents. Some are secular, others are religious, but they are thought to be more firmly embedded in the social fabric of their home country.

Migration Within the EU

In some ways, it is migration within the European Union, not immigration from outside, that brings more challenges—because the numbers are larger and because this phenomenon is more recent. With the inclusion of a number of less developed countries, mainly from the east, the EU has become ever more disparate and economically uneven. Combined with the free movement of people and labor inside the single market, this has induced substantial movement across member-states. From Poles moving to the United Kingdom, Romanians to Spain, or Ukrainians to Poland, many are looking for better opportunities to work, study, or live in other parts of the EU. Since the cultural differences (language, to start with) are so substantial, these new migrants are sometimes viewed as foreign whether they are from inside the EU or not. In this way, internal migration in the EU has also contributed to political division and a nationalist turn in many countries. In the EU countries of origin, this migration has a more direct negative effect: the ones leaving tend to be young and educated, precisely the demographic cohorts needed to advance these relatively less developed regions. As a consequence, and notwithstanding efforts from the EU to support poorer areas, the regional inequalities remain or become even more pronounced.

Regions of Europe

In the past, Europe divided rather easily into Western, Northern (Nordic), Mediterranean (Southern), and Eastern regions, and countries were grouped accordingly. This traditional configuration remains relevant today, but in the era of European unification, a *core–periphery* framework has been superimposed on that regional scheme, defined in terms of how central—or peripheral—countries are to the workings of the EU. In other words, Europe can still be understood as a set of formal, more or less homogeneous cultural regions; yet at the same time, the entire realm increasingly functions as a tightly interdependent economic and political regional system, a spatial network that revolves around the EU. We distinguish four major regions that constitute the European realm (**Fig. 5-15**): (1) *Western Europe*, including Britain and Ireland; (2) *Northern Europe*, (3) *Mediterranean Europe*, including Slovenia

EUROPEAN REGIONS
- ▪▪▪ European Core Boundary
- Western Europe
- Northern (Nordic) Europe
- Mediterranean Europe
- Eastern Europe

© J. Nijman, P. O. Muller, and John Wiley & Sons, Inc.

FIGURE 5-15

and Croatia; and (4) *Eastern Europe*, the broad eastern tier of the realm that extends to the Russian border.

Take a close look at Figure 5-15 and note that the European Core (bounded by the red dashed line) in many places extends beyond Western Europe and rarely coincides with national borders. Thus, the European Core is primarily defined on the basis of regional economic performance, not political territorialization. For instance, northern Italy (but not the Italian south) and southern Sweden (but not the Swedish north) are both part of this Core; also, note the inclusion of Estonia in the Core, despite its relatively remote location. This regionalization underscores how complicated and challenging it can be to understand the complexity of spatial organization. Even though Europe typifies that challenge like no other realm, the goal of this chapter is to bring clarity to what at first glance appears to be a chaotic regional arena.

Region Western Europe

Western Europe consists of 11 countries including the **microstate** [25] of Liechtenstein (**Fig. 5-16, Box 5-8**). This region also lies almost entirely within the European Core that forms the heart of the EU. The large offshore islands of Britain and Ireland (referred to in the past as the British Isles) have been closely involved in Western Europe's history as well as its ongoing economic and political development. Let us begin on the mainland with Germany, which is both the realm's most populous country and largest economy.

Germany

Twice during the twentieth century Germany plunged Europe and the world into war. At the end of World War II in 1945, the

FIGURE 5-16

©J. Nijman, P. O. Muller, and John Wiley & Sons, Inc.

Box 5-8 From the Field Notes ...

"It has been a long day, driving north from Italy through the Swiss Alps. I just crossed the border into the ministate of Liechtenstein to look for a place to spend the night. From my hotel in the town of Triesenberg, I have a panoramic view of an upper stretch of the Rhine River. It flows from left to right in the picture. This is Western Europe's leading river and one of its defining geographic features—as a means of transportation, sometimes as a barrier, and frequently as a natural boundary. Here it forms the ministate's border with Switzerland, located on the far side. Even though we are looking at what is a political boundary, the physical and cultural landscapes on both sides of the Rhine have much in common."

© J. Nijman

defeated and devastated German state was divided into two parts: West Germany and East Germany (their former border is delimited by the red line in **Fig. 5-17**), the latter of which was forced to surrender some important territory to Poland.

West and East Germany then set forth on widely different economic and political trajectories. The Soviet governance of East Germany was established on the USSR's communist model, and as repayment for the extreme hardships the Soviet Union had suffered at German hands during the war, this rule was unremittingly punitive. The U.S.-led authority in West Germany was far less strict and aimed at rehabilitation rather than retribution. West Germany was included when the Marshall Plan was instituted in 1948, and the massive infusion of aid stimulated rapid economic recovery. At the same time, West Germany was reorganized politically into a modern federal state based on democratic foundations.

West Germany's economy soon began to thrive. Between 1949 and 1964, its gross national income (GNI) tripled as industrial output rose by 60 percent. Simultaneously, its political leaders participated enthusiastically in the negotiations that created the six-member Common Market in 1957 (and thereafter evolved to become today's EU). Geography worked in West Germany's favor: it had common borders with all but one of the original member-states. Its swiftly rebuilt transportation infrastructure was second to none in Europe. The reinvigorated spatial economy focused on heavy industrial production in the Ruhr Basin (in the hinterland of the gigantic Dutch port of Rotterdam) in addition to the leading-edge manufacturing complexes centered on Hamburg in the north, Frankfurt (the leading financial and air transport hub as well) in the center, and Stuttgart in the south. West Germany now exported huge quantities of iron, steel, motor vehicles, machinery, textiles,

and farm products. Until just a few years ago, Germany had the largest export economy by value in the world; although it has now been surpassed by China and the United States, Germany's exports per capita are still by far the highest on Earth.

At the time of Germany's reunification in 1990, West Germany had a population of about 62 million and East Germany 17 million. By 2015, former East Germany's population was down to less than 13 million, and its shrinkage is expected to continue in the coming years. Decades of comunist misrule in the

© J. Nijman

Leipzig is eastern Germany's biggest city after Berlin. It has a rich cultural history and has been home to, among others, Johann Sebastian Bach, Johann Wolfgang von Goethe, Richard Wagner, Friedrich Nietzsche, and Felix Mendelssohn. But during communist times Leipzig was dragged down along with the rest of East Germany. Although this city is now finally experiencing something of a revival, there are still many struggling neighborhoods and outlying communities.

STATES (*LÄNDER*) OF GERMANY

GDP PER CAPITA, IN EUROS, 2016
- Over 35,000
- 30,000–35,000
- 24,000–30,000

City population
- 50,000–250,000
- 250,000–1,000,000
- 1,000,000–5,000,000

National capitals are underlined

- - - - Railroads
——— Roads

0   50   100 Kilometers
0   25   50 Miles

Data source: Eurostat, 2016

© J. Nijman, P. O. Muller, and John Wiley & Sons, Inc.

FIGURE 5-17

east had yielded outdated factories, crumbling infrastructures, polluted environments, drab cities, inefficient farming, and inadequate legal and other institutions. Thus the 1990 reunification was more of a rescue than a merger, and the cost to the reconstituted German state since then has been gargantuan. Figure 5-17 maps the 16 States (or *Länder,* in German) that constitute today's reunified Federal Republic of Germany and also makes a key point: regional disparity in terms of gross domestic product (GDP) per person remains a serious problem. Note that all of former East Germany's six States (except Berlin) still rank in the lowest income category, whereas nine of the ten former West German States rank in the two higher-income categories.

Another challenge facing Germany, along with other Western European countries, is the integration of sizeable immigrant minorities. Germany now has nearly 16 million immigrants, about 20 percent of the total population of 83.5 million. They not only are culturally different but also worse off economically: more than a third of immigrant households live below the official poverty line, largely because the unemployment rate of immigrants far exceeds that of native Germans.

Nevertheless, Germany navigated the recent global recession better than any other European country and retains its position as the realm's economic powerhouse. Today, this country is developing closer ties with Eastern Europe, and from the Balkans to the Baltic Sea, Germany is regarded as the region's most desirable trading partner. Germany has also played a leading role in EU bailouts of Greece and other countries, and—in an ironic historical twist, given the initial reasons for European integration—has been a vital force in maintaining a sufficient degree of EU unity.

France

Territorially, France is larger than Germany, and the map suggests that France has a superior relative location, with coastlines on the Mediterranean, the Atlantic Ocean, and, at Calais, even a window on the North Sea. But France does not have any good natural harbors, and oceangoing ships cannot navigate its rivers and other waterways very far inland.

The map of mainland Western Europe (Fig. 5-16) reveals a significant contrast between France and Germany in terms of their **urban systems [26]**, the spatial distribution and interrelationships among cities of various sizes. France has one dominant city, Paris, and no other city comes close either in size or centrality: Paris has 11 million residents, whereas its closest rival, Lyon, contains only 1.7 million. Germany has no city to match Paris, but it does have a wider range of medium-sized cities: in the 100,000-plus category, Germany has more than 80, whereas France has less than 40.

Paris is an example of a **primate city [27]**, one that is disproportionately large compared to all others in the national urban system, exceptionally expressive of its country's culture,

and (almost always) the capital. Paris grew so large and dominant within the French urban system for reasons of geography and history. Geographically, Paris has advantages in terms of both **site [28]** (the physical attributes of the place it occupies) and **situation [29]** (its location relative to surrounding areas). The *site* of the original settlement at Paris lay on an island in the Seine River, a defensible spot where the river was often crossed. This island, the Île de la Cité, was a Roman outpost 2000 years ago; for centuries, its security ensured continuity. Eventually the island became overcrowded, and the city expanded along the banks of the river (**Fig. 5-18**). Paris's advantageous *situation* lay in its fertile agricultural hinterland and the presence of key waterways. The Seine River is joined near Paris by several navigable tributaries (the Oise, Marne, and Yonne). When canals extended these waterways even farther, Paris was linked to the Loire Valley of central France; the Rhône–Saône Basin of the southeast; Lorraine, a key industrial area in the northeast; and the northern border zone with Belgium. When Napoleon reorganized France and built a radial system of roads (duplicated later by railroads) that focused on Paris from all parts of the country, the city's primacy was assured (Fig. 5-18).

Historically, France early on developed the strong tradition of a highly centralized state in which the central government in Paris maintained tight control over 96 subnational *départements*. It remained so for nearly two centuries, but France decentralized its administrative structure in 1982 with the introduction of a new layer of governance, consisting of 26 larger *régions* (22 in France itself and 4 overseas). These régions

SITE AND SITUATION OF PARIS, FRANCE

- ▪▪▪▪ 12th Century Wall
- ══ Wall of 1840s
- —— Roads
- —— Railroads
- ▧ Forests, Parks
- ▢ Built-up area

A SITE | SITUATION **B**

© J. Nijman, P.O. Muller, and John Wiley & Sons, Inc.

FIGURE 5-18

were inserted at the level between Paris and the *départements* both to accommodate devolutionary forces and to seize the opportunities for regional and local growth throughout the country. In 2016, the number of régions inside France itself was further consolidated from 22 to 13. **Figure 5-19** shows today's régions and the regional variation in GDP per capita—with the Paris région leading far ahead of the other régions.

France has long had one of the most diversified economies in the world. Northern French agriculture remains intensive and varied, exploiting the country's wide range of soils and microclimates and enjoying state subsidies and protections. In the southeast, Lyon, France's second city and headquarters of the région named Rhône-Alpes, has become a focus for growth industries and multinational firms. This

région is evolving into a stand-alone economic powerhouse that is becoming a driving force in the European economy. Indeed, it is one of the *Four Motors of Europe* (discussed earlier) with its own global business connections to countries as far away as China and Chile. More broadly, France's economic geography is marked by new high-tech industries, and it is a leading producer of high-speed trains, aircraft, fiber-optic communications systems, and space-related technologies. France is also the world leader in nuclear power, which currently supplies about 75 percent of its electricity and thereby reduces dependence on foreign oil imports; but rising public opposition today all but prohibits the construction of new generating facilities given the national goal of reducing nuclear reliance to 50 percent by 2025.

© J. Nijman, P. O. Muller, and John Wiley & Sons, Inc.

FIGURE 5-19

France is another European state with a rapidly aging population (totaling just over 65 million), and that poses some major challenges for the years ahead. Natural population growth is close to zero and must be compensated for through immigration. However, most immigrants hail from poor Islamic countries with very different cultures, and their integration and acceptance into French society has been a painful process. The graying of the French population means that in the future more and more elderly people must be cared for with the incomes earned by a shrinking productive workforce.

The French economy has already exhibited signs that it might encounter problems of the kind that have bedeviled countries such as Greece and Spain—which would raise serious alarms because France's economy is far bigger and of much greater importance to the EU as a whole. Government debt is high; the public sector is very large (nearly twice the size of Germany's), draining expenditures while youth unemployment exceeded 20 percent in early 2019.

France is a politically divided country, in part because of growing economic inequalities and in part because of polarized public opinion regarding immigrants and the Muslim minority. The right-wing *Front National* has been a fast-growing French political party, espousing an extremist, ultranationalist agenda that is highly critical of Muslim and other immigration as well as EU policies in general.

Benelux

Three small countries are crowded into the northwestern corner of mainland Western Europe—Belgium, the Netherlands, and Luxembourg—and are collectively referred to by their first syllables (*Be-Ne-Lux*). Their total population of 29 million reminds us that this is one of the most densely peopled corners of our planet. Not coincidentally, the *Low Countries*, as they are also sometimes called, are situated at the **estuary [30]** (mouth) of the great Rhine and Scheldt rivers and have access to the seas that, throughout modern history, has been the envy of the Germans and the French. They are a highly productive trio: both the Netherlands and Belgium generally rank among the top 20 economies on Earth, and tiny Luxembourg has one of the world's highest per capita gross national incomes.

The Netherlands, one of Europe's oldest democracies and a constitutional monarchy today, has for centuries been expanding its territory—not by warring with its neighbors but by wresting land from the sea. About a quarter of the country lies below sea level. The urban, demographic, and economic center of gravity lies within a **conurbation [31]** (a large urban complex formed by the coalescence of two or more cities) in the west of the country. This so-called *Randstad* ("Ring City"), roughly diamond in shape, is anchored by Amsterdam, the constitutional capital; Rotterdam at the mouth of the Rhine, Europe's largest port; The Hague, the national seat of government; and Utrecht.

Belgium also has a thriving economy as well as a leading port in the city of Antwerp. With 11.5 million people, Belgium's regional geography is dominated by a cultural fault line that cuts diagonally across the country, separating a Flemish-speaking majority (60 percent) centered on Flanders in the northwest from a French-speaking minority in southeastern Wallonia (40 percent). Brussels, the predominantly French-speaking capital, forms a linguistic island in the Flemish-speaking sector; but the city also is one of Belgium's greatest assets because it serves as the headquarters, and in many ways as the *de facto* capital, of the European Union (though it is important to note that the EU does not have an official capital city). Nonetheless, devolution is a persistent problem for Belgium, with political parties that agitate for Flemish separatism constantly roiling the social landscape.

Luxembourg, one of the realm's numerous ministates, lies between Germany, Belgium, and France, with a Grand Duke as head of state (its official name is the Grand Duchy of Luxembourg), a territory of only 2600 square kilometers (1000 sq mi), and a population of just over 600,000. Luxembourg has translated sovereignty, relative location, and stability into a haven for financial, services, and information-technology industries. There are more than 150 banks in this tiny country and about 14,000 holding companies (corporations that hold controlling stock in other businesses in Europe and worldwide). With its unmatched GDP per capita (more than U.S. $104,000 in 2017), Luxembourg is in some ways the greatest beneficiary of the advent of the European Union. And in no country in Europe is support for the EU stronger than it is here.

The Alpine States

Switzerland, Austria, and the microstate of Liechtenstein on their border share an absence of coasts and the mountainous terrain of the Alps and not much else (Fig. 5-16). Austria speaks one language; the Swiss speak German in their north, French in the west, Italian in the southeast, and even a bit of Rhaeto-Romansch in the remote central highlands (Fig. 5-7). Austria has a large primate city; multicultural Switzerland does not. Austria has a substantial range of domestic raw materials; Switzerland does not. Austria is twice the size of Switzerland and has a larger population, but far more trade moves across the Swiss Alps between Western and Mediterranean Europe than crosses Austria.

Switzerland, not Austria, is in most ways the leading state in the Alpine subregion of Western Europe. Mountainous terrain and a **landlocked location [32]** can constitute crucial barriers to economic development, but, as is so often the case, geography poses constraints in one way while offering opportunities in another. That is why Switzerland is such an important lesson in the complexities of human geography. Through the skillful maximization of their opportunities (including the transfer needs of their neighbors), the Swiss have transformed their seemingly restrictive environment into a prosperous state. They deftly utilized the waters cascading from their mountains to generate hydroelectric power needed to develop highly specialized industries. Swiss farmers perfected methods to optimize the productivity of mountain pastures and valley soils. Swiss leaders converted their country's isolation into stability, security, and neutrality, making it a world banking behemoth, a global magnet for money. Zürich, in the German-speaking sector, is the financial center; Geneva, in the French-speaking sector, is one of the world's most internationalized cities. The Swiss feel that they do not need to join the EU—and they have no plans to apply in the foreseeable future.

Austria, which joined the EU in 1995, is a remnant of the old Austro-Hungarian Empire and has a historical geography that is far more reminiscent of unstable Eastern Europe than that of Switzerland. In fact, even Austria's physical geography seems to demand that the country look eastward: it is at its widest, lowest, and most productive in the east, where the River Danube links it to Hungary, its old ally in the anti-Muslim wars of the past.

Vienna, by far the Alpine subregion's largest city, also lies on the country's eastern perimeter. One of the world's most expressive primate cities with magnificent architecture and monumental art, Vienna today is Western Europe's easternmost major city, but its relative location has changed dramatically with EU enlargement since 2000. Even though Vienna found itself in a far more centralized position when the EU border shifted eastward, most Austrians doubted that their neighbors to the east could ever become EU members of potentially equal standing. One result is that Austrian public support for the European Union today has fallen to a record low level.

The Czech Republic

The Czech Republic is focused on the historic Bohemian Basin, the mountain-encircled national core area centered on the capital, Prague. This is another classic primate city, its cultural landscape highly reflective of Czech traditions. Prague is also an important industrial hub and collection center, surrounded by mountains that contain many valleys with small towns that specialize, Swiss-style, in fabricating high-quality manufactures. In the old Eastern Europe, even during the communist

A view of Lake Lucerne in central Switzerland, with the town of Stansstad in the foreground and the city of Lucerne in the distance. In this country, connectivity has triumphed over the challenging terrain. Near the opposite end of this lake, Switzerland in 2016 opened the Gotthard Rail Tunnel (the world's longest)—a 57-kilometer (35-mi) marvel running below the entire massif of the Alps.

period, the Czechs always were leaders in technology and engineering; their products could be found on markets in foreign countries near and far. Moreover, Bohemia always was cosmopolitan and Western in its exposure, outlook, development, and linkages; Prague lies in the upper basin of the Elbe River, its traditional outlet through northern Germany to the North Sea. Not long ago, the Czech Republic would have been considered part of "Central Europe," but today it has shifted its gaze entirely to the west, with its solid economic development fully warranting inclusion in the European Core (Fig. 5-15).

The United Kingdom (UK)

It has often been observed that the geographical position of the United Kingdom, just off the European mainland, reflects its traditional hesitation to consider itself part of Europe. It did join the EU in 1973, but was always reluctant to move toward tighter integration. When most member-states in 2002 adopted the new euro in favor of their national currencies, the British kept their pound sterling. Meanwhile, within the UK, opposition to continued EU membership steadily mounted, culminating in the 2016 Brexit referendum—whose close vote to break away from the EU deeply divided the country.

With an area about the size of Oregon and a population of 67.5 million, the UK is by European standards a fairly large country. Based on a combination of physiographic, historical, cultural, economic, and political criteria, the United Kingdom can be subdivided into four subregions (numbered in **Fig. 5-20**):

1. **England.** So dominant is this subregion of the United Kingdom that the entire country is sometimes referred to by this name. England is anchored by the sprawling London metropolitan area, which by itself contains nearly one-sixth of the UK's total population. Indeed, along with New York, London is regarded as one of the top-tier **world cities [33]**, with financial services, high technology, communications, engineering, and associated industries reflecting the momentum of its long-term growth and agglomeration. Within England, there is a notable difference between the crowded, fast-paced, and globally connected southeast centered on London, and the far less prosperous north and west (see inset map in Fig. 5-20). The latter areas, once the global hearth of the Industrial Revolution, have now completely lost their dynamism as well as segments of their population in our age of deindustrialization. Spatial inequality has also been rising, and today's widening gap between rich and poor within England's residential mosaic shows no sign of abating (photo next page).

2. **Wales.** This nearly rectangular, rugged territory was a refuge for ancient Celtic peoples, and in its western counties more than half the inhabitants still speak Welsh. Because of the high-quality coal reserves in its southern tier, Wales, too, was engulfed by the Industrial Revolution, and Cardiff, the Welsh capital, was once the world's leading coal

© J. Nijman

Since the mid-2010s, London has been ranked (by the high-end real estate company Knight Frank) as the best place for the world's wealthiest people to reside and/or own a home. Not surprisingly, it is also one of the most expensive cities in the world. To reside here in the Royal Borough of Kensington and Chelsea, a high-cachet neighborhood in central London close to Buckingham Palace, be prepared to pay more than U.S. $2 million for a two-bedroom apartment (houses in this area start at around $5 million). Foreign investors from Russia, China, and the oil-rich Persian Gulf States are most prominent among the buyers. London may be an extreme case, but it is indicative of a twenty-first-century trend among select cities (New York, Paris, and San Francisco are others) that are experiencing skyrocketing real estate values due to foreign interest—increasingly to the detriment of established middle-class residents who can no longer afford to remain in their longtime neighborhoods. It remains to be seen if and how a Brexit will affect London's business climate, its socio-economic fabric, and its overall standing as a leading world city.

exporter. But the fortunes of Wales also declined, and many Welsh emigrated. In several cultural and economic spheres, Wales is closely tied to England, and independence is not in the cards. The situation in Scotland, however, is entirely different.

3. **Scotland.** With a population of 5.5 million, Scotland is a major component of the UK—but perhaps for not much longer. It is also well removed from Britain's thriving southeastern core area. Most of the population resides in the Scottish Lowlands, anchored by Edinburgh, the capital, in the east and Glasgow in the west. Attracted there by the labor demands of the Industrial Revolution (coal and iron reserves lay nearby), the Scots developed a world-class shipbuilding industry. Subsequent decline and obsolescence were followed by high-tech development (notably around Glasgow) and Scottish participation, beginning in the 1970s, in the exploitation of oil and gas reserves beneath the adjacent North Sea. Nevertheless, a growing number of Scots feel that they are increasingly disadvantaged within the UK and should play a bigger role in the EU. In 2014, a referendum on the question of full independence won the support of 45 percent of the electorate. Since then, the separatist cause has gained momentum—only fueled

further by the Brexit decision (opposed strongly here) and its new opening for independence followed by Scotland rejoining the EU.

4. **Northern Ireland.** Northern Ireland was home not only to Protestants from Britain but also to a substantial population of Irish Catholics who found themselves on the wrong side of the border when Ireland became independent nearly a century ago. Ever since, conflict has intermittently engulfed Northern Ireland and spilled over into Britain. A declining share of the overall population of 1.8 million is Protestant and traces its ancestry to Scotland or England; meanwhile, the proportion of Roman Catholics is rising, and in the capital (Belfast) they achieved majority status in 2014. Catholics share their religion with the Irish Republic on the southern side of the border. Although Figure 5-20 suggests that there are majority areas of Protestants and Catholics in Northern Ireland, no such clear separation exists; mostly they reside in clusters throughout the territory, including the walled-off neighborhoods that still mark the major cities of Belfast and Londonderry.

Republic of Ireland

Hard-won independence from Britain in 1921 did not bring real economic prosperity until the final decades of the twentieth century, when Ireland became known as the *Celtic Tiger* (likening it to the miraculous, Hong Kong-style economic development of the *Asian Tigers* discussed in Chapter 10). Participation in the EU (since 1973), adoption of the euro (in 2002), business-friendly tax policies, comparatively low wages, an English-speaking workforce, and an advantageous relative location combined for a time to produce the highest rate of economic growth in the EU. During the 1990s, Ireland's booming, service-based economy, accompanied by burgeoning towns and cities (especially the capital, Dublin), skyrocketing real estate prices, mushrooming industrial parks, and bustling traffic, transformed a country long known for emigration into a magnet for postindustrial workers, producing new social challenges for a closely knit, long-isolated society. It also brought deep inequalities, between income classes and between various parts of the country, with Dublin and a few other bigger cities becoming much more wealthy and expensive, and the east and south economically much better off than the rest (see Fig. 5-20). After a painful intermezzo during the Great Recession (2007-2011), Ireland today (population: 4.9 million) continues its role as a major EU locus of high-technology investment from the United States.

Region Northern Europe

Northern Europe's remoteness, isolation, and challenging environment, especially at latitudes higher than 60°N, have had positive binding effects for much of this domain and seem to have

BRITAIN AND IRELAND

POPULATION

—— Railroad
—— Road
⊢⊣ Canal
●—● Gas pipeline
●—● Oil pipeline
◆ Gasfield
⚒ Oilfield
▨ Coalfield

● 50,000–250,000
● 250,000–1,000,000
● 1,000,000–5,000,000
● Over 5,000,000

National capitals are underlined

Maritime Boundaries
Area of Catholic majority in N. Ireland
Area with a majority of Welsh speakers
European Core Boundary

0   50   100   150   200   250 Kilometers
0   25   50   75   100   125   150 Miles

GDP PER PERSON, IN EUROS, BY SUBNATIONAL EU REGION, 2016

Over 35,000
30,000–35,000
18,000–30,000

Data Source: Eurostat

© J. Nijman, P. O. Muller, and John Wiley & Sons, Inc.

FIGURE 5-20

fostered similar cultures. The Scandinavian Peninsula carrying Sweden and Norway lay removed from most of the wars of mainland Europe and developed in relatively tranquil circumstances (even though Norway was overrun by Nazi Germany during World War II). Three of its major languages (Swedish, Norwegian, and Danish) are mutually intelligible, and in terms of religion there is overwhelming adherence to the same Lutheran church. Democratic and representative governments emerged early on, and individual rights and social welfare have been carefully protected for centuries. Women participate more fully in government and politics here than in any other part of the world.

But consider the implications of Figure 5-15 and **Figure 5-21**: the southern, coastal, and more urbanized areas of the region form part of the European Core, but the sparsely populated remainder does not. In the aggregate, economic indicators for most of this region are quite strong, but almost all development is concentrated within that smaller Core zone. Territorially, northern Norway and Sweden as well as most of Finland belong to the European periphery. The five countries of this higher-latitude domain of Europe have a combined population of only 27 million, yet their core areas make up in prosperity and external linkages what they lack in size.

Interestingly, some of the countries in the region have stopped short of full participation in the EU: Norway is not even a member, and only Finland has joined the Euro zone to date.

Sweden

Sweden is northern Europe's largest country in terms of both population (10 million) and territory. Most Swedes live south of latitude 60°N (which passes through Uppsala just north of the capital, Stockholm) in what is climatically the most moderate part of the country (Fig. 5-21). Here lie the primate city, core area, and the main industrial districts; here, too, are the main agricultural areas that benefit from the lower relief, better soils, and somewhat milder climate (**Box 5-9**).

Sweden long exported raw or semifinished materials to industrial countries, but today the Swedes make finished products themselves, including motor vehicles, electronics, stainless steel, furniture, and glassware. Much of this production is based on local resources, including a major iron ore reserve at Kiruna in the far north (there is a steel mill at Luleå). Swedish

FIGURE 5-21

© J. Nijman, P. O. Muller, and John Wiley & Sons, Inc.

Box 5-9 From the Field Notes …

"From the water, we have a view of outer Stockholm's prosperous suburban landscape. We are cruising along Lilla Värtan, the narrow strait that separates the Swedish capital from the adjacent island of Lidingö. Stockholm is home to 1.5 million people, and its long history dates back to the thirteenth century. Its layout is quite unusual in that it has been built on a cluster of no less than 14 islands. Sweden is a wealthy country, and much of that affluence is concentrated in and around this capital city. Which is not to say that everybody lives like this. The myriad waterways and islands have allowed for considerable segregation, with drab, low-income, high-rise residential areas located well away from these obviously well-to-do suburbs."

© J. Nijman

manufacturing, in contrast to that of several Western European countries, is based in dozens of small and medium-sized towns specializing in particular products. Energy-poor Sweden was a pioneer in the development of nuclear power, but a national debate focused on the risks involved has reversed that course. Sweden joined the EU in 1995 but chose to stay out of the Euro zone. Sweden has a long tradition of neutrality in Europe and it is not a member of NATO but, with a wary eye on Russia, in recent years it has intensified relations with the Western alliance, and has been actively involved in joint military exercises.

As Nordic Europe's biggest and most influential country, Sweden has historically dominated both Norway and Denmark. Culturally, Sweden has time and again made its presence known on the international scene, from popular music to films to literary works to the annual awarding of the Nobel Prizes.

Norway

Norway does not need a nuclear power industry to supply its energy needs. It has found its economic opportunities on, in, and beneath the sea. Norway's fishing industry, now augmented by highly efficient fish farms, has long been a cornerstone of the economy, and its merchant marine spans the world. But over the past four decades, Norway's economic life has been transformed by the bounty of oil and natural gas discovered beneath the floor of its sector of the North Sea.

With its limited patches of cultivable soil, high relief, extensive forests, and spectacular fjords, Norway has nothing to compare to Sweden's agricultural or industrial development. Its cities, from the capital Oslo and the North Sea port of Bergen to the historic national focus of Trondheim as well as Arctic Hammerfest, lie on narrow coastal lowlands and have difficult overland connections. The isolated northern province

of Finnmark has even become the scene of a devolutionary movement among the reindeer-herding indigenous Saami (Fig. 5-10). The distribution of Norway's population of 5.4 million has been described as a necklace, its beads linked by the thinnest of strands, but this has not constrained national development. Economically, Norway has one of the lowest unemployment rates in Europe, and in terms of income per capita it is one of the world's richest countries. Interestingly, Norway uses almost none of the oil it produces and exports nearly all of it. The country's own energy consumption is virtually entirely from hydroelectric, renewable sources. It is no coincidence that the largest Tesla plant in the world outside the U.S. is located here, and the Norwegian government provides major financial incentives to buyers of electric vehicles. In 2018, half of all cars sold in Norway were electric, the highest proportion in the world.

Norwegians have a strong national consciousness and spirit of independence. In 1995, when Sweden and Finland voted to join the European Union, the Norwegians again said no. They had little desire to exchange their economic independence for the regulations of a broader, even possibly safer, Europe. They did, however, join the Schengen Area in 2010.

Denmark

Territorially small by Scandinavian standards, Denmark has a population of 5.8 million, the North's second-largest country after Sweden. It consists of the Jutland Peninsula and several islands to the east at the gateway to the Baltic Sea; it is on one of these islands, Sjaelland, that the capital of Copenhagen is located. Copenhagen has long been a port that collects, stores, and transships large quantities of goods. This **break-of-bulk** [34] function exists because many oceangoing vessels cannot enter the shallow Baltic Sea, making the city an **entrepôt [35]** where transfer facilities and activities prevail. The completion of the Øresund bridge-tunnel link to southern Sweden in 2000 further enhanced Copenhagen's situation (Fig. 5-21).

Denmark remains a kingdom, and in centuries past Danish influence extended far beyond its present confines. Remnants of that period continue to challenge Denmark's governance. *Greenland* came under Danish rule following union with Norway (1380) and continued as a Danish possession when that union ended in 1814. In 1953, Greenland's status changed from colony to province, and in 1979 its 56,000 inhabitants were granted home rule under the new Inuit name of *Kalaallit Nunaat*. They promptly exercised their rights by withdrawing from the EU, of which they had become a part when Denmark joined in 1973. Greenland was dependent on financial support from Denmark, but this has changed in recent years following the discovery of oil off Greenland's west coast, north of the Arctic Circle. Through Denmark, Greenland forms a part of NATO, which has a military base on Greenland's northwest coast, deep inside the Arctic.

Another restive Danish dependency is the Faroe Islands, located between Scotland and Iceland (Fig. 5-15). These 17 minuscule islands and their 50,000 inhabitants were awarded self-governance in 1948, complete with their own flag and currency, but Denmark remains sovereign, and debates continue about full independence.

Finland

Finland, territorially almost as large as Germany, has only 5.5 million residents, most of them concentrated in the southern triangle formed by the capital, Helsinki, the textile-producing complex of Tampere, and the shipbuilding center of Turku (Fig. 5-21). A land of evergreen forests and glacial lakes, Finland's national income has long been sustained by wood and wood-product exports. But the Finns have now developed a more diversified economy with staple agricultural crops and the manufacture of precision machinery and telecommunications equipment. Finland's environmental challenges and relative location have created cultural landscapes similar to those of northern Sweden and Norway, but the Finns are not a Scandinavian people—their linguistic and historical affinities are with the Estonians across the Gulf of Finland to the south (Fig. 5-7). In fact, other ethnic groups speaking these Finno-Ugric languages are also widely dispersed across what is today western Russia. Finland shares a long border with Russia, and during the Cold War it had to tread very carefully in international affairs. But after the collapse of the Soviet Union it joined the EU in 1995 and was one of the first countries to adopt the euro in 2002. In the twenty-first century, Finland has decidedly positioned itself within Europe. It did not, however, join NATO, a move considered too provocative toward Russia.

Iceland

Iceland is a volcanic, glacier-studded island in the frigid waters of the North Atlantic just south of the Arctic Circle. Inhabited by people with Scandinavian ancestries (2019 population: 340,000), Iceland is of special scientific interest because it lies astride the

Strokkur Geyser, about a 90-minute drive east of the capital (Reykjavik), is one of Iceland's famous attractions. Geysers are vents in the Earth's surface that periodically eject a column of boiling water and steam. Iceland's geysers, volcanoes, and frequent earthquakes are products of its hyperactive geology, the result of its location astride the Mid-Atlantic Ridge.

Mid-Atlantic Ridge. Here, the Eurasian and North American tectonic plates of the Earth's crust are diverging (see Fig. 1-5), new land can be seen forming, and spectacular volcanic eruptions are periodically on display (as occurred memorably in 2010 when the Eyjafjallajökull volcano brought trans-Atlantic and domestic air traffic to a standstill for days on end).

Iceland's population is almost completely urban, and the capital, Reykjavik, is home to fully half of the country's inhabitants. The nation's economic geography is traditionally oriented to the frigid surrounding waters, whose seafood harvests have given Iceland a high standard of living. In the 1990s, Iceland embarked on economic liberalization policies, and its financial industries and banks grew rapidly. Iceland's landscapes formed the backdrop for movies like *Lord of the Rings* and TV shows like *Game of Thrones*, and in the last decade or so the country has become an important tourist destination.

Region Mediterranean Europe

As Figure 5-15 reveals, the northern sectors of two major countries of southern Europe—Italy and Spain—form parts of the European Core. Portugal, on the western flank of the Iberian Peninsula, remains outside the Core, far less urbanized, much more agrarian, and not strongly integrated with Western Europe. The remaining Mediterranean countries include Greece, the island-states of Cyprus and Malta, Croatia, and Slovenia, of which only the last can be considered part of the European Core in recognition of its rapid economic development. Slovenia's recent upward trajectory is similar to that of the Czech Republic (and, as we shall presently see, Estonia).

Spain and Portugal

At the western end of Southern Europe lies the broad Iberian Peninsula, separated from France and Western Europe by the rugged mountain wall of the Pyrenees and from North Africa by the narrow Strait of Gibraltar. Spain (population: 47 million) occupies most of this compact Mediterranean landmass, and peripheral Portugal lies in its southwestern corner.

Spain followed the leads of Germany and France and decentralized its administrative structure as a concession to devolutionary pressures. These pressures were especially strong in Catalonia, the Basque Country, and Galicia; in response, the Madrid government created a subnational regional framework of 17 so-called *Autonomous Communities* (ACs) (**Fig. 5-22**). Even so, the new AC mosaic has not defused any secessionist-minded movements. Relations between Madrid and the Basque Country remain tense, and in Catalonia the drive toward independence is strengthening as the Spanish economy struggles while Catalonia continues to forge ahead. In late 2014, the Catalan AC government held a referendum in which, reportedly, about 80 percent of the population voted in favor of outright independence from Spain; however, the Spanish High Court swiftly voided this vote, declaring it to be in violation of Spain's constitution.

Centered on the dynamic, productive coastal city of Barcelona, the Catalonia AC is not only Spain's leading industrial agglomeration but has also become one of Europe's *Four Motors*. Catalonia is endowed with its own distinctive language and culture that find vivid expression in Barcelona's vibrant urban landscape. In recent years, Catalonia—with 6 percent of Spain's territory and 16 percent of its population—has annually produced at least a quarter of all Spanish exports and more than one-third of its industrial exports.

As Figure 5-22 shows, Spain's capital and largest city, Madrid, lies near the geographic center of the state. It also lies astride an economic–geographic transition zone. In terms of annual GDP per capita, the Spanish north is far more affluent than the south, a reflection of the country's inequitable distribution of resources, development opportunities, and environmental assets (the south suffers from drought as well as inferior soils). The most prosperous ACs, apart from Madrid, are in the northeastern tier closest to the French border.

Economically, Spain has been going through a difficult stretch. Economic decline has plagued most of the country since 2008, when it was hit hard by recession and a housing market crash. By the mid-2010s, a modest recovery took hold. In early 2019, overall unemployment was among the highest in Europe, at 14 percent; youth unemployment hovered around 32 percent.

Portugal (10.2 million) occupies the southwestern corner of the Iberian Peninsula. Because EU agricultural policies entail transfers from rich to poor areas in the Union and because the EU funds major infrastructural projects in more isolated areas, Portugal at first benefited substantially from its 1986 admission to the EU (in more recent years, most of these subsidies went to the new members in the East). Nevertheless, it still remains far behind Europe's leading economies. Unlike Spain, which

FIGURE 5-22

© J. Nijman, P. O. Muller, and John Wiley & Sons, Inc.

has major population clusters on its interior plateau as well as its coastal lowlands, the Portuguese are mainly concentrated within their Atlantic coastal zone. The country has several excellent natural harbors, including those of primate city Lisbon and Sines (Fig. 5-22), with considerable potential to attract a larger share of container shipping (especially for transfer onto freight trains bound for Spain), but these ports need to become more efficient and competitive. The best farmlands lie in the moister western and northern zones of the country; but farms here are small and inefficient, and even though Portugal remains dominantly rural, it must import as much as half of its food. In recent years, Portugal has seen growing tourism and residential real estate investment from more prosperous EU countries.

Italy

Centrally located within Mediterranean Europe, most populous of the realm's southern states, best connected to the European Core, and economically most advanced is Italy (population: 60 million), a charter member of what is now the EU. The northern half of Italy stands in strong social, economic, and political contrast to such southern regions as Calabria (the "toe" of the Italian "boot") and Italy's two major Mediterranean islands, Sicily and Sardinia. Not surprisingly, Italy is often described as two countries—a progressive north and a stagnant south (known as the *Mezzogiorno*). The urbanized, industrialized north is very much a part of the European Core.

North and south are bound by the ancient (and current) capital, Rome (photo below), which lies within the narrow transition zone between Italy's contrasting halves. This clear manifestation of Europe's core–periphery distinction is referred to in Italy as the *Ancona Line*, named after the city on the Adriatic coast where it reaches the eastern side of the peninsula (**Fig. 5-23,** blue line). Whereas Rome remains Italy's governmental and cultural focus, the functional core area of the country has shifted northward into Lombardy in the basin of the Po River. Here lies Mediterranean Europe's leading industrial complex, the Milan–Turin–Genoa triangle. Metropolitan Milan itself personifies the new, modern Italy—not only as its primary manufacturing center (making Lombardy one of Europe's *Four Motors*) but also as the headquarters of the country's financial and service industries. Although the Milan area contains only 9 percent of Italy's population, it now accounts for more than one-third of the entire country's national income.

The inset in Figure 5-23 shows the results of Italy's general election in 2018 for the three major coalitions. As usual, the country's north–south divide is expressed in its electoral geography. The Center-Right coalition, which includes the Northern League party and is pro-EU, receives most of its support in the economically advanced areas of the north; the Five Star Movement is mostly supported by lower-income classes in southern Italy and many of its voters are anti-EU; the Center-Left coalition is a more old-fashioned labor/socialist party, and it maintained some of its support in select areas in the north. It was a tight race, with the Five Star Movement winning most seats in parliament—but not enough to form a government, so that a number of parties that made up the Center-Right had to be engaged to form a coalition government.

The Italian economy also has been slow to recover from the global recession, and here too austerity measures have been a hard sell. Beyond these challenges, Italy has been burdened by the flow of undocumented migrants who illegally enter southern Italy from the northern coast of Africa. Between 2013 and 2015, these boat people, desperately seeking political and/or economic refuge in Europe, streamed by the thousands across the Mediterranean in the flimsiest of vessels toward Sicily, many of these clandestine voyages ending tragically. It is estimated that several thousand drowned. The disintegration of Libya, and subsequently Syria (see Chapter 7), caused an upsurge in this immigration flow and intensified the crisis. As noted earlier in this chapter, the EU made a deal with Turkey to stem the flow of refugees coming from the Middle East; a similar deal was brokered with Libya, to turn the tide of migrants coming from Africa.

Slovenia and Croatia

These two countries, both lying in close proximity to Italy (Fig. 5-22), were the most progressive to emerge from the 1990s disintegration of former Yugoslavia. **Slovenia**, wedged against Austria and Italy in the hilly terrain near the head of the Adriatic Sea, has just over 2 million inhabitants, a nearly homogeneous ethnic complexion, and a productive economy. Slovenia joined the European Union in 2004 and entered the Euro zone three years later. **Croatia** (population: 4.1 million), an EU member since 2013, has a long coastline on the Adriatic that has recently become a major European tourist destination. In the long and bitter aftermath of the collapse of Yugoslavia, relations with eastern neighbor Serbia are still strained, and the presence of a Croatian minority of about 600,000 in neighboring Bosnia continues to stir a range of political sensitivities. Both Slovenia and Croatia joined NATO in 2009.

© J. Nijman

A view down the Via dei Fori Imperiali, one of the Rome's main avenues. It passes by several famous sites of ancient Rome, including the Colosseum (in the background), and was constructed under the fascist government of Mussolini in the late 1920s as part of his glorification of the Roman Empire.

FIGURE 5-23

Greece, Cyprus, and Malta

Greece is the birthplace of European civilization, a major historic presence in the Mediterranean region, and a longtime member of the EU and NATO. Geographically, it is also relatively far removed from the European Core. Seeming to dangle from the southern end of Eastern Europe, Greece was the wellspring of one of the ancient world's greatest civilizations, its scientists and philosophers still cited to this day, its famous tragedies still staged in the amphitheaters built more than 2000 years ago. Its familiar peninsulas and islands are bounded by Turkey to the east and by Bulgaria, North Macedonia, and Albania to the north

(Fig. 5-15). Note that some of Greece's islands in the Aegean Sea lie on the very doorstep of Turkey; in addition, Greeks represent the majority of the population of Cyprus in the remote northeastern corner of the Mediterranean Sea.

But EU membership had other, far less favorable consequences. When the Union expanded to include poorer states such as Bulgaria and Romania, Greece lost much of its EU subsidy as Brussels had to divert its equalization funds to needier members. Then the global financial crisis hit Greece especially hard in the late 2000s, and it became increasingly difficult for the government to continue to pay for such personal frills as guaranteed lifetime employment in the public sector and generous (early) retirement, and to reward businesses with protection against competition and the loose enforcement of tax regulations. Suddenly, this proud and historic nation of 10.5 million found itself at the center of an unfolding economic and social calamity as the overmatched Greek government—mired in mountainous debt and facing massive street riots that greeted every effort at urgently needed reform—was forced to conform to the painful bailout terms imposed by the EU agencies providing the indispensable loans. By 2019, Greece was doing somewhat better, especially Athens, indicative of an urban–rural divide that seems to be widening across Europe and elsewhere in the world (elaborated in Chapter 2).

This southeastern corner of Europe also contains the island country of **Cyprus**, an EU member since 2004, but one whose fractious political geography continues to cause problems for the Union. As Figure 5-15 shows, Cyprus lies closer to Turkey than it does to any part of Europe, and for more than three centuries it was ruled by the Turkish Ottoman Empire. But today's population of 1.2 million is predominantly Greek, the people who have been there the longest. In 1983, the 40 percent of Cyprus under Turkish control, with about 100,000 inhabitants (plus 30,000 Turkish soldiers), declared itself the independent *Turkish Republic of Northern Cyprus*. Only Turkey recognizes this ministate (which today contains a population of about 330,000), whereas the international community recognizes the government on the Greek side as legitimate. Things got even more complicated when only the Greek side of the island was admitted to the EU in 2004. Resentment was high on both the Turkish side and in Turkey itself, a reminder that Cyprus's "Green Line" separating its Greek and Turkish communities constitutes not just a regional border but a boundary between neighboring geographic realms.

The final component of Mediterranean Europe is the ministate of **Malta**, a cluster of five islands located just to the south of Sicily (Fig. 5-23). Although Malta contains merely 450,000 inhabitants, its history is that of a crossroads at the very center of the Mediterranean Sea, a cultural nexus that received infusions from Britain (as a former dependency), Italy, North Africa, and Southwest Asia. As an EU member since 2004, Malta enjoys a comparatively high standard of living reinforced by a thriving tourist industry.

Region Eastern Europe

Europe has changed fundamentally in the three decades since the Cold War ended. The old Eastern Europe, its western edge sharply defined by the Iron Curtain, no longer exists, and to a certain extent we have seen the revival of an older cultural region known as Central Europe or, in its prevailing German terminology, *Mitteleuropa*. Many of these changes came about as NATO and the European Union expanded eastward. Several Eastern European states joined the EU, and the Czech Republic and Slovenia have already formed part of the European Core. More are knocking on the EU door. Some

are unlikely to be admitted because of their very poor economic circumstances (e.g., Albania and Moldova), whereas others—especially Ukraine—would pose serious geopolitical challenges.

The region defined as Eastern Europe covers roughly half of the realm's territory and encompasses no fewer than 17 countries. Our coverage is organized within the framework of four subregions: the Balkans, East-Central Europe, the Baltic States, and the Eastern Periphery.

The Balkans

This subregion of Eastern Europe **(Fig. 5-24)** was deeply impacted by the violent dismantling of communist Yugoslavia during the final decade of the twentieth century. It was the latest chapter in a particularly volatile history as the Balkans constitute a classic example of a **shatter belt [36]**—a zone of persistent political splintering and fracturing. Geographers employ many terms to describe the breakup of an established order, and one of the best is **balkanization [37]**—the recurrent division and fragmentation of a region. That term is rooted in this part of Europe, derived from the east–west trending Balkan Mountains that stretch across Bulgaria. Today, none of the Balkan subregion's countries can be considered part of the European Core, and none are EU members.

The key state on the current map is **Serbia**, the name of what is left of a much larger domain once ruled by the Serbs, who were dominant in the former Yugoslavia. Centered on the historic capital of Belgrade on the Danube River, Serbia (with a population of 8.8 million) is the largest and potentially most important new country in the Balkan subregion. But Serbia faces some tough challenges. First, more than one million Serbs live in neighboring Bosnia and Herzegovina, where they have an uneasy relationship with the local Muslims and Croats. Second, the coastal province named **Montenegro** broke away in 2006, when voters there decided to convert it into an independent state. Third, the Muslim-majority, southernmost province of **Kosovo** declared independence in 2008, its sovereignty immediately recognized by the United States and a majority of (but not all) European governments. And fourth, Serbia still incorporates a Hungarian minority of more than 250,000 in its northern province of *Vojvodina*. Clearly, the Balkans contain a virtual kaleidoscope of ethnic identities that severely complicates any kind of orderly territorial arrangement.

When former Yugoslavia collapsed, **Bosnia and Herzegovina** (*Bosnia* in our shorthand form) was the cauldron of calamity. No ethnic group was numerically dominant here, and this multicultural, effectively landlocked triangle of territory, lying between the Serbian stronghold to the east and the Croatian republic to the west and north, fell victim to disastrous conflict among Serbs, Croats, and *Bosniaks* (now the official name for Bosnia's Muslims, who comprise just over half of the population of 3.3 million). At least a quarter of a million civilians perished in concentration camps associated

FIGURE 5-24

with *ethnic cleansing* practices. In 1995, the U.S. brokered a truce—the Dayton Accords—that partitioned the country (Fig. 5-24). In 2015, Bosnia reached an initial understanding with the EU as to a future candidacy for EU admission.

The southernmost "republic" of the former Yugoslavia is *Macedonia*, renamed **North Macedonia** in 2019. About two-thirds of the population of two million are Macedonian Slavs. As the map shows, North Macedonia adjoins Muslim Albania and Kosovo, and its northwestern corner is home to the other third who are Muslims. The remainder of this culturally diverse population are Turks, Serbs, and Roma (Gypsies). North Macedonia is one of Europe's poorest countries, landlocked and powerless. Why the name change? One of Greece's northern provinces is also named Macedonia, and it figures prominently in ancient Greek history. For that reason, Greece objected from the start to the naming of the new country of Macedonia after Yugoslavia fell apart. These protests were in vain until Macedonia wanted to join the EU and NATO, because Greece, as an established member of both, blocked Macedonia's admission until it would change its name.

The only other dominantly Muslim country in Europe is **Albania**, where 60 percent of the population of 2.9 million are adherents of Islam. Albania shares with one other country—Moldova—the status of being Europe's poorest state. It has one of Europe's higher rates of natural population growth, and

many Albanians try to emigrate to the EU by crossing the Adriatic Sea to Italy. Most Albanians subsist on livestock herding and farming. They also are eager to join the EU—which is not likely in the foreseeable future.

East-Central Europe

As Figure 5-15 and **Figure 5-25** show, the three states of this subregion—Poland, Hungary, and Slovakia—all adjoin or are partially integrated into the European Core. All are EU and NATO members, and Slovakia has also joined the Euro zone. Considering that these countries were tightly bound into the Soviet bloc before its collapse in 1991, these changes have been truly remarkable.

The largest and most populous of these countries is **Poland**. With 38 million people and lying between two historic enemies, Poland's borders have shifted time and again; but its current, post–1945 reconfiguration appears to be more durable than previous territorial layouts. As the map shows, the historic and once-central capital of Warsaw now lies closer to Russia in the east than to Germany in the west, but the country today has decidedly shifted its gaze westward. After entering the EU in 2004, Poland saw hundreds of thousands of workers depart for jobs in the European Core. But many have since returned as

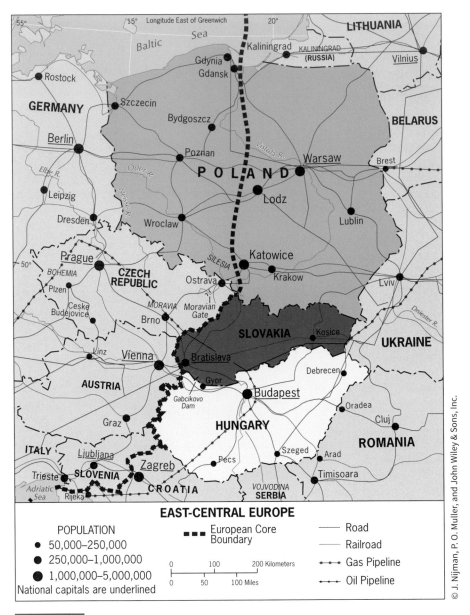

FIGURE 5-25

the national economy grew robustly during the 2010s (Poland may now be in its best shape ever), and economic relations with Germany are at their closest in history. Overall, Poland has benefited substantially from joining the EU, and the country's westernmost provinces, which border and are more closely tied to Germany, are much better off than those in the east. As we noted in the section on Baltic geopolitics, Poland provides an important corridor overland to the Baltic states (the Suwalki Gap), and as such plays an important part in NATO. Politically, Poland has been on the receiving end of criticism from established EU members about its authoritarian penchant and its defiance of common EU agreements regarding the shared admission of refugees.

Hungary's post-communist economic transition went surprisingly smoothly. By the mid-1990s, most government-owned companies had been privatized, and Hungary was attracting significant foreign investment. It joined NATO in 1997 and the

EU in 2004. But soon thereafter its situation began to deteriorate as the Hungarian economy was hit especially hard by the global recession of the late 2000s. Politically, heightening discord has centered on Hungary's persistent problem of **irredentism [38]**, which refers to a government's open support for fellow ethnics residing in nearby countries. Ethnic Magyars (Hungarians) are found in relatively large numbers in parts of neighboring Romania, Slovakia, and Serbia (**Fig. 5-26**). As for domestic politics, Hungary has recently taken a turn to the far right, unleashing both nationalistic and anti-Semitic tendencies that have been roundly criticized by EU leaders. In early 2019, some Western observers were questioning if Hungary was "going rogue."

Slovakia (5.5 million people), sitting north of Hungary and sharing a long border, has followed a trajectory similar to that of Hungary: a relatively smooth post-communist transition and rapid accession to the EU, followed by a severe

FIGURE 5-26

economic downturn and rising political controversies that also drew rebukes from Brussels. In Slovakia, those controversies revolved around the alleged poor treatment of ethnic Germans, Hungarians, Roma (formerly known as Gypsies), and LGBTQ populations.

The Baltic States

Estonia, Latvia, and Lithuania are positioned at the northeastern extremity of Eastern Europe (Figs. 5-15, 5-21). They could be grouped with the Northern European region or just with

Finland, around the Baltic Sea. In our regional scheme of the European realm, we consider the Baltic States a subregion of Eastern Europe because their historical geography is so closely tied to Russia, with which they share a border and a common recent history as Soviet Republics. The geopolitical sensitivities regarding Russia are palpable here, yet the pro-European orientation is crystal clear. All three countries have joined the EU and NATO. See the section on Baltic geopolitics (pages 155-156) for an overview of the subregional political context of these countries.

Estonia is the northernmost of the three Baltic states, and it has longstanding ethnic and linguistic ties to Finland.

During the period of Soviet control from 1940 to 1991, Estonia's demographic structure transformed drastically: today about a quarter of its 1.3 million inhabitants are Russians, most of whom came here as colonizers. After an initially difficult period of adjustment, Estonia has forged ahead of its Baltic neighbors, Latvia and Lithuania. Busy traffic links Tallinn, the capital, with its nearby counterpart in Finland, Helsinki, and a free-trade zone at Muuga Harbor facilitates commerce with Russia. But more important for this country's future was its entry into the European Union in 2004. Since then, Estonia has gained attention as a result of its economic advances, particularly its creative entrepreneurship in the high-tech and software industries (Skype, for instance, was originally an Estonian company). Having entered the Euro zone in 2011, Estonia has developed an economy that has remained vigorous, and we classify it here as a stable component of the European Core.

As Figure 5-21 reminds us, the boundary of the European Core that traverses Europe's north includes southern Norway, Sweden, and Estonia but excludes the other two Baltic states. **Latvia**, the middle Baltic state centered on the port and capital city of Riga, was tightly integrated into the Soviet system throughout Moscow's long domination. Here, as in Estonia, about one-quarter of its population of 1.9 million is ethnic Russian. Consider this: 25 years ago, virtually all of Latvia's trade was with the Soviet Union. Today its principal trading partners are Germany, the United Kingdom, and Sweden. In 2014, Latvia once again asserted its European reorientation by joining the Euro zone.

Lithuania is the southernmost Baltic state, with a population of 2.8 million and a residual Russian minority of only about 9 percent. But relations with this gigantic neighbor are tense—despite a continuing dependence on Russia as a trading partner. One reason for this strained relationship has to do with neighboring *Kaliningrad*, Russia's **exclave [39]** facing the Baltic Sea (Fig. 5-21). When Kaliningrad became a Russian territory after World War II (Russia still regards this exclave as an important outpost), Lithuania was left with only about 80 kilometers (50 mi) of Baltic coastline and a small port that was not even connected by rail to the interior capital of Vilnius. Significantly, Lithuania joined NATO in 2004, and a year later it called for the demilitarization of Kaliningrad as a matter of national security. Russia did not oblige, nor will it in the future. Of vital importance to Lithuania is the shared border with fellow NATO member Poland. Following in the footsteps of Estonia and Latvia, Lithuania joined the Euro zone in 2015.

The Eastern Periphery

This subregion consists of the five remaining countries of Eastern Europe: Romania, Bulgaria, Ukraine, Moldova, and Belarus (**Fig. 5-27**). The first two joined NATO in 2004 and were admitted to the EU three years later. For the other three, such membership is not even on the horizon. Ukraine and Belarus find themselves on Russia's doorstep, and separatist-minded eastern Ukraine (with

its substantial population of more than 8 million ethnic Russians) descended into a *de facto* state of civil war in 2014.

Just how **Romania** managed to persuade EU leaders to endorse its 2007 accession still remains a mystery to many (Western) Europeans. Its economy is weak, its incomes are low, its governance has not been sufficiently upgraded from communist times, and political infighting and corruption are endemic. But this is a strategically important country, located in the lower basin of the Danube River and occupying much of the heart of Eastern Europe. And like a number of other countries in this region, it was incorporated into NATO before it was allowed to join the EU. With 19 million inhabitants and a pivotal location on the Black Sea, Romania is a bridge between Central Europe and the realm's southeastern corner, where EU member Greece faces its historical adversary, Turkey, across both land and water (Fig. 5-27).

Across the Danube River lies Romania's southern neighbor, **Bulgaria**. The rugged Balkan Mountains form Bulgaria's physiographic backbone, separating the Danube and Maritsa basins. As Figure 5-27 shows, Bulgaria has five neighbors, several of which are in political turmoil. In 2007, Bulgaria also became an EU member, once required judicial and other social reforms were finalized. But today its 6.9 million citizens constitute one of the Union's poorest countries, and development prospects are not encouraging. Bulgaria fronts the Black Sea and has a major port at Varna, but the main advantage it derives from its seaside location is the tourism its beaches generate. The capital, Sofia, lies near the opposite, western border with Serbia, and in its core-area hinterland modest foreign investment is slowly modernizing parts of the economic landscape. A substantial number of Bulgarians have used their newfound freedom of movement in the EU to seek out opportunities in Europe's core. According to the UN, Bulgaria has the world's fastest-shrinking population, due to outmigration and declining birthrates. It does not help that most of those who move away are the young and the educated.

Territorially, **Ukraine** is the realm's largest country. Demographically, however, its population of 44 million only qualifies Ukraine for second-tier ranking among European states, on a par with Poland and Spain. And, like Bulgaria, it is shrinking. As Figure 5-27 underscores, Ukraine's relative location is crucial. Not only does it link the core of Russia to the periphery of the European Union: it also forms the northern shore of the Black Sea from Russia to Romania, in the center of which lies the strategically prominent Crimea Peninsula. Perhaps most important of all, oil and gas pipelines connect Russian fields to European markets across Ukrainian territory. Once a land of farmers tilling its famously fertile soils, Ukraine emerged from the Soviet era with the huge (and decaying) Donbas heavy-industrial complex in its east as well as with a substantial (18 percent) Russian minority population. Most of the Russians are concentrated in the zone that stretches eastward from the Dnieper Valley (Fig. 5-27)—which is why that part of the country is mapped as a transition zone in Figure 5-15.

The vital Dnieper River forms a valuable geographic reference for comprehending Ukraine's division that creates

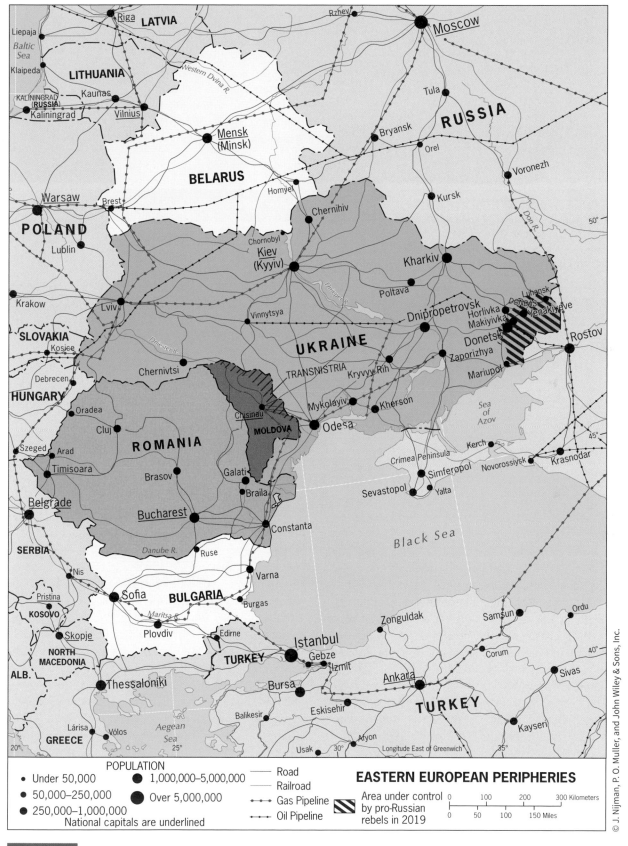

POPULATION

- Under 50,000
- 50,000–250,000
- 250,000–1,000,000
- 1,000,000–5,000,000
- Over 5,000,000

National capitals are underlined

— Road
— Railroad
•–•–• Gas Pipeline
•+–•+ Oil Pipeline

EASTERN EUROPEAN PERIPHERIES

Area under control by pro-Russian rebels in 2019

0   100   200   300 Kilometers
0   50   100   150 Miles

© J. Nijman, P. O. Muller, and John Wiley & Sons, Inc.

FIGURE 5-27

the transition zone (Fig. 5-27). To its west lies agrarian, rural, mainly Roman Catholic Ukraine; in the Dnieper's great southern bend and eastward lies industrial, urban, Russified (and Russian Orthodox) Ukraine. For the most part, Ukraine's electorate is divided between a pro-Western (and pro-EU) north and west and a pro-Russian south and east (see **Box 5-10**).

Violent protests broke out in Kiev (Kyyiv), the Ukrainian capital, in February 2014 over major political disagreements concerning relations with Russia and possible membership in the European Union. Pro-Russian groups inside Ukraine raised their profile by protesting against EU membership. The Crimea was swiftly annexed by Russia in March 2014 amid vociferous objections from the Kiev government and most of the international community. Within days, the Russian border zone of eastern Ukraine was engulfed by an escalating conflict that pitted Ukrainian government forces against those of the separatists, now aided directly by Russian military power. As of early 2020, much of eastern Ukraine is controlled by Russian rebels, with both open and clandestine support from Moscow (Fig. 5-27, Box 5-10, Fig. 5-28).

Moldova, Ukraine's minuscule, impoverished southwestern neighbor, was a Romanian province seized by the Soviets in 1940 and converted into a landlocked Soviet Socialist Republic. A half-century later, along with the USSR's 14 other "republics,"

Box 5-10 Geographic Information Analysis

Ukraine: A Linguistic and Economic Transition Zone

Maps are quite helpful in demarcating and analyzing regional transition zones. **Figure 5-28** places the ongoing Ukrainian political crisis in spatial context. For each oblast (province), excluding Crimea, it shows GRP per capita as well as the percentage of Russian speakers. Carefully examine the spatial distribution of both variables, particularly their east–west gradients. Then closely track the Russian–Ukrainian border and consult the relevant maps in this chapter to discover which of Ukraine's neighboring countries are members of the EU and/or NATO. What are the key geographic factors that explain why EU membership is so strongly supported in the western half of Ukraine?

Access your WileyPLUS course to interact with a dynamic version of this map and to engage with online exercises and questions.

Russian language speakers (%)

80
50
20

Gross regional product (GRP) per capita, 2017 (US dollars)

No Data | Below $1,500 | $1,500 - $2,000 | $2,000 - $3,000 | > $3,000

FIGURE 5-28 **Interactive**

© J. Nijman, M. Shin, P.O. Muller and John Wiley & Sons, Inc.

Moldova gained independence when the Soviet Union disintegrated. Romanians remain in the majority among its 4.1 million (and shrinking) population, but many of the Russians and Ukrainians (each about 6 percent) have migrated across the Dniester River to an industrialized strip of land between that waterway and the Ukrainian border, proclaiming there a Republic of Transnistria (see Fig. 5-27). This separatist movement constitutes only one of Moldova's multiple problems. Its economy, dominated by farming, is in decline; an estimated 40 percent of the population works at jobs outside its borders because unemployment rates in Moldova typically hover around 30 percent; smuggling and illegal arms trafficking are rife. These many misfortunes translate into weakness, and Russia's influence in the form of support for Transnistria's separatists keeps the country in turmoil.

Belarus is in many ways the most peripheral European country of all. Landlocked, autocratic, isolated from Western Europe (in part by choice), and sustained in no small measure by the transit of energy supplies from Russia to countries of the European Core, Belarus itself has few functional ties with the rest of the realm and little prospect of progress. Nearly 85 percent of Belarus's 9.5 million inhabitants are Belarussians ("White" Russians), a West Slavonic people; only about 8 percent are (East Slavonic) Russians. The economy of Belarus is almost that of a bygone era: no less than four-fifths of the workforce is employed by the state, which also allocates most housing. Except for a small contingent of oligarchs, private property does not exist, and neither do labor unions nor chambers of commerce. This is a society stuck in the old Soviet mode in a realm that continues to undergo profound change. In effect, Belarus trades energy subsidies for serving as a stable and willing buffer against further European encroachment on Russia's shrinking western doorstep.

Europe's Future Prospects

The future of the European realm, involving both its internal dynamics and external relations, has become ever more tightly intertwined with integration—in the European Union, but also in the security structure of NATO. The EU and NATO today have penetrated deeply into Eastern Europe, encompass a growing number of members, and have facilitated economic progress as well as political stability as they have evolved over the past six decades. This amounts to a tremendous achievement. The EU now has a combined population of more than 500 million that constitutes one of the world's richest markets.

But Europe is under significant stress. First, internally, the EU has to balance its policies and priorities among an ever more diverse and disparate number of member-states. It must also demonstrate the value and legitimize the costs of integration to the European electorates. The problem of the so-called *democratic deficit* is a serious obstacle: from Germany to Greece, citizens often feel that they are not being heard, yet national governments must have their votes to support the Union. The Brexit saga illustrates as much. Second, Europe in recent years has been facing challenges of immigration and refugee flows. This aging realm surely needs migrants, in the long term, to remain demographically stable, but most countries have a culture that needs to adapt to the challenges of multiculturalism. Third, relations with Russia will continue to be critical to the stability of this realm, but they are far from cordial and are unpredictable. The expansion of NATO onto Russia's doorstep has brought in a number of countries that badly needed security and stability, but it has only compounded the difficult relations with the mighty neighbor to the east. We turn to the Russian realm next.

Points to Ponder

- Per capita income in Norway or Luxembourg is nearly twice as high as in the United States.

- With the UK intent on leaving the EU, there is a good probability that Scotland will leave the UK.

- Brussels is the main seat of the European Union, yet it is the capital of a country faced with strong divisive forces that at times threaten its national unity.

- In Northern European countries, the south is more developed than the north; in Southern European countries, the north is more developed than the south.

- In Europe, nearly half of the population say they do not believe in God, compared to 10 percent in the United States.

FIGURE 6-1

In Central Asia, hunting with eagles has a long tradition but it is said to be a dying art. This photo was taken in January in the snowy steppes of eastern Kazakhstan, about 150 miles northeast of the city of Almaty, during the yearly National Competition of "Berkutchi" (Golden-Eagle Hunters). Kazakhs are renowned for their extraordinary skills in using these mighty birds to hunt small prey. Eagles are captured as chicks from their nests in the wild. The training takes three or four years and has to be done by just one person in order to develop the necessary bond between master and eagle.

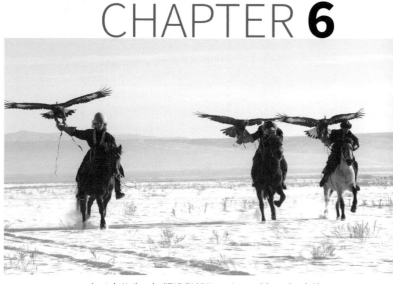

Anatoly Ustinenko/ITAR-TASS News Agency/Alamy Stock Photo

The Russian/Central Asian Realm

IN THIS CHAPTER

- Military buildup in the Arctic
- Shrinking populations in eastern Russia
- Siberia's thawing permafrost
- Russia in Ukraine
- A New Silk Road across Central Asia
- Astana: Kazakhstan's shining new capital

CONCEPTS, IDEAS, AND TERMS

Monocentric geographic realm	**[1]**	Satellite state	**[14]**
Exclave	**[2]**	Near Abroad	**[15]**
Irredentism	**[3]**	NATO	**[16]**
Geopolitics	**[4]**	Distance decay	**[17]**
Continentality	**[5]**	Unitary state system	**[18]**
Permafrost	**[6]**	Population implosion	**[19]**
Tundra	**[7]**	BRICs	**[20]**
Taiga	**[8]**	Centrality	**[21]**
Northeast Passage	**[9]**	Shi'ite Islam	**[22]**
Forward capital	**[10]**	Sunni Islam	**[23]**
Federation	**[11]**	Belt and Road Initiative (BRI)	**[24]**
Russification	**[12]**	Eurasian Customs Union	**[25]**
Command economy	**[13]**		

Every geographic realm has a dominant, distinguishing feature: Europe's jigsaw-puzzle map of 40 countries, South America's familiar triangle, Southeast Asia's thousands of islands and peninsulas. The Russian/Central Asian realm stands out because of its vast areal extent, particularly its enormous, nine-time-zone-wide east–west dimension. The country of Russia by itself is twice the territorial size of Canada, the world's second-largest country, and three times as big as neighboring Europe. Not surprisingly, Russia dominates this part of the world, making this a classic example of a **monocentric geographic realm** [1]. But not all of this realm is Russian: it also includes the regions of Central Asia and Transcaucasia, which now consist of independent states that were once part of Czarist Russia and its successor Soviet Union.

Defining the Realm

As **Figure 6-1** reveals and as a globe shows even better, Russia's entire northern coast—from its border with Norway in the west to the Bering Strait across from Alaska in the east—faces the Arctic Ocean, with all the environmental consequences you would expect. And for all of its stupendous dimensions, much of this realm lies at higher latitudes, shut off by mountains and deserts from the warmer and moister parts of Eurasia. That has posed a geographic challenge for Russian rulers and governments for as long as Russia has existed as a state. Russia's ports on the Pacific Ocean in the east lie about as far from where most Russians live as they could be. And in the west, where Russia's core area is situated, every maritime outlet is restricted in some way. The Arctic ports operate on seasonal access. The Baltic ports face a shallow sea far from the open ocean. The Black Sea ports require maneuvering through narrow straits across Turkish territory—simply to reach the eastern Mediterranean Sea.

The Russia/Central Asia realm may be gigantic in size, but its human numbers (that total about 231 million) are modest and not much larger than those of such individual countries as Brazil and Pakistan. Moreover, much of the realm's population has been declining since the 1991 collapse of the Soviet Union in response to changed political, economic, and social conditions that even now continue to produce negative demographic effects.

This realm, as defined by our global framework, consists of nine political entities: the giant Russian state itself, three small nations in Transcaucasia, and five countries in Central Asia. Transcaucasia, the mountainous zone between the Black and Caspian seas, contains Georgia, Armenia, and Azerbaijan. Population-wise, all are small in size, with a combined population of 17 million (7.5 percent of the realm's total); they also lie in a historically turbulent arena of conflict between Russian and numerous non-Russian peoples. The five Central Asian countries are Kazakhstan, Uzbekistan, Turkmenistan, Kyrgyzstan, and Tajikistan—sometimes referred to collectively as the "Stans" (although they do not include Afghanistan, which we assign to the South Asian realm in Chapter 9). In contrast to Transcaucasia, Central Asia covers a comparatively large area. Territorially, Kazakhstan is the biggest Central Asian country (about five times the size of France) and also has a lengthy border with Russia. Central Asia's overall population is about 75 million, accounting for just about one-third of the realm's total.

Box 6-1 Major Geographic Features of Russia/Central Asia

1. Russia/Central Asia is an enormous realm that is almost entirely lacking in warm-water access to the world ocean.

2. This is by far the "widest" geographic realm on Earth, stretching from the Baltic Sea in the west to the Pacific Ocean in the east for some 10,000 kilometers (6200 mi) and spanning no less than nine time zones.

3. Much of Russia/Central Asia is cold and/or dry. Extensive rugged mountain zones separate the realm from warmer subtropical air, and most of it lies open to Arctic air masses.

4. Global climate change is having a massive impact on this realm, in different ways: from the opening up of navigable waters in the Arctic Sea to the melting of permafrost, the latter posing significant ecological and economic challenges.

5. Russia/Central Asia borders four other realms: Europe, North Africa/Southwest Asia, South Asia, and East Asia. These boundaries have shifted throughout history, and today some of them must be mapped as transition zones.

6. For so large a territory, the population size of Russia/Central Asia (just over 230 million) is relatively small. In Russia itself, the population has been declining in recent decades, especially in the country's east and in most of the rural areas.

7. This realm's core area is located in western Russia. New development is now mostly concentrated to the west of the Ural Mountains; here are most of the major cities, leading industrial complexes, densest transport networks, and most productive farming areas. Outside Russia, certain parts of Central Asia (most notably Kazakhstan) have experienced rapid development based on their rich natural resources.

8. For a monocentric realm, this one is decidedly multicultural and has a complex political geography. Russian domination has long inflamed tensions within this ethnic mosaic, especially Muslim concentrations along the realm's southern periphery.

9. Central Asia is a region that highly interests China, for both its natural resources and its geopolitical role on China's western frontier. China's Belt and Road Initiative is transforming relations with this region.

10. The country's powerful position as a leading exporter of oil and natural gas has been a springboard in Russia's quest to reclaim its Soviet-era role as a major player on the international stage.

The political geography of this realm is complex: it contains myriad contested boundaries and territories; **exclaves [2]** from (Russian) Kaliningrad in northwest Europe to (Azerbaijani) Nakhchivan in southern Armenia; and a fair number of **irredentist [3]** claims that are at the root of intractable conflicts. The implosion of the Soviet Union in 1991 still resonates across many parts of the realm and beyond, especially near the Russian border. This is a realm, perhaps more than any other, where **geopolitics [4]** is at the forefront (**Box 6-1**).

What Transcaucasia and Central Asia have in common is that both were part of Czarist Russia and the Soviet Union. Today they remain firmly within Russia's sphere of influence (the same could be said for parts of Eastern Europe, such as Belarus and eastern Ukraine). It is also worth noting that these southern peripheral components of the realm display characteristics of a *transition zone*: Transcaucasia is positioned in a historically volatile corridor between Russia and Southwest Asia, whereas the "Stans" of Central Asia form a broad arena that has for centuries functioned as a crossroads linking Russia, Southwest Asia, South Asia, and East Asia.

Physical Geography of Russia/Central Asia

Physiographic Regions

One of the first features to observe on the realm's physiographic map is a prominent north–south trending mountain range in western Russia that extends from the Arctic Ocean southward to Kazakhstan (Fig. 6-1 and **Fig. 6-2**), dividing Russia into two unequal halves: the Russian Plain to the west and the subdivisions of Siberia to the east. This range, the *Ural Mountains* (numbered ① in Fig. 6-2), is sometimes designated as the eastern boundary of Europe. That is a mistake because Russia is not Europe: with both sides of the range equally Russian, the Urals cannot form a cultural boundary. Note, too, that the island of Novaya Zemlya

(Fig. 6-1) is the northern extension of the Urals, which at their opposite end also extend southward into Kazakhstan.

The *Russian Plain* ② to the west of the Ural Mountains is the eastward continuation of the North European Lowland, and here lies Russia's *core area*. Travel northward from the centrally located capital of Moscow, and the countryside soon changes to coniferous (needleleaf) forests like those of Canada. The Volga River, Russia's Mississippi, flows southward in a wide arc across this plain from the forested north—but unlike the Mississippi, the Volga drains into a landlocked lake, the Caspian "Sea." Travel southward from Moscow today, and the land is draped in grain fields and pastures. Eastward, the Urals are not high enough to impede surface transportation, but they retain their prominence because this range separates two vast lowlands. At the southern end of the Russian Plain, the *Caucasus Mountains* ③ rise like a wall between Russia and the lands beyond—a multi-tiered barrier over which Russians have fought with their southern neighbors for centuries.

The *West Siberian Plain* ④ to the east of the Urals is often described as the world's largest unbroken lowland, and here all the rivers flow northward. In fact, this plain is so flat that over the last 1600 kilometers (1000 mi) of its course to the Arctic Ocean, the Ob River (the lowland's largest) falls less than 90 meters (300 ft) in enormous meanders flanked by forests until, in the near-Arctic cold, the trees yield to mosses and lichens. Follow an imaginary trail eastward, and in just about the middle of Russia, at the Yenisey River, the West Siberian Plain gives way to the higher relief of the massive *Central Siberian Plateau* ⑤, one of the most sparsely populated areas in the habitable world.

To the south, beyond the Russia–Kazakhstan border, the West Siberian Plain transitions into the *Caspian–Aral Basin* ⑥. This physiographic region contains almost all of Central Asia and is dominated by windswept plains of moisture-starved steppes and desert. Poorly executed irrigation schemes along the rivers draining into the Aral Sea marked much of the Soviet era, and the price was paid after 1990 as an environmental disaster unfolded that has now resulted in the nearly complete drying up of the Aral Sea and the waterways that supply it. To the southeast, lining the long border with China and Mongolia,

Contrasting landscapes of the Russia/Central Asia realm. On the left is a coniferous forest in early autumn near where the Arctic Circle crosses the West Siberian Plain; to the right are the Tian Shan Mountains that parallel the Chinese border in northern Kyrgyzstan.

Natalya Onishchenko/Alamy Stock Photo

Speedskater/Deposit Photos

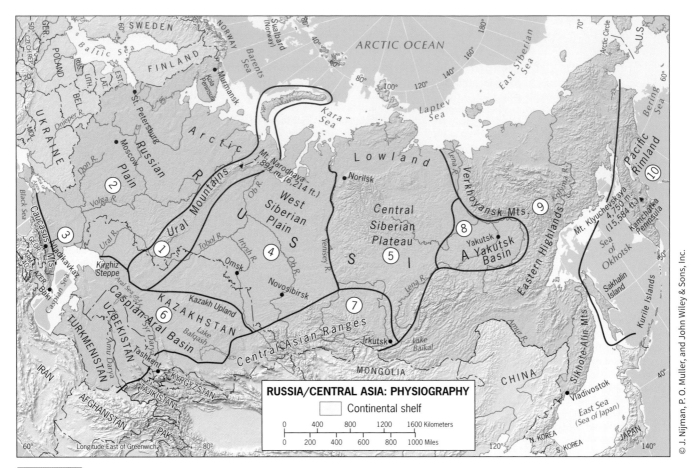

FIGURE 6-2

lie the *Central Asian Ranges* ⑦ that stretch northeastward from Tajikistan and Kyrgyzstan to Lake Baikal.

Continuing our eastward transect of Russia toward the Pacific, we first encounter the *Yakutsk Basin* ⑧ drained by the Lena River, and then we approach the realm's mountainous eastern perimeter (look again at Figs. 6-1 and 6-2), where the few roads and railroads must wend their way through tunnels and valleys full of hairpin turns. We map this jumble of ranges collectively as the *Eastern Highlands* ⑨, which blanket one of the most spectacular, diverse, and still-remote landscapes on the face of the Earth.

Finally, our survey of the realm's physiographic regions ends at the farthest extremity of Russia, the partially offshore *Pacific Rimland* ⑩ that consists of the huge Kamchatka Peninsula and the long, narrow island of Sakhalin to its southwest across the Sea of Okhotsk. Here, this immense geographic realm makes contact with the Pacific Ring of Fire (see Fig. 1-5). There are no erupting volcanoes in Siberia, and earthquakes are rare, but the Kamchatka Peninsula has plenty of both (see **Box 6-2**). This is one of the most volatile segments of the Pacific Ring of Fire, with more than 60 volcanoes, many of them active, forming the spine of the peninsula. Nothing here resembles Siberia (the climate is moderated by the Pacific waters offshore; the vegetation is mixed, not coniferous), and Kamchatka is one of the world's most difficult places to inhabit, effectively separated from the rest of Russia by the absence of overland connections. On the island of Sakhalin, earthquakes rather than volcanism are

the leading environmental hazard. Long a battleground between Russians and Japanese, Sakhalin has remained in Russian hands since the end of World War II. In subsequent decades, the island and its maritime surroundings proved to contain substantial reserves of oil and natural gas, making it a most valuable asset in Russia's energy-based export economy.

Climate and Vegetation

As the northernmost populous country on Earth, Russia has virtually no natural barriers against the onslaught of Arctic air masses. Moscow lies farther north than Edmonton, Canada, and St. Petersburg lies at the latitude of the southern tip of Greenland. Winters are long, dark, and bitterly cold in most of Russia; summers are short and growing seasons limited. Precipitation, even in western Russia, ranges from modest to minimal because the warm, moist air carried eastward across Europe from the North Atlantic Ocean loses both heat and moisture by the time it reaches Russia. **Figure 6-3** reveals the consequences. Russia's climatic **continentality [5]** (inland climatic environment remote from moderating and moistening maritime influences) is expressed by its prevailing *Dfb* and *Dfc* conditions. Compare the Russian map to that of North America (Fig. 1-6), and you will note that, except for a small corner off the Black Sea, Russia's climatic conditions resemble those of the U.S. Upper Midwest and interior Canada. Along its entire

Box 6-2 From the Field Notes . . .

"On a flight from Seattle to Kyoto (Japan), I get a perfect view of the rugged, mostly snow-covered east coast of Kamchatka. The peninsula is littered with volcanoes, as it is positioned on one of the most volatile segments of the 'Pacific Ring of Fire,' where tectonic plates meet (see Fig. 1-5 in Chapter 1). The photo shows a pair of volcanoes about half-way down the coast. It is a spectacular landscape and of growing interest among more adventurous travelers and some tourists. But it is barely inhabited (see Fig. 6-5 later in the chapter), and its remoteness stands in the way of commercial exploitation."

© J. Nijman

northern edge, Russia has a zone of **E** climates, the most frigid on the planet. In these Arctic latitudes originate the polar air masses that dominate its environments.

The Russian realm's harsh northern climates affect people, animals, plants—and even the soil. **Figure 6-4** shows the geographic extent of **permafrost** [6], wherein subsoils are permanently frozen underneath surface layers that thaw in summer. Its southern boundary starts at the shore of the Barents Sea, crosses Siberia eastward (while bulging to the south in central Siberia), and reaches the coast of the Pacific Ocean

RUSSIA/CENTRAL ASIA: CLIMATE

B DRY CLIMATE   **C HUMID TEMPERATE CLIMATE**   **D HUMID COLD CLIMATE**   **E COLD POLAR CLIMATE**   **H HIGHLAND CLIMATE**

| BS | Semiarid | Cf | No dry season | Df | No dry season | E | Tundra and ice | H | Unclassified highlands |

| BW | Arid | Cw | Dry winter | Dw | Dry winter |

h=hot
k=cold

Cs Dry summer

a=hot summer
b=cool summer

c=short, cool summer
d=very cold winter

Taiga (snowforest)

0   500   1000   1500 Kilometers
0   500   1000 Miles

© J. Nijman, P. O. Muller, and John Wiley & Sons, Inc.

FIGURE 6-3 **Interactive**

north of the neck of the Kamchatka Peninsula. Under conditions of 'continuous' permafrost (Fig. 6-4), water in the ground is permanently frozen, creating an even more formidable obstacle to settlement and infrastructure than the severe weather alone—but it is the thawing of permafrost due to global climate change that causes even greater problems (**Box 6-3**).

Examine Figure 6-3 carefully, and you will see two ecological terms: *tundra* and *taiga*. **Tundra [7]**, as the map suggests, refers to both climate and vegetation. The blue area mapped as *E* marks the coldest ice-affected environmental zone in Russia and elsewhere in the high Arctic: this is frigid, treeless, wind-swept, low-elevation terrain where bare ground and rock prevail and mosses, lichens, patches of low grass, and a few hardy shrubs are all that grow. The stippled area on the map is the **taiga [8]**—a Russian word meaning "snowforest" (also called boreal forest)—which extends across vast reaches of Eurasia as well as northern North America and is dominated by coniferous (as in pinecone) trees. As the map shows, the taiga extends southward into more moderate environments, and there it becomes a mixed forest of coniferous and deciduous trees. Although the taiga prevails from northern Scandinavia to the Russian Far East, this is the vegetative landscape most often associated with Siberia: endless expanses of rolling countryside draped in dense stands of evergreen pine trees (see left photo on page 187).

Climates and Peoples

Climate and weather have always challenged the peoples of this realm. The fierce winds that drive the bitter Arctic cold southward deep into Eurasia, the blizzards of Siberia, the aridity of Central Asia, the temperature extremes, rainfall variability, and short and unreliable growing seasons even in the more moderate western parts of the realm have always made farming difficult, to the point that grain often had to be imported into Russia.

As was pointed out in Chapter 1 and will be repeatedly observed throughout this book, humankind's long-term dependence on agriculture remains etched on the population map, even as our planet becomes ever more urbanized and its economy more globalized. By examining the climates of the Russian/Central Asian realm, we can begin to understand what the map of population distribution communicates (**Fig. 6-5**). The overwhelming majority of the realm's inhabitants remain concentrated in western and southwestern Russia, where environmental conditions were least difficult at a time when farming was the mainstay of most of the people. To the east of the Urals, the population is sparser and tends to cluster like beads on a string along the southern margin of the realm, becoming even more thinly distributed east of Lake Baikal. Central Asia's population distribution is very closely tied to local water supplies, which are most readily available in the hilly terrain that lines the rugged mountains of the southeast. And let us also keep in mind that nearly three-quarters of this realm's population now resides in cities and towns, so it is not difficult to imagine just how empty vast stretches of countryside must be—especially in those frigid northerly latitudes.

Climate Change and Arctic Prospects

As every map of this high-latitude realm shows, its northern coast lies entirely poleward of the Arctic Circle (latitude 66.5°N). Such a lengthy coastline on the Arctic Ocean is not exactly an advantage: most of that ocean's surface is frozen much of the

---

Box 6-3 Global Climate Change in Russia

The Perils of Thawing Permafrost

Permafrost refers to a permanently frozen layer of subsoil that is generally overlain by a layer of soil that thaws in summer. In the coldest areas of the Arctic, permafrost can reach hundreds of meters in depth and has been undisturbed for centuries. The southern boundary of the permafrost zone is not that easy to map because the summer melting season of the topsoil varies along with the depth of the permafrost. Figure 6-4 shows the extent of continuous and discontinuous (i.e., with seasonal melting down to certain depths) permafrost in the Northern Hemisphere. In most of these areas, the topsoil will thaw for a short period during the summer, but the duration and depth of thawing have both increased in recent times. Global warming is estimated to have caused a poleward shift of permafrost averaging 100 kilometers (65 mi) over the past three decades, both in Eurasia and North America. Although the somewhat milder temperatures may seem appealing for inhabitants, thawing permafrost in fact presents major challenges. First, the permafrost stores the carbon remains of plants and animals that froze before they could decompose; thus when it thaws, this carbon (carbon dioxide and methane, to be precise) is released into the atmosphere. It is estimated that the world's permafrost contains 1500 billion tons of carbon, nearly half the quantity that currently circulates in the atmosphere. In other words, melting permafrost is not just the result of global warming but could significantly accelerate it. Second, and of more immediate concern, is that melting permafrost causes major problems for surface infrastructure and construction. Permafrost, as long it remains frozen, provides a stable foundation for buildings, roads, railroads, pipelines, and the like. But when it melts and becomes soft and soggy, it destabilizes whatever is built on top of it. For the Russian government, this means constant repairs and redesign of high-latitude infrastructure, a very costly affair. For individual Russians, it affects the homes in which they live, the buildings where they work, and the roads in their locality (see photo). These mounting challenges aside, one perceived advantage of global warming in the permafrost zone is the heightened melting of sea ice in the Arctic Ocean. As this polar icecap recedes, waterways open up and widen along its surrounding shores, thereby enhancing opportunities for forging new long-distance shipping lanes as well as accessing fossil-fuel deposits and other rich mineral resources currently buried beneath the seafloor.

PERMAFROST IN THE NORTHERN HEMISPHERE

- Continuous permafrost
- Discontinuous permafrost

0   500   1000   1500 Kilometers
0   500   1000 Miles

ATLANTIC OCEAN

PACIFIC OCEAN

ARCTIC OCEAN
North Pole

Source: The International Permafrost Association, 2009. See https://ipa.arcticportal.org/news/106-the-ipa-releases-the-first-globa-snapshot-on-permafrost-temperatures

FIGURE 6-4

One of many sinking older houses in the south-central Siberian city of Irkutsk (Figs. 6-2, 6-4). Many of the older wooden houses here date back to the early twentieth century. They were built on pilings resting on deeper icy layers of permafrost that began thawing, during the summers, in recent decades. Many such homes had already become uninhabitable years ago, and most are demolished to make way for new construction, such as the mid-rise apartment building seen at the left. This particular structure is now used for storage but even for that it probably doesn't have many years left. Irkutsk is located along the edges of the permafrost zone, making it particularly vulnerable to the effects of global warming.

© J. Nijman

POPULATION DISTRIBUTION: RUSSIA AND CENTRAL ASIA

Population
per Square Kilometer

0 to 5
5 to 10
10 to 50
50 to 100
100 to 500
Over 500

© J. Nijman, P. O. Muller, and John Wiley & Sons, Inc.

FIGURE 6-5 Interactive

year, and only limited warmth from the North Atlantic keeps the western ports of Murmansk and Arkhangelsk open a bit longer (Fig. 6-1). Ports like these (and even St. Petersburg on the Gulf of Finland, an arm of the Baltic Sea) would never have developed the way they did if Russia had better access to the oceans (see **Box 6-4**).

These days it looks as though nature may be about to give Russia a helping hand, at least in one respect. As global warming causes long-term melting of sizeable portions of the Arctic Ocean's ice cover, that body of water may come to play a new and different role in this realm's future. Warmer water may even keep Arctic ports open year-round, and the so-called **Northeast Passage [9]** will open up to directly connect the Bering Strait with the North Sea (see Chapter 13), shortening certain international sea routes by thousands of kilometers and ushering in a new era of long-distance trade.

That, at least, is how many Russians like to envision global warming—as a potential advantage nature has long denied them. And indeed, the Arctic provides much evidence for such warming: the melting of parts of the Greenland Ice Sheet and the shrinking of the average extent of permanent Arctic Ocean ice, opening up navigable waterways. Russian economic planners look at the shallow offshore waters along the Arctic Ocean coastline (see the white and light-blue areas in Fig. 6-1) and hope that oil and gas reserves beneath the seafloor will come within reach of exploitation. These developments may even have the effect of expanding the realm far into the Arctic. In 2007, the Russian government even placed a metal Russian flag on the seafloor at the North Pole under the permanent ice of the Arctic Ocean to symbolize its intentions (see final photo in Chapter 13).

Ecologies at Risk

If, as most computer models predict, global warming persists and perhaps intensifies in the polar latitudes, these ecologically sensitive environments are sure to be severely affected. Flora and fauna have long adapted to prevailing climatic conditions, and such adaptations would be significantly disrupted by rapid environmental change. The intricate web of relationships among species and their environments on the one hand, and among species themselves on the other, can be quickly damaged by temperature shifts. The polar bear is just one prominent example: it depends on ample floating sea ice to hunt as well as raise cubs. When sea ice contracts and wider stretches of open water force polar bears to swim greater distances during the cub-rearing season, fewer infants make it to adulthood. If ice-free Arctic summers are indeed in the offing, the polar bear could become extinct in your lifetime. Rapid ecosystem change would also endanger seal, bird, fish, and other high-latitude wildlife populations.

Such changes affect human populations as well. For example, indigenous communities still pursuing traditional lives in parts of the Arctic domain are similarly adapted to the harsher environments of the past. Their traditions, already under pressure from political and economic forces resulting from their incorporation into modern states, would be further imperiled by environmental changes that are likely to irreversibly modify the ways of life they developed over thousands of years.

And if a new cycle of oil and natural gas exploration and exploitation is indeed about to begin, already-fragile offshore environments face even greater danger. As the technologies of recovery put ever more of these reserves within reach, oil

---

Box 6-4 Technology & Geography

Russia's Nuclear Icebreakers in the Arctic Ocean

Nuclear-powered icebreaker ships were first built in the 1970s by Russia. They are specialized vessels whose design, weight, and fuel system allow them to crush sea ice up to three meters (10 ft) thick. The icebreakers facilitate access to the frozen seas for commercial, scientific, and military purposes. They can be on dedicated missions, or they may carve a pathway through the ice for other ships to follow (see photo). The introduction of icebreakers has greatly facilitated access to the Arctic Ocean, and their importance has only increased with discoveries of oil, gas, and other natural resources below the ice. Icebreakers are also used more routinely to keep Russia's waterways and ports at lower latitudes ice-free throughout the year. Notwithstanding the momentous impact of global warming on the Arctic (discussed further above and elaborated in Chapter 13), these ships continue to be a vital technology, especially during the winters.

These icebreakers have a double hull of reinforced steel, each about two inches thick at the bow, with a thick synthetic coating that minimizes friction with the ice. When the ship ploughs into the ice, the smooth bow causes it to ride up on the ice, which then gets crushed under its weight. The vessel's relatively wide body ensures that it creates a pathway that is broad enough for other ships to follow in its wake. Icebreakers require massive thrusting power, equivalent to 75,000 horsepower, therefore consuming huge amounts of fuel. Earlier-generation icebreakers ran on diesel oil: they had enormous fuel tanks, but their mileage and reach were very limited. In those days, an icebreaker would burn 100 tons of diesel fuel in a day; today, for the same amount of work, their nuclear-powered engines use up less than a pound of uranium and, with nuclear reactors on board, it could keep going for a year without refueling. Thus, in the absence of nuclear power, they could never reach the North Pole or get anywhere near it.

In 2019, the United States had only two icebreakers, but it had begun planning for the construction of a new state-of-the-art model. This has everything to do with the new geopolitics and economics of the Arctic Basin.

VIT AP-ACC/Newscom

The Russian nuclear icebreaker *Yamal* carving a channel for ships in the Arctic Ocean just north of the Eurasian coast. This vessel was launched in 1992 and crushed its way to the North Magnetic Pole (ca. 79°N, 105°W—1250 kilometers [775 mi] from the geographic North Pole) for the first time in 1994. It has been used for commercial, military, scientific, and even touristic purposes.

---

platforms, drills, pumps, and pipelines will materialize—along with risks of oil spills, pollution, and collateral ecological damage of the kind we have seen all over the lower-latitude world. Climate change and technological development are allowing the economic penetration of a part of the world long protected from it by distance and nature.

Natural Resources of Russia and Central Asia

Given the gigantic territorial layout of the realm, it is not surprising that its natural resource base is vast and varied. Indeed, it is already one of the world's leading producers and exporters of oil and natural gas. **Figure 6-6** maps the growing number of producing fields dispersed across the realm: from western and central Russia eastward to Sakhalin in the Pacific Rimland, and from the Arctic shore of Siberia in the north to the Caspian–Aral Basin in the far south (Fig. 6-6).

Petroleum and natural gas are not the only underground riches of this realm. In fact, it is so well endowed with minerals that virtually all of the raw materials required by modern industry are extractable. Nor are oil and gas the only energy reserves. Major coalfields lie both east and west of the Urals as well as in Siberia's more southerly latitudes (the string of mines producing high-quality coal along the corridor of the all-important Trans-Siberian Railroad, completed in the early twentieth century, were essential to Russia's industrial development and its successful war against Nazi Germany). Large deposits of iron ore, too, are widely scattered across Russia, from the so-called Kursk Magnetic Anomaly near the border with Ukraine to the remotest corners of Siberia's Arctic north. And when it comes to other metallic ores, this realm has everything from gold to lead and from aluminum to zinc. In Central Asia, Kazakhstan has led the way in expanding the exploitation of its rich deposits of oil, natural gas, uranium, chromium, and titanium.

Russia's Czarist Roots

Grand Duchy of Muscovy

During the fourteenth century, the Grand Duchy of Muscovy rose to preeminence under the rule of its princes and dukes (**Fig. 6-7**). They extended Moscow's trade connections from the Baltic to the

FIGURE 6-6

shores of the Black Sea and forged enduring religious ties with the leadership of the Eastern Orthodox Church in what is now Istanbul, Turkey. Then, beginning around 1450, there began a period of more than three centuries during which powerful Russian rulers—now called czars—etched Muscovy's imprints ever more widely on the map of Russia, Eurasia, and, by extension, the world.

By the sixteenth century, during the reign of Czar Ivan IV (better known as Ivan the Terrible), the Grand Duchy had been transformed into a military powerhouse and an imperial state. Ivan expanded Moscow's empire first by conquering the Islamic areas of what is now western Russia and then by gaining control over present-day Estonia and Latvia on the Baltic Sea. It was a time of almost constant warfare, and Ivan acquired his reputation from the reign of terror he unleashed in his pursuit of military discipline, centralized administrative control, and retaliation against many members of the nobility.

Eighteenth-Century Czarist Russia

When Peter the Great became czar (he ruled from 1682 to 1725), Moscow already lay at the center of a great empire—great, at least, in terms of the territories it controlled. The Islamic threat had been ended, and the growing influence of the Russian Orthodox Church was represented by its distinctive religious architecture and powerful bishops. Peter consolidated Russia's gains and endeavored to make a modern, European-style state out of his loosely knit country. He built St. Petersburg as a

forward capital [10] on the doorstep of Swedish-held Finland, fortified it with major military installations, and made it Russia's leading port (Fig. 6-7).

Peter the Great was an extraordinary leader and in many ways the founder of modern Russia. In his desire to remake the country—to pull it from the forests of the interior to the seas of the west, to open it to outside influences, and to relocate its population—he left nothing to chance. Prominent merchant families were forced to move from other cities to St. Petersburg. Ships and wagons entering the city had to bring building stones as an entry toll. The czar himself, aware that to become a leading power Russia had to be strong at sea as well as on land, went to the Netherlands to work as a laborer in the famed Dutch shipyards to learn the most efficient methods for building them. Meanwhile, the czar's forces continued to conquer peoples and territory: Estonia was incorporated in 1721, widening Russia's window to the west; and in Siberia steady expansion soon began in the area south of the city of Tomsk (Fig. 6-7).

Under Czarina Catherine the Great, who ruled from 1760 to 1796, Russia's empire in the Black Sea Basin grew at the expense of the Ottoman Turks. The Crimean Peninsula, the port city of Odesa, and the entire northern coastal zone of the Black Sea fell under Russian control. Also during this period, the Russians made a fateful move: they penetrated the corridor between the Black and Caspian seas, whose spine is the mountainous Caucasus with its dozens of ethnic and cultural homelands, many of which were Islamized. The cities of Tbilisi, in Georgia, Baki (also spelled Baku),

CZARIST RUSSIA, SOVIET UNION, AND RUSSIA TODAY
- Grand Duchy of Moscow, 1462
- Czarist Russia (Russian Empire), 1914
- Soviet Union, 1989
- Former Soviet republics
- Russia today

FIGURE 6-7

in Azerbaijan, and Yerevan, in Armenia, were captured. Eventually, the Russian push toward an Indian Ocean outlet was thwarted by the British, who held sway in Persia (modern Iran), and also by the Turks. Under Catherine the Great, St. Petersburg gained in stature, accumulating more and more wealth (see **Box 6-5**).

Around the same time, Russian colonists were advancing far to the east, settling along southern Siberia's frontier, crossing the Bering Strait, and even entering Alaska in 1784. But in less than a century the Russians gave up on their North American outposts (which at their height had reached as far south

Box 6-5 From the Field Notes...

"The Catherine Palace, now a museum here on the outskirts of St. Petersburg, emerges from the greenery like a fairy tale castle. Not only the city of St. Petersburg itself, but also its surrounding areas display the architectural and artistic splendor of Czarist Russia. The czars built spectacular palaces in these outlying districts (then some distance from the built-up center), among

which the Catherine Palace, begun in 1717 and completed in 1723 followed by several expansions, was especially majestic. These photos show the magnificent exterior and one of the famous pieces on display in the part that is now a museum: a dress and robe worn by Catherine the Great herself. The former palace exudes opulence in the extreme."

as central California). The sea-otter pelts that had attracted the early pioneers were running out, European and white American hunters were cutting into the profits, and indigenous American resistance was growing. When the United States offered to purchase Russia's Alaskan holdings for $7.2 million (about $125 million in today's dollars) in 1867, the Russian government quickly agreed, setting Alaska on a new course that culminated in its joining the United States as the forty-ninth State in 1959. From the U.S. point of view, besides the $24 allegedly paid to acquire Manhattan Island, this transaction is often regarded as the most lucrative real estate deal in world history.

Nineteenth-Century Expansion

Although Russia had withdrawn from North America, Russian expansionism during the nineteenth century forged ahead in Eurasia. As they were extending their empire southward, the Russians also took on the Poles, old enemies to the west, and succeeded in conquering most of what is now Poland, including the capital of Warsaw. To the northwest, Russia took over Finland from the Swedes in 1809.

Throughout most of the nineteenth century, however, the Russians were preoccupied with Central Asia, where Tashkent and Samarqand (Samarkand) came under St. Petersburg's control (Fig. 6-7). The Russians here were still bothered by raids of nomadic horsemen, and they sought to establish their authority over the Central Asian drylands as far southeast as the high mountains that marked China's western frontier. As it made headway here, Russia gained many Muslim subjects because it was now penetrating Islamic Asia. Nonetheless, under czarist rule local peoples retained some autonomy.

Thus, the Russian state, like the states of Britain, France, and other European powers, expanded through *imperialism*. Whereas those European mother countries expanded their domains overseas, Russian influence spread overland into Central Asia, Siberia, China's inland peripheries, and the far eastern Pacific coastlands. What emerged in the end was not the greatest empire but the largest territorially contiguous empire in the world.

The czars embarked on their imperial conquests in no small part because of Russia's relative location—most important, the lack of warm-water ports. Had the early-twentieth-century revolution not intervened, their southward push might have reached the Indian Ocean or even the Mediterranean Sea. Czar Peter the Great envisaged a Russia open to trading with the entire world, and he made sure his new capital of St. Petersburg on the Baltic Sea developed into Russia's leading port. But in truth, throughout the long arc of its history, Russia's geography has mostly been one of remoteness from international mainstreams of change and progress, repeatedly punctuated by episodes of self-imposed isolation.

A Multinational Empire

Centuries of Russian expansionism did not confine themselves to empty land or unclaimed frontiers. The Russian state became an imperial power that annexed and incorporated numerous nationalities and cultures. This was done by employing force of arms, by overthrowing uncooperative rulers, by seizing and annexing territory, and by stoking the fires of ethnic conflict. By the time the ruthless Russian regime had begun to confront revolution among its own citizens, Czarist Russia controlled peoples representing more than 100 nationalities. The winners in the revolutionary struggle that ensued—the communists who forged the Soviet Union—did not liberate these subjugated peoples. Rather, they changed the empire's framework, binding the peoples colonized by the czars into a new system that would in theory give them autonomy as well as identity. In practice, however, it doomed those peoples to bondage and, in certain cases, extinction. The Soviet Union, which arose in the wake of the communist revolution of 1917, expeditiously undertook another effort to integrate all these peoples and regions into this vast multinational state—notwithstanding the mantra-like socialist condemnations of imperialism.

The Soviet Experiment (1922–1991)

Communism found fertile ground in the Russia of the 1910s and 1920s. In those days, Russia was infamous for the wretched serfdom of its peasants, the cruel exploitation of its workers, the excesses of its nobility, and the ostentatious palaces and riches of the czars. Ripples from Europe's Industrial Revolution during the 1890s had introduced a new age of misery for those laboring in factories. There were workers' strikes and ugly retributions, but when the czars finally tried to better the lot of the poor, it was too little too late.

There was no democracy, and the people had no way to express or channel their grievances. Europe's democratic revolution had bypassed Russia, and its economic revolution touched the czars' domain only slightly. Most Russians, as well as tens of millions of non-Russians under czarist control, faced exploitation, corruption, starvation, and harsh subjugation. When the people finally began to rebel in 1905, there was no hint of what lay in store. Over the next two decades, it would result in nothing less than a full-scale revolution that transformed Czarist Russia into a behemoth communist state. In the process, the lives of nearly 150 million now-Soviet citizens were fundamentally reorganized, and the seeds sown for what would eventually become one of the twentieth century's superpowers on the global political stage.

The Soviet Territorial Framework

Russia's great expansion had brought myriad nationalities under czarist control; now the revolutionary government set out to restructure this heterogeneous ethnic mosaic into a smoothly functioning state, the *Union of Soviet Socialist Republics (USSR)*—or *Soviet Union* in short form. The czars had conquered, but they had done little to bring Russian culture to the peoples they ruled. When the old order was overthrown in 1917, ethnic Russians themselves constituted only about half of the population of their empire.

The political framework for the new Soviet Union was based on the ethnic identities of its numerous incorporated peoples. Given the size and cultural complexity of the now Soviet Empire, it was impossible to allocate territory of equal political standing to each of the nationalities (keep in mind that the communists now controlled the destinies of well over 100 peoples, ranging from large nations to small isolated ethnic pockets). It was decided to subdivide this gargantuan realm into 15 *Soviet Socialist Republics* (*SSRs*), each of which was delimited to correspond broadly to one of the major nationalities. (As **Figure 6-8** shows, much later these SSR territories morphed into the framework of the national state borders that emerged after the 1991 dissolution of the USSR.) Although Russians constituted barely 50 percent of the nascent Soviet Union's population, they were (and still are) the most widely dispersed ethnic group in the realm (Fig. 6-8). Thus, the Russian Republic was by far the largest designated SSR, encompassing just over three-quarters of the Soviet Empire's territory.

As designed by Vladimir Lenin, the Soviet Union's founder, the new political system claimed to be a **federation [11]**, which suggests power sharing among the various republics—but the reality was quite different. The Russian leadership in Moscow maintained absolute control over the SSRs, and they made it Soviet policy to relocate entire peoples from their homelands in order to better fit the grand design as well as rewarding or punishing localities—in many instances capriciously. The overall effect, however, was to move minority peoples eastward and replace them with Russians.

In time, this **Russification [12]** of the Soviet Empire produced substantial ethnic Russian minorities in each of the 14 other republics.

The centerpiece of the tightly controlled Soviet "federation" was the Russian Soviet Federative Socialist Republic. Containing half the USSR's population, the capital city, the realm's core area, and 76 percent of all Soviet territory, Russia was unmistakably the Empire's nucleus. Outside the Russian Republic, "Soviet" almost always meant "Russian"—the reality with which the lesser republics lived. Russians came to these 14 republics to teach (Russian was taught in the colonial schools), to organize (and frequently dominate) the local Communist Party, and to implement Moscow's economic decisions. Clearly, this was colonialism, but somehow the communist disguise—how could socialists, as the communists called themselves, be colonialists?—and the contiguous spatial nature of the Empire made it appear to the rest of the world as something else. Indeed, on the international stage the Soviet Union became a champion of oppressed peoples, a force in the post-1950 decolonization process. It was an astonishing contradiction that would, over time, be fully exposed.

The Soviet Economic Framework

The geopolitical changes that resulted from the formation of the Soviet Union were accompanied by a gigantic economic experiment: the conversion of the Empire from a czarist autocracy

RUSSIA/CENTRAL ASIA: ETHNOLINGUISTIC AREAS

Legend: Armenian, Azerbaijani (Azeri), Buryat, Caucasian, Finno-Ugric, Kalmyk, Paleo-Siberian, Russian, Tajik, Turkic, Ukrainian, Sparsely populated or uninhabited

FIGURE 6-8

with a capitalist veneer to communism. From the early 1920s onward, the country's economy would now be *centrally planned*—the communist leadership in Moscow would make all decisions regarding economic planning and development. Soviet planners had two principal objectives: (1) to accelerate industrialization and (2) to collectivize agriculture. For the first time ever on such a scale, an entire country was organized to work toward national goals prescribed by a central government.

But Soviet agriculture never came close to achieving its intended level of productivity because farmers were inadequately motivated. Most opposed collectivization, but any who obstructed the communists' grand design suffered a dreadful fate. It has been estimated that between 30 and 60 million people lost their lives through imposed starvation, constant political purges, Siberian exile, and/or forced relocation. Even though this historic Soviet "experiment" produced an incalculable human tragedy, the secretive nature of Soviet officialdom made it possible to hide this disaster from the world.

The USSR operated a **command economy [13]**, in which state planners assigned the production of particular manufactures to particular places, often disregarding the basic rules of economic geography. For instance, the manufacture of railroad cars might be assigned (as indeed it was) to a factory in Latvia. No other factory anywhere else would be permitted to produce this equipment—even if supplies of raw materials would make it cheaper to build them near, say, Volgograd 2000 kilometers (1250 mi) away. Yet, despite an expanded and improved transportation network (see Fig. 6-14), such policies made manufacturing in the USSR extremely expensive, and the absence of competition made managers complacent and workers far less productive than they might have been.

The Downfall of the Soviet Union

On December 25, 1991, the inevitable came to pass: the Soviet Union imploded. The centrally planned economy went into terminal structural failure; the arms race with the United States during the Cold War had drained too many resources; and the long effort at Russification had finally backfired, fueling the drive for independence among most of the peoples who had been forcibly incorporated into the USSR. The last Soviet president, Mikhail Gorbachev, resigned, and the Soviet hammer-and-sickle flag flying atop the Kremlin in Moscow was lowered for the final time and was immediately replaced by the white, red, and blue Russian tricolor. Yet even though a new and turbulent era had opened literally overnight, deeply entrenched Soviet institutions and systems will long continue to cast their shadows over transforming Russia.

When the Soviet system collapsed and the 15 Soviet Socialist Republics became independent states in 1991, Russia was left without the empire that had taken centuries to build and consolidate—and that contained crucial agricultural as well as industrial resources. No longer did Moscow control the farms of Ukraine or the oil and natural gas reserves of the Caspian Basin and Central Asia. The former Russian Soviet Federative Socialist Republic had suddenly become the newborn country of Russia, still a state of enormous territorial size but no longer able to claim sovereignty over the valuable lands beyond its new borders.

robertharding/Alamy Stock Photo

The spirit of the communist revolution found expression in "socialist-realist" art and political propaganda. This famous landmark at the entrance to Moscow's Mosfilm Studios dates from 1937 and glorifies the working class in typical fashion: a heroic statue of individual male and female workers hoisting the iconic hammer and sickle above their heads. Russians call this sculpture Rabochiy i Kolkhoznitsa (Worker and Kolkhoz Woman—a kolkhoz being a collectivized farm). Even today, statues like this stud urban landscapes all across Russia.

Post-Soviet Russia and the Near Abroad

The Soviet Union's implosion signaled a major reorientation of the Russian Republic as well as the other 14 former SSRs that became independent countries. Central planning came to an end, and free markets had to be introduced; national industries and corporations had to be privatized; and the communists lost their political monopoly to a new multiparty system that (at the outset) offered real elections. Although individual freedoms increased, state support in a variety of forms, from pensions to rent subsidies, mostly came to a halt. As all the components of the former Soviet Union went through an initial phase of shock, the overriding hope was that all this would be transitional, and that temporary dislocation and occasional chaos would soon be followed by a better life for most citizens. Before we examine the new political and economic forces unleashed at the regional level, we need to establish their broader context by reviewing two important developments reshaping the realm's human geography in the twenty-first century: (1) the continuing evolution of its complex cultural mosaic, and (2) Russia's geopolitical repositioning in its "Near Abroad" (the former non-Russian SSRs) as well as the world beyond.

A Complex Cultural Mosaic

For centuries, this realm has been a culturally and ethnically diverse part of the world whose traditions and customs spill over into neighboring realms even as neighbors have come to live here. Ethnic Russians still form the overall majority, but sizeable parts of this realm are home to a growing number of non-Russians—and not just in Transcaucasia and Central Asia. As Figure 6-8 indicates, the realm contains clusters of Finnish, Turkic, Armenian, Kazakh, and dozens of other minorities, but the scale of this map is too small to reflect the complexity of the total ethnic mosaic (especially within Russia itself). Here is an interesting example: in the southwest, facing the Caspian Sea, lies a small "republic" named Dagestan that is part of modern Russia. It is half the size of Maine, has a population of 3 million, and contains nearly 30 ethnic communities that speak their own languages, most of which are variants of languages spoken in the nearby Caucasus, Turkey, and Iran.

Figure 6-8 shows that the Slavic people collectively known as the Russians form the majority of the realm's population and also are the most widely dispersed across it. Although the large lowland west of the Urals is the historic hearth and modern core area of the Russian state, the distribution of Russians today still extends from the shores of the Arctic Ocean southward to the Black Sea coast and from St. Petersburg on the Gulf of Finland all the way east to Vladivostok on the Pacific's Sea of Japan. Discontinuous nuclei and ribbons of Russian settlement are even scattered across parts of Siberia. Nonetheless, another glance at Figure 6-5 reminds us that the population as a whole decidedly concentrates in the realm's southwestern sector.

Traces of the penetration of the Islamic Ottomans hundreds of years ago can clearly be discerned on today's cultural map (Fig. 6-8): the large swath of Turkic peoples (shaded tan) extending from near Nizhniy Novgorod southeastward across Kazakhstan to the fringes of Central Asia is a solid piece of evidence. Contemporary Russia has a relatively higher percentage of Muslims in its population (roughly 13 percent) than Western European countries do. As for Central Asia, the combined percentage of Muslims living in its five countries today is more than 75 percent.

Long before Islam appeared on the scene, Slavic peoples from one end of their domain to the other had accepted the teachings of the Eastern Orthodox Church. It stayed that way until the communists overthrew the czar and promptly began the disestablishment of the Church. For the next seven decades, atheism was official policy; but following the demise of the USSR in the early 1990s, the Russian Orthodox Church has made a powerful comeback, today increasingly accompanied by nationalistic propaganda.

The Near Abroad

The ending of the Cold War upon the breakup of the USSR implied an abrupt termination of Soviet geopolitical strategies that had been aimed at confronting U.S.-led alliances in Europe, the Middle East, and South Asia as well as Southeast Asia. Soviet foreign policy and power had been projected well beyond Soviet borders to other geographic realms from Africa to Middle America. The downfall of the Soviet Union meant that the new Russia, shorn of its **satellite states [14]** in Eastern Europe and abandoned by all the former SSRs, had to completely reinvent its position in the world, and most immediately within Eurasia. Moreover, it had to meet this challenge in the knowledge that a total makeover of Russia's economy was the foremost priority.

In 1992, the new Russian government introduced the term **Near Abroad [15]** to refer to the newly formed countries that surround Russia today. For the most part, these encompass the 14 former Soviet republics, stretching from Estonia to Tajikistan. Those states may now be located "abroad," but seen from Moscow they are so "near"—and collectively constitute such a pronounced encirclement of Russia—that the Russian government feels entitled to take a special interest in them. What this really means is that Russia will not permit any of these countries to mount a threat of any kind and that it reserves the right to intervene if it deems that is necessary. It is not difficult to envision this sense of territorial encirclement in Russia, particularly with the eastward expansion of the EU and **NATO [16]** (now almost at Russia's western doorstep), heightened Islamic militancy in the southern periphery, and the swift rise and growing assertiveness of China in the Far East as well as Central Asia.

And then there is the issue of ethnic Russian minorities who now find themselves living "abroad." Although many Russian nationals have returned from the former SSRs since 1991, millions more stayed on. A large proportion of these expatriates reside in areas where Russians were in the majority (for example, in parts of eastern Ukraine, northern Kazakhstan, and eastern Moldova) and where post-Soviet life was generally unchanged from what it was during Moscow's rule. But elsewhere, Russians who remained in the erstwhile Soviet republics were instantly reduced to minority status; many were mistreated in various ways, and they often appealed to Moscow for assistance. Thus, the concept of a Russian Near Abroad found its way into Russia's national discourse. Beyond Russia's borders, but of growing concern to its leadership and citizens, the Near Abroad became geographic shorthand for an enduring Russian sphere of influence, whereby Moscow stood ready to help its fellow ethnics in case of trouble.

So far, the most acute and dangerous confrontation continues to unfold in Ukraine following Russia's unexpected, forcible annexation of the Crimean Peninsula in 2014. Within weeks, parts of eastern Ukraine had descended into *de facto* warfare as separatist Russian rebels, strongly supported by Moscow (both openly and surreptitiously), battled the Kiev-based Ukrainian government in their struggle to secede from Ukraine and achieve annexation by Russia (also see Chapter 5, especially Box 5-10).

After the Crimea was annexed, Russia began construction of the Kerch Strait bridge, which connects Kerch, the easternmost city of Crimea, with Taman on the Russian side, close to the city of Krasnodar. This gives Russia direct access to the Crimea without having to traverse Ukrainian territory (**Figure 6-9**). The bridge opened to cars in mid-2018 while the part of the bridge for trains was still under construction. Importantly, the Kerch Strait is a narrow sea lane connecting the Sea of Azov to the Black Sea, and in turn to the Mediterranean. Ukraine has accused Russia of building the bridge so low that large cargo ships

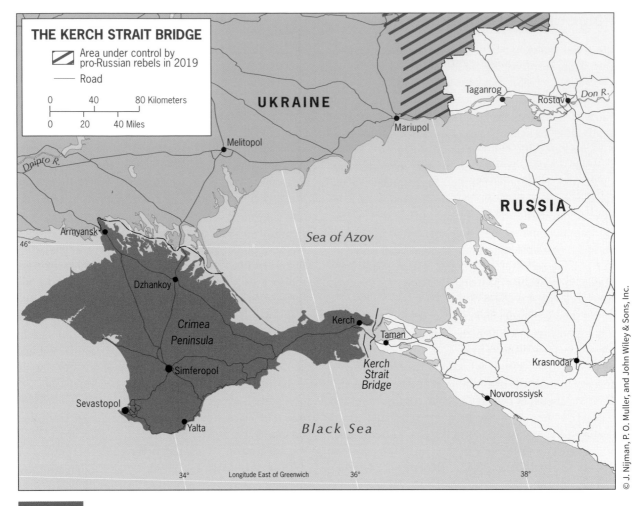

© J. Nijman, P. O. Muller, and John Wiley & Sons, Inc.

FIGURE 6-9

AP Images/Vitaliy Timkiv

Aerial view of the Kerch Strait Bridge, just after it opened for automobile traffic on May 16, 2018, looking northwestward from Taman toward Kerch (see Fig. 6-9). The bridge is the only direct overland connection from Russia to the Crimean Peninsula without having to first traverse Ukrainian territory. Near the center of the Strait, the bridge sits 35 meters above the water, allowing ships to enter and leave the Sea of Azov, essential to several Russian and Ukrainian ports. But the passageway is controlled by Russia, allowing it to further tighten its grip on Ukraine.

cannot pass under it and that this will deprive the Ukrainian port of Mariupol, located on the Sea of Azov, of about 25–30 percent of its trade. The Ukrainians seem unable to do much about it. When Russia annexed the Crimea, it also effectively confiscated about three-quarters of Ukraine's naval ships.

Playing Russia's Energy Card

During Soviet times, the non-Russian republics all became dependent to some extent on Russia's prodigious oil, natural gas, and mineral supplies. Today, this dependency on Russian energy reserves in particular plays a critical role in the evolving geopolitics of the realm. In parts of Central Asia, however, Russia has actually been losing ground because a number of countries contain their own abundant fuel supplies. Most of all, Kazakhstan has begun to exploit its huge reserves of oil and gas—attracting growing interest as well as substantial investments from around the world.

In the last 15 years or so, Kazakhstan's exports to China have increased rapidly, especially oil and gas. The Chinese have been working with the Kazakhs to complete a pipeline clear across western China and Kazakhstan in order to tap the latter's high-yielding northern Caspian Sea oilfields, and China is investing heavily in new infrastructure in this part of the world. Similarly, oil- and gas-rich Turkmenistan has in recent years intensified relations with

China at Russia's expense: 81 percent of Turkmen exports (mostly natural gas) now flow to China. The remaining (energy-poorer) Central Asian countries—Uzbekistan, Kyrgyzstan, and Tajikistan—continue to be tightly tethered to the Russian economy, and Moscow's political influence in all three remains considerable.

As for the European countries to Russia's west, dependence on Russian energy supplies is enormous, implying significant leverage for Moscow in its political dealings with that realm. The EU has to import about two-thirds of its consumed natural gas, and about 37 percent of it comes from Russia. For some individual countries, the Russian share of gas imports is considerably higher: Bulgaria, the Czech Republic, Estonia, Latvia, Hungary, Austria, Poland, Romania, Slovenia, Slovakia and Finland all imported more than 75 percent of their natural gas from Russia in 2018.

In the late 2010s as criticism of Russia accelerated in Europe during the deepening crisis in Ukraine, there was much talk about the need for Europe to reduce its energy reliance on Russia. But this cannot happen overnight because, quite apart from the economic and political issues involved, energy infrastructures are complex, very expensive, and can take years to build. **Figure 6-10** maps both existing and proposed gas pipelines connecting Russia and Europe. Note that a great deal of Russian gas being pumped to Europe passes through Ukraine—yet another potentially critical factor in the evolving crisis

FIGURE 6-10

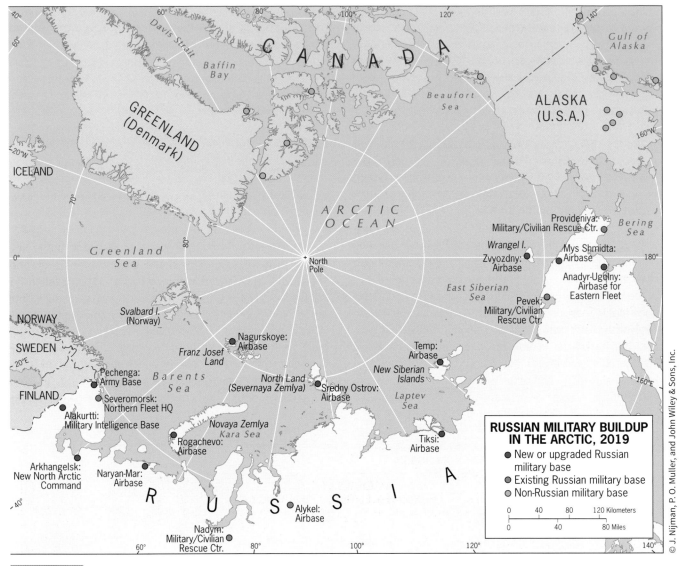

FIGURE 6-11

there. Ukraine itself is hurriedly trying to decrease its reliance on Russia by seeking new suppliers in the West (also see the section on Geopolitics in the Baltics on pages 155-156).

Militarizing the Arctic

Unlike the Antarctic, the Arctic Ocean Basin is not officially divided up among countries claiming sovereignty (compare Figs. 13-6 and 13-7). Yet claims are being made by the United States, Canada, and Denmark (all three members of NATO) as well as Russia. Clearly, Russia has a longstanding interest in the Arctic because its northern coast fronts nearly half of it.

With increasing evidence of major mineral deposits below the seafloor and with global warming improving Arctic accessibility, Russia has in recent years stepped up its military presence there, and the Russian Defense Ministry in early 2019 announced it had built nearly 500 new military "infrastructures" in the Arctic since 2006. **Figure 6-11** shows only known Russian existing and new/upgraded military bases, along with those of Canada, Denmark, and the United States (in Alaska). Seen from Moscow, a strong military presence in the Arctic is paramount, especially if the *Northeast Passage* increasingly opens up to international shipping in the coming years (see Chapter 13).

Regions of Russia and Central Asia

Few other single states dominate their geographic realm to the extent that Russia does, but we also need to keep in mind all those recent changes that have occurred within and beyond the realm's

peripheries. After the USSR imploded nearly three decades ago and all the former SSRs had gone the way of independent statehood, Russia was left by itself—still of unmatched proportions territorially, but a nation that had lost control of bordering regions in Europe and Southwest Asia, from the Baltic Sea to the Tian Shan Mountains lining China's westernmost boundary. Today, we refer to the Russian/Central Asian realm, which consists of Russia, the three comparatively tiny countries of Transcaucasia, and, to the east of the Caspian Sea, the five Central Asian countries that were all once part of the Soviet Union.

As we are aware, Russia remains a gigantic country, one so diverse in terms of its human and physical geography that it could be regionally subdivided in a variety of ways and at different scales. We will keep things simple by delimiting what can be called six first-order regions, recognizing that each of them contains its own notable subdivisions. Four of these six regions to be discussed, which are mapped in **Figure 6-12**, lie entirely inside Russia, and the other two are located in separate sectors of Russia's southern frontier zone that is almost transcontinental in east–west length:

1. The Russian Core
2. Southern Siberia
3. Northern Siberia
4. The Russian Far East
5. Transcaucasia
6. Central Asia

Before we turn to a discussion of these regions of the realm, we must outline some important politico-geographical and economic changes that affect Russia as a whole.

The New Russia

Russia's Political-Administrative Structure

When Russia's new government was installed after the disintegration of the USSR, it immediately faced an old Soviet problem in the form of an exceptional geographical challenge—the sheer size of the country, its vast distances, and the remoteness of many of its far-flung spatial components. Geographers refer to the principle of **distance decay [17]** to explain how increasing distances between places tend to reduce interactions among them. Because Russia is the world's largest country, distance is a significant factor in the relationships between the capital and outlying areas. Moreover, Moscow and St. Petersburg, decidedly the two biggest cities of the realm, lie in the far west of this behemoth country, literally half a world away from Russia's Pacific coast. Still another problem is the

FIGURE 6-12

© J. Nijman, P. O. Muller, and John Wiley & Sons, Inc.

astonishing size variation among the country's myriad administrative units, as can be seen in **Figure 6-13**. Whereas the (territorially) smallest divisions are concentrated in the Russian core area, the largest lie far to the east, where the Republic of Sakha is nearly 1000 times the size of the Republic of Ingushetiya. On the other hand, the populations of the enormous eastern divisions are minuscule compared to those of the smaller ones in the Russian Core. Such diversity spells administrative difficulty.

Figure 6-13 reflects this especially complex Russian administrative system, which in large part is a legacy of the Soviet Union. There are eight Federal Districts (newly created in 2000) that are subdivided into 21 Ethnic Republics and Autonomous Regions, and 35 Russian Regions. These administrative units function with varying degrees of power and autonomy from the federal government. The 21 Republics, established to recognize and accommodate the largest ethnic minorities, include a cluster in the far south, another lying to the east of Moscow, and a third that parallels the far-off Mongolian border.

In 1992, almost all of Russia's administrative units signed a document known as the Russian Federation Treaty, which committed them to cooperate within the new *federal system*. At first, a few holdouts refused to sign, including a Republic in the remote Caucasus periphery, Chechnya-Ingushetiya, where Muslim rebels had launched a campaign for independence (Fig. 6-13). Chechnya-Ingushetiya subsequently split into two separate Republics. Eventually, only Chechnya refused to sign the Russian Federation Treaty; Russian military intervention soon followed and triggered a prolonged, violent conflict, producing disastrous consequences for Chechnya's people and infrastructure (the capital, Groznyy, was completely destroyed).

In 2000, the newly elected administration of President Vladimir Putin moved to diminish the influence of the Regions by creating a new geographic framework that combined the 83 Regions, Republics, and other entities into a set of eight new Federal Districts—not to enhance their clout in Moscow, but to increase Moscow's authority over them. As Figure 6-13 shows, each of these new Federal Districts has a capital, elevating cities such as Rostov and Novosibirsk to a status secondary only to Moscow. In a related move supported by Russia's parliament, Putin decreed that regional governors would be appointed rather than elected—thereby concentrating even more power in Moscow. Thus Russia's federal system began its steady shift toward a **unitary state system [18]**—one dominated by a centralized government that exercises power equally throughout the country. This administrative reorganization reinforced the position of western Russia as the unrivaled political focus of the state; this heartland is also the demographic center of gravity (Fig. 6-5) and by far Russia's most accessible as well as urbanized region (**Fig. 6-14**).

Russia's Shrinking Population

Some of these administrative problems are exacerbated by an ominous trend: the Russian population has been shrinking steadily since 1990. In fact, certain parts of the country have witnessed a veritable **population implosion [19]**. When the Soviet Union disintegrated in 1991, Russia's population was about 149 million. Today, that total has dropped nearly 3 percent and is down to 146 million—despite the in-migration of several million ethnic Russians from the old non-Russian SSRs. Undoubtedly, the transition away from Soviet rule is one reason: uncertainty is known to cause families to have fewer children. The fertility rate in 2018 was 1.82 births per woman, far below the replacement reproduction rate (2.1).

Russian males are the ones most affected. Male life expectancy dropped from 71 in 1991 to 67 in 2018 (over the same period, the rate for females dipped only slightly to 78). Males are more likely to be afflicted by alcoholism and related diseases, by AIDS (chronically underreported in Russia), by heavy smoking, and by suicide, accidents, and murder. This may, however, not be true anymore for the younger generation. Russian millennials have grown up long after the Soviet Union fell, and they have little interest in its glorious revival or in mourning its demise. They grew up with the Internet, and many of them are seeking a good education and a future as entrepreneurs or well-paid professionals. Whether their aspirations can be met remains to be seen.

Demographically, at least, things may have taken a slightly positive turn in 2011 when, for the first time in 20 years, the overall population expanded, albeit minimally. Between 2011 and 2017, the overall population increased annually by 0.15 percent—minimal, indeed, but at least growth turned positive. It is too early to tell if this turnaround will last. Whatever the outcome, the situation is grimmest at the regional level, where a few Federal Districts have been hit hard by very substantial population losses. Out-migration unquestionably plays a role because people are no longer forced to stay in place by a totalitarian regime. Villages in particular continue to empty out all across Russia, and at least 20,000 have turned into "ghost villages"—more than 15 percent of all such settlements in the country two decades ago. **Box 6-6** and **Figure 6-15** provide further insights into the spatial dimension of Russia's ongoing population decline.

A Volatile and Uneven Economy

At the beginning of this century, Russia's economy appeared to have turned completely around from what it was under its communist predecessor. Private property, start-up companies, trade, foreign investment, and functioning stock exchanges—all the things we regard as vital components of a free market economy—seemed to have found their way into post-Soviet Russia. The city of Moscow changed beyond recognition with a new high-rise skyline and unprecedented linkages to the global economy. After some major (and for many people painful) adjustments were made, Russia's GDP grew robustly as the price of shares on the Moscow stock exchange skyrocketed. In common parlance, this performance catapulted Russia into membership in the exclusive club of the so-called **BRICs [20]**, the world's four biggest emerging market economies (Brazil, Russia, India, China—though in subsequent years the BRICs lost much of their shine as momentum slowed in China and Brazil, and Russia

RUSSIA'S ADMINISTRATIVE STRUCTURE

Ethnic Republics and Autonomous Regions

38.4 Population, in millions, 2010 census

• Federal District capital

— Federal District boundary

Russian Regions

National capital is underlined

0 100 200 300 400 500 Miles
0 200 400 600 800 Kilometers

RUSSIAN REGIONS
(named after their capitals)

1. Astrakhan
2. Belgorod
3. Bryansk
4. Chelyabinsk
5. Ivanovo
6. Kaluga
7. Kemerovo
8. Kostroma
9. Krasnodar
10. Kurgan
11. Kursk
12. Lipetsk
13. Mariy-el
14. Moscow
15. Novosibirsk
16. Omsk
17. Orenburg
18. Orel
19. Penza
20. Pskov
21. Rostov
22. Ryazan
23. St. Petersburg
24. Samara
25. Saratov
26. Smolensk
27. Stavropol
28. Tambov
29. Tula
30. Tver
31. Ulyanovsk
32. Vladimir
33. Volgograd
34. Voronezh
35. Yaroslavl

FIGURE 6-13

FIGURE 6-14

Box 6-6 Geographic Information Analysis

Population Change Across Russia

Between the dismantling of the Soviet Union in 1991 and 2018, Russia's total population dropped from 149 million to just above 146 million. Because this population decline has varied markedly from region to region, its geographic impact needs to be more closely examined. Figure 6-15 shows population loss from 1989 to 2018 in Russia's eight Federal Districts. Can you identify the spatial pattern of population shrinkage? This map also shows the size of the economy (GRP) in each District for 2017. Is it possible to determine which is the better predictor of population decline in a District, geographic remoteness or economic robustness?

Access your WileyPLUS course to interact with a dynamic version of this map and to engage with online exercises and questions.

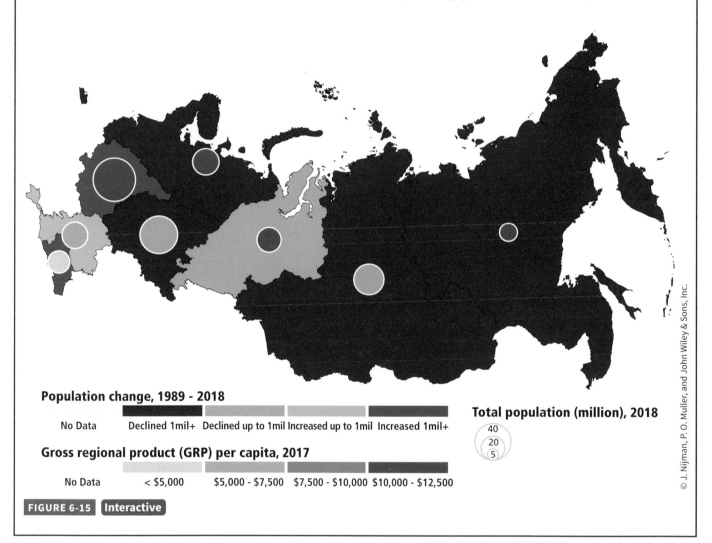

Population change, 1989 - 2018

No Data   Declined 1mil+   Declined up to 1mil   Increased up to 1mil   Increased 1mil+

Total population (million), 2018

40
20
5

Gross regional product (GRP) per capita, 2017

No Data   < $5,000 $5,000 - $7,500 $7,500 - $10,000 $10,000 - $12,500

FIGURE 6-15   Interactive

© J. Nijman, P. O. Muller, and John Wiley & Sons, Inc.

itself stagnated). Russia's *nouveaux riche*, nonetheless, became well known for their extravagant consumption on shopping trips to Paris, Milan, and New York; for sending their children to the highest-cachet Swiss and British boarding schools; and even for purchasing premier soccer-team franchises in Europe.

However, Russia's advancement had been precarious at best. Too much of it is based primarily on exports of oil and gas. Think of it: when is the last time you bought something that was made in Russia? Incoming foreign direct investment is anemic in comparison with the other emerging markets; the get-rich-quick schemes of a handful of oligarchs (highly successful businessmen—but almost never women) rarely benefit the masses; and the gargantuan state apparatus left by the Soviets is responsible for widespread corruption as well as close connections to organized crime syndicates. The value of the "bribes market" was consistently estimated to be an alarming 20 percent of the national economy. Even though statistics on the Russian economy since 2000 suggested abundant growth, ample evidence has also accumulated to indicate that most citizens were unable to better their standard of living.

Frustrations with the economic situation have boiled over into the political arena. The rigged parliamentary elections of 2011 that handed victory to Putin's United Russia Party triggered massive demonstrations on Moscow's streets. Nonetheless, a few months later Putin himself easily won election to serve as president until 2018—when his previously engineered constitutional change kicked in to allow him to run for reelection for yet another six-year term. He won that election as well (his fourth), leaving him in power from 2018 to 2024. Prospects for true democracy in Russia, therefore, seem likely to be deferred until at least the mid-2020s.

A view of the Kremlin with the statue of St. Vladimir (also known as Vladimir the Great) in the foreground, to the left. The statue was erected by his namesake, Vladimir Putin, in 2016, to honor one of Russia earliest rulers and the father of Orthodox Christianity. It was viewed by many as a political gesture by Putin to the large community of religious Russians.

Region The Russian Core

The first of Russia's regions—the Russian Core—extends from the western border of the realm to the Ural Mountains in the east (Fig. 6-12). This is the Russia of Moscow and St. Petersburg, of the Volga River and its industrial cities. This is the most developed component of this enormous realm and the focus of its transportation network (Fig. 6-14).

Moscow is the megacity hub of a wider area encompassing some 50 million inhabitants (more than one-fifth of the realm's entire population), most of them concentrated in cities that, during the communist period, specialized in assigned industries such as auto making in Nizhniy Novgorod (the "Soviet Detroit") and textile fabrication in Ivanovo. But today, Russia is less a diversified manufacturing country than a commodity producer and exporter (dominated by oil and natural gas). The tall steel-and-glass towers you see rising above the CBDs of the biggest cities mainly belong to energy companies such as Gazprom. And the cars being built are more likely to display foreign brand names such as Fiat or Volkswagen.

The Moscow metropolitan area itself has grown rapidly in recent years and today has a population of nearly 17 million, of which an estimated 2 million are undocumented migrants. It is the economic and business opportunities that continue to draw people in, but steadily worsening congestion and an increasingly brutal social environment mean that many dream of leaving the city as soon as they can afford to. The Moscow metro system (now the world's sixth longest) carries more than nine million passengers daily, and new route extensions and metro stations are built nearly every year. To handle all this overcrowding, the government announced several years ago plans for a massive expansion of the capital to the southwest over the next two decades, but implementation is still awaited.

St. Petersburg (formerly Leningrad) remains Russia's second city, with a population of 5.2 million. In the late seventeenth and early eighteenth centuries, Czar Peter the Great and his architects transformed this city at the head of the Baltic Sea's Gulf of Finland into the "Venice of the North." St. Petersburg's palaces, churches, waterfront façades, bridges, and monuments still give the city a decidedly European look unlike that of any other city in Russia. It was proclaimed the Russian capital in 1712, with Moscow serving as the distant second city. The Imperial Winter Palace and the adjoining Hermitage (one of the biggest art museums on Earth) at the heart of the city are among a host of surviving architectural treasures (see Box 6-5). Today, St. Petersburg is again important to Russia's trade with the rest of the world, and the city has become a major destination for many cruise ships.

But it possesses none of Moscow's political stature (photo above) or Moscow's locational advantages in the domestic market. With Moscow these days increasingly defined in terms of global finance and producer services, St. Petersburg's niche involves trade, culture, and tourism.

Central Industrial Region

At the heart of the Russian Core lies the Central Industrial Region (**Fig. 6-16**), sometimes also referred to as the Moscow Region, thereby emphasizing that for over 400 kilometers (250 mi) in all directions from the capital, everything is oriented toward this historic focus of the state. As Figure 6-14 underscores, Moscow has maintained its decisive **centrality [21]**: roads and railroads converge from all directions—from Ukraine in the south; from Mensk (also spelled Minsk) in Belarus and the rest of Eastern Europe in the west; from St. Petersburg and the Baltic coast in the northwest; from Nizhniy Novgorod and the Urals in the east; from the cities and waterways of the Volga Basin in the southeast (a canal links Moscow to the Volga, the country's single most important navigable river); and even to the subarctic northern periphery that faces the Barents Sea, where the strategic naval port of Murmansk and the lumber-exporting outpost of Arkhangelsk lie.

Povolzhye: The Volga Region

Another major subregion within the Russian Core is the *Povolzhye*, the Russian name for an area that extends along the middle and lower valley of the Volga River as well as its tributary, the Kama, above the city of Kazan. It would be appropriate to call this the Volga Region, for this greatest of Russia's rivers is its lifeline, and most of the Povolzhye's cities lie on its banks (Fig. 6-14). The city of Volgograd is centrally located within this subregion, on the west bank of the Volga. It was named Stalingrad from 1925 to 1961, and that name will always be

FIGURE 6-16

associated with the decisive battle here that became the turning point of World War II on Nazi Germany's Eastern Front. The nearby city of Rostov, located where the Don River flows into the Sea of Azov (Fig. 6-9), has in recent years been one of southern Russia's most entrepreneurial cities, combining a good economic location with mild weather and a developing waterfront.

The elongated Povolzhye (the *vol* refers to the Volga) forms the eastern flank of the Central Industrial Region, its cities and farms sustained by this Mississippi River of Russia. In the 1950s, the crucial Volga-Don Canal (connecting to Rostov) was opened, creating new linkages and opportunities. Next, the full dimensions of the oil- and gasfields in and near the Povolzhye became clear (see Fig. 6-6), and for some time these reserves were the largest in the entire Soviet Union. And finally, the Volga Region's northwestward linkages, via the Moscow and Mariinsk canals to the Baltic Sea, were improved by upgraded rail and road connections. Today, the industrial composition of the Volga's riverside cities continues to evolve—but if the Central Industrial Region is Russia's heart, then the Povolzhye is its aorta.

The Urals Region

The Ural Mountains form the eastern edge of the Russian Core. They are not particularly high; in the north they consist of a single range, but southward they broaden into a hilly zone. Nowhere are they an obstacle to east–west surface transportation. An enormous storehouse of metallic mineral resources located in and near the Urals has made this subregion a natural

place for industrial development. Today, the Urals Region, well connected to the nearby Volga and Central Industrial Regions, extends from Serov in the north to Orsk in the south (Fig. 6-16).

Region Southern Siberia

From the southeastern flanks of the Ural Mountains to the headwaters of the Amur River, and from the latitude of Tyumen to the northern zone of neighboring Kazakhstan, lies Russia's vast region of Southern Siberia (Fig. 6-12), the product of a gigantic Soviet experiment in the eastward extension of the Russian Core along the axial Trans-Siberian Railroad. As the maps of Russia's cities and transportation (Fig. 6-14) as well as population distribution (Fig. 6-5) suggest, this region is more densely peopled and more fully developed in the west than in the east. To the east of the Yenisey River, linear settlement prevails, marked by ribbons and clusters along the Trans-Siberian Railroad and the newer Baikal–Amur Mainline Railroad that parallels it to the north. Two subregions dominate the human geography: the Kuznetsk Basin in the west and the Lake Baikal area in the east (**Box 6-7**).

The Kuznetsk Basin (Kuzbas)

Some 1500 kilometers (950 mi) east of the Urals lies another of Russia's primary regions of heavy manufacturing resulting from the communist period's national planning: the Kuznetsk Basin, or *Kuzbas* (Fig. 6-16). The leading city, located just outside

---

Box 6-7 From the Field Notes …

"It is a warm day in June as I arrive at the shores of Lake Baikal, having driven down about 40 miles from the city of Irkutsk. This is the heart of Southern Siberia, but it can actually get quite warm here on summer days—a huge contrast to the very cold, long winters. In the middle of winter, the lake freezes, which is quite amazing, considering that it is the largest, deepest, freshwater body in the world. It contains more water than all the Great Lakes of North America combined. Geologically, Lake Baikal is part of a rift valley, similar to the East African Rift Valley: an area

where tectonic plates are moving apart, creating (over very long periods of geological time) valleys and lakes. Further down the shore, I encounter a Buryat shaman (a person believed to have contact with the spiritual world of good and evil), said to be performing a traditional ritual invoking the spirit of the lake. The Buryats, a local 'tribe' with an animistic religion and little or no cultural resemblance to Russians (genetically, they are closer to the Mongolians), have inhabited the shores of the lake for many generations."

© J. Nijman

© J. Nijman

---

the Kuzbas, is Novosibirsk, which lies at the intersection of the Trans-Siberian Railroad and the Ob River as the symbol of Russian enterprise in the vast eastern interior. To the northeast lies Tomsk, one of the oldest Russian towns east of the Urals,

founded in the seventeenth century and now caught up in the modern development of the Kuzbas. Southeast of Novosibirsk lies Novokuznetsk, a city that produces steel for this subregion's machine and metal-working plants as well as aluminum products from Urals bauxite.

LM3311/iStock/Getty Images

The river port of Krasnoyarsk, where the south-to-north-flowing Yenisey River meets the Trans-Siberian Railroad, Russia's east–west transport axis, near the center of Southern Siberia (see Fig. 6-12). This is the northwestern extremity of the Baikaliya manufacturing region (Fig. 6-16). Supplies brought by rail from the west are shipped downriver by boat and barge to settlements in Northern Siberia, whose southward-flowing raw material exports arrive in Krasnoyarsk for transshipment to Russia's factories and markets. Major storage facilities can be seen across the river.

The Lake Baikal Area (Baikaliya)

East of Lake Baikal lies Southern Siberia's most rugged, remote, and forbidding country. North of the central segment of the Mongolian border zone and eastward around Lake Baikal, larger and smaller settlements cluster along the two railroads to the Pacific coast (Fig. 6-16). West of the lake, these rail corridors lie in the headwater zone of the Yenisey River and its tributaries. A number of dams and hydroelectric projects serve the Angara Valley, particularly the city of Bratsk. Mining, lumbering, and some farming sustain life here, but isolation dominates it. The city of Irkutsk, near the southern end of Lake Baikal, is the principal service center for the enormous Northern Siberia region to the north as well as for a lengthy east–west stretch of southeastern Russia.

Beyond Lake Baikal, settlements are rare, with many being mere camps. The Republic of Buryatiya (see Fig. 6-13) is part of this zone; the territory bordering it to the east was taken from China by the czars and may yet become an issue in the future. Where the Russian–Chinese boundary turns southeastward along the Amur River, the region labeled Southern Siberia peters out and Russia's Far East begins.

Region Northern Siberia

Before we assess the potential of Russia's Pacific Rim, we should remember that the ribbons of settlement just discussed hug the southern perimeter of this gargantuan country, avoiding the huge region of Northern Siberia to the north (Fig. 6-12). Northern Siberia extends from the Ural Mountains to the Kamchatka Peninsula—a vast, frigid, forbidding land. Larger than the conterminous United States but inhabited by fewer than 20 million people, Northern Siberia quintessentially symbolizes the Russian environmental plight: immense distances, cold temperatures aggravated by ferocious Arctic winds, difficult terrain, unproductive soils, and limited options for survival.

But Northern Siberia does possess valuable resources. From the days of the first Russian explorers, Siberia's riches have beckoned. Gold, diamonds, and many other precious minerals were found. Later, metallic ores including iron and bauxite were discovered. Still more recently, this region proved to contain massive deposits of oil and natural gas (Fig. 6-6) that now contribute significantly to Russia's energy supply and exports.

As the physiographic map (Fig. 6-2) reveals, major rivers—the Ob, Yenisey, and Lena—flow gently northward, traversing Northern Siberia and the Arctic Lowland before emptying into the Arctic Ocean. Hydroelectric power development in the basins of these rivers has generated electricity used to extract and refine local ores, and to run the lumber mills that have been built to exploit the vast Siberian forests.

Understandably, the population geography of Northern Siberia is extremely fragmented. Most of the region is devoid of inhabitants, with hundreds of square kilometers of empty territory separating the relatively few islands of settlement (Fig. 6-5). Nonetheless, some small clusters of ethnic Russians have developed. Accordingly, the Yenisey River can be traced on the map of the realm's peoples (Fig. 6-8) as a thin ribbon of Russian settlements north of Krasnoyarsk (photo page 210); farther east, the upper Lena Valley can similarly be recognized.

Region The Russian Far East

Officially, the huge Far Eastern Federal District incorporates six Russian administrative regions and four residual ethnic republics (Fig. 6-13). But many Russian citizens have a different view of what constitutes the geographic region they call the Far East. Justifiably, they see most of the northern zone of the official Far East as a continuation of Siberia. To them, the real Far East is constituted by the territory beyond Southern Siberia to the Pacific coast, the island of Sakhalin, the Kamchatka Peninsula, and a narrow strip of land along the northern shore of the Sea of Okhotsk (Fig. 6-12 and **Fig. 6-17**).

In Soviet times, citizens who agreed to move to this most distant region to work in factories or for the administration

Oymyakon is known as the coldest town on Earth. It is located north of the city of Yakutsk in Northern Siberia. The average daily temperature during winter hovers around −46° Celsius (−51° Fahrenheit), and the record low recorded here in 1933 was −68°C (−90°F). A few hundred people live here, and there is even a school—which does not open if the temperature drops below −52°C (−62°F). Cell phones would not work here, even if there was satellite coverage, and visitors are advised not to wear glasses—they freeze to your face.

Dean Conger/Corbis Historical/Getty Images

were rewarded with special privileges to compensate for their hardship. Make no mistake—even though the Pacific Ocean's proximity tends to moderate natural environments here, life anywhere in the Russian Far East is tough. Moreover, Moscow's favors were withdrawn early in the post-communist era, so under the new democracy hundreds of thousands of inhabitants voted with their feet and simply departed. Petropavlovsk, the largest urban center on the Kamchatka Peninsula, exemplifies the devastating consequences of this exodus (see **Box 6-8**).

The Far East, too, has significant reserves of what underpins today's Russian economy—oil and natural gas (the most important deposits are located on and around the island of Sakhalin). As Figure 6-6 also reminds us, the southern sector of

A Russian landing ship releases armored personnel carriers during a 2019 naval exercise of Russia's Pacific Fleet near its home base of Vladivostok. Russia's growing visibility and military strength in the Far East in recent years suggest a widening geostrategic posture in the East Asian arena—an impression reinforced by warming relations with China.

Dmitry Yefremov/ITAR-TASS News Agency/Alamy Live News/Alamy Stock Photo

FIGURE 6-17

this region lies between additional reserves and potential markets: China next door and Japan across a narrow sea. We are not used to thinking about Russia as an Asia–Pacific power, but it certainly qualifies. The distance to Moscow notwithstanding, the Far East provides Russia with highly valuable geopolitical as well as economic assets. In 2012, Russia hosted the Asia–Pacific Economic Cooperation (APEC) summit in Vladivostok (the home base of the Russian Pacific Fleet), spotlighting its strengthening military presence in this part of the world (photo page 211). And by 2016, the Putin regime was implementing new policies in line with its latest mantra that developing the Far East "must be a 21st century national priority." Its centerpiece is a free land program offering one-hectare (2.5-acre) properties to new rural settlers. There have not been too many takers.

Region Transcaucasia

Now let us focus on the smallest but most complicated region of all—the area where Russia's leaders, communists as well as czars, pushed the expanding state's influence across and beyond the Caucasus Mountains. On regional maps (Fig. 6-12 and **Fig. 6-18**), it looks as though the Russian Core extends to the international boundary (which coincides, more or less, with the crestline of this towering range). However, things are not that simple. When you travel southward from Volgograd or Rostov, the landscape—physical as well as cultural—does not change; the Russian Plain's gently rolling countryside with its villages and farms is very much like that south of Moscow. But

Box 6-8   From the Field Notes . . .

"The Russian Far East is losing people at the fastest rate among all Russian Federal Districts, and my field trip to Petropavlovsk-Kamchatsky on the Kamchatka Peninsula only reinforced what I had learned in Vladivostok some years earlier. During Soviet times, Russians willing to move to the Far East were given special privileges (an older man told me that 'being as far away from Moscow as possible was reward enough'), but today locals here feel abandoned, even repressed. In 1990, Petropavlovsk had about 270,000 residents; in 2018 the total was down to 180,000. It is not just the bleak environment—the ubiquitous ash and soot from dozens of active volcanoes on Kamchatka, covering recent snowfalls and creating a pervasive black mud—it is also a loss of purpose that people feel here. There was a time when the Soviet windows on the Pacific were valuable and worthy; now the people scramble to make a living, and when they find a way, for example, by selling used cars bought cheaply in Japan and brought to the city by boat, Moscow sends law enforcement to halt the illegal trade. Petropavlovsk has no surface links to the rest of the world, and this springtime view of the city at the foot of the volcano Koryakskaya (left) and a street scene near the World War II museum (right) underscore why the Far East is depopulating."

taking the train south from Astrakhan in the Volga River Delta, you will soon see the Caucasus wall looming in the distance, but you may not realize that you have entered a different world until you get off near its base at Groznyy, the capital of Chechnya, one of Russia's 21 Ethnic Republics (Fig. 6-18). Just observing the mix of peoples here makes you aware that you have left the heartland and reached Russia's periphery.

Figure 6-18 is important enough to spend some serious time studying it. Here the communist leaders who succeeded the czars confronted multiple challenges. First, inside the Russian state itself, there were many minority peoples (such as the Chechens) who had fought the czars' armies and who now faced new rulers. Second, there were peoples who had sided with the czars and who wanted to cooperate with the Soviet regime, such as the Ossetians. To recognize the aspirations of these peoples, some (but not all) of whom were Muslim, the communist administration designated ethnic "republics" for them. On our map, these are colored dark orange to distinguish them from the rest of Russia, which is shown in light orange.

Russia's Internal Periphery and the Case of Chechnya

From Buddhist-infused Kalmykiya on the perimeter of the Caspian Sea westward to mainly Orthodox-Christian Adygeya near the Black Sea, the *Internal Periphery* of southern Russia is defined by eight ethnic Republics, each with its own identity (Fig. 6-18).

Physiographically and culturally divided Chechnya lies astride a transition zone between plains and mountains, with the mountains serving as a refuge for those opposed to Moscow's rule. When the post-Soviet Russian government asked all of the country's Regions and Republics to sign its proposed Russian Federation Treaty in 1992, Chechnya's leaders refused, believing that they now had the opportunity to terminate Russian control. The Chechens mounted a campaign for independence that led to a bitter and costly war that destabilized a number of neighboring republics as well. Chechen terrorists eventually carried this campaign to the very heart of Moscow, killing hundreds of civilians there and forever altering Russia's political landscape. When presidential candidate Vladimir Putin promised victory, the Russians elected him overwhelmingly in 2000. Subsequent military intervention did overpower the Chechens, and the war finally came to an end in 2007; Chechnya now has a pro-Russian regime that receives substantial economic support from Moscow in exchange for fierce military loyalty. Its western neighbor, Ingushetiya, got caught up in the anti-Russian activism, and the ensuing drawn-out conflict also took a terrible toll in adjacent North Ossetia. Still farther west, things remained calmer, and no problems at all arose in generally pro-Moscow Adygeya.

This tier of ethnic Republics along Russia's southernmost border displays all the properties of a disadvantaged periphery: not only do they share a history of subjugation by the powerful Russian Core to the north, but they also lag far behind the rest of the Federation by almost any measure of social progress. In terms of health, education, income, opportunity, and other social indicators, the peoples of the Internal Periphery, whether they are pro-Russian or not, have a long way to go to achieve current Russian living standards. The Internal Periphery as a whole remains restive and barely stable.

Russia's External Periphery

Russian (and later Soviet) manipulation of the ethnic map did not terminate at the international boundary we see in Figure 6-18, but it did take a different turn. Beyond the Russian border, the Soviet Empire extended its control over the three adjoining Transcaucasian entities where the Russians had for centuries exerted influence and waged war (**Box 6-9**). These three political units were **Georgia** (mapped in yellow in Fig. 6-18), facing the Black Sea; **Azerbaijan** (green), bordering on the Caspian Sea; and **Armenia** (lavender), landlocked in the middle. During Soviet times, Georgia was a loyal partner (Stalin was a Georgian by birth). Christian Armenia appreciated membership in the Soviet Union because of the security it provided next door to hostile Islamic Turkey. And Islamic Azerbaijan, whose ethnic majority is constituted by a people known as the *Azeris* (with close kinship ties to Iranian Azeris directly across their southern border with Iran), was a valued member of the USSR because of its rich oil reserves.

But note in Figure 6-18 that Azerbaijan has a large exclave on the Iranian border (Nakhchivan), separated from the main part of the country by Armenian territory—and that the Armenians hold another exclave named Nagorno-Karabakh inside what would appear to be Azerbaijan's territory (which is home to more than 150,000 Christian Armenian citizens who were placed under Islamic Azerbaijan's jurisdiction by Soviet mandate).

Soviet rule kept the lid on potential conflict in these three Transcaucasian countries. After the collapse of the USSR, Russia sought to retain its influence and maintain stability in this vital sector of the Near Abroad. It wanted to keep Azerbaijan's oil flowing to and through Russia (see the pipeline connecting Baki [Baku] to the Black Sea terminal at Novorossiysk in Fig. 6-18); Russia also wanted to keep Georgia in the fold, and it looked for ways to mitigate the long-running clash between Azerbaijan and Armenia over Nagorno-Karabakh.

But things have not gone well for Moscow in post-Soviet Transcaucasia. Driven by Western investments, Azerbaijan in 2006 began exporting oil via a new pipeline across Georgia and Turkey to the latter's Mediterranean port of Ceyhan, providing it with a more favorable economic as well as political alternative to Russian transit. Meanwhile, Armenia and Azerbaijan have not been able to resolve their sometimes violent quarrel over Nagorno-Karabakh. And things really fell apart for Russia in Georgia, where the government pursued pro-European policies and allegedly mistreated its pro-Russian Ossetian minority, and where Russia openly supported devolutionary initiatives in the northwestern corner of the country, a small ethnic entity known as Abkhazia (Fig. 6-18).

Soon after the USSR fell apart in 1991, the disagreement between Russia and Georgia escalated into serious discord. The Russians closed their border with Georgia, shutting Georgian farmers out of their crucial market. Then the Russians started issuing Russian passports to Abkhazian citizens of Georgia. Next, the Russians signaled even stronger support for their Ossetian allies, and in a still-disputed sequence of events during 2008 Russian forces entered South Ossetia and engaged the Georgian military in a brief but costly conflict. After a cease-fire was brokered by international mediators, Moscow took the extraordinary step of recognizing the two pro-Russian autonomous regions of Georgia—Abkhazia and South Ossetia—as independent states! Russian military intervention here sent shudders across Europe and much of the rest of the world—and turned out to be the prelude to similar Russian aggression in 2014, which saw the forcible annexation of Crimea followed by the outbreak of hostilities in nearby eastern Ukraine.

Unlike Azerbaijan, predominantly agricultural Georgia has no energy card to play—although it did agree to permit the new pipeline from Azerbaijan to Turkey to traverse its territory, thereby angering the Russians even further. In addition, the Georgian government has courted the European Union and even NATO, seeking the security Georgians have lacked for centuries. By 2018, it appeared that Georgian membership in the EU could be a possibility in the next several years.

The landlocked Armenians (who total 3 million) live in a world of unsettled scores—with the Turks over their decimation by Turkish Ottoman Muslims during World War I; with the Azeris over their exclave of Nagorno-Karabakh; and with the international community over the proper recognition of their plight. The Azeris of Azerbaijan are adherents of **Shi'ite Islam [22]**, the religion that also dominates neighboring Iran. As we know, this small country of 10 million is a major oil producer. It also aspires to become a major hub on China's "New Silk Road" (the Belt and Road Initiative discussed at the end of this chapter), linking Central Asia to Turkey and Europe beyond. Over the past decade, Azerbaijan's economy has expanded steadily, and a growing number of its citizens appear to be benefiting (the government claims the poverty rate fell from 50 to 5 percent between 2000 and 2018). Baki, the growing, thriving capital, is both the leading port city on the Caspian Sea as well as the leading oil and gas transport node in this part of Eurasia (photo below).

Eric Nathan/Alamy Stock Photo

The modern skyline of Baki, the oil-rich capital of Azerbaijan, with the beach and promenade on the Caspian Sea in the foreground. This city of nearly three million has modernized rapidly since the turn of the century.

Box 6-9 Geographic Information Analysis

Ethnic Geographies of Southern Russia and Transcaucasia

Geopolitically, southern Russia and Transcaucasia form a highly complex part of this realm. This applies to areas inside Russia but also to Georgia, Armenia, and Azerbaijan. Take a close look at the layout of the national borders and the location of various ethnic groups and exclaves south of the Russian border. Geopolitically, which of these three countries do you think is the hardest to govern? Why?

Access your WileyPLUS course to interact with a dynamic version of this map and to engage with online exercises and questions.

FIGURE 6-18 Interactive

Region Central Asia

The sixth and final region of the realm is Central Asia, which consists of five states that simultaneously became independent upon the disintegration of the USSR in 1991: Kazakhstan, Uzbekistan, Turkmenistan, Kyrgyzstan, and Tajikistan. Until quite recently, Central Asia was regarded as both isolated and remote; today Central Asia's relative location in the very heart of Eurasia is being translated into an increasingly strategic regional position. Geopolitically, this region lies directly between Russia and China as well as bordering Transcaucasia, Iran, and portions of South Asia. In terms of economic geography, Central Asia contains huge deposits of oil and natural gas, and key parts of it are being developed as a bridge between China and western Eurasia in the form of major new overland transport corridors that are being likened to a twenty-first-century version of the historic Silk Road.

In Central Asia today, Turkic cultural influences blend with Islam and Soviet legacies. During its hegemony over this region, the USSR tried to suppress Islam and install secular regimes. Nevertheless, Islam reasserted itself strongly after the Soviet withdrawal and has now become one of the defining qualities of Central Asia. From Almaty to Samarqand (the region's most densely populated area as shown in Fig. 6-5), mosques are being restored and reactivated, and Islamic dress has again become part of the cultural landscape. National leaders make high-profile visits to Mecca, and most are sworn into office with a hand on the holy Quran.

Central Asia is marked by enormous ethnic complexity and diversity. Figure 6-8 is only a generalization of the highly intricate mosaic of peoples and cultures, because every cultural territory on this map contains clusters of minorities that cannot be captured at this scale. Thus, people of different faiths, languages, and lifestyles rub shoulders everywhere, and this often produces friction that can escalate into ethnic conflict. Therefore, it should not come as a surprise, in this post-Soviet geographic context, that democracy is hard to come by and that autocratic government is common.

Kazakhstan is the region's giant but lies between two even greater territorial titans: Russia and China (**Fig. 6-19**). During the Soviet era, the northern tier of Kazakhstan was heavily Russified, and in 2018 ethnic Russians still accounted for about 22 percent of the total national population of 18.8 million. To stay on good terms with post-Soviet Russia, the Kazakhs carefully maintain the railway and highway links that connect their north to Russia—and they even moved their capital to Astana in the heart of Russified Kazakhstan (see photo). Today, this country is in the forefront of Central Asia's energy boom: it already ranks among the world's leading producers of uranium, and in 2018 it ranked 16th among the oil-producing countries. Figure 6-19 further reveals Kazakhstan's intermediate position between the huge oil reserves of the northern Caspian Basin and China. New oil and gas pipelines operating across Kazakhstan are serving to reduce China's dependence on oil imports carried by long-haul tankers along distant international sea lanes. Kazakhstan's economic growth has averaged 8 percent annually over the past decade, a development thrust that is certain to continue as the country becomes

Astana, the capital of Kazakhstan since 1998, was designed almost from scratch. As Figure 6-19 shows, it is located closer to Russia (and to concentrations of Russians living inside Kazakhstan) than Almaty, the former capital. Astana is the country's political and economic center and much of its new wealth comes from oil. The photo shows Bayterek Tower, which depicts a mythical giant poplar tree containing a golden egg in its branches.

even more of an oil power as well as a leading beneficiary of China's Belt and Road Initiative (described in the next section).

Uzbekistan occupies the heart of the region and borders every state within it. Uzbeks not only make up more than 80 percent of the population of 34 million but also form substantial minorities in a number of neighboring countries. The capital, Tashkent, lies in the eastern core area of the country, where most of the people live in towns and farm villages, and where the densely settled Fergana Valley is the focus. In the desert northwest, straddling the border with Kazakhstan, lies what is left of the still-shrinking Aral Sea, whose feeder rivers were diverted into cotton fields and croplands during the Soviet occupation. Heavy use of pesticides contaminated the groundwater of the Aral Basin, and countless thousands of local people continue to suffer from cancers and other catastrophic medical problems as a result.

Turkmenistan, the autocratic desert republic that extends all the way from the Caspian Sea to the borders of Afghanistan, has a population of 6 million, of which nearly 90 percent are ethnic Turkmen. During the days of Soviet control, communist planners initiated work on a massive project—the Garagum (Kara Kum) Canal designed to transfer water from Turkestan's eastern mountains into the heart of the desert. Today the canal is 1100 kilometers (700 mi) long, and it has enabled the cultivation of more than one million hectares (2.5 million acres) of cotton, vegetables, and fruits. The plan is to extend the canal all the way to the Caspian Sea. Meanwhile, Turkmenistan has greatly expanded its oil and gas extraction from Caspian coastal and offshore reserves, most of which is transported to China via a new pipeline that runs through Kazakhstan to northwestern China.

Kyrgyzstan's topography and political geography are reminiscent of the Caucasus. The agricultural economy is generally weak, consisting of pastoralism in the highlands and farming in the valleys. Mapped in yellow in Figure 6-19, Kyrgyzstan lies intertwined with Uzbekistan and Tajikistan to the point of having

FIGURE 6-19

exclaves and enclaves along their common borders. Ethnic Kyrgyz constitute about 75 percent of the population of 6.5 million. More than three-fourths of the people profess allegiance to Islam, and *Wahhabism*, an orthodox form of **Sunni Islam [23]**, has achieved a strong foothold here as well; the town of Osh is often referred to as the Central Asian headquarters of this movement. In 2010, ethnic rioting erupted in Osh between the Kyrgyz majority and the Uzbek minority, killing hundreds and putting the country in danger of breaking apart. Tens of thousands of Uzbeks, who make up roughly 14 percent of the total population and form a prosperous merchant class, fled their homes.

Tajikistan's mountainous scenery is even more spectacular than Kyrgyzstan's, and here too topography is a barrier to the integration of a multicultural society. The Tajiks, who constitute around 80 percent of the population of just under

10 million, are ethnically Persian (Iranian), not Turkic, and speak an Indo-European language related to Persian. Most Tajiks, despite these cultural affinities, are Sunni Muslims, not Shi'ites. Territorially minute though it is, regionalism to the point of state failure plagues Tajikistan: the government in Dushanbe is repeatedly at odds with the barely connected northern part of the country (Fig. 6-19), a boiling cauldron not only of Islamic fundamentalism but also of anti-Tajik, Uzbek activism.

A New Silk Road?

Central Asia is still very much a region in evolution—redefining its relations with Russia, searching for new ways to tame its ever-turbulent ethnic mosaic, and working to lure the interest of investors worldwide in the quest to more fully develop its bountiful resource base. As noted previously, once-dormant Central Asia has now become a key crossroads linking the different realms and powerful countries that surround it. Of the utmost importance today is China's growing challenge to Russia's longstanding regional influence here. The centerpiece of that effort is an enormous project the Chinese announced in 2013: the **Belt and Road Initiative (BRI) [24]**. The historic Silk Road, of course, was the ancient linear trade network that connected Mediterranean Europe to Asia, stretching eastward from Constantinople (modern Istanbul) to Xian and beyond in eastern China—and named after the prized commodity that had been perfected by the Chinese: silk.

The geographic idea that forms the cornerstone of this Chinese initiative is to construct an ultramodern overland routeway extending from East Asia to Europe via Central Asia—a new "Eurasian land bridge." The Chinese claim that it will benefit everyone, particularly the Central Asian countries hungry for investment and cutting-edge industrial technology. China's funding of the project will mainly go into building infrastructure—railroads, highways, and freight terminals—but Chinese companies would also become a driving force in the industrialization of Central Asia. Because of its relative location, Kazakhstan is of supreme importance, and the Kazakh and Chinese governments already have completed deals said to be worth tens of billions of dollars.

China's motives are fourfold. The first entails the insatiable Chinese appetite for raw materials and energy supplies, and Central Asia is richly endowed with both. Second, as industrial production in China shifts westward (discussed in Chapter 10), Europe-bound goods become costlier to ship if they must be first sent east to the Pacific coast and then transshipped via the far longer, circuitous oceanic route around the entire southern rim of Eurasia. Besides shipping costs, significant time would be saved as well: an existing railroad that moves freight from China to western Kazakhstan and connects to a rail line that crosses Russia and Belarus on its way to Germany is said to require only 14 days versus the 60 days needed to sail the much slower and indirect oceanic route. The third reason for the BRI involves a geostrategic imperative—the need to secure open overland access in case of maritime blockades (especially in the hotly disputed South China Sea [discussed in Chapter 11]). And fourth, the BRI will propel the economic development of China's far western periphery, particularly the restive Autonomous Region of Xinjiang, through which all routes to Kazakhstan are channeled.

As new freight-rail lines open (see photo of one inaugural run), along with oil and gas pipelines that connect Central Asia to China, trade between the two—at the expense of Russia—is steadily increasing. If China and Russia are seen to be competing for influence in Kazakhstan, then China has taken a major step forward. **Figure 6-20** is a map redrawn from a Chinese government publication that first appeared in 2014, and it reveals a most interesting perspective on this contest: the proposed main route of the BRI veers southward from Kazakhstan to Uzbekistan before turning westward toward Europe. That not only avoids Russian territory in the layout of the main east–west corridor, but also puts Moscow at the end of a branch line that hives off from Eastern Europe.

Russia's latest response is a half-hearted attempt to strengthen the so-called **Eurasian Customs Union [25]**, a supranational organization it created in 2010 with Belarus and Kazakhstan that was subsequently joined by Armenia and Kyrgyzstan in 2015. Even so, Russia may already have lost out to China's initiative in Central Asia—perhaps providing another reason why Russia is strengthening its presence in the Far East as it keeps an ever-watchful eye on China all around their common, 4200-kilometer (2600-mi)-long border.

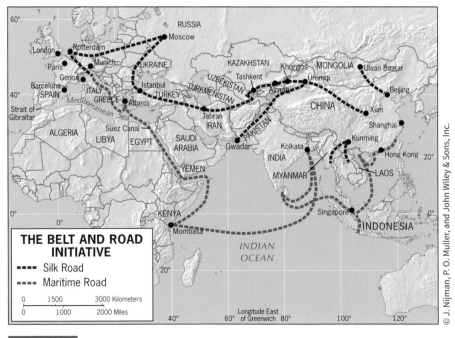

THE BELT AND ROAD INITIATIVE
- - - - Silk Road
- - - - Maritime Road

© J. Nijman, P. O. Muller, and John Wiley & Sons, Inc.

FIGURE 6-20

The first train (on February 25, 2015) pulls out of the new portside container terminal in Lianyungang on China's Yellow Sea coast to begin its 3750-kilometer (2300-mile) westbound run to Almaty, Kazakhstan's largest city. The freight-rail corridor it will traverse is the pioneering component of the Belt and Road Initiative, the Chinese high-profile Eurasian transportation program that may fundamentally change relations between China, Russia, and Central Asia.

Points to Ponder

- Russia has land borders with 14 countries and three other geographic realms. Compare that to the conterminous United States, which borders only two countries and only one other realm.

- High-speed rail travel, so urgently needed in this far-flung realm, has been extremely slow to develop. The first such route opened in 2010, linking the capital, Moscow, with the leading port and second city, St. Petersburg; but since then, not a kilometer of new, dedicated high-speed track has been laid.

- One out of seven Russian villages has been abandoned, with many turning into ghost towns.

- Climate change may create new opportunities for shipping in the Arctic Ocean, but its effects on thawing permafrost could significantly accelerate global warming.

- Landlocked, once-remote Central Asia could soon become a first-order global geopolitical hotspot involving Russia, China, India, and a host of lesser regional powers.

FIGURE 7-1

Sunset over the ancient city of Yazd in central Iran. The city dates back to the 5th century AD and its historic core was named a UNESCO World Heritage Site in 2017. Besides its architectural beauty, Yazd is famous for its silk-weaving industry and as the last remaining center of Zoroastrianism, an ancient Persian faith that predates Islam.

Konstantin Kalishko/Zoonar GmbH/Alamy Stock Photo

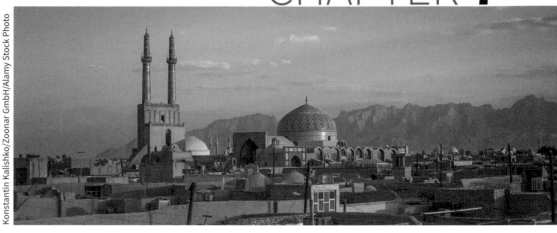

The North African/ Southwest Asian Realm

IN THIS CHAPTER

- The demise of ISIS as a territorial insurgent state
- Women's rights in Tunisia and Saudi Arabia
- The steady rise of the Gulf States
- A late Arab Spring in Sudan and Algeria?
- Jihadism comes to the African Transition Zone
- Desalination as an answer to water needs—for those who can afford it

CONCEPTS, IDEAS, AND TERMS

Cultural geography	**[1]**	Domino effect	**[14]**
Culture hearth	**[2]**	Religious revivalism	**[15]**
Cultural diffusion	**[3]**	*Jihad*	**[16]**
Cultural landscape	**[4]**	Insurgent state	**[17]**
Hydraulic civilization theory	**[5]**	Caliphate	**[18]**
Spatial diffusion	**[6]**	Megacity	**[19]**
Expansion diffusion	**[7]**	Basin irrigation	**[20]**
Relocation diffusion	**[8]**	Externality effect	**[21]**
Stateless nation	**[9]**	Rain shadow effect	**[22]**
Choke point	**[10]**	Islamic Frontier	**[23]**
Fragmented modernization	**[11]**	Failed state	**[24]**
Desalination	**[12]**	Environmental determinism	**[25]**
Arab Spring	**[13]**		

From Morocco on the shores of the Atlantic to the Iranian Plateau, and from the Horn of Africa to the Turkish Straits, lies a vast geographic realm of enormous cultural complexity. It stands at the crossroads where Europe, Asia, and Africa meet, and it is part of all three (**Fig. 7-1**, **Box 7-1**). Throughout history, its influences have radiated across Eurasia and Africa and reached most other parts of the world as well. This is one of humankind's primary source areas. On the Mesopotamian Plain between the Tigris and Euphrates rivers (in modern-day Iraq) and on the banks of the Egyptian Nile arose several of the world's earliest civilizations. In its soils, plants were domesticated that are now grown from the Americas to Australia. Along its paths walked prophets whose religious teachings are still followed by hundreds of millions. And in this first quarter of the twenty-first century, the heart of this realm has been engulfed by political turbulence, religious strife, and some of the most dangerous conflicts on Earth.

Defining the Realm

As you may conclude from its long and somewhat cumbersome name—North Africa/Southwest Asia—this is a sprawling, geographically complex realm. In our era of high-speed communication, we will also refer to it with the acronym NASWA. Sometimes, references to this realm resort to certain kinds of geographic shorthand and it is important to point out, from the start, that this can be misleading.

So, here's what NASWA is not. First, this is *not* just "the Middle East"—that label is reserved for one of the realm's constituent regions; in fact, the realm extends far beyond it. Second, this realm should not be equated with "the Arab World." The label "Arab" connotes a more or less homogeneous regional culture (expressed, for instance, in the Arabic language) but it does not apply to large populations inside the realm such as Turks, Iranians, most Israelis, or the Berbers in Northwest Africa. Finally, this realm is sometimes referred to as "the Islamic World" and that, too, is misleading because there are major populations inside the realm that adhere to other religions (e.g., Judaism, Christianity) and because the Islamic faith extends far beyond this realm (e.g., the largest Muslim country in the world today is Indonesia).

Although Islam has a dominant presence in this realm, its political and social geographies are divided and often fractious. We encounter states with strong internal divisions (Iraq, Lebanon), stateless nations (Palestinians, Kurds, Berbers), and territories still in the process of integration (Western Sahara, Palestine). Of its two dozen states and territories, the three largest, and in many ways the most important, are Egypt in North Africa, Turkey on the threshold of Europe, and Iran at the margins of Central Asia. All three have populations between 84 and 102 million, modest by world standards. But at the other end of this continuum are mini-states such as Bahrain, Qatar, and Djibouti, with populations between 1 and 3 million.

Box 7-1 Major Geographic Features of North Africa/Southwest Asia

1. North Africa and Southwest Asia were the scene of several of the world's great ancient civilizations, based in its river valleys and basins.

2. The boundaries of the North African/Southwest Asian realm consist of volatile transition zones in several sectors in both Africa and Asia. Afghanistan forms a transition zone vis-à-vis South Asia (i.e., Pakistan), and the African Transition Zone connects to Subsaharan Africa.

3. Islam dominates the cultural geography of this realm, and religion plays a leading role in political developments and conflicts.

4. The conflict-prone Middle East, one of six constituent regions, lies at the heart of the realm. It includes Israel, but also contains Syria and Iraq—the backdrop to some of the most destructive wars in this realm in the past couple of decades.

5. Drought and unreliable precipitation dominate natural environments, and population concentrations occur where water supply is adequate to marginal. Conflict over water sources and supplies looms as a steadily rising threat all across this realm, where population growth rates are high by current world standards.

6. Certain countries of the realm have enormous reserves of oil and natural gas, creating great wealth but also regional disparities.

7. Democracy and individual freedoms are rare in this realm, and the high hopes that accompanied the short-lived Arab Spring of 2011 have vanished almost everywhere.

8. Several countries in the realm are ravaged by warfare, especially Syria, Iraq, Yemen, and Libya. Terrorist organizations such as ISIS and Al Qaeda have proven to be powerful forces. External powers, including the United States, have been repeatedly involved in regional conflicts.

9. The challenges that afflict parts of this realm are expressed in a massive and expanding refugee crisis. Most pressing is the plight of Syrian war refugees.

POPULATION DISTRIBUTION: NORTH AFRICA / SOUTHWEST ASIA

Population
per Square Kilometer

0 to 5
5 to 10
10 to 50
50 to 100
100 to 500
Over 500

© J. Nijman, M. Shin, P.O. Muller and John Wiley & Sons, Inc.

FIGURE 7-2   **Interactive**

It is useful to take a careful look at **Figure 7-2** because it shows how differently populations are distributed across and within the countries we will be examining. The co-varying relationship between population distribution and available water dominates, and many of this realm's most prominent physiographic features—river valleys, basins, and deltas; moist coastlines; adequately watered mountain basins and plateaus—are virtually defined by the clusters you see on this map.

Finally, this is not only a highly complex and volatile realm but also one that lacks hard external boundaries. It is characterized by transition zones to the east (Afghanistan, see Chapter 9), to the northeast (Central Asia, see Chapter 6), to the north (Turkey, see Chapter 5), and to the south (Subsaharan Africa, see pages 259-263 and Chapter 8). This means that, sometimes, in order to understand events within NASWA, we must consider its relations to adjacent realms.

Ancient Hearths of Cultures

This geographic realm occupies a pivotal part of the world: here Africa, the source of humankind, meets Eurasia, crucible of human cultures. Two million years ago, the ancestors of our species walked from East Africa into North Africa and Arabia and then spread all across Asia. Less than 100,000 years ago, *Homo sapiens* crossed these lands on their way to Europe, Australia, and, eventually, the Americas. Ten thousand years ago, human communities in what we now call the Middle East began to domesticate plants and animals.

They subsequently learned to irrigate their fields, enlarged some of their settlements into towns, and formed the earliest states. The world's dominant monotheistic religions originated here: first came Judaism, followed by the rise of Christianity; most recently the heart of the realm was stirred and mobilized by the teachings of Muhammad and the *Quran* (or Koran), and Islam arose.

Dimensions of Culture

In Chapter 1, we discussed the concept of culture and its regional expression in the cultural landscape. **Cultural geography [1]**, we observed, is a wide-ranging and comprehensive field that studies spatial aspects of human cultures, focusing not only on cultural landscapes but also on **culture hearths [2]**—the crucibles of civilization, the sources of ideas, innovations, and ideologies that transformed regions and realms. Those ideas and innovations spread far and wide through a set of processes that we study under the rubric of **cultural diffusion [3]**. Because we understand these processes better today, we can reconstruct many of the ancient routes by which the knowledge and achievements of culture hearths spread—that is, *diffused*—to other areas.

Another aspect of cultural geography, particularly noteworthy in the context of the North Africa/Southwest Asia realm, is the study of **cultural landscapes [4]** that a dominant culture creates. Human cultures exist in long-term accommodation with (and adaptation to) their natural environments, exploiting opportunities that these habitats present and coping with the extremes they can impose. In the process, they fuse their physical landscape and cultural landscape into an interacting unity between nature and society.

Rivers and Communities

In the basins of the major rivers of this realm (the Tigris–Euphrates system of modern-day Turkey, Syria, and Iraq as well as the Nile of Egypt) lay two of the world's earliest culture hearths (**Fig. 7-3**). Mesopotamia (which means land between the rivers) possessed fertile alluvial soils, abundant sunshine, ample water, and animals and plants that could be domesticated. Here, in the Tigris–Euphrates lowland between the Persian Gulf and the uplands of present-day Turkey, arose one of humanity's first culture hearths, a cluster of communities that grew into larger societies and, eventually, into the world's first cities containing tens of thousands of people. Each of the ancient cultural hearths in Figure 7-3 had two critical geographic features. First, they were all located in arid or semi-arid climates where rainfall was scarce and where desert conditions prevailed. But, second, they were also situated along major rivers that enabled the discovery of irrigation techniques and intensive farming. It is a fascinating illustration of how geography simultaneously imposes constraints and provides opportunities; depending on human ingenuity, the outcomes are not always easy to predict. None of the early civilizations of this era (around 2500 BC) were located in more lush, temperate, zones.

Mesopotamia   Mesopotamians were innovative farmers who domesticated cereal grains such as wheat and knew when to sow and harvest various other crops, water their fields, and store their surpluses. Their knowledge spread to villages near and far, and a *Fertile Crescent,* a region of significant agricultural productivity, evolved, curving westward from Mesopotamia across Syria, the eastern Mediterranean coastlands, and terminating at the lower Nile Valley (Fig. 7-3).

Irrigation was the key to prosperity and power in Mesopotamia, and urbanization was its major reward. Among many settlements in the Fertile Crescent, some thrived, grew, enlarged their hinterlands, and diversified socially as well as occupationally; others failed. What determined success? One theory, the **hydraulic civilization theory [5]**, holds that cities that could manage irrigated farming across large hinterlands were able to produce a sufficient agricultural surplus to feed the growing urban populations, and grew more powerful. One of these cities, Babylon on the Euphrates River, endured for nearly 4000 years. A busy port, its walled and fortified center endowed with temples, towers, and palaces, Babylon for a time was the world's largest city, its population evolving from about 35,000 to 200,000 people.

FIGURE 7-3

Egypt and the Nile   Egypt's cultural evolution may have started even earlier than Mesopotamia's, and its focus lay upstream from (south of) the Nile Delta and downstream from (north of) the first of the Nile's series of rapids, or cataracts. This part of the Nile Valley is surrounded by inhospitable desert, and unlike Mesopotamia (which lay open to all comers), the Nile provided a natural fortress here. The ancient Egyptians converted their security into progress. The Nile was their leading avenue of trade and interaction; it also supported agriculture through irrigation (**Box 7-2**). The Nile's cyclical ebb and flow was much more predictable than that of the Tigris–Euphrates river system. By the time Egypt fell victim to outside invaders (around 1700 BC), a full-scale urban civilization had emerged. Ancient Egypt's artist-engineers left a magnificent legacy of huge stone monuments, some of them containing treasure-filled crypts of god-kings called pharaohs. These tombs have enabled archeologists to reconstruct the ancient history of this culture hearth.

Today, the world continues to benefit from the accomplishments of the ancient Mesopotamians and Egyptians. They domesticated cereals (wheat, rye, barley), vegetables (peas, beans), fruits (grapes, apples, peaches), and several animals (horses, pigs, sheep). They also advanced the study of the calendar, mathematics, astronomy, government, engineering, metallurgy, and a host of other skills and technologies. In time, many of their innovations were adopted and then modified by other cultures in the Old World and eventually in the New World as well. Europe was the greatest beneficiary of these legacies of Mesopotamia and ancient Egypt, whose achievements constituted the foundations of Western civilization.

Stage for Islam

Long after the ancient civilizations of Egypt and Mesopotamia had faded away, this realm became the backdrop to an entirely new religion and culture. In a remote place on the Arabian Peninsula, where the foreign invasions of the Middle East had had little effect on the Arab communities, an event occurred early in the seventh century that was to change history and affect the destinies of people in many parts of the world. In a town called Mecca (Makkah), about 70 kilometers (45 mi) from the Red Sea coast in the Jabal Mountains, a man named Muhammad in the year AD 611 began to receive revelations from Allah (God). Muhammad (571–632) was then in his early forties. Arab society was in social and cultural disarray, but Muhammad forcefully taught Allah's lessons and began to transform his culture. His personal power soon attracted enemies, and in 622 he fled from Mecca to the safer haven of Medina (Al Madinah), where he continued his work until his death. This moment, the *hejira* (meaning migration), marks the starting date of the Muslim era, Year 1 on Islam's calendar. Mecca, of course, later became Islam's holiest place.

The Faith

The precepts of Islam in many ways constituted a revision and embellishment of Judaic and Christian beliefs and traditions. All of these faiths have but one god, who occasionally communicates with humankind through prophets; Islam acknowledges that Moses and Jesus were such prophets but considers Muhammad to be the final and greatest prophet.

Box 7-2   From the Field Notes . . .

"Flying north over central Egypt from Luxor's ruins toward Cairo along the Nile's ribbon of green, I have to think of the famous remark by the ancient Greek historian Herodotus that 'Egypt is a gift of the Nile.' The river brings water from the interior of Africa that crosses the vast northern deserts and is crucially important to a country with as little rainfall as Egypt. About 95 percent of the Egyptian population lives within 20 kilometers (12 mi) of the river's banks. Almost all of Egypt's agriculture has always depended on irrigation, which is how this narrow strip of desert turned fertile. Notice the sharp boundaries between irrigated and nonirrigated land. Undoubtedly, the Nile was the lifeline that supported one of the ancient world's leading civilizations. And it is equally important to Egypt today."

© J. Nijman

Islam brought to Arabian culture not only the unifying religious faith it had lacked but also a new set of values, a new way of life, and a new individual and collective dignity. Islam dictated observance of the Five Pillars: (1) repeated expressions of the basic creed, (2) daily prayer, (3) a month each year of daytime fasting (Ramadan), (4) the giving of alms, and (5) at least one pilgrimage to Mecca in each Muslim's lifetime (the *hajj*). Islam prescribed and proscribed in other spheres of life as well. It forbade the use of alcohol, smoking, and gambling. It tolerated polygamy, even though it acknowledged the virtues of monogamy. Mosques appeared in Arab settlements, not only for the Friday prayer, but also as social gathering places to knit communities closer together. Mecca became the spiritual center for a divided, widely dispersed people for whom a collective focus was something new.

The Arab-Islamic Empire

Muhammad provided such a powerful stimulus that Arab society was mobilized almost overnight. The prophet died in 632, but his faith and fame spread like wildfire. Arab armies carrying the banner of Islam formed, invaded, conquered, and converted wherever they went. As **Figure 7-4** shows, by AD 700 Islam had reached far into North Africa, into Transcaucasia, and into most of Southwest Asia. In the centuries that followed, it penetrated both southern and eastern Europe, Central Asia, West Africa, East Africa, South Asia, and Southeast Asia, even reaching into China by AD 1000.

The spread of Islam provides a good illustration of a set of processes known as **spatial diffusion** [6] that focus on the way ideas, inventions, and cultural practices propagate through a population in space and time. Diffusion takes place in two forms: **expansion diffusion** [7], when propagation waves originate in a strong and durable source area and spread outward, affecting an ever-larger region and population; and **relocation diffusion** [8], in which migrants carry an innovation or an idea (or belief). Islam spread through expansion as well as relocation diffusion, reaching such faraway places as the Ganges Delta in South Asia, what is now Indonesia in offshore Southeast Asia, and East Africa.

The heart of Islam, however, remained in Southwest Asia. There, Islam became the cornerstone of an Arab Empire with Medina as its first capital. As the Empire expanded, its headquarters shifted from Medina to Damascus (in what is now Syria) and later to Baghdad on the Tigris River (in modern-day

**DIFFUSION OF ISLAM
630–1600 AD**

FIGURE 7-4

Iraq). And it prospered. In architecture, mathematics, and science, the Arabs overshadowed their European contemporaries (now mired in their medieval standstill). The Arabs established institutions of higher learning in many cities, including Cairo, Baghdad, and Toledo (Spain), and their distinctive cultural landscapes united their far-flung domain (**Box 7-3**). Non-Arab societies in the path of the Muslim drive were not only Islamized, but also Arabized, adopting other Arab traditions as well. Islam had spawned a culture; it still continues to function at the heart of that culture today.

Islam and Other Religions

Two other major faiths also originated in the *Levant* (the lands bordering the Mediterranean's eastern coast, extending inland for about 300 kilometers [200 mi])—Judaism and Christianity—with both predating the founding of Islam. Islam's rise submerged many smaller Jewish communities, but the Christians, not the Jews, waged centuries of holy war against the Muslims, seeking, through the Crusades, not only to drive Islam back but to reestablish Christian communities where they had dominated before the Islamic conquest. The aftermath of that lengthy campaign still marks the realm's cultural landscape today. A substantial Christian community remains in Lebanon, and Christian minorities also survive in Israel, Egypt, Syria, and Jordan. Since World War II, the most intense conflict to date has pitted Israel, the Jewish state formed in 1948, against its Islamic neighbors near and far. Jerusalem—holy city for Judaism, Christianity, and Islam—lies squarely in the middle of this confrontation.

Islam Divided

For all its vigor and success, Islam is still fragmented into sects. The earliest and most consequential division arose after Muhammad's death between those who believed that only a blood relative should follow the prophet as leader of Islam (*Shi'ites*) and others who felt that any devout follower of Muhammad was qualified to take over (*Sunnis*). Overall, the Sunnis prevailed, and, in the centuries that followed, the great expansion of Islam was mainly propelled by Sunnis; the Shi'ites survived as minorities scattered throughout the realm. Today, even though roughly 90 percent of all Muslims are Sunnis, the Shia–Sunni schism has strongly resurfaced to become the root of a number of the realm's ongoing conflicts.

The minority Shi'ites have always vigorously promoted their version of the faith. In the early sixteenth century, their work finally paid off: the royal house of Persia (present-day Iran) made Shi'ism the only legal religion throughout its vast empire. That domain extended from Persia into lower Mesopotamia, into Azerbaijan, and into western Afghanistan and Pakistan. As the world map of religions (Fig. 1-9) shows, this created for Shi'ism a large culture region and gave the faith unprecedented strength. Iran remains the bastion of Shi'ism in the realm today, and its appeal continues to radiate into neighboring countries and even farther afield.

The Ottoman Empire and Its Aftermath

If we want to understand this sprawling realm today, we must appreciate the historical role of the Ottoman Empire, its enormous expansion, and the ways it was undone by other

---

Box 7-3 From the Field Notes . . .

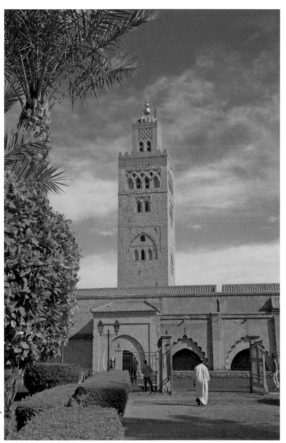

© J. Nijman

"Walking through the medina (old city) of Marrakech, Morocco, I come upon the famous Koutoubia Mosque. Its minaret is the city's signature landmark, standing 77 meters (250 ft) tall. It was built in the twelfth century and reminds me of the masterful architecture and cosmopolitanism that flourished in parts of the Arab-Islamic Empire while European culture lay dormant during the depths of the so-called Dark Ages. In fact, this Islamic cultural imprint also extended into southern Europe, and comparable architectural wonders still stand in the Spanish cities of Seville and Granada. When, in the late fifteenth century, Spain expelled its Jewish population, some found refuge across the Strait of Gibraltar in Morocco. I tried to take a look inside the Mosque, but non-Muslims are refused entry. Today, Morocco is a relatively stable outpost on the western edge of a turbulent and expansive realm."

great powers. It is ironic that Islam's last great advance into Europe eventually led to European occupation of Islam's very heartland. The Ottomans (named after their leader, Osman I), based in what is today Turkey, conquered Constantinople (now Istanbul) in 1453 and then pushed into southeastern Europe. Soon Ottoman forces were on the doorstep of Vienna; they also invaded Persia, Mesopotamia, and North Africa. At its height, the Ottoman Empire under Suleyman the Magnificent (ruled 1522–1560) was the most powerful state in western Eurasia.

Although the Ottoman Empire survived for more than four centuries until its dissolution in 1923, it lost territory as time went on. First came losses to the Hungarians, then to the Russians, and later to the Greeks and Serbs until, after World War I, the European powers took over its provinces and turned them into colonies—colonies we now know by the names of Syria, Iraq, Lebanon, and Yemen (**Fig. 7-5**). As the map shows, the French and the British took several possessions; even the Italians annexed the part of the Ottoman domain that is now Libya.

The boundary framework that the colonial powers created after World War I to delimit their new holdings was unsatisfactory. As Figure 7-2 illustrates, this realm's population is clustered, fragmented, and strung out in river valleys, coastal zones, and crowded oases. The colonial powers laid out long stretches of boundary as ruler-straight lines across uninhabited territory; they saw no need to make those boundaries conform to cultural or physical features in the landscape. Other borders, even some in desert zones, were poorly defined and never marked on the ground. Later, when the colonies had become independent states, such boundaries led to quarrels, even armed conflicts, among neighboring Muslim states. In some instances, they left entire nations, such as the Kurds and the Palestinians, without any territory of their own, thereby rendering them **stateless nations** [9].

Economic Geography

The Power and Peril of Oil

About 30 of the world's countries have significant oil reserves, with six of the nine leading countries located in this realm: Saudi Arabia, Iran, Iraq, Kuwait, the United Arab Emirates, and Libya. In general terms, NASWA's deposits of oil (and associated natural gas) are concentrated within two discontinuous zones (**Fig. 7-6**). The most productive of these zones extends from the southern and southeastern part of the Arabian Peninsula northwestward around the rim of the Persian Gulf,

EXTENT OF OTTOMAN EMPIRE AND SUBSEQUENT TERRITORIAL CONTROL

— - — Maximum Extent of Ottoman Empire

Subsequent Territorial Control by:

Austria   France   Persia

Britain   Italy   Russia

Countries that became independent

1912 Date province lost

FIGURE 7-5

© J. Nijman, P.O. Muller and John Wiley & Sons, Inc.

FIGURE 7-6

NORTH AFRICA/SOUTHWEST ASIA: OIL AND NATURAL GAS, 2018

BARRELS OF OIL PRODUCED ANNUALLY

- Over one billion
- 100 million–one billion
- Under 100 million
- Little or no oil production

— Oil and gas production region
◗ Oilfield
◗ Gasfield
•–•–• Oil pipeline
- - - - Proposed oil pipeline
—•—•— Gas pipeline
IRAN OPEC member
Algiers• Oil shipping terminal

ANNUAL OIL PRODUCTION

TOP COUNTRIES	BARRELS (in billions)
United States	4.00
Russia	3.93
Saudi Arabia	3.81
Iraq	1.68
Canada	1.55
Iran	1.55
China	1.38
United Arab Emirates	1.18
Kuwait	1.03
Brazil	0.95

TOTAL OIL RESERVES

TOP COUNTRIES	BARRELS (in billions)
Venezuela	302.3
Saudi Arabia	266.2
Canada	170.5
Iran	157.2
Iraq	148.8
Kuwait	101.5
United Arab Emirates	97.8
Russia	80.0
Libya	48.4
United States	42.0

ANNUAL NATURAL GAS PRODUCTION

TOP COUNTRIES	CUBIC FEET (in billions)
United States	32,914
Russia	22,728
Iran	9,098
Canada	6,752
Algeria	6,492
Qatar	6,001
Norway	5,763
China	4,560
Saudi Arabia	4,232
United Arab Emirates	3,179

TOTAL NATURAL GAS RESERVES

TOP COUNTRIES	CUBIC FEET (in billions)
Russia	1,688,200
Iran	1,194,000
Qatar	842,600
Saudi Arabia	307,700
Turkmenistan	265,000
United Arab Emirates	215,100
China	212,600
Venezuela	201,500
Nigeria	198,700
Algeria	159,100

reaching into Iran and continuing northward into Iraq, Syria, and southeastern Turkey, where it peters out. The second zone lies across North Africa and extends from north-central Algeria eastward across northern Libya to Egypt's Sinai Peninsula.

Winners and Losers

Saudi Arabia is one the world's largest oil exporters, and the combined oil production of the NASWA countries far exceeds that from all other sources. Throughout recent times, the United States has been the world's leading importer, followed closely by China. Today, the United States is both the biggest consumer of oil as well as the top producer, making it less dependent on imports than it would otherwise be.

As Figure 1-10 in Chapter 1 indicates, the production and export of oil and gas have vaulted a number of this realm's countries into the higher-income categories. But petroleum wealth also has entangled these Islamic societies and their governments in global strategic affairs. When regional conflicts create instability in producing countries that have the potential to disrupt international supply lines, powerful consumers are tempted to intervene and have done so.

When the colonial powers laid down the boundaries that partitioned this realm among themselves, no one knew about the riches that lay beneath the ground. A few wells had been drilled, and production in Iran had begun as early as 1908 and in Egypt's Sinai Peninsula in 1913. But the major discoveries came later, in certain cases after the colonial powers had already withdrawn. Some of the newly independent countries, such as Libya, Iraq, and Kuwait, found themselves endowed with wealth undreamed of when the Ottoman Empire collapsed. As Figure 7-6 shows, however, others were less fortunate. A few countries had (and still have) potential. The smaller, weaker emirates and sheikdoms on the Arabian Peninsula always feared that powerful neighbors would try to annex them. The unevenly distributed oil wealth, therefore, created yet another source of division and distrust among Islamic neighbors.

A Foreign Invasion

The oil-rich countries of the realm found themselves with a coveted energy source but lacking in the skills, capital, or equipment to exploit it. Those resources had to come from the Western world and entailed what many tradition-bound Muslims feared most: a strong foreign presence on Islamic soil, outside intervention in political as well as economic affairs, and penetration of Islamic societies by the vulgarities of Western ways. The transport of oil from these countries to destinations in other geographic realms also heightened the importance of strategic arteries that connect NASWA to the rest of the world. Most significant are **choke points [10]**, the narrowing of international waterways that increase the risk of ship collisions and vulnerability to piracy and other forms of attack. This realm contains no less than five of the world's most critical choke points: the

Suez Canal and Bab-el-Mandeb Strait at opposite ends of the Red Sea; the Strait of Hormuz at the mouth of the Persian Gulf; the Turkish Straits that connect the Black and Mediterranean seas; and the Strait of Gibraltar at the Mediterranean's outlet to the Atlantic Ocean (Fig. 7-1). At the end of this chapter, we discuss the virtual invasion of a number of foreign powers in Djibouti, which occupies the crucially important west shore of the Bab-el-Mandeb Strait. In the case of Djibouti, it is not about oil but all about geostrategic location.

Foreign intervention has often aggravated clashes between the traditional and the modern. In several countries, the social chasms between the rich, well-connected, Westernized elites and the far less fortunate local citizenries bred resentment. In Iran, these issues played a role in the alienation of the ruling shah (king) from his own people, fueling a religious revival that ultimately led to the monarch's ouster in 1979. Thus oil, foreign involvement, economics, and religion are often inextricably intertwined.

The Impact of Oil Across the Realm

As often occurs with exploited natural resources, oil has completely transformed cultural landscapes in certain parts of the North African/Southwest Asian realm and left others virtually unchanged. For hundreds of millions of this realm's inhabitants, a patch of tillable soil and a source of water still mean more to daily life than all the oil in OPEC.[1] The countryside in Arab as well as non-Arab territories in this part of the world continues to carry the imprints of centuries of cultural tradition, not decades of oil-driven modernization. Nonetheless, oil and natural gas—their location, production, transportation, marketing, and sale—are vital to national economies. Just consider the estimated share of the economy that is generated in the oil sector for these countries, in 2019: United Arab Emirates, 33 percent; Saudi Arabia and Kuwait, 50 percent; Qatar, 60 percent; Iraq, 65 percent.

The impacts of oil across the realm include the following:

1. ***Urban Transformation.*** Undoubtedly, the most visible manifestation of oil wealth is the modernization of cities. The tallest building on Earth (as of 2019) rises above Dubai (photo page 231) in a mini-state named the United Arab Emirates (UAE), but it is only one in a forest of new glass-encased skyscrapers, many of which test the limits of design and engineering. Besides the capitals (such as Saudi Arabia's Riyadh), the landscapes of a number of other cities on the Arabian Peninsula also reflect the riches oil has brought—most of all Mecca.

2. ***Volatile Incomes.*** Oil and natural gas prices fluctuate on world markets. In 2008, oil sold for more than $140

---

[1]OPEC, the Organization of Petroleum Exporting Countries, is the international oil cartel (syndicate) formed by 14 producing countries to promote their common economic interests through the setting of joint pricing policies and the limitation of market options for consumers. Its seven NASWA members are Algeria, Iran, Iraq, Kuwait, Libya, Saudi Arabia, and the United Arab Emirates.

per barrel before plunging to less than $40; by 2013, the price of a barrel had climbed back to nearly $120, only to drop to less than $50 by 2015. As 2020 opened, the price had edged back up to about $60. When energy prices are high, several states in this realm rank among the highest-income societies in the world, but the countries without oil have to pay dearly for their energy. Even when prices decline, many petroleum-exporting countries manage to remain at least in the upper-middle-income category (Fig. 1-10).

3. **Infrastructure.** Massive spending on airports, seaports, bridges, tunnels, four-lane highways, public buildings, shopping malls, recreational facilities, and other components of national infrastructure creates an image of comfort and affluence quite unlike that prevailing in countries without significant oil or gas revenues. Saudi Arabia, for instance, is now engaged in a gigantic modernization project extending from coast to coast.

4. **Economic Diversification.** A number of far-sighted governments among those with oil wealth, realizing that reserves will not last forever, are investing some of their income in industries that will survive the oil-exporting era. Petrochemical manufacturing using domestic supplies, aluminum, steel, and fertilizers are among those industries, along with forward-looking high-technology fields. Countries like Qatar, the UAE, and Saudi Arabia are also actively promoting their roles as international trade hubs, convention centers, and regional hubs for shopping and tourism.

5. **Regional Disparities.** Oil wealth, like other high-value resources, tends to create strong regional contrasts both within and among countries. Saudi Arabia's ultramodern east coast is a world apart from most of its interior, which remains a land of barren deserts, widely scattered oases, enormous distances, and isolated settlements. Among countries, the contrasts can also be striking, for example, between Qatar and Yemen.

6. **Foreign Investment.** Governments and private entrepreneurs have invested enormous amounts of oil-generated wealth in foreign countries, buying financial securities as well as acquiring prestigious hotels, upscale retailers, and other high-cachet properties. These investments have created a network of international linkages that not only connects NASWA states and individuals to the national economies of other realms but also links them to the growing Islamic communities in those countries.

7. **Foreign Involvement.** To many inhabitants of this realm, particularly those with strong Islamic-revivalist convictions (a topic to be discussed shortly), the inevitable presence of foreigners (including businesspeople, politicians, architects, engineers, and even armed forces) on Islamic soil is an unwelcome byproduct of the energy era. In fact, public opinion in Saudi Arabia forced its ruling regime to negotiate the 2003 departure of U.S. troops from the country's territory. Still, Saudi Arabia has remained a close (though controversial) U.S. ally in the realm.

8. **Intra-Realm Migration.** Petroleum wealth enables governments, industrialists, and private individuals to hire workers from less-favored parts of the realm to labor in the oilfields, ports, and numerous menial service occupations. This has brought many Shi'ites to the countries of the eastern Arabian Peninsula, where they now form significant sectors of national populations; hundreds of thousands of Palestinian Arabs also sought temporary employment in local industries here. In 2019, it was estimated that Saudi Arabia, with a permanent population of 35 million, hosted more than 11 million foreign workers (no less than 80 percent of the private sector workforce).

9. **Migration from Other Realms.** The labor market in such burgeoning places as the UAE's Dubai and Abu Dhabi has also attracted hordes of workers from well beyond the realm's borders during recent periods of rapid growth. Wages in South Asian and African countries are considerably lower than those paid by the building industry or private employers in oil-boom-driven NASWA states. Besides low-level construction jobs, these workers serve as domestics, gardeners, trash collectors, and the like. In 2019, almost 90 percent of the total population of the United Arab Emirates and Qatar consisted of foreign employees! Over the past few years, substandard working conditions and wage issues have also led to protests that have exposed the harsh conditions faced by many guest workers throughout this realm.

10. **Diffusion of Revivalism.** Islamist regimes utilize oil and gas revenues to support Muslim communities as well as their mosques and cultural centers throughout the world. No NASWA country has spent more money on such causes than Saudi Arabia, and thousands of mosques from England to Indonesia prosper as a result. This example of relocation diffusion creates myriad nodes of recruitment for the faith—and ensures the dissemination of revivalist principles.

Markus Mainka/Alamy Stock Photo

The Burj Khalifa, centerpiece of the ultramodern city of Dubai on the Persian Gulf in the UAE. Completed in 2010, the 163-story Burj is still the world's tallest building, topping out at 828 meters (2716 ft)—just over half a mile high, almost twice the height of New York's Empire State Building.

Box 7-4 Technology & Geography

Global Water Needs and Desalination

No single natural resource defines a world geographic realm as oil does for North Africa/Southwest Asia. Yet it is access to fresh water that permits oil-rich economies, some of which are located in the extremely arid climates of the Sahara and Arabian deserts, to endure and even thrive. The fate of the ancient civilizations, cities, and communities of this realm were once solely tied to river systems like the Nile and Tigris–Euphrates. Today, the sustainability and future of developed economies and societies of this realm depend on desalination technology. **Desalination [12]** refers to the process of obtaining fresh water from saline (salt) water. There are two primary methods of desalination. The first, *multistage flash distillation,* uses heat to evaporate water, leaving the salt behind. The second, *reverse osmosis,* entails the forcing of salt water through specialized filters at extremely high pressures to purify the liquid. Although simple in theory, desalination is very expensive because it is such an energy-intensive process.

Desalination technology has advanced in this century, thereby cutting the cost, but it remains expensive. For that reason,

it is not uniformly available across this particularly uneven realm. Geographic variations in access to desalination, and more generally to potable water, are not merely a matter of having a location near the sea. Desalination capacity is also related to standards of living, economic development, and levels of health found across the realm. **Figure 7-7** shows the global variation in water stress levels (measured as available renewable fresh water per person) and the leading countries in desalination capacity (also on a per capita basis). Note the exceptional position of the rich oil states in the NASWA realm. For Saudi Arabia, one of the world's wealthiest countries and leading producers of oil, it is estimated that 15 percent of annual oil income is allocated to desalination operations. The combination of volatile oil prices, regional political instability, and rapidly increasing demand for both electrical power and water is pushing the kingdom to construct nuclear power plants to supply the energy for desalination and to wean itself from oil dependency. It remains to be seen whether or not, and how, the development of nuclear power will contribute to greater access to fresh water, economic growth and sustainability, and realm-wide security.

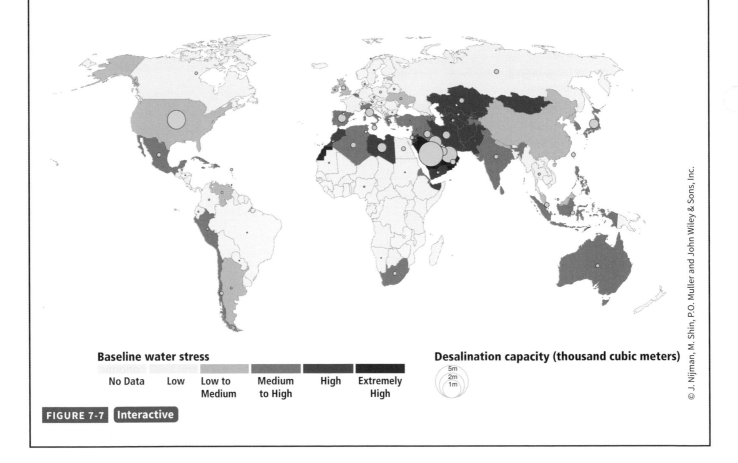

Baseline water stress

No Data | Low | Low to Medium | Medium to High | High | Extremely High

Desalination capacity (thousand cubic meters)

5m
2m
1m

© J. Nijman, M. Shin, P.O. Muller and John Wiley & Sons, Inc.

FIGURE 7-7 Interactive

Oil-Generated Inequalities

Perhaps more than any other realm, this one is marked by a human spatial mosaic of **fragmented modernization [11]**. To a certain degree NASWA is highly modern and prosperous,

but in many more respects it remains traditional, stagnant, and quite poor. These disparities are best explained by the uneven impact of oil, the lack of democracy, the uneven availability of technology (see **Box 7-4**), and the mercurial role of religion.

Oil brought this realm into contact with the outside world in ways unforeseen just a century ago. It has strengthened and empowered some of its peoples but has dislocated and imperiled others. It has truly been a double-edged sword. We should remind ourselves that the great majority of NASWA's inhabitants are not, in their daily lives, directly affected by the changes the energy era has brought. Most Moroccans, Tunisians, Egyptians, Jordanians, Yemenis, Lebanese, and countless millions of others—Kurds, Palestinians, Berbers, Tuaregs—make ends meet by trading or farming or working at jobs their parents and grandparents performed. Take the case of Iran, which as a country earns about two-thirds of its income from oil and natural gas. Only one-half of 1 percent of that country's workforce (just over 100,000 out of more than 20 million workers) is employed in the energy-related sector. By far the largest number engaged in any single occupation are Iran's five-million-plus farmers. And for all their oil, Iranians in 2018 earned an average yearly income of $5,500, just over half the earnings in neighboring, energy-poor Turkey. While it is important to consider the fundamental economic geography of the realm, the regional dynamics are in large part shaped by cultural and political forces.

The two dozen states that constitute NASWA often display stunning degrees of variety and diversity. This difference is accentuated by the severe disparities that result from the uneven distribution of the realm's oil wealth. And even within those countries possessing large energy reserves, glaring inequalities have arisen between their have and have-not citizens. Ultramodern urban skylines on the Arabian Peninsula stand in sharp contrast to upstream Nile Valley villages virtually unchanged for millennia. Modern expressways cross deserts still traversed by nomadic traders riding camels. Social geographies vary just as greatly—for instance, the female literacy rate in Lebanon is 92 percent but only 55 percent in Yemen.

Pervasive Political Geographies

The Absence of Democratic Traditions

Democracy has been a rarity in this realm. Although specific reasons vary, the majority of states share a history of autocratic, conservative regimes, usually backed strongly by their military forces. Most countries were part of the Ottoman Empire before World War I (see Fig. 7-5) and only became independent after a period of (attempted) European rule that was, in some cases, sanctioned by the League of Nations (the predecessor of the United Nations that functioned from 1920 to 1946). Accordingly, Egypt, Iraq, Palestine, Transjordan, and parts of Yemen were administered by Britain; Lebanon, Syria, Tunisia, Morocco, and Algeria were ruled by France; and Libya was controlled by Italy.

In most cases, the Europeans faced persistently tenacious opposition from local populations but were reluctant to let go. Independence was usually preceded by intense conflict or even war; at any rate, none of these countries can be said to have been prepared to govern themselves as democracies. Indeed, virtually everywhere the newly independent countries fell into the hands of the military (such as Egypt) or restored monarchs with strong military support (such as Yemen).

After World War II, these autocratic regimes were often cemented into place through foreign involvement. Mostly, this was about oil: foreign powers needed oil, showed little interest in domestic politics, and provided incumbent regimes with abundant revenues to build up their military forces, secure their political position, and maintain order. Saudi Arabia, the world's richest petrostate, is the quintessential example. In other cases, it was Cold War politics that effectively supported the realm's nondemocratic governments.

Political regimes in this realm differ, but at least until the so-called **Arab Spring [13]** movement emerged in late 2010, most heads of government had been in power for a substantial number of years, and their top priority was to preserve their domination and the status quo. Overall, their rule had been generally conservative and based on top-down control, with minimal concern for the "people on the street" (**Box 7-5, Fig. 7-8**).

The Arab Spring of 2011

On December 17, 2010, in the Tunisian capital of Tunis, a 26-year-old fruit-and-vegetable vendor named Mohammed Bouazizi became so distraught when police, following months of constant harassment, confiscated his unlicensed produce stand, that he set himself on fire. His act of desperation resonated with countless Tunisians whose frustration with their government had reached the breaking point. Bouazizi died soon afterward, but his self-immolation quickly came to symbolize Tunisia's "Jasmine Revolution." Within a month, popular opposition to the regime had grown so fierce that Tunisia's president was forced to resign after 23 years in power.

In Arab countries, according to conventional wisdom, not only democratic governance but even the desire for such a political system was nonexistent. That myth exploded as Tunisians of all backgrounds, including legions of women as well as youths, took to the streets to demand an end to official corruption, repression, cronyism, and economic mismanagement. The Tunisian people were insisting that they be heard—and their voices swiftly and loudly echoed across much of the realm.

What originated in Tunisia rapidly diffused eastward across North Africa and into adjacent parts of Southwest Asia. By the spring of 2011, popular uprisings had spread into Egypt, Libya, Syria, Yemen, and Bahrain. The conditions in all these countries were similar: they were ruled by long-established autocratic regimes that had failed to bring economic progress, repressed their own people, and had lost touch (most especially with the younger generation). Throughout the realm,

Box 7-5 Geographic Information Analysis

Development and Democracy

Worldwide, high levels of national development tend to be associated with democratic political systems that offer a full range of personal freedoms. Conversely, it is thought that the poorer the country, the less likely it is to be a democracy (for example, think of North America versus Subsaharan Africa). But all this can range considerably from one realm to another or within realms. Figure 7-8 maps GDP per capita against level of democracy for selected NASWA countries in 2017. Can a clear relationship between wealth and democracy be observed in this realm? Can you discern any regional patterns? Comparing this map with the one in Figure 7-6, do you think oil and gas production is associated with the spatial variations you observe?

Access your WileyPLUS course to interact with a dynamic version of this map and to engage with online exercises and questions.

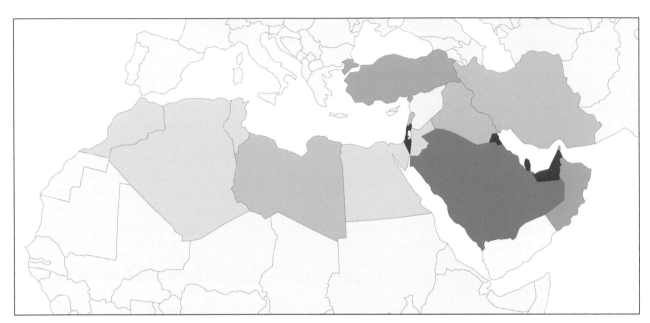

Gross domestic product (GDP) per capita, 2017 (inflation-adjusted U.S. dollars)

| No Data | < $2,500 | $2,500 - $5,000 | $5,000 - $10,000 | $10,000 - $20,000 | $20,000 - $30,000 | > $30,000 |

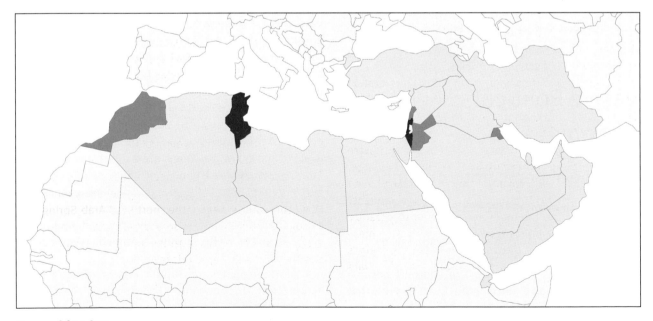

Personal freedom

| No Data | Not Free | Partially Free | Free |

FIGURE 7-8 **Interactive**

AP Images/Riadh Dridi

Tunisians celebrate the fifth anniversary of the Arab Spring on January 14, 2016, in Tunis. Teachers, students, activists, and political parties joined to celebrate five years since protesters drove out their autocratic president. Tunisia is one of a few countries where the Arab Spring ushered in what seem to be lasting democratic reforms.

people now closely followed these events via television and the Internet, and grassroots revolutionary fervor spread like wildfire. It may well have been the most spectacular **domino effect [14]** in recent history.

The pivotal tile in this row of falling dominoes was Egypt. The most heavily populated, historically prominent, and centrally positioned of the Arab countries, Egypt had for decades suffered from stunted economic growth and the stagnation of an out-of-touch, repressive government. In Cairo's Tahrir Square and elsewhere, the masses boldly took to the streets and soon forced the ouster of President Mubarak, even though protesters were repeatedly confronted by harsh, bloody encounters with the police and military forces; nearly 1000 demonstrators were killed in these clashes. The events in Egypt generated an even more powerful round of shockwaves that reverberated throughout the realm. If the tide could be turned here at the Arab epicenter, reasoned the fearless throngs of protesters in several other countries, it could happen anywhere.

In most NASWA countries, more than half the population is under 25 years of age. This enormous demographic cohort has grown up in a world of social media, so they know only too well that their home countries deny them many opportunities that exist elsewhere. Not surprisingly, the 2011 uprisings were predominantly (though not exclusively) led by the younger generation, ranging from teenagers to thirty-somethings—and included many women. In Tunisia, Egypt, and Syria, students took the lead, using the Internet's social media to organize demonstrations. This emphasized another critical dimension of the Arab Spring: the disconnect between the realm's aging, entrenched autocrats and their young, globally aware subjects.

Unfulfilled Arab-Spring Promises

Tunisia, where it all began in December 2010, emerged as the one success story of the Arab Spring uprisings. Here, a long-reigning dictator was ousted and replaced by a democratically

elected government. This fragile democracy was still in place in 2019, though thousands of people seemed to have become disillusioned with the failure of the political system to provide improved economic conditions.

Elsewhere, things looked worse. If the very term *Arab Spring* expressed the hopefulness of a new political dawn across this realm, such romantic notions were rapidly dispelled. First of all, even though the spate of uprisings resonated and jangled nerves across the Arabian Peninsula, they did not seriously challenge any of that region's governments (except in peripheral Yemen). It was here—as well as in Jordan, Morocco, and Iran—that the established monarchies and/or theocracies prevailed, with only modest responses to popular demands for reform.

Second, the more overtly tyrannical regimes in Libya and Syria responded with stubborn resistance and iron fists. Libya quickly degenerated into civil war as President Muammar Qadhafi unleashed his military forces against his unarmed people. That attack was so vicious that a makeshift alliance of France, the United Kingdom, and the United States (later subsumed by NATO) intervened with aerial bombing raids to help protect the Libyans from annihilation by their own government. By October 2011, the tide had turned, the capital of Tripoli fell, and Qadhafi was cornered and killed in his hometown of Sirte on the central coast. But that did not end the conflict. The country's power structure soon imploded, the civil war intensified, and by 2015 Libya had become a failed state.

In Syria, protesters were met with nothing less than state-sanctioned violence perpetrated by the army and the air force, aimed at all civilians who opposed the dictatorial regime of President Bashar al Assad. This ruthless campaign of repression not only triggered a brutal civil war in 2011 that still continues, but also fractured the country in a manner that lured in some of the most radical extremists, who quickly turned the conflict into a regional war. By mid-2019, the death toll resulting from this unrelenting carnage was estimated to have exceeded half a million.

And third, where it had appeared that a transformation of government had been achieved—most notably in Egypt—democratic prospects steadily faded as the military conspired to retain *de facto* control over the country. In retrospect, the Arab Spring was fueled in large part by the aspirations of masses of young people yearning for economic progress and democracy; however, without the institutions for a successful transition (such as an independent judiciary, a free press, and a multiple-party electoral system), they faced an uphill struggle. Five years after it began, the short-lived Arab Spring had run its course—its initial promise and excitement gradually eroding into crushed hopes, with the inexorable restoration of the business-as-usual autocratic regimes.

Politics and Religion

The Arab Spring was also driven by populist grievances concerning the suppression of religious freedoms. Keep in mind that almost all of the governments in this realm were (and are) secular militaristic regimes that view Islamic parties and organizations with suspicion and as threats to political stability. Religious leaders and activists saw their opportunity as part of the wider revolt that was primarily aimed at the removal of repressive regimes. Thus, the protest movement, overall, was both complex and potentially conflicted. The notions of democracy that stirred so many of the young protesters ran counter to the fundamentalist views of reactionary Islamists, a divide that added yet one more obstacle to much awaited social change in this realm.

To the extent that the Arab Spring exposed the vulnerability of some long-entrenched secular regimes, it may have unintentionally fueled Islamic fundamentalism—or, more broadly **religious revivalism [15]**. That resurgence had already been in the making for decades. During the 1970s, the *imams* (mosque worship leaders) in Shi'ite Iran sought to reverse the shah's moves toward liberalization and secularization: they wanted to (and soon did) recast society in a traditional, revivalist Islamic mold. An *ayatollah* (leader under Allah) replaced the deposed shah in 1979; Islamic rules and punishments were instituted. Urban women, many of whom had been considerably liberated and educated during the shah's regime, were forced to return to more traditional roles.

The rise of Islamic revivalism was neither limited to Iran nor confined to Shi'ite communities. Many Muslims—Sunnis as well as Shi'ites—in various parts of the realm disapproved of the erosion of traditional Islamic values, the corruption of society by European colonialists and later by Western modernizers (which brought economic benefits only to the elites), and the declining power of the faith within the secular state. As noted earlier, most governments in the NASWA realm continue to be both secular and repressive, and they do not provide the economic progress their citizens yearn for. Put in those terms, it is not that difficult to understand why many people seek solace in religion as a way to regain hope and dignity.

Some have taken their return to the faith much farther and imbued it with a fanaticism that is alien to most Muslims—and this has set Muslim against Muslim in various parts of the realm. Some revivalists fired the faith with a new militancy, aggressively challenging the status quo from Iran to Algeria. These militants forced their governments to ban "blasphemous" books, to resegregate schools according to gender, to enforce traditional dress codes, to legitimize religion-based political parties, and to heed the wishes of the *mullahs* (teachers of Islamic ways). Militant Muslims also proclaimed that Western ideas about modernization, inherited from colonialists and adopted by Arab nationalists, were incompatible with the dictates of the Quran.

Terrorism in the Name of Islam

The great majority of Muslims are not fundamentalists; not all fundamentalists are militants; and not all militants are terrorists. Most religious scholars point out that the Islamic faith is no more predisposed toward militancy than other religions. And we should not forget that terrorism is a tool of war that has been used around the world, and indeed throughout modern history, by people of several different beliefs and ethnic backgrounds.

But nowadays the religion of Islam has proven to be a significant vehicle for political mobilization and a powerful source of inspiration for militants with terroristic agendas. Certainly, many Muslim adherents contest this extreme interpretation of the Quran; yet, on the other hand, there is little doubt that militants justify their actions as expressions of their particular fundamentalist beliefs. Theirs is a "holy war"—a *jihad* [16]—and as such it is a deeply reactionary movement that looks to the past rather than to the future.

Broadly speaking, two extremist Islamist organizations are dominant today in spreading terrorism within the realm—and sometimes beyond. The first is *Al Qaeda* (the name means "base" or "foundation"), an internationally networked organization whose goal is to establish Islamic rule across NASWA. They originated in Afghanistan in the late 1980s and have since masterminded terrorist acts in many other countries, from Kenya to Indonesia to the United States (where Al Qaeda was responsible for the 9/11 attacks). After their leader, Osama Bin Laden, was killed by U.S. special forces in Pakistan in 2011, the organization was weakened and forced to decentralize. Currently, Al Qaeda's main "branch" is constituted by *AQAP*—shorthand for *Al Qaeda in the Arabian Peninsula*. It is especially active in Yemen, where it controls key territory and worked to destabilize that country to the point of its 2015 collapse into chaotic civil war.

The other leading Islamist terrorist organization is ISIS, which has rapidly gained importance since 2013. The acronym *ISIS* stands for *Islamic State of Iraq and Syria*; the organization is also known by the acronyms *IS* (*Islamic State*) and *ISIL* (*Islamic State of Iraq and the Levant*). ISIS is notoriously violent and employs unprecedented tactics of intimidation, its videos of decapitations of "infidels" sending shockwaves around the globe. This highly militarized organization emerged in Iraq after the U.S. invasion of 2003 and initially declared allegiance to Al Qaeda. Thereafter, it steadily assumed a more independent role in Iraq and then expanded into unguarded, sparsely populated eastern Syria following the outbreak of civil war in 2011. Exploiting the chaos of that war and Sunni–Shi'ite schisms in both countries, ISIS effectively rallied Sunnis against secular governments that were dominated by Shi'ites.

ISIS: From International Network to Territorial State, and Back

By 2015, ISIS had become an established force that controlled sizeable swaths of territory in eastern Syria as well as in northern and western Iraq (**Fig. 7-9**). At its greatest extent, it covered about 88,000 square kilometers (34,000 sq mi), though much of it noncontiguous. ISIS evolved over a short period of time from a dispersed terrorist organization into a territorial quasi-state (perhaps even a formative **insurgent state [17]**,

FIGURE 7-9

with the Syrian city of Raqqa serving as its capital). Its armed forces were estimated by the U.S. government to total somewhere between 20,000 and 30,000 fighters (ISIS claimed it had 200,000). Strict Islamic (*sharia*) law prevailed in its territories; common people and shopkeepers were taxed; and control of oilfields and infrastructure provided sufficient revenues to fund its military operations.

As ISIS tried to hold and consolidate its territorial gains, its ambitions in this realm were clearly not limited to parts of Iraq and Syria. The designs of ISIS reached far beyond NASWA: it declared itself to be nothing less than a **caliphate** [18]—an imperial-scale Islamic government led by a direct successor to the Prophet Muhammad who rules and exerts moral authority over Muslims worldwide. History's most recent caliphate was the Ottoman Empire, whose caliphs ruled from 1299 until abolition in 1922. Were the ISIS caliphate to succeed, the notorious terrorist leader Abu Bakr al-Baghdadi was long designated as its 'Caliph'—but he was killed by U.S. special forces in October of 2019. Since the mid-2010s, ISIS has claimed responsibility for terrorist attacks outside of the realm, most notably in Europe but also, for example, in Sri Lanka.

The threat posed by ISIS in Syria intensified so rapidly that it essentially dissipated much of the international

opposition to the Assad regime and its war crimes against hundreds of thousands of Syrian civilians. By the end of 2015, ISIS came to be regarded as the greater danger, and the U.S.-led allied coalition (which included Saudi Arabia,

The city of Raqqa, Syria, on August 12, 2017. Raqqa was the "capital city" of ISIS between 2013 and 2017. At the time of this photo, much of the city was destroyed by coalition air strikes, and it was surrounded on all sides by Syrian government forces who would soon retake the city.

the UAE, the UK, and France) started intensive campaigns of air strikes on ISIS strongholds. Russia, which had emphatically declined to join the U.S. coalition and which had continued its close relationship with the Syrian regime, initiated its own air strikes in 2015 but in a way that clearly supported the Syrian government forces. The Russians bombed both ISIS strongholds as well as other Syrian rebel forces, including Kurdish positions.

In Iraq, the Shi'ite-dominated government forces, with support from the U.S. in the form of air strikes and military training, aligned with Kurdish forces in the northeast. Here, too, the ISIS strongholds were destroyed. In March 2019, ISIS lost its final stronghold around the city of Al Bukamal on the Syrian–Iraqi border (Fig. 7-9).

If this meant the end of ISIS as a kind of territorial state, the organization itself survived, as it regrouped as an international terrorist network, loosely connected in many ways but with a strong ideological core and with continued appeal to militants inside and outside of the realm. For example, the bombings of several churches in Sri Lanka in 2019, killing hundreds and wounding many more, were claimed by ISIS-inspired terrorists, and ISIS leader al Baghdadi praised the perpetrators in a later video recording. Within the realm, there has also been a regional shift, since around 2017, of jihadist activity from Southwest Asia to Africa's Transition Zone. This is also where U.S./allied forces have turned their attention, with an estimated 15,000 troops deployed mainly in Mali, Niger, and Burkina Faso. The African Transition Zone is one of the regions discussed in more detail toward the end of this chapter.

Ironically, a parallel result of the focus on ISIS by allied forces and by Russia was the strengthening of the—much condemned—Syrian government's position. And for most Syrian people, little had changed. About a quarter of all Syrians had fled their country and remained abroad in 2020 (Fig. 7-9).

The Refugee Problem

NASWA's political instabilities, raging conflicts, social dislocations, and associated economic problems have generated a rising tide of refugee flows in recent years. No longer limited to internal, short-distance, cross-border population shifts, these migratory movements were increasingly directed toward Europe and reached real crisis proportions in 2015 and 2016 (also see Chapter 5). Most refugees from the Syrian war are still housed in camps in neighboring countries that have been overwhelmed, especially Lebanon, where refugees now account for more than one out of every six inhabitants (Fig. 7-9). Additional hundreds of thousands of Syrian refugees have taken flight and attempted to gain entry to Europe. And they were hardly alone, joining broad streams of Europe-bound migrants that originate in a wide range of troubled areas in Southwest Asia, North Africa, and Subsaharan Africa (**Fig. 7-10**).

The number of detected illegal border crossings into Europe rose from around 400,000 in 2014 to more than 2 million in 2015. The two main refugee routes were either to cross the waters of the central and eastern Mediterranean Sea or to proceed overland through Turkey into Greece or Bulgaria. Most attempts to cross the Mediterranean were made from the coasts of Libya and, to a lesser extent, Tunisia and Egypt; the distance to Lampedusa, Italy's southernmost island south of Sicily, is just over 100 kilometers (65 mi) from Tunis and about 250 kilometers (150 mi) from Tripoli (Fig. 7-10). These desperate illegal migrants, however, were largely in the hands of smugglers, who charged exorbitant fees to cram as many as possible into their rickety boats for the dangerous voyage north.

For most of these refugees, whatever their source, the greatest challenge of their journey begins at the Mediterranean shore. Among those fortunate to have completed the treacherous maritime passage, thousands are refused asylum and are forced to return to their country of origin. Nobody knows how many undocumented migrants perish at sea: in 2015, about one million were known to have crossed the Mediterranean, and the most realistic estimates tell us that many tens of thousands drowned. This colossal human tragedy continued unabated into 2016 (see photo below), and started to decrease only in 2017. The reason for the decrease was the drastic tightening of borders around Europe and deals made between EU countries and Turkey and Libya, inducing the latter two to discourage and halt the massive refugee streams passing through their jurisdictions (Fig. 7-10; also see Chapter 5).

A Realm in Turmoil

The Arab Spring movement failed to achieve the diverse goals of the protesting masses in the streets of Cairo and most other cities, but it did destabilize those parts of the NASWA realm in and around the countries that experienced the greatest violence. A good example is Libya, whose political breakdown unleashed a wide spectrum of radical Islamist groups across

May 25, 2016: Migrants in a capsizing overcrowded boat in the Mediterranean Sea between Libya and Italy. The Italian navy, on this occasion, reported saving about 500 migrants and recovering seven dead bodies, with the total number of deaths probably considerably higher.

Anadolu Agency/Getty Images

FIGURE 7-10

northern Africa, especially in Mali. And what began as an internal conflict in Syria soon escalated into a dangerous regional war when it spread into neighboring Iraq—drawing in a wide range of combatants including ISIS, Turkey, Kurdish militias, Iran (traditional supporter of the Shi'ite cause in Syria and Iraq), bordering states swamped by refugee flows (such as Lebanon and Jordan), and the outside powers of Russia and the U.S.-led,

anti-ISIS coalition. NASWA, and its Middle East region in particular, have a long history of conflict, and the present turmoil is as intractable as ever. Recently, one Arab author gloomily observed that "history in this part of the world does not move forward: it slips sideways." Clearly, for many young people in this realm, such a view is not really an option, and many continue to aspire to living in peace, freedom, and prosperity.

Regions of North Africa/Southwest Asia

Identifying and delimiting regions in the sprawling North Africa/Southwest Asia realm is quite a challenge. Populations are widely scattered in some areas but highly concentrated in others. Cultural transitions and landscapes—internal as well as peripheral—make it difficult to discern a regional framework. Furthermore, this is a highly changeable realm and always has been. Nonetheless, we can distinguish the following six regions (**Fig. 7-11**):

1. **Egypt and the Lower Nile Basin.** This region in many ways constitutes the heart of the realm as a whole. Egypt

(together with Iran and Turkey) is one of the realm's three most populous countries. It is the historic focus of this part of the world as well as a major political and cultural force. Also included is Sudan, the truncated, northern portion of the much larger former Sudan that split apart in 2011.

2. **The Middle East.** This pivotal region includes Israel, Jordan, Lebanon, Syria, and Iraq. In effect, it is the crescent-shaped zone of countries that extends from the eastern coast of the Mediterranean to the head of the Persian Gulf.

FIGURE 7-11

3. **The Arabian Peninsula.** Dominated by the enormous territory of Saudi Arabia, the Arabian Peninsula also encompasses the United Arab Emirates (UAE), Kuwait, Bahrain, Qatar, Oman, and Yemen. Here lies the source and focus of Islam, the holy city of Mecca; here, too, lie many of the world's greatest oil deposits.

4. **The Empire States.** Consisting of two of the realm's giants—Turkey and Iran—we refer to this region as the Empire States because each functioned as the core of a major historic empire. They are located on the northeastern edges of the realm.

5. **The Maghreb and Its Neighbors.** Western North Africa (known as the *Maghreb*) and the lands that border it also form a region, consisting of Algeria, Tunisia, and Morocco at the center, and Libya, Chad, Niger, Mali, Burkina Faso, and Mauritania along the volatile African Transition Zone where the Arab-Islamic realm of North Africa merges into Subsaharan Africa.

6. **The African Transition Zone.** This area is the complicating factor on the regional map of Africa. In Figure 7-11, note that this striped zone of increasing Islamic influence to the north fully incorporates certain countries (e.g., Somalia in the east and Senegal in the west) while cutting across others, in the process creating Islamized northern sectors and non-Islamic southern ones (e.g., Nigeria, Chad, Ivory Coast). Note, too, that an elongated strip of Islamic dominance extends southward from the Horn of Africa, lining the Indian Ocean coastlands of both Kenya and Tanzania.

Region Egypt and the Lower Nile Basin

Egypt occupies a pivotal location in the heart of a realm that extends more than 9600 kilometers (6000 mi) longitudinally and some 6400 kilometers (4000 mi) latitudinally. At the northern end of both the Nile and the Red Sea, at the eastern end of the Mediterranean Sea, in the northeastern corner of Africa across from Turkey to the north and Saudi Arabia to the east, adjacent to Israel, to Sudan,[2] and to Libya, Egypt unmistakably lies in the crucible of the NASWA realm (**Fig. 7-12**). Because it also extends into the

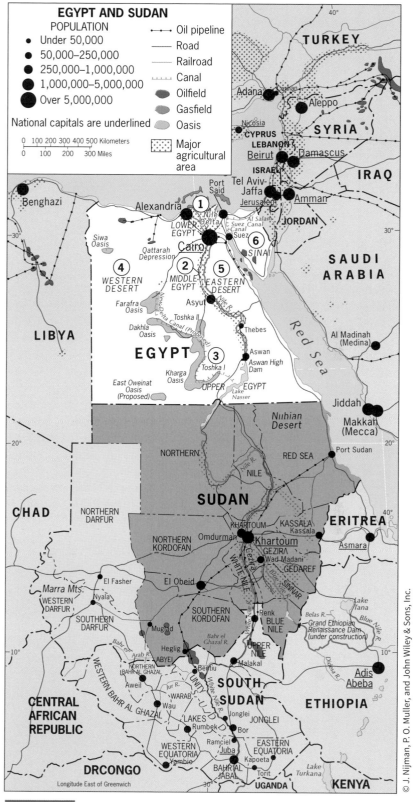

FIGURE 7-12

© J. Nijman, P. O. Muller, and John Wiley & Sons, Inc.

---

[2]*Sudan* refers to the truncated, northern portion of the former country of Sudan that remained after South Sudan (discussed in Chapter 8) achieved independence and split away in 2011. The much larger country that existed before this breakup is referred to in this book as *former Sudan*.

Sinai Peninsula, Egypt, alone among countries on the African continent, has a foothold in Asia. This foothold gives it a coast overlooking the strategic Gulf of Aqaba (the northeastern arm of the Red Sea). Egypt also controls the Suez Canal, the vital artificial waterway that links the Indian and Atlantic oceans and forms one of Europe's lifelines. The capital, Cairo, is the realm's largest city and a leading center of Islamic civilization.

Egypt has six subregions, mapped in Figure 7-12. Most Egyptians live and work in Lower (i.e., northern) and Middle Egypt, subregions ① and ②. This is the country's core area, anchored by Cairo and flanked by the leading port and second-largest manufacturing center, Alexandria. The other four subregions are remote and sparsely populated, especially Upper Egypt ③, the Western Desert ④, and the Sinai ⑥. The northern stretches of the Eastern Desert ⑤ are now in the path of Cairo's encroaching urban sprawl (photo below in **Box 7-6**).

Gift of the Nile

Egypt's Nile is the aggregate of the two great branches upstream: the White Nile, which originates in the streams that feed Lake Victoria in East Africa, and the Blue Nile, whose source lies in Lake Tana in Ethiopia's highlands. The two Niles converge at Khartoum, capital of present-day Sudan. About 95 percent of Egypt's 102 million people live within 20 kilometers (12 mi) of the great river's banks or in its delta (Figs. 7-12, 7-2).

Before dams were constructed on the Nile, the ancient Egyptians used **basin irrigation [20]**, building fields with earthen ridges and trapping the annual floodwaters with their fertile silt, to grow their crops. That practice continued for thousands of years until, during the nineteenth century, the construction of permanent dams made it possible to irrigate Egypt's farmlands year-round. These dams, with locks for navigation, controlled the Nile's annual flood, expanded the country's cultivable area, and allowed farmers to harvest more than one crop per year on the same field. The largest of these dams, the Aswan High Dam completed in 1970, created Lake Nasser (the reservoir that extends southward 150 kilometers [100 mi] into northern Sudan) and increased Egypt's irrigable land by nearly 50 percent. Today it still provides a sizeable proportion of the country's electricity.

Much farther upstream, in the western highlands of Ethiopia where the Blue Nile originates, its government is constructing the Grand Ethiopian Renaissance Dam, with much of the funding coming from China (Fig. 7-12). Note that the location of this dam is near the Sudanese border. It is a common strategy that minimizes possible negative consequences for Ethiopia itself, such as reduced water flow and entrapment of fertile sediments. Instead, such possible consequences, also called **externality effects [21]**, will be felt downstream in Sudan and Egypt. After completion in (probably) 2022, the dam is expected to boost electrical power production to meet all domestic demands as well as earn new revenues from sales to Ethiopia's energy-deficient neighbors. But the project has also attracted strong opposition (and raised the potential for conflict) in Sudan and Egypt.

Egypt's elongated oasis along the Nile, just 5 to 25 kilometers (3 to 15 mi) wide, broadens north of Cairo across a delta anchored in the west by the historic city of Alexandria and in the east by Port Said, gateway to the Suez Canal. The delta contains extensive farmlands, but it is a troubled area today. The ever more intensive use of the Nile's water and silt upstream is depriving the delta of desperately needed replenishment. The low-lying delta is also at risk from geological subsidence and sea-level rise, increasing the danger of saltwater intrusion from the Mediterranean that can severely damage soils.

Box 7-6   From the Field Notes . . .

"Taking off from Cairo International Airport at the northeastern edge of this megacity of nearly 17 million, we flew over one of Cairo's newest housing developments at the edge of the northern sector of the Eastern Desert. Look at this residential density—in a desert! With a birth rate more than twice that of the United States, Egypt's critical problem is the provision of a sufficient water supply—which right now is steadily decreasing. These are middle- and upper-middle-class residential complexes. All those new apartment towers will require water, and so will anything green planted here. Where will all these people work, and how will they get there? Is this sustainable urban development?"

© J. Nijman

Political Upheaval and Economic Challenges

Before 2011, Egypt had experienced steady economic growth for almost a decade at an annual rate of around 5 percent. At the same time, its annual rate of population growth declined while literacy and per capita consumption increased. Egypt became an exporter of fruits, vegetables, rice, and textiles in addition to its longtime staple, cotton. The country also nurtured a thriving tourist industry that accounted for about one-tenth of GDP. And the exploitation of (modest) oilfields in the Sinai and adjacent Red Sea (Fig. 7-12) resulted in reduced expenditures on fuel imports.

But then came the tumultuous events of the Arab Spring and its aftermath, which decimated the tourist industry and damaged much of the rest of the economy. Politically, it was hoped that Egypt might be moving toward multiparty democracy, but instead it returned to the old ways of repression of both Islamists and more Western-oriented democratic movements.

Once political stability was restored, however, the economy regained momentum as the government started to actively pursue pro-growth policies. This included economic reforms along with major infrastructural projects to spur employment and reduce popular discontent. In 2015, the Suez Canal was enlarged to accommodate two-way traffic, double the canal's capacity (unlike the Panama Canal, there is minimal relief and locks are unnecessary), as well as increase traffic and revenues. A further boost came with the discovery in 2018 of additional massive natural gas fields in Egyptian territorial waters (Fig. 7-12), which will soon provide enough for all Egyptian gas consumption plus growing energy exports. By 2019, economic growth had returned to the previously high levels. Newly constructed suburbs have started to dominate Cairo's outer metropolitan ring (see Box 7-6). It seems that economic progress has calmed some of the voices demanding political reforms. And, by 2019, tourism was picking up again. From an all-time high of 14.7 million tourists in 2010, the number had dropped to 4.8 million in 2015, then climbed back up to an estimated 9 million in 2018 with further growth expected in 2019 and 2020.

Sudan

As Figure 7-12 indicates, Egypt is bordered by two countries that have recently gone through even more turbulent times than Egypt itself: Sudan to the south, and Libya to the west. Sudan, the northern remnant of the 2011 breakup of former Sudan, is almost twice the size of Egypt (but with less than 45 percent of the population) and lies centered on the confluence of the White Nile (from Uganda) and the Blue Nile (from Ethiopia). Here the capital, Khartoum, anchors a sizeable agricultural area where cotton was planted during colonial times. Almost all of Sudan is desert, with irrigation-based farming along the banks of the White Nile and the Blue Nile. The country also has a 500-kilometer (300-mi) coastline on the Red Sea, where Port Sudan lies almost directly across from Jiddah and Makkah (Mecca) in Saudi Arabia.

Note that a large part of Sudan is situated in the African Transition Zone (Figs. 7-11 and 7-20), which helps to explain some long-running regional conflicts. The triad of southwestern provinces named *Darfur* (Fig. 7-12) are inhabited by the *Fur*, a mix of Arabized pastoralists and settled farmers who in past years apparently sympathized with anti-Muslim rebels in the far south of former Sudan. This pretext led to a gruesome conflict with the Khartoum government's military forces in the early 2000s, and as many as 400,000 (mostly civilians) may have died. The southern part of former Sudan, in contrast to the north, was in many ways a typical Equatorial African entity, dominated culturally by Christianity and animist African religions. Ongoing conflicts between north and south were exacerbated by the discovery of oil on contested land. A long, bloody war ensued, with an estimated 1.5 million deaths. The war finally ended in the breakup of former Sudan, with the south gaining independence in 2011 as the newborn country of South Sudan (discussed separately in Chapter 8).

Sudan today is a far more homogeneous Islamic state, with a Muslim population of more than 95 percent (compared to the 68 percent of pre-breakup Sudan). Even though the still unofficial international boundary puts the majority of former Sudan's oil reserves in South Sudan, the pipelines for exports run northward across post–2011 Sudan. Both Sudan and South Sudan remain involved in costly military operations and internal as well as cross-border conflicts with each other—while their populations remain in critical need of economic and political progress. In 2018 and early 2019, popular protests against the Sudanese government escalated and on April 11, 2019, with the help of members of the armed forces, the 75-year-old president Omar al-Bashir, was forced out after 30 years in power. After the ouster of the 82-year-old Algerian president one week earlier, these were signs that the Arab Spring lives on in some parts of the realm, after all.

Region The Middle East

The regional term *Middle East*, as noted at the beginning of this chapter, is not satisfactory, but it is so common and generally used that avoiding it creates more problems than it solves. It originated when Europe was the world's dominant realm and when places were *near*, *middle*, and *far* from Europe: hence a Near East (Turkey), a Far East (China, Japan), and a Middle East (Egypt, Arabia, Iraq).

Five countries constitute the Middle East (**Fig. 7-13**): Iraq, largest in population and territorial size, facing the Persian Gulf; war-torn Syria, next in both categories and fronting the Mediterranean; Jordan, linked by the narrow Gulf of Aqaba to the Red Sea; Lebanon, whose survival as a unified state is a perennial question; and Israel, a Jewish nation in the crucible

FIGURE 7-13

of the Islamic domain. The complexity of this region today is heightened by the presence and aftermath of two nonstate entities: Kurdistan and ISIS (though the latter had lost its territory by late 2019). Because of the extraordinary importance of this region in world affairs, we focus on its cultural, economic, and political geography.

Iraq's Enduring Importance

Even a glance at the regional map (Fig. 7-13) underscores why Iraq is crucial among the states of the Middle East. With about 60 percent of the region's total area, 48 percent of its predominantly Arab population, and most of its valuable energy and agricultural resources, Iraq is key to the Middle East's fortunes.

Having no less than six neighbors, it is no surprise that Iraq is practically landlocked, and this map shows just how narrow and congested its single outlet to the Persian Gulf is. As a result, a network of pipelines across Iraq's neighbors must carry much of Iraq's oil export volume to coastal terminals in other countries.

Iraq is divided into three cultural domains in which religion, ethnicity, and tradition form the basis of its political geography (Fig. 7-13). The largest population sector is composed of about 20 million Shi'ites (about 63 percent of the national total of 40 million) in the southeast, whose religious affinities are with neighboring Iran. The next largest is the Sunni minority of some 12 million (about 32 percent) in the north and west. Perhaps the most striking feature on this map is the gray striped zone that blankets northeastern Iraq, large parts of Turkey and Iran, and a few pockets of northern Syria. In this territory, the majority of inhabitants are not Arabs but Kurds—regarded by many as the world's biggest stateless nation. More specifically, about 6 million Kurds (16 percent of the Iraqi population) live in the northern highlands of Iraq; adjacent southeastern Turkey is the traditional home of at least 15 million more; another 6 million or so inhabit northwestern Iran; and 2 million congregate in the scattered Kurdish clusters of northern Syria. The large majority of Kurds, by the way, are also Sunni Muslims.

The U.S.-Led Invasion and Its Aftermath
The United States invaded Iraq in 2003, claiming that it possessed nuclear and other weapons of mass destruction and that Al Qaeda had important bases there. Neither charge proved to be true, and the invasion has been heavily criticized ever since. When American combat troops finally departed in 2011, the dictator Saddam Hussein had been eliminated, a new democratically elected government was functioning, and U.S.-trained police and military forces were in place.

But the U.S. incursion had also unleashed sectarian violence between majority Shi'ites and minority Sunnis as well as enabling Al Qaeda terrorists (who had not previously been active inside Iraq) to undermine the country's security situation. Even more ominously, these events empowered the Sunni jihadists of ISIS (as discussed earlier), who established themselves in the north and west, challenged the government in Baghdad, and effectively plunged Iraq into civil war (Fig. 7-9). After 2011, U.S. military involvement, as part of a multinational coalition, was confined to air strikes on ISIS military targets in northern Iraq and Syria. But in early January of 2020, the U.S. government executed a surprise drone attack and assassinated Iran's top-ranked general, Suleimani, who was in

Thousands of Iranians take to the streets to mourn the death of Iranian Revolutionary Guard Commander Qassem Soleimani during a demonstration after Friday prayer in Tehran, Iran, on January 3, 2020. The day before, Soleimani was killed by a U.S. drone strike in Baghdad, Iraq. See Box 9-3 (Chapter 9) for a Technology & Geography feature about military drones.

Maryam Rahmanian/UPI/Alamy Live News/Alamy Stock Photo

Iraq at that time. Within days, Iraq's parliament voted for the removal of all U.S. military personnel, and Iran's government vowed revenge. At the time of writing, the full ramifications of the assassination were not yet clear (photo below).

A Kurdish State in the Making?
The Kurdish homeland in northern Iraq is independent in all but name. The Kurdish nation is the product of diverse ethnic influences. Its language has Persian rather than Arabic roots. The majority adhere to Sunni Islam, although the Kurds are known to be rather pluralistic in their religious beliefs and practices. Like many other Sunni communities in Iraq and Syria, the Kurds had no sympathy for ISIS, whose violent jihadist offensives were met with fierce resistance from Kurdish armed forces supported by the air power of the U.S.-led coalition. Kurdish aspirations benefited from American support (both financial and diplomatic) since the U.S. invasion of Iraq in 2003. The nation became politically stabilized and more or less democratic, and its economy was fueled by the significant revenues from its major oilfields. By 2014, it seemed the process aimed at achieving statehood (including a popular referendum) was well under way.

However, the notion of a Kurdish state is a most delicate political matter because "Kurdistan" extends well beyond Iraq into Turkey and Iran—strongly organized states that have no interest in setting a precedent that might force them into territorial concessions. Moreover, the defeat of ISIS as a territorial power in Iraq and Syria was likely to render the Kurds less relevant to U.S./allied interest, and support for their cause of statehood seemed to be fading.

Syria

Like Lebanon and Israel, Syria possesses a Mediterranean coastline where crops can be raised without irrigation. Behind this densely populated coastal zone, Syria has a much larger interior than these neighbors, but its areas of productive capacity are widely dispersed, and many have been unusable during the ongoing civil war (Fig. 7-13). Damascus, the capital located in the southwestern corner of the country, was built on an oasis and is considered to be one of the world's oldest continuously inhabited cities.

Although Syria's population of 17 million (which has dwindled due to the war) is 74 percent Sunni Muslim, the ruling elite has long come from a small Shi'ite Islamic sect based in the country—the Alawites—who account for only about 12 percent of the population. Leaders of this powerful minority have retained control over the state for decades by ruthlessly suppressing dissent. In 2000, president-for-life Hafez al-Assad died and was succeeded by his son, Bashar.

When the Arab Spring movement reached Syria in 2011, it was immediately met with a savage military crackdown by the Assad regime. But popular resistance did not disappear, and the violence swiftly escalated into a full-fledged civil war marked by regime-inflicted atrocities on civilian populations in cities as well as the countryside. By early 2019, over half a million people had died in the violence, and countless additional thousands were reported to be incarcerated or missing. This conflict has also displaced some 13 million people—around 7 million forced to relocate within Syria and nearly 6 million others who fled the country for the squalid refugee camps of neighboring Turkey, Jordan, and Lebanon (Fig. 7-9). Internationally, even though the murderous Assad regime has been almost universally condemned for perpetrating the worst humanitarian crisis of our time, it continues to receive steady (if tacit) support from Russia, Iran, and China.

As discussed earlier, it did not take ISIS long to enter the fray in Syria, and soon its own brutalities reached a level that undercut international criticism and opposition to the Syrian regime. Figure 7-9 starkly illustrates the fragmentation of Syria by 2015. With the destruction of ISIS as a territorial power in 2019, the Syrian regime of Bashar al-Assad seemed to emerge as a winner, notwithstanding the almost universal condemnation of that regime. But strong internal opposition (including that of the Kurds) remained. Even more than in Iraq, it is difficult to imagine a future restoration of Syria as an integrated national territorial state.

Jordan

With a lack of oil reserves and possession of only a small and remote outlet to the Gulf of Aqaba, Jordan (population 10 million) has survived with U.S., British, and other aid. Remittances from around 800,000 Jordanian workers in the Arabian Gulf states are also vital to the economy. Jordan lost its West Bank territory (i.e., the area west of the Jordan River) in a 1967 war with Israel, including its sector of Jerusalem (then the kingdom's second-largest city). No third country has a greater stake in a settlement between Israel and the Palestinians than does Jordan, especially because Palestinians form the majority here.

Jordan did not experience the kind of upheaval that shook many other Arab countries in 2011 because the monarchy had enjoyed widespread support and government policies lacked the level of repression found in Syria or Tunisia. Yet there were some demonstrations to air grievances, and King Abdullah II had to dismiss his cabinet to quell the protests. Jordan's long-term challenges are primarily economic. The wider regional turmoil has brought in nearly 700,000 refugees across Jordan's northern border with war-ravaged Syria. It is also important to recognize the remarkable long-term stability of this country within this most volatile of political environments—perhaps comparable to Morocco in this regard, and suggesting that responsible NASWA monarchies have the ability to sense and respond to the pulse of the people. Jordan has exhibited considerable generosity toward Syrian refugees, but the country has suffered a severe decline in tourism (e.g., to the historic site of Petra—see photo) due to its close proximity to the region's wars.

Lebanon

The map suggests that Lebanon has significant geographic advantages in this region: a lengthy coastline on the Mediterranean Sea; a well-situated capital, Beirut, on its shoreline; oil terminals along its coast; and a major capital (Syria's Damascus) in its hinterland. The scale of Figure 7-13 cannot reveal yet another asset: the fertile, agriculturally productive Bekaa Valley in the eastern interior. Nevertheless, if Lebanon is well endowed from a physical-geographic perspective, its chaotic cultural geography and fragile geopolitical circumstances are problematic. As a result, the country has gone back and forth between good times and bad. Lebanon (population nearly 7 million) is sometimes aptly described as a "garden without a fence" because of its vulnerability to outside interference (especially from neighboring Syria), its latest burden being the more than one million refugees it has taken in from Syria's civil war. This means that in 2019 one in five persons in Lebanon was a refugee. With its economy facing severe challenges and so many external powers and influences involved in its affairs, Lebanon is unlikely to achieve stability and significant progress in the foreseeable future.

Israel and the Palestinian Territories

Israel, the Jewish state, lies in the center of the Arab domain (Fig. 7-11). Since 1948, when Israel was created as a homeland for the Jewish people on the recommendation of a United Nations commission, the Arab–Israeli conflict has cast a deep shadow across this entire realm. **Figure 7-14** helps us to understand the complex issues involved here. In 1946, the British, who had administered this area in post-Ottoman times, granted independence to what was then called Transjordan, the kingdom to the east of the Jordan River. In 1948, the orange-colored area became the UN-sponsored state of Israel—including, of course, land that had long belonged to Arabs in this territory named Palestine.

As soon as Israel proclaimed its independence, it was attacked by neighboring Arab states. Israel, however, not only held its own but pushed the Arab forces back beyond its borders, gaining the green-colored areas in 1949 (Fig. 7-14). Meanwhile, Transjordanian armies crossed the Jordan River and annexed the yellow-colored area named the West Bank (of the Jordan River), including part of the city of Jerusalem. The king called his newly enlarged country Jordan.

Further conflict was to follow. In 1967, a six-day-long war produced a major Israeli victory: Israel took the Golan Heights from Syria, the West Bank from Jordan, and the Gaza Strip from Egypt; it also conquered the entire Sinai Peninsula all the way to Egypt's Suez Canal (Fig. 7-14). In later peace agreements, Israel returned the Sinai to Egypt but not the Gaza Strip.

All this strife produced a huge outflow of Palestinian Arab refugees and displaced persons. The Palestinian Arabs, therefore, constitute another of this realm's stateless nations; nearly 2 million continue to live as Israeli citizens within the borders of Israel, but more than 2.8 million now reside in the West Bank

FIGURE 7-14

and another 1.7 million in the Gaza Strip. Many Palestinians also live in neighboring and nearby countries, though exact numbers are hard to know since some are registered as refugees while others are not. According to estimates, there are about 3 million in Jordan, half a million in Syria, nearly half a million in Lebanon, 300,000 in Saudi Arabia, and around another half million scattered around the rest of the Middle East (there are also about 250,000 Palestinians in the United States).

Israel is approximately the size of Massachusetts and has a population of 8.7 million (including its nearly 2 million Arab citizens), but because of its location, its powerful military, and its strong international links, these data do not reflect Israel's

importance. Israel has been a recipient of massive U.S. financial aid, and American foreign policy has long been to seek an accommodation between Jews and Palestinians, as well as between Israel and its Arab neighbors. However, in 2018 the U.S. government recognized Jerusalem as Israel's capital, and moved the American embassy from Tel Aviv to Jerusalem. This pleased the Israelis as much as it aggrieved the Palestinians (who have always laid partial claim on East Jerusalem, which is why previous U.S. governments and other countries always kept their embassy in Tel Aviv). According to most observers, this emphatically pro-Israel symbolic move by the U.S. would stand in the way of a two-state agreement.

Geographic Obstacles to a Two-State Solution

1. **The West Bank.** This would be the centerpiece of a future Palestinian state, but its 2.8 million Palestinians are being joined by growing numbers of Israeli Jewish settlers—who in early 2019 numbered around 450,000, dispersed across a discontinuous network of around 200 settlements fortified by walls and/or security fences (**Fig. 7-15**). Moreover, the West Bank is almost entirely bounded by the Israeli-built Security Barrier.

2. **The Security Barrier.** In response to the infiltration of suicide terrorists, Israel has walled off of the West Bank along the Security Barrier border (photo, top p. 249) mapped in Figure 7-15. This reinforcement, which in places penetrates deeply into former Palestinian areas, imposes much inconvenience for Palestinians as well as outright hardship for many where it runs between their homes and farm fields. Palestinians demand that the fences and walls be taken down; Israelis reply that the Palestinian (National) Authority (the interim self-government body) must control the terrorists.

3. **Jerusalem.** The United Nations intended Tel Aviv to be Israel's capital, and Jerusalem an international city (photo, bottom p. 249). But the 1948–1949 war ended with Israel

FIGURE 7-15

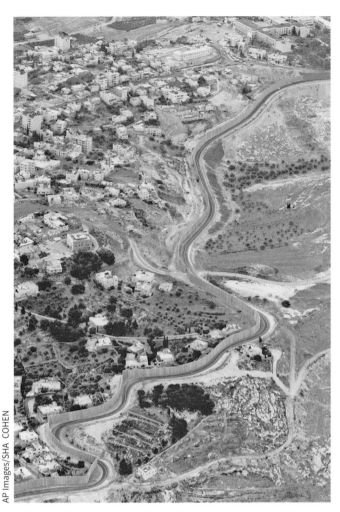

AP Images/SHA COHEN

Israel's Security Barrier takes several forms—a concrete wall, a spiked metal fence—and its effect also has multiple dimensions. This section of the barrier separates the village of Abu Dis on the outskirts of Jerusalem (left) from the West Bank (right). Israel defines this demarcation as a security issue, but the Arabs, citing its location inside Palestinian territory, argue that it is motivated primarily by politics.

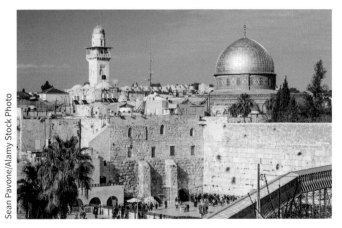

Sean Pavone/Alamy Stock Photo

The old city of Jerusalem, home to holy sites for Jews, Muslims, and Christians. The photo shows Jews praying at the Wailing Wall of Temple Mount and, above it, the Islamic Dome of the Rock and Al Aqsa Mosque.

holding the western part of the city and Arab forces the eastern sector. Then, in the 1967 war, Israel conquered all of the West Bank, including East Jerusalem; in 1980, the Jewish state reaffirmed Jerusalem's status as its capital, calling on all countries to relocate their embassies from Tel Aviv (some did for a while; the U.S. first did so in 2018). Meanwhile, the government redrew the map of the ancient city, building Jewish settlements in a ring around East Jerusalem that would obliterate the old distinction between a Jewish west and an Arab east. This enraged Palestinian leaders, who still view Jerusalem as the eventual headquarters of a future Palestinian state. Renewed U.S. efforts to persuade Israel to halt the construction of new settlements in East Jerusalem have so far been futile. Note, in Figure 7-15, that the Security Barrier runs well to the east of Jerusalem today.

4. ***The Gaza Strip.*** Israel in 2005 decided to withdraw from the Gaza Strip, removing all Jewish settlements there and yielding control of this territory to the Palestinian Authority (Fig. 7-14, inset map). Gaza has few natural resources, and its near-encirclement by Israel puts it in a most precarious position. Relations between Israel and Hamas (the extremist Palestinian faction that rules Gaza) are severely strained and repeatedly lapse into open conflict—usually in the form of rocket attacks from Gaza into Israel, followed by punishing Israeli retaliation.

Israel lies at the center of an intricate geopolitical puzzle. The four obstacles just discussed above are only part of the overall predicament: others involve water rights, compensation for expropriated land, Arabs' "right of return" to pre-refugee abodes, and the exact form a Palestinian state would take.

Region The Arabian Peninsula

The regional identity of the Arabian Peninsula is clear: south of Jordan and Iraq, the entire peninsula is encircled by water. This is a region of (mainly) old-style sheikdoms made fabulously wealthy by oil, modern-looking emirates, and the locale where Islam originated. But it also includes, at its southern periphery on the Gulf of Adan, the economically deprived country of Yemen, which in 2014 collapsed into political chaos, warfare, and state failure.

As a region, the Arabian Peninsula is environmentally dominated by a harsh desert habitat and politically dominated by the Kingdom of Saudi Arabia (**Fig. 7-16**). More than twice the territorial size of Egypt, Saudi Arabia is the realm's second-biggest state after Algeria. Saudi Arabia's neighbors (proceeding clockwise from the head of the Persian Gulf) are Kuwait, Bahrain, Qatar, the United Arab Emirates (UAE), the Sultanate of Oman, and the Republic of Yemen. Together, these countries on the eastern and southern fringes of the peninsula contain 53 million inhabitants; the largest by far is Yemen, with more than 29 million.

FIGURE 7-16

Saudi Arabia

Saudi Arabia itself has only 35 million residents (plus about 8 million expatriate workers, at least one-fourth of whom are undocumented) in its vast territory, but we are already aware of the kingdom's importance on the region's oil map (Fig. 7-6): according to some measures, the Arabian Peninsula contains the world's largest concentration of known petroleum reserves. As Figure 7-16 shows, the Saudi reserves lie in the eastern part of the country, particularly in the vicinity of the Persian Gulf, and extend southward into the Rub al Khali (Empty Quarter).

Figure 7-16 also reveals that most of the economic activities in Saudi Arabia are concentrated in a wide belt across the "waist" of the peninsula, from the oil boomtown of Dhahran on the Persian Gulf through the national capital of Riyadh in the interior to the Mecca–Medina anchors of the far west near the Red Sea. A modern transportation and communications network has recently been completed, but in the more remote zones of the interior, Bedouin nomads still ply their ancient caravan routes across the vast trackless deserts.

Efforts to reduce Saudi Arabia's regional economic disparities have been impeded by the enormous cost of bringing water from deep underground sources to the desert surface to stimulate agriculture, by the sheer size of the country, and by a population growth rate that is still far above the global average. Nonetheless, housing, healthcare, and education have experienced major improvements, and industrialization also has been stimulated in such newer cities as Jubail on the Persian Gulf and Yanbu on the Red Sea. The latter is part of the historic Mecca–Medina corridor, which is now modernizing rapidly.

Saudi Arabia's archconservative monarchic rule, official friendship with the West, and enormous social contrasts stemming from ongoing economic growth have raised political opposition, for which no adequate channels exist. Consider this: women were not allowed to drive until 2018, and they could not vote until 2015 (in municipal elections); women cannot even travel today without a male guardian. Even though things remained relatively quiet here, the Saudi regime watched nervously as the Arab Spring uprisings reverberated across the realm in 2011. By no coincidence, the government was quick to announce a new jobs program involving billions of dollars while beefing up police presence on the streets. But Saudi Arabia's planners have also learned to think much bigger than this, and their grandiose plans now call for nothing less than a complete economic transformation of the country.

The Saudis have good reason to invest in their economic future. Today, nearly three quarters of the population is under the age of 30, and unemployment among these young people runs around 25 percent. To accelerate the generation of the millions of jobs that will be needed, the Saudi royal rulers are spending their country's oil money on four new *economic cities* that by the mid-2020s should create well over a million new jobs (see Fig. 7-16; photo below). Overall, in anticipation of the end of the oil era, the ultimate goal is to transform Saudi Arabia from a petrodollar-based economy into a world-class industrial power.

The four new cities are (see Fig. 7-16):

- *King Abdullah Economic City (KAEC)*, the biggest EC, is located near Mecca and Jiddah and is planned to be a hub for global trade as well as a center for industry and investment.
- *Prince Abdulaziz bin Mousaed Economic City (PABMEC)*, located near the northernmost city of Ha'il, will be a center of logistics, agribusiness, minerals, and construction materials.
- *Jazan Economic City (JEC)*, in the southwestern corner of Saudi Arabia, will serve as a manufacturing and raw materials complex.
- *Knowledge Economic City (KEC)*, the smallest of the four, is located outside the holy city of Medina and near the modern Red Sea port and industrial city of Yanbu. It will be more like a science park, focusing on knowledge-based industries such as biotechnology and higher education.

Saudi Arabia is an important player in the geopolitics of this realm and a longtime American ally. Frequently, this partnership involves the writing of checks by Saudi Arabia and the provision of military power by the U.S.—as in Syria, or simply taking the form of Saudi purchases of American weapons. But the United States and Saudi Arabia make a strange pair at times, given the latter's highly imperfect democratic record and their clashing interests—for instance, regarding Iran. The Saudis are mostly orthodox Sunnis, and they eye the rising regional influence of Shi'ite Iran with great trepidation. Since 2015, the

dpa picture alliance archive/Alamy Stock Photo

A foreign business delegation, visiting in nearby Rabigh, Saudi Arabia, takes a look at a model of the planned King Abdullah Economic City.

Saudis have intervened in the downward-spiraling conflict in neighboring Yemen (see below) with air strikes to push back Houthi rebels, a Shi'ite militia that was advancing against the Sunni forces of the beleaguered Yemeni government.

The Gulf States

Five of Saudi Arabia's six neighbors on the Arabian Peninsula face the Persian and Oman gulfs: Kuwait, Bahrain, the United Arab Emirates, Qatar, and Oman (Fig. 7-16). All are known for their oil reserves, from which they earn a consequential foreign income. On a GDP per capita basis, these five are all high-income countries; they are also highly urbanized and quite modern in certain ways but traditional in others; and all are monarchies in the (Sunni) Islamic tradition. Their response to popular demands during the short-lived Arab Spring was comparable to that in Saudi Arabia: overly modest political concessions combined with increased public spending to appease the masses. Another similarity is that they each contain a substantial population of foreign workers.

The Gulf States, overall, maintain cooperative relations with the United States. They are generally anti-jihadist, and their economies are becoming ever more globalized. The rise of these states over the past decade or so is the result of a planned diversification away from a singular reliance on oil revenues—similar to what Saudi Arabia is doing. They seek various niches in the global economy, such as Bahrain's emergence as a center of international finance. Most spectacular is

Foreign laborers on the construction of Qatar's Ras Abu Aboud Football Stadium, December 12, 2018. This is one of nine new stadiums being built for the 2022 World Cup tournament. The decision to award this highest-profile sports event to Qatar has been widely criticized, not least because summer temperatures here are among the hottest on Earth (an average July day in Doha, the capital, records a high of 42°C [108°F] and a low of 31°C [88°F]). The Qataris have responded by promising to build covered, air-conditioned stadiums—yet another extravagance in a Gulf State that seems to have unlimited money to spend. A far more serious criticism, however, has also been leveled at Qatar: the consistently poor treatment of its (overwhelmingly foreign) workforce.

the UAE's burgeoning Dubai. This ultramodern city has turned itself into a world-scale air transportation hub (its airport traffic surpassed London's Heathrow in 2014 to become the busiest on Earth) as well as tourist destination with elaborate theme parks and shopping malls. Visitors cannot help but be impressed with the city's ultramodern high-rises, much of them built in the last 15 years (see photo page 231). Moreover, these countries are particularly adept at promoting themselves through global public relations campaigns—think of the high visibility of the UAE's state-supported airlines (Emirates and Etihad), or the 2022 FIFA World Cup tournament being awarded to Qatar (see photo), home of the global media giant Al Jazeera. All of the Gulf States, it should be noted, are still highly dependent on oil and sensitive to fluctuations in world oil prices.

These five states have learned how to take advantage of their location on the energy-rich Persian Gulf, whose busy oil-tanker traffic must navigate the treacherous Strait of Hormuz outlet, one of the most strategic choke points in the world ocean (Fig. 7-16). At the regional scale, the leading geographic asset of the Gulf States is their central location in Southwest Asia, providing a stable political and economic haven in an otherwise turbulent realm; moreover, none of these small countries even has a hostile neighbor today. Not a single one of these advantages, however, applies to Yemen, the country that occupies the southern corner of the Arabian Peninsula.

Yemen

With a population of 30 million—approximately 56 percent Sunni and 43 percent Shi'ite—Yemen is plagued by grinding poverty, centrifugal political forces, and incessant terrorist activity. In the chaotic mid-2010s, an ongoing Shi'ite rebellion against the central government destabilized the north while a secessionist movement continued to gain ground in the south, where oilfields encourage such ambitions.

Figure 7-16 highlights the geographic appeal of Yemen for Al Qaeda and other Islamic terrorist groups. Not only does Yemen lie at the back door of Saudi Arabia: its southern tip overlooks another of the world's most crucial and busiest choke points at the Indian Ocean entrance to the Red Sea—the Bab-el-Mandeb ("Gate of Grief") Strait—where ships tightly converge and run the risk of capture by pirates. Across the Gulf of Adan to the south lies another terrorist haven (and pirate stronghold), the failed state of Somalia, and just across the Red Sea to the west lies Eritrea, a likely target for extremists of the Al Qaeda stripe.

The mass demonstrations of Yemen's Arab Spring in 2011 against the 32-year-old regime of President Ali Abdullah Saleh led to his resignation a few months later. But here, like mostly elsewhere, the removal of an autocratic government was not followed by a democratic turn. On the contrary, Yemen experienced a political transition that mirrored Libya's—a brief spasm of destabilization, an ensuing power vacuum, and the swift emergence of opposing, armed factions.

By 2014, Yemen's government had been rendered powerless and the country descended into anarchy as three opposing forces began their battle for supremacy. One is the *Houthi* tribe, based in the country's far west, which is dominated by Shi'ites; another contender is the powerful jihadist branch of *Al-Qaeda of the Arabian Peninsula (AQAP),* which is expanding its territory westward from the central Gulf of Adan coast; the third force is Sunni-dominated government troops. The conflict is exacerbated and complicated by foreign involvement: Iran actively supports the Shi'ite Houthis, whereas Saudi Arabia and the United Arab Emirates (with support from the U.S. government) have conducted massive air strikes against them to shore up what is left of the country's Sunni-dominated government. By late 2019, more than 100,000 people had died in the conflict and the UN reported that 14 million people were at risk of starvation. While the Saudi-led air strikes had wreaked havoc on Houthi forces, the country remained hopelessly fractured (see inset, Fig. 7-16). San'a, the capital, was still in the hands of the Houthis while the main oilfields and pipelines were in the hands of government forces or AQAP. Like Syria, Yemen had become a failed state.

Region The Empire States

Two major states, both with long imperial histories, dominate the region that lies immediately to the north of the Middle East and Persian Gulf (Fig. 7-11), where Arab ethnicity gives way but Islamic culture endures—Turkey and Iran. Although they share a short border and are both Muslim countries, they display significant differences as well. Even their versions of Islam are different: Turkey is an officially secular but dominantly Sunni state, whereas Iran is the heartland of Shi'ism. Turkey's leaders have worked to establish satisfactory relations with Israel; on several occasions, Iran's leadership promised to "wipe Israel off the map." Turkey has been involved in negotiations aimed at entry into the European Union. Iran's goals have been quite different as that country has defied Western efforts to constrain its nuclear ambitions.

These parts of Southwest Asia are a long way from North Africa, and given their location, they are geographically linked to regions outside the realm, particularly Central Asia and westernmost South Asia. For centuries, Turkey and Iran have been involved in what are now, respectively, Turkmenistan and Afghanistan. Iran's southeastern Baluchi nation extends well into southwestern Pakistan, whereas its far northwestern Azeri minority spills over the border into Azerbaijan. Most important, Afghanistan constitutes a transition zone between NASWA and South Asia because that war-ravaged country has been tightly enmeshed in the geopolitics and cultures of both realms (see Fig. 7-11). Today, however, it can be argued that Afghanistan is more firmly connected to South Asia, and as such it will be discussed separately in Chapter 9.

Turkey

As **Figure 7-17** indicates, Turkey is a mountainous country with generally moderate relief; it also exhibits considerable environmental diversity, ranging from steppe to highland (see Fig. 1-6). On central Turkey's dry Anatolian Plateau, villages tend to be small, and subsistence farmers grow cereals and raise livestock. Coastal plains are not large, but they are productive as well as densely populated. Textiles (from home-grown cotton) and farm products dominate the export economy, but Turkey also has substantial mineral reserves, some oil in the southeast as well as major dam-building projects on the Tigris and Euphrates rivers, and a modest steel industry based on domestic raw materials.

The ancient capital of Turkey was Constantinople (long renamed Istanbul), located on the Bosphorus, a narrow waterway component of the strategic Turkish Straits that connect the Black and Mediterranean seas. But the struggle for Turkey's survival had been waged from the heart of the country, the Anatolian Plateau, and it was here that Kemal Atatürk (modern Turkey's founding leader who ruled from 1923 to 1938) decided to place his seat of government (Fig. 7-17). Ankara, the new capital, possessed some unique advantages: it would remind the Turks that they were (as Atatürk always said) Anatolians, it lay much closer to the center of the country than Istanbul, and it could therefore act as a stronger unifier.

Even though Atatürk moved the capital eastward and inward, his orientation was westward and outward. To implement

A view of the historic center of Istanbul, located on the western (European) side of the Bosphorus, with Süleymaniye Mosque visible to the right in the background.

YILMAZ SAVAS KANDAG/Alamy Stock Photo

FIGURE 7-17

his plans for Turkey's modernization, he initiated reforms in almost every sphere of life within the country. Islam, formerly the state religion, lost its official status, and Turkey became a secular state whose armed forces ensured that the Islamists would not take over again. The state took over most of the religious schools that had controlled education. The Roman alphabet replaced the Arabic. A modified Western code supplemented Islamic law. Symbols of old—growing beards, wearing the fez—were prohibited. Monogamy was made law, and the emancipation of women was initiated. The new regime emphasized Turkey's separateness from the Arab World, and in many ways it has remained aloof from the affairs that engage other Islamic states.

In present-day Turkey, the divide between traditional Muslims and secular Westerners is quite active. This can be seen in Figure 7-17, which maps support for the country's three leading political parties in the general election of 2015. The AKP (Justice and Development Party) holds a narrow parliamentary majority, has been the governing party since 2002, and is led by President Recep Tayyip Erdoğan; it is a conservative Muslim party, with its power base located in the rural provinces of central Turkey. The CHP is the secular Republican People's Party that was founded by Atatürk in 1923; its support

is based in several pockets of western Turkey that lie close to Europe. The HDP is the People's Democratic Party, dominated by Kurds in the southeastern margin of the country, and emphasizes equal rights for minorities. These internal divisions dramatically surfaced in July 2016, when a faction of the military attempted (and failed) to ignite a *coup d'état* to overthrow the government. This rebellion was largely aimed at President Erdoğan, whom they (and others) criticize for his increasingly autocratic behavior and appeasement of conservative Islamic groups.

Turkey and Its Neighbors   From before Atatürk's time, Turkey had a history of intolerance of minorities. Soon after the outbreak of World War I, the (pre-Atatürk) regime decided to expel the Armenians, concentrated in the country's northeast. Nearly 2 million Turkish Armenians were uprooted and brutally forced out; well over a million died in an eight-year genocidal campaign, the centennial of whose beginning was widely marked in 2015.

In modern times, Turkey also has been widely criticized for its harsh treatment of its large and regionally concentrated Kurdish population. About one-fourth of Turkey's population of 85 million is Kurdish, and successive Turkish governments have mishandled relationships with this minority nation. The historic Kurdish homeland lies in the southeast of Turkey, centered on Diyarbakir (Fig. 7-17), but millions of Kurds have moved to the outskirts of Istanbul—and to jobs in EU countries. With Kurdish nationalism rising just across the border with embattled Iraq, Turkey has responded by suppressing Kurdish insurgent movements while awarding more rights and freedom to those Kurds not involved in armed resistance.

Turkey's overriding foreign policy concern today is Syria. By 2019, 3.6 million Syrian refugees had settled on Turkish soil, and the alleged cross-border movements of Syrian rebel forces had drawn fire from Assad-regime troops inside Syria. While it was clear that Turkey opposed jihadist Islamic forces and formally participated in the U.S.-led anti-ISIS coalition, Turkey was also intent on the weakening of the Kurds, even though the latter fought side-by-side with the coalition against ISIS.

Turkey and the European Union

With its diversified and growing economy over the last couple of decades, secular government, and independent posture in international affairs, Turkey is the key predominantly Islamic state eligible for admission into the European Union. However, while formal talks on admission were launched in 2005, progress has been very slow because Europeans have been fundamentally divided over Turkey's qualifications. Whereas some focus on the country's hugely important strategic value (Turkey is also a member of NATO—see Chapter 5), others consider it "not European enough," particularly when it comes to so-called European democratic values. Some Turks take offense at the unrestrained public debate in Europe concerning their candidacy, often including criticisms of Turkish customs and traditions, not only in Turkey but also in existing Turkish communities already established in Europe. Still, about three out of four Turks say they are in favor of joining the European Union.

Events took an interesting turn in late 2015, against the backdrop of the refugee crisis that was overwhelming Europe in which Turkey occupied a crucial geographic position in the pathway of the refugee flow (especially of Syrians and Afghans). In return for Turkish efforts to block refugees from leaving Turkey for Greece or Bulgaria, the EU provided major financial support for the refugee camps in Turkey, the scrapping of visa requirements for Turks in the EU, and the promise of accelerated negotiations for Turkey's EU membership (also see Chapter 5).

For some time, during the Arab Spring, protesters across the realm espoused the *Turkish Model* as the ideal future scenario for their countries—a "mildly" Islamist government that does not alienate the religiously devout, that walks a fine line on democratic reform, and that prioritizes economic growth. In more recent years, as the government grew increasingly autocratic, such praise for Turkey has dissipated. Nonetheless, Turkey is a crucial state in a unique geographical position within a highly volatile realm.

Iran

Iran has its own history of conquest and empire, collapse and revival. Iran is the remnant of a once-vast regional empire that waxed and waned, from Europe's Danube Basin to the Indian subcontinent's Indus Valley, over more than two millennia. Long known to the outside world as *Persia* (although the people had called themselves Iranians for centuries), the state was officially renamed *Iran* by the reigning shah (king) in 1935. His intention was to stress his country's Indo-European cultural heritage as opposed to its Arab and Mongol infusions. But for all the rhetoric about modernization, the shahs failed to advance the lot of the common people, and by 1979 a revivalist revolution had engulfed the country. The monarchy was replaced by an Islamic republic ruled by an ayatollah (leader under Allah), and Islamists continue to tightly control this Shi'ite bastion today.

A Crucial Location and Dangerous Terrain

As Figure 7-11 shows, Iran occupies a critical position in this tempestuous realm. It controls the entire corridor between the Caspian Sea and the Persian Gulf. To the west it adjoins Turkey and Iraq, both historic adversaries. To the north (west of the Caspian Sea) Iran borders both Azerbaijan and Armenia, where Islam meets Christianity. To the east Iran borders Afghanistan and Pakistan, and to the northeast lies similarly volatile Turkmenistan.

As **Figure 7-18** reveals, Iran is a country of mountains and deserts. The heart of it is an extensive upland, the Iranian Plateau, that lies surrounded by even higher mountains, including the Zagros in the west, the Elburz in the north along the Caspian Sea coast, and the mountains of Khorasan to the northeast. This mountainous terrain also signals geologic danger: here the Eurasian and Arabian tectonic plates actively converge, causing major and often devastating earthquakes (Fig. 1-5, Chapter 1). The Iranian Plateau is actually a huge highland basin marked by salt flats and wide expanses of sand and rock. The highlands wrest some moisture from the air, but elsewhere only oases interrupt the arid monotony—oases that for countless centuries have been stops on the area's caravan routes.

The Ethnic-Cultural Map

Today, Iran's population of nearly 84 million is 76 percent urban, and the capital, Tehran (population 14 million), lies far to the north on the southern slopes of the Elburz Mountains, at the heart of modern Iran's core area.

Iranians dominate Iran's affairs, but the country is no more ethnically unified than Turkey. Figure 7-18 shows that three corners of Iran are inhabited by non-Iranians, two in the west and one in the southeast. Iran's northwestern corner is part of greater Kurdistan, the multistate territory in which Kurds are in the majority and of which southeastern Turkey also is a part (Fig. 7-13). Between the Kurdish area and the Caspian Sea the majority population is Azeri, who number more than 10 million in Iran and also form the ethnic majority in the neighboring

FIGURE 7-18

state of Azerbaijan to the north. In the southwest, concentrated in Khuzestan Province at the head of the Persian Gulf, an Arab cluster predominates. A significant part of Iran's huge oil reserve is located here (Abadan is the main export terminal, shown in Fig. 7-18), but this Arab community is among Iran's poorest and most restive—in a province whose stability could not be more essential to the country's economy. The third ethnically non-Iranian corner lies in the southeast, where a coastal Arab minority (remember, Iranians are not considered Arabs) focused on the key port of Bandar-e-Abbas on the Strait of Hormuz gives way eastward to a scattered and still-nomadic population of Baluchis, who have strong ties to fellow ethnics

in Balochistan Province just across the border of neighboring Pakistan. The Iranian component of the Baluchi homeland is a remote and inconsequential area, but in Pakistani Balochistan that country's government faces a persistent rebellion by tribes who feel slighted by official neglect.

Iran's Choices
Iran is a crucial player in the geopolitics of Southwest Asia. Its regional ambitions are hardly surprising. Tehran has a major stake in developments in Iraq and Afghanistan, both neighbors; it has an already-nuclear and unstable Pakistan on its southeastern border; it has a strong interest in the fate of Shi'ite majorities and minorities elsewhere in the

realm; it has long been an avowed enemy of Israel and a strong supporter of Palestinian causes; and it has in the past funded organizations labeled as terrorist whose actions have even reached across the Atlantic. In Iraq, it played an important role in providing ground military support to (Shi'ite-dominated) government forces in their battles with ISIS (Sunni Muslims).

But Iran has also suffered for decades from the effects of Western economic sanctions because of its efforts to acquire nuclear capabilities. It possesses the realm's second-largest concentration of oil reserves. Iran is overly dependent on oil (around 80 percent of its exports) and lacks a diversified economy. Slow economic development and the traditional disposition of the Islamic state (even under the ayatollahs) have kept most Iranians in a state of poverty and backwardness. And yet, young urban Iranians can be surprisingly modern (and Western) in their outlook; there is a palpable yearning among them for openness, democratization, and economic progress.

Iran's worsening economic plight and rising domestic discontent made the regime more receptive to down-sizing its nuclear ambitions in exchange for an end to sanctions. In 2015, such a pact was agreed by Iran and the United States (partnered by China, Russia, Germany, France, and the UK). But the arrangement soon derailed: first, as a result of the U.S. government rescinding the agreement in 2018 citing violations by the Iranians; and then by the U.S. assassination of Iran's top-ranking general in early 2020 (see p. 245), which took U.S.-Iran relations to their lowest point in recent memory.

Region The Maghreb and Its Neighbors

The countries of northwestern Africa are collectively called the *Maghreb,* but the Arab name for them is more elaborate than that: *Jezira-al-Maghreb,* or "Isle of the West," in recognition of the great Atlas Mountains rising like an enormous island from the Mediterranean Sea in the north and the sandy flatlands of the immense Sahara to the south (see Fig. 7-1).

Whereas Egypt is the gift of the Nile, the Atlas Mountains facilitated the settlement of the Maghreb. These high ranges wrest from the rising air enough orographic rainfall to sustain life in the intervening valleys, where good soils support productive farming. From the vicinity of Algiers eastward along the coast into Tunisia, annual rainfall averages greater than 75 centimeters (30 in), a total more than three times as high as that recorded for Alexandria in Egypt's delta. Even 240 kilometers (150 mi) inland, the slopes of the Atlas still receive over 25 centimeters (10 in) of rainfall. The effect of this topography can even be read on the world climate map (Fig. 1-6, Chapter 1): where the highlands of the Atlas terminate, aridity immediately takes over.

The Atlas Mountains are aligned southwest–northeast and begin inside Morocco as the High Atlas, with elevations over 4000 meters (13,000 ft) (**Box 7-7**). Eastward, two major ranges

dominate the landscapes of Algeria proper: the Tell Atlas to the north, facing the Mediterranean, and the Saharan Atlas to the south, overlooking the great desert. Between this pair of mountain chains, with each consisting of several parallel ranges and foothills, lies a series of intermontane basins markedly drier than the northward-facing slopes of the Tell Atlas. Within these valleys, the **rain shadow effect [22]** of the Tell Atlas is reflected not only in the semi-arid, steppe-like natural vegetation but also in land-use patterns: pastoralism replaces cultivation, and stands of short grass and shrubs blanket the countryside.

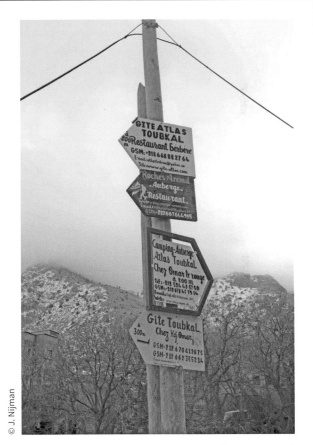

Box 7-7 From the Field Notes . . .

© J. Nijman

"It is hard to believe this scene is in North Africa. I am at the village of Imlil, Morocco, near the base of Mount Toubkal, the highest peak in the Atlas Mountains at 4167 meters (almost 14,000 ft). It is February, and snow can be seen on the north-facing slopes. Most of the people residing in these small villages have simple, hard lives, and keeping warm can be a challenge. Ethnically, they are Berbers and speak a language quite different from Arabic, even though Arab cultural influences are strong throughout the country. But these directional signs, for shelter and food, are in French, a colonial vestige aimed at the few foreign tourists who come through this area. At this moment, it looks like it might soon be snowing again. What a stark contrast to the bone-dry, sizzling hot, endless desert flatlands that cover nearly all of the rest of northern Africa."

The Maghreb Countries

The countries of the Maghreb include Morocco, the last remaining North African kingdom; Algeria, a secular republic besieged by religious-political problems; and Tunisia, smallest and most Westernized of the three (**Fig. 7-19**). **Morocco**, a relatively stable and rather conservative kingdom in a revolution-scarred region, is tradition-bound and economically weak. Like some of the other Arab monarchies in this realm, Morocco's government introduced some modest reforms in the early 2010s to preempt growing popular discontent. Its core area lies in the north, anchored by four major cities—Casablanca, Marrakech, Tangier, and the capital, Rabat. But the political attention of the Moroccan government is focused on the south, where it seeks to incorporate its neighbor, *Western Sahara* (a former Spanish dependency with almost 600,000 inhabitants, many of them immigrants from Morocco). Even if this campaign is successful, it will do little to improve the lives of most of Morocco's 37 million people, of whom hundreds of thousands have migrated to Europe. Of greater value to Morocco would be the free access of their (mostly agricultural) products to European markets.

Algeria, France's one-time colony whose major agricultural potential drew more than one million European settlers to its arable coastal strip known as the *Tell*, now has an economy based primarily on its substantial oil and natural gas reserves. The country's primate city, Algiers, is centrally situated within the densely settled corridor lining the Mediterranean that contains most of Algeria's 43 million inhabitants (Fig. 7-2). But a sizeable contingent of Algerians—officially estimated at about 2 million—has emigrated to France, where they form one of Europe's largest Muslim population sectors. Hard-nosed military control combined with generous social spending kept the Arab Spring movement at bay—at first; but in early 2019, after persistent and massive popular protests, the country's 82-year-old ruler, who had been in office for more than two decades, was forced to resign. Challenges remain: a transition toward greater democracy is in doubt, and jihadists affiliated with ISIS have become active in the remote Sahara of southeastern Algeria, even more so since the defeat of ISIS in Syria and Iraq.

The smallest of the Maghreb states, **Tunisia**, lies at the eastern end of the region. Tunisia in many ways outranks surrounding countries: it has a higher urbanization level, more favorable social indicators, and a much lower growth rate (among its population of 12 million) than elsewhere in the Maghreb. Most of the country's productive capacity lies in the north in the hinterland of its historic capital, Tunis. As noted previously, this country took center stage as the source of the Arab Spring uprisings that diffused across NASWA. Of all the revolutions sparked by this movement, Tunisia's seems to have been one of the more successful, with an ongoing vibrant civil society and, for example, movement toward gender equality that are unmatched in this realm. Still, preserving its democratic gains remains a major challenge, and being the region's most Westernized country also attracts trouble in the form of terrorism:

FIGURE 7-19

© J. Nijman, P. O. Muller, and John Wiley & Sons, Inc.

attacks in 2015 on a Tunis museum and a nearby beach killed a total of 59 international tourists. On a more positive note, on National Woman's Day on August 13, 2018, the government announced legislation that would give women equal inheritance rights across the country.

Libya

The Maghreb's eastern neighbor, Libya, is unlike any other North African country: an oil-rich desert state whose population is almost entirely clustered in settlements along the coast—and suffering through a chaotic, multiethnic civil war that broke out following the 2011 Arab Spring revolution that led to the overthrow of the Qadhafi dictatorship (Fig. 7-19).

Almost rectangular in shape, Libya (population: 6.9 million) faces the Mediterranean Sea between the Maghreb states and Egypt. What limited coastal-zone agricultural possibilities exist lie in the two subregions known as Tripolitania in the northwest, centered on the capital of Tripoli, and Cyrenaica in the northeast, focused on the city of Benghazi. But it is oil, not farming, that propels the economic activity that remains in this highly urbanized country. The oilfields are located well inland from the central Gulf of Sidra, linked by pipelines to coastal terminals. Libya's two interior corners, the desert Fezzan district in the southwest and the Kufra oasis in the southeast, are connected to the coast by two-lane roads subject to frequent sandstorms.

Muammar Qadhafi had been one of the realm's most violent despots, clinging to power for more than four decades. His brutal ground and air attacks on anti-regime demonstrations in several Libyan cities in early 2011 were hardly unexpected. Within a few months, however, advancing rebel forces captured a number of cities including Benghazi, which became their main base of operations. Assisted by NATO air support, which enforced a no-fly zone that grounded the government's military aircraft, the rebels prevailed; they soon tracked down and killed Qadhafi, who had fled Tripoli to take refuge in his hometown of Sirte.

But even though the Libyan people had rid themselves of a despised tyrant, their country collapsed into political chaos in late 2011 because a new democratic government could not be formed. Local tribes, ethnic minorities, and jihadists of various stripes, long suppressed under Qadhafi, swiftly reemerged to contest each other's turf, but only succeeded in fragmenting the country. In 2019, the country was carved into three main areas controlled by competing factions (Fig. 7-19 inset). The UN-backed government, a coalition of groups endorsed by the UN in 2015, controls a relatively small area in the northwest that includes the former capital Tripoli. The largest territory, to the east and south, and comprising most of the oilfields and installations, is in the hands of the former military under the Qadhafi regime. Finally, a stretch of territory against the border with Algeria is controlled by the *Tuareg* and allied tribal factions.

The Tuareg are a nomadic people of about 3 million who live across the Sahara in an area covering parts of Mali, Niger, Libya, Algeria, and Burkina Faso—but in none of these countries do they form an ethnic majority. Culturally and linguistically they are related to the Berbers of the Maghreb. Politically, they can be considered a nation without a state, in some ways like the Kurds or the Palestinians. The Tuareg in Libya were courted by the Qadhafi regime and held a special role in the military forces. The area under their control inside Libya and across the Sahara is indicated by the dashed line in Figure 7-19.

Adjoining Saharan Africa

Between the chain of North Africa's coastal states and the southern margins of the Sahara lies an east–west tier of five states, all but one of them landlocked, dominated by the world's greatest desert, sustained by modest ribbons of water, and thoroughly under the sway of Islam: Mauritania, Mali, Niger, Burkina Faso, and Chad (Fig. 7-19). Although their huge overall territory is sparsely populated (Fig. 7-2), the populations of these countries total around 80 million, not insignificant by African standards. Furthermore, all five are at least partially located within the *African Transition Zone*—a region bedeviled by awesome political and environmental challenges (as we shall shortly see).

The destabilizing effects of Libya's breakdown are especially felt in **Mali**, territorially the size of Texas and California combined, with a population of 20 million that mostly inhabits the country's southern half. The country descended into chaos after a military coup in 2012. The military was said to be frustrated with the government's handling of the Tuareg insurgency in the sparsely populated far north (Fig. 7-19). But the coup sparked a rebel surge, and shortly thereafter Mali was effectively split in two. A Tuareg–Jihadist alliance took control of the fabled city of Timbuktu—located near the center of this hourglass-shaped country—and declared that the north was now an independent state named *Azawad*. In response, France sent thousands of its troops to help its former colony recapture Timbuktu, did so, and then turned over the effort to restore central-government rule over northern Mali to a UN-mandated, all-African military force. That approach has not proved particularly effective, and much of the country remains unstable and prone to violent clashes between government and rebel forces.

▣ Region The African Transition Zone

Mali's well-populated southern half, which contains its core area, lies in the final region to be discussed in this chapter—the African Transition Zone (Fig. 7-11). From the Atlantic coast of Mauritania and Senegal in the far west to the easternmost Horn of Africa that protrudes into the Indian Ocean, a semi-arid belt of land, about 500 kilometers (300 mi) wide, lies between the bone-dry Sahara to the north and the more humid savanna zone to the south. This is the *Sahel* (meaning "shore"—with the bordering Sahara analogous to an ocean), where steppe (**BS**) climates separate desert (**BW**) from tropical savanna (**Aw**) climates (see Fig. 1-6).

Life in this environmentally stressed belt of steppelands is quite unpredictable. During the nine-month dry season, there is virtually no rain and temperatures can rise to 50°C (122°F). When the erratic midsummer rains finally come, some areas receive a modest amount while others are flooded. In wet years, the harvest can be good, but at other times it is almost nonexistent, particularly during the Sahel's recurrent droughts. Thus, human–environment relationships here are always fragile, and it is not surprising that, according to UNICEF, malnutrition kills about a quarter-million children annually in the five westernmost Sahelian countries. But the problems of the Sahel go far beyond climate and food provision.

As both Figure 7-11 and **Figure 7-20** indicate, the African Transition Zone is unlike the five other regions of this realm. This is where Subsaharan Africa's cultures to the south intersect with those of the Muslim domain to the north. This Islamic influence into Subsaharan Africa engulfs some countries in their entirety (such as Senegal and Burkina Faso) and divides others into Muslim and non-Muslim sectors (Nigeria, Ivory Coast, and Chad). As elsewhere in the world where certain geographic

© J. Nijman, P. O. Muller, and John Wiley & Sons, Inc.

FIGURE 7-20

realms meet and overlap, complications abound within the African Transition Zone. In some countries, the geographic transition from Muslim to non-Muslim society is gradual, as in Sierra Leone and Ivory Coast. In other areas, the cultural divide is far more abrupt, as in eastern Ethiopia and the two halves of former Sudan, where the border between Christian/animist African cultures and Islamic African communities is quite sharp. The southern border of the African Transition Zone, the religious boundary we refer to as Africa's **Islamic Frontier [23]** (mapped in Fig. 7-20), is therefore neither static nor advancing uniformly.

Conflict marks most of the Islamic Frontier: in former Sudan, where a decades-long war for independence by non-Islamic Africans in the south against the Arabized, Muslim north—which finally resulted in the 2011 creation of South Sudan—cost millions of lives; in northern Nigeria, where violent Islamic revivalism continues to severely threaten that country's prospects; more recently in Mali, northern Burkina Faso, and

FIGURE 7-21

© J. Nijman, P. O. Muller, and John Wiley & Sons, Inc.

Chad, where jihadism has been on the rise and political rivalries have been reawakened; and in the Horn of Africa—nowhere more than in Somalia, the fifth in the category of failed states in this realm (also see the discussion of certain countries in the African Transition Zone in Chapter 8, which covers Subsaharan Africa).

The Horn of Africa

The most potentially explosive subregion of the African Transition Zone is the Horn of Africa, which consists of Somalia, Djibouti, Ethiopia, and Eritrea (**Fig. 7-21**). You will find some discussion of the Horn that is more pertinent to Subsaharan Africa in Chapter 8. Here, we focus on the first three countries. Regarding the fourth, the small state of **Eritrea** (about the size of Ohio), suffice it to say that it came into existence in 1993 when it separated from Ethiopia, with whom it still engages in boundary disputes and clashes that have damaged the economies of both countries. Eritrea's early hopes for representative government and economic progress were dashed by war, dictatorial rule, and, most recently, its involvement in Islamic sectarian strife. Eritrea is sometimes referred to as "Africa's North Korea" because of it rogue politics and economically dismal situation.

Somalia

The key component in the eastern sector of the African Transition Zone is Somalia, where 16 million people, almost all of them Muslim, live at the mercy of a desert-dominated environment that requires cross-border migration into Ethiopia's easternmost Ogaden zone in pursuit of seasonal pastures. As many as 5 million Somalis reside permanently on the Ethiopian side of the border, but this is hardly the only such division the Somali "nation" faces. Somalia's cultural geography comprises an assemblage of five major ethnic groups fragmented into hundreds of clans—all engaged in an endless struggle for power as well as survival.

In the early 2000s, Somalia fragmented into three major subdivisions that still exist today (Fig. 7-21). In the north, the sector known as *Somaliland*, which had proclaimed its independence in the 1990s and which remains by far the most stable of the three, essentially functions as a state even though the international community will not recognize it as such. In the east, a conclave of local chiefs declared their territory to be distinct from the rest of Somalia, gave it the name of *Puntland*, and asserted varying degrees of autonomy. In the south, where the official capital, Mogadishu, is located on the Indian Ocean coast, local secular warlords and Islamic militias continue their grim, long-running contest for supremacy. In 2006, the Islamic militias stormed into the capital and took control, ousting the warlords and proclaiming their determination to create an Islamic state. Six years later, an internationally backed government regained control of Mogadishu, but the Al Qaeda–linked *al-Shabaab* Islamist terrorist movement it ousted reemerged strongly in southern Somalia and continues to use it as a base to launch its bold attacks. The government receives American

support, and a U.S. drone strike killed al-Shabaab's leader in 2014. Nonetheless, the jihadist mosaic continues to blanket and control large territorial expanses of Somalia, political stability is as hard to imagine as ever today, and the country continues to be regarded as a classic **failed state [24]**. As we have seen in this chapter, besides Somalia, four other failed states mark the NASWA realm: Syria, Yemen, Libya, and Mali.

Djibouti

This ministate counts less than a million people and covers a territory about the size of New Jersey, but it is increasingly significant because it lies directly across from Yemen as it overlooks the narrow entry to the Red Sea, the Bab-el-Mandeb Strait (see Fig. 7-21 and the satellite image below). Djibouti is fully 97 percent Muslim, but it has escaped the kinds of entanglements with jihadist groups that have afflicted Somalia. Economically, Djibouti is very closely tied to

Jacques Descloitres, MODIS Land Rapid Response Team, NASA/GSFC

The narrow Bab-el-Mandeb Strait connects the Red Sea (upper left) to the Gulf of Adan arm of the Indian Ocean (lower right). This classic choke point—the only one located within the African Transition Zone—accommodates one of the world's most crucial sea lanes, especially for oil shipments. With the unstable African Horn on the western shore and failed-state Yemen on the eastern side, it is not hard to envision the Strait's surrounding turmoil escalating into attempts to disrupt traffic here. That would trigger outside intervention, not only by such NASWA countries as Saudi Arabia and Egypt, but also by the United States, Europe's powers, and very likely China. It also explains the strong interest of foreign powers in gaining a foothold in Djibouti—which surrounds the narrow inlet on the western shore just to the southwest of where the Strait meets the Gulf.

Ethiopia—90 percent of the shipments through its main port come from or go to Ethiopia. Given that country's rapid economic growth, Djibouti has benefited as well. But it is Djibouti's geostrategic position that has attracted a significant international presence. This country now contains military bases operated by the United States (4000 troops are permanently stationed there), France, Japan, the EU, and China (it is China's only overseas military base in the world); and Saudi Arabia and India are also believed to be interested. These land leases bring in considerable revenues: "We don't have anything else but location," said a government official recently—but it has been turned into a very significant economic asset. The broader challenge for Djibouti is to diversify its economy: nearly three quarters of its GDP is directly associated with the port, and unemployment in 2019 was estimated at around 44 percent.

Ethiopia   Ethiopia, the last country covered in this chapter, allows us to end on a fairly positive note. The highlands of Ethiopia, where the country's core area, its capital (Adis Abeba), and its historic Christian heartland lie, are sandwiched between dominantly Muslim societies (Fig. 7-21). And Figure 7-20 reminds us that Ethiopia's eastern Ogaden is traditional Somali country and almost entirely Islamic. Today, one third of Ethiopia's population of 115 million is Muslim, and here the Islamic Frontier is sharply defined. With the continent's second-largest population, Ethiopia is also an important African cornerstone state and is discussed more fully in Chapter 8 as part of the region of East Africa. It not only faces some of the usual challenges of the African Transition Zone but has been landlocked as well since it lost Eritrea in 1993.

Yet Ethiopia's highland climate and soils, quite conducive for productive agriculture, strongly contrast with the less encouraging conditions that prevail in the stressed, semi-arid environments across so much of the African Transition Zone. Indeed, Ethiopia exports fruits, vegetables, flowers, coffee, and even tea. Moreover, this country has in recent years attracted considerable investment from China to finance badly needed infrastructure, construction, and agribusiness ventures (see Chapter 8); and in terms of economic growth, Ethiopia has ranked among the world's top 25 countries for most of the past decade. Politically, Ethiopia remains stable despite some questionable recent elections managed by a hard-nosed, oppressive regime. But economically, at least, Ethiopia's recent performance far exceeds that of other countries in the African Transition Zone. This not only reminds us of the importance of geography but also of the pitfalls of **environmental determinism [25]**: geography explains a great deal because it poses constraints and offers opportunities—but geography never dictates.

Points to Ponder

- The Arab Spring movement of the early 2010s did not produce major advances toward democracy in most countries, but it did contribute to dangerous political destabilization in the realm's regions and a number of individual countries.

- In 2018, Tunisia's women received rights equal to those of men regarding family inheritance laws; in that same year, laws were changed in Saudi Arabia to allow women to drive a car.

- Since 2014, Dubai International Airport (DXB) in the United Arab Emirates has been the world's busiest airport measured in terms of international passengers (London Heathrow came in second each of these years).

- In 2020, the North Africa/Southwest Asia realm counted five failed states (Syria, Yemen, Libya, Mali, and Somalia)—more than in any other realm.

- In Lebanon in 2019, a small country the size of Connecticut, one in five people were refugees, mostly from Syria.

FIGURE 8-1

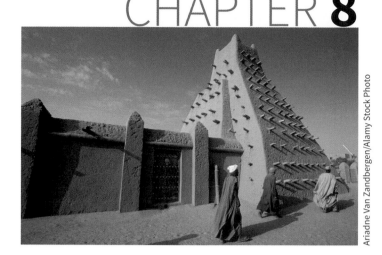

The city of Timbuktu in present-day Mali was founded in the early 12th century by Tuareg nomads. Located on important African trading routes (especially of gold and salt), it quickly grew and developed into a cosmopolitan city that included the first African universities or scholastic centers. These were based in mosques but scholarship went far beyond learning the Quran and covered the sciences at large. The photo shows Sankoré Mosque, built in the 15th century. It is one of Timbuktu's famous landmarks, where the university was based.

Ariadne Van Zandbergen/Alamy Stock Photo

The Subsaharan African Realm

IN THIS CHAPTER

- Africa's growing economies
- Southern Africa's vulnerability to global climate change
- China's foothold in Africa
- Boko Haram in Nigeria's north
- The paradox of export manufacturing in Ethiopia
- "Day Zero:" Cape Town running out of water

CONCEPTS, IDEAS, AND TERMS

Human evolution **[1]**	Green Revolution **[11]**	Boko Haram **[21]**
Cold War **[2]**	Agribusiness **[12]**	Anthropogenic **[22]**
Rift valley **[3]**	Medical geography **[13]**	Mobile money **[23]**
Continental drift **[4]**	Endemic **[14]**	Cash economy **[24]**
State formation **[5]**	Epidemic **[15]**	Exclave **[25]**
Ethnicity **[6]**	Pandemic **[16]**	Apartheid **[26]**
Nation-state **[7]**	Multilingualism **[17]**	Xenophobia **[27]**
Indirect rule **[8]**	Formal economy **[18]**	State capture **[28]**
Land tenure **[9]**	Informal economy **[19]**	
Land alienation **[10]**	*Sharia* law **[20]**	

The African continent occupies a special place in the physical as well as the human world. This is where **human evolution [1]** began. In Africa, we formed our first communities, spoke our first words, and created our first art. From this realm, our ancestor hominins spread outward into Eurasia more than 2 million years ago. From Africa, our species emigrated, beginning perhaps 95,000 years ago, northward into present-day Europe and eastward via southern Asia into Australia and, much later, farther afield into the Americas. Disperse our forebears did, but we should remember that, at the source, we are all Africans.

Defining the Realm

For millions of years, Africa served as the cradle for the emergence of humankind. For tens of thousands of years, Africa served as the source of human cultures. Yet Africa has not experienced economic development on a par with other world regions. Its underdevelopment is difficult to separate from foreign involvements: from imperialism to colonialism to the geopolitical impact of the **Cold War [2]**. Today, there are hopeful signs that this catastrophic era is ending as a number of states are being propelled forward by the new global economic relationships they are building.

The focus in this chapter will be on Africa south of the Sahara, also known as *Subsaharan Africa*, an unsatisfactory but popular name that applies to the extensive landmass south of the great desert (**Fig. 8-1**; **Box 8-1**). The African continent contains two geographic realms: the African, extending from the southern margins of the Sahara to the Cape of Good Hope, and the western flank of the North African/Southwest Asian realm. Although the great desert constitutes a formidable barrier between the two, the powerful influences of Islam crossed the Sahara from North Africa centuries before the first Europeans set foot in West Africa. By that time, the African kingdoms lining the southern margins of the Sahara had been converted, creating an Islamic foothold in the African Transition Zone (the final region discussed in Chapter 7) that now marks the northern periphery of the realm we are about to investigate.

Africa's Physiography

Africa accounts for about one-fifth of the Earth's entire land surface. Territorially, it is as big as China, India, the United States, Mexico, and Europe combined. The north coast of Tunisia lies 7700 kilometers (4800 mi) from the southernmost coast of South Africa. Coastal Senegal, in West Africa, lies 7200 kilometers (4500 mi) from the tip of Africa's Horn in easternmost Somalia. These distances have critical environmental implications. Much of Africa lies far from maritime sources of moisture. In addition, as Figure 1-6 shows, large parts of the landmass lie in latitudes where global atmospheric circulation systems produce arid conditions. As a result, water supply is one of Africa's foremost problems.

As the physiographic map (**Fig. 8-2**) reveals, Africa's topography is marked by several properties that are not replicated on other landmasses. Alone among the continents, Africa does not possess a mountain backbone; neither the northwestern Atlas nor the far southern Cape Ranges are in the same league as the Andes or Himalayas. Where Africa does have mountainous terrain, as in Ethiopia and South Africa, it takes the form of deeply eroded plateaus or, as in East Africa, high snowcapped volcanoes. Furthermore, Africa is one of only two continents containing a cluster of great lakes. These lakes (with the exception of Lake Victoria) lie in

---

Box 8-1 Major Geographic Features of Subsaharan Africa

1. Physiographically, Africa is a massive plateau continent without a major "spinal" mountain range but with a set of great lakes, several major river basins, variable rainfall, soils of generally inferior fertility, and dominantly savanna and steppe vegetation.

2. Hundreds of distinct ethnic groups make up Subsaharan Africa's culturally rich and varied population mosaic. They far outnumber the states in this realm, and rarely do state and ethnic boundaries coincide.

3. Approximately three-fifths of Subsaharan Africa's peoples still depend on farming for their livelihood.

4. The realm is rich in raw materials vital to industrialized countries, and many economies continue to rely on primary activities—the extraction of resources—and not the greater income-generating activities of manufacturing and assembly.

5. The realm is famous for its wildlife, but many species are threatened. Combining bioconservation with sustainable development is a big challenge when the outside world covets elephant tusks and rhinoceros horns.

6. Severe dislocation still affects a number of Subsaharan African countries, from South Sudan to Zimbabwe.

7. Foreign interest in the realm's natural resources, ranging from commodities to agricultural land, has expanded steadily in this century.

8. Subsaharan Africa experienced rapid economic growth in the 2000s, but growth rates after 2009 became subdued as commodity prices declined and China's economic slowdown took hold. Nevertheless, certain countries still record strong growth.

FIGURE 8-2

deep trenches called **rift valleys [3]**, which form when huge parallel fractures appear in the Earth's crust and the strips of crust between them sink to form great, steep-sided, linear valleys. These rift valleys, stretching more than 9600 kilometers (6000 mi) from the Red Sea to South Africa, are mapped by the red lines in Figure 8-2.

Africa's rivers, too, are unusual: their upper courses frequently bear landward, seemingly unrelated to the coast toward which they eventually flow. Several rivers, such as the Nile and the Niger, have inland as well as coastal deltas. Major waterfalls, notably Victoria Falls on the Zambezi, or lengthy

systems of cataracts separate the upper from the lower river courses.

Finally, Africa may be described as the *plateau continent*. Except for some comparatively limited coastal plains, almost the entire continent lies above 300 meters (1000 ft) in elevation, and half of it lies over 800 meters (2500 ft) high. As Figure 8-2 indicates, the plateau surface has sagged under the weight of accumulating sediments into six major basins (three of them in the Sahara), but the margins of Africa's plateau are marked by escarpments, often step-like, such as the Great Escarpment of southeastern South Africa that marks the eastern edge of the Drakensberg Mountains.

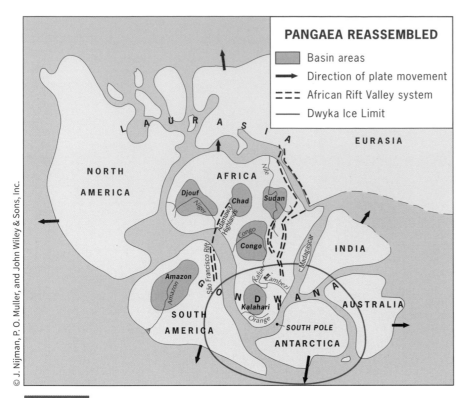

FIGURE 8-3

© J. Nijman, P. O. Muller, and John Wiley & Sons, Inc.

Continental Drift and Plate Tectonics

Africa's remarkable and unusual physiography informed geographer Alfred Wegener's hypothesis of **continental drift [4]**. All of the present-day continents, Wegener reasoned, lay assembled into one giant landmass called *Pangaea* not very long ago (ca. 220 million years) in geologic time. The southern part of this supercontinent was *Gondwana*, of which Africa formed the core (**Fig. 8-3**). When, approximately 200 million years ago, tectonic forces began to split Pangaea apart, Africa and the other landmasses acquired their present configurations. That process, now known as *plate tectonics*, continues across the world and is marked by earthquakes and volcanic eruptions where it is most intense (also see Fig. 1-5 in Chapter 1).

Africa's ring of escarpments, its rifts, its river systems, its interior basins, and its lack of significant mountains all relate to the continent's central position within Pangaea—all pieces of the puzzle that led to an explanation based on plate tectonics theory.

Africa's Historical Geography

Africa is the cradle of humankind. Archeological research has chronicled 7 million years of transition from Australopithecenes to hominins to *homo sapiens*. It is, therefore, ironic that we know comparatively little about Subsaharan Africa from 5000 to 500 years ago—that is, before the onset of European colonialism. This is partly due to the colonial period itself, during which African history was neglected, numerous African traditions and artifacts were destroyed, and many misconceptions about African cultures and institutions became entrenched. It is also a result of the absence of an accurate and detailed written history about the region until the contemporary period.

African Genesis

Africa on the eve of the colonial period was far from a static realm. In interior West Africa, cities were developing; in central and southern Africa, peoples were moving, readjusting, frequently struggling with each other for territorial supremacy. African cultures had been established in all the environmental zones shown in Figure 1-6 for thousands of years and thus long before the intrusions of Islamic and/or European influences. One of these, the Nok culture, endured for over eight centuries (north of the Niger–Benue confluence in what is now central Nigeria) from about 500 BC to the third century AD (**Fig. 8-4**). The Nok people made stone as well as iron tools, and they left behind a treasure of art in the form of clay figurines representing humans and animals.

Early Trade

West Africa, over a north–south span of a few hundred kilometers, displayed a wide range of contrasting environments, economic opportunities, modes of life, and products. The peoples of the tropical forest in the south produced and needed goods that were different from the products and requirements of the peoples of the dry, distant north. For example, salt was a prized commodity in the forests, where humidity precluded its formation, but it was plentiful in the semi-arid steppe and the desert beyond. This enabled the desert-margin peoples to provide salt to the forest peoples in exchange for ivory, spices, and dried foods. Accordingly, there evolved a degree of *regional complementarity* between the peoples of the forest and those of the drylands. And the savanna peoples—those located in between—found themselves in a position to channel and handle this profitable trade.

The markets in which these goods were exchanged prospered and grew, some into urban centers. One of these old cities, now an epitome of isolation, was once a thriving center of commerce and learning and one of the leading crossroads in the world—Timbuktu (see photo page 265). In fact, its university is one of the oldest anywhere, with a library that holds irreplaceable documents, thankfully now being preserved. Others, predecessors as well as successors of Timbuktu, declined, a few of them into oblivion. Still other savanna cities, such as Kano in northern Nigeria, remain important today.

Early States

Strong and durable premodern states arose in the interior of West Africa (Fig. 8-4). The oldest known state was ancient Ghana, located to the northwest of present-day Ghana and centered on parts of what are now Mali, Burkina Faso, and Ivory Coast. Ancient Ghana managed to weld various groups of people into a stable territory between the ninth and twelfth centuries AD. Taxes were collected from its citizens, and tribute was extracted from subjugated peoples along Ghana's periphery; tolls were levied on goods entering the state, and an army maintained control. Muslims from the drylands to the north invaded Ghana in 1067, when it was in decline, hastening the state's demise as it shattered into smaller units.

In the centuries that followed, the focus of politico-territorial organization in this West African *culture hearth* shifted almost continuously to the east—first to ancient Ghana's successor state of Mali, which was centered on Timbuktu and the middle Niger River Basin, and then to the state of Songhai, whose focus was Gao, a city farther down the Niger that still exists (Fig. 8-4). This eastward movement may have been the result of the growing influence and power of Islam. Traditional animist religions prevailed in ancient Ghana, but Mali and its successor states sent massive, gold-laden pilgrimages to Mecca along the savanna corridor south of the Sahara, then turning northward to follow a Nile route that passed through present-day Khartoum and Cairo.

FIGURE 8-4

© J. Nijman, P. O. Muller, and John Wiley & Sons, Inc.

Beyond West Africa

West Africa's savanna zone undoubtedly experienced momentous cultural, technological, and economic developments, but other parts of ancient Africa made progress as well. Early states emerged in present-day Sudan, Eritrea, and Ethiopia. Influenced by key innovations from the Egyptian culture hearth to the north, these kingdoms were stable and durable: the oldest, Kush, lasted 23 centuries (Fig. 8-4). The Kushites built elaborate irrigation systems, forged iron tools, and created impressive structures as the ruins of their long-term capital and industrial center, Meroe, reveal. Nubia, to the southeast of Kush, became Christianized until the Muslim tide overtook it in the eighth century. And Axum was the richest market in northeastern Africa, a powerful kingdom that controlled Red Sea trade and endured for more than six centuries.

The process of **state formation** [5] diffused throughout Africa and was still in progress when the first European contacts occurred in the late fifteenth century. Large and effectively organized states developed on the equatorial west coast (notably Kongo) and on the southern plateau from the southern margin of the Congo River Basin southeastward to Zimbabwe. East Africa, as well, gave rise to a number of city-states, including Mogadishu, Mombasa, and Sofala.

At a broader scale, a crucial series of events commencing about 5000 years ago affected virtually all of West, Equatorial, and Southern Africa: the Great Bantu Migration from present-day Nigeria and Cameroon southward and eastward across the continent. This epic expansion appears to have occurred in waves that populated the Great Lakes area and ultimately penetrated South Africa, where it culminated in the formation of the powerful Zulu Empire in the nineteenth century (Fig. 8-4).

All this reminds us that, prior to European colonization, Africa was a realm of rich and varied cultures, diverse lifestyles, technological advances, mobility, and trade. But it was also a highly fragmented realm with a cultural mosaic (**Box 8-2**; **Fig. 8-5**) constituted by myriad **ethnicities** [6] sharing a common ancestry, culture, language, and/or religion. After the Europeans intervened and redrew the social and political map, the modern African state emerged as an amalgamation of multiple ethnic groups that almost entirely disregarded the preexisting cultural-territorial layout. Thus, the European idea of the **nation-state** [7], where state borders coincide with ethnic boundaries, is a rarity within the patchwork of pluralistic countries that marks this realm (Fig. 8-5).

Box 8-2   Geographic Information Analysis

Despite the common legacies of European colonization, Subsaharan Africa is marked by a mosaic of cultural, linguistic, and ethnic diversity. The coexistence and/or integration of different groups can be a major challenge anywhere, especially when the ethnic patterns become so intricate. Closely observe and compare the realm's state borders with its myriad ethnic boundaries. Which states have particularly complex ethnic configurations?

Access your WileyPLUS course to interact with a dynamic version of this map and to engage with online exercises and questions.

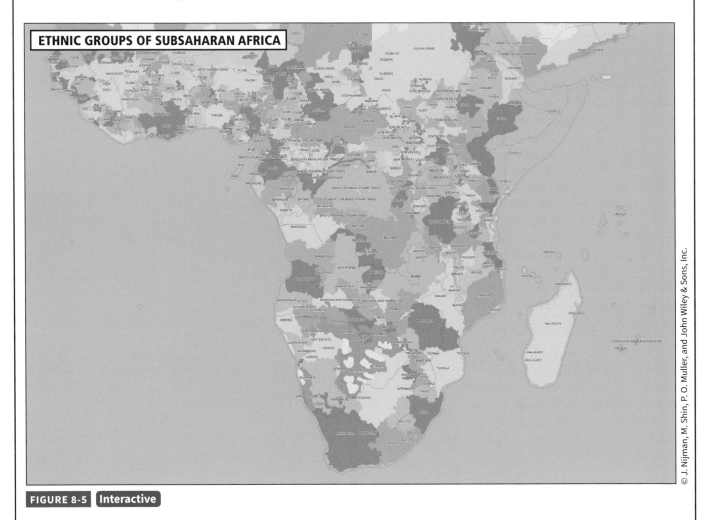

ETHNIC GROUPS OF SUBSAHARAN AFRICA

© J. Nijman, M. Shin, P. O. Muller, and John Wiley & Sons, Inc.

FIGURE 8-5   Interactive

The Colonial Transformation

European involvement in Subsaharan Africa began in the fifteenth century. It disrupted the path of indigenous African development and irreversibly altered the entire cultural, economic, political, and social complexion of the continent. Starting quietly in the late 1400s, Portuguese ships groped their way all along the west coast and by 1488 had rounded the Cape of Good Hope. Their goal was to find a sea route to the spices and riches of India and points east. Soon other European countries were dispatching their vessels to African waters, and a string of coastal stations and forts sprang up. In West Africa, the part of the continent closest to the European spheres in Middle and South America, the initial impact was strongest. At their coastal control points, the Europeans bartered with African intermediaries for the slaves who were destined to work the New World plantations, for the gold that had been flowing northward across the Sahara, and for highly prized ivory and spices.

Suddenly, the centers of activity lay not with the inland cities of the savanna belt but with the foreign stations on the Atlantic coast. As the interior declined, the coastal peoples thrived. Small forest states gained unprecedented wealth, transferring and selling slaves captured in the interior to the European traders on the coast. Dahomey (now called Benin) and Benin (now part of neighboring Nigeria) were two modern states built on the slave trade.

Impact of the Slave Trade

Millions of Africans were forced to migrate from their homelands to the Americas, especially Brazil, the Caribbean, and the United States. The slave trade was a catastrophe that was facilitated, in part, by the peril of proximity. The northeastern tip of Brazil, by far the largest single destination for the millions of Africans forced from their homes in bondage (**Fig. 8-6**), lies about as far from the nearest West African coast as South Carolina lies from Venezuela. This was a much shorter maritime intercontinental journey because the distance from West Africa to South Carolina is more than twice as far.

Slavery was not new to West Africa. In the interior of Africa and within city-states, kings, chiefs, and prominent families traditionally took a few slaves, but the status of these slaves was relatively benign. Indeed, large-scale slave trading was first introduced to East Africa long before the Europeans brought it to western Africa. African intermediaries from the coast raided the interior for able-bodied men and women of various ethnic backgrounds and marched them in chains to the Arab markets on the coast (the island of Zanzibar was one such slave trading hub). There, packed into specially built sailboats, they were carried off to Arabia, Persia, and India.

When the European slave trade took hold in West Africa, however, its volume and impact were far greater than anything seen before, and so were the horrors. Europeans, Arabs, and collaborating Africans now ravaged the continent, forcing perhaps as many as 30 million persons away from their homelands in captivity (Fig. 8-6). Families, entire communities, and even cultures were destroyed; those surviving in exile suffered unfathomable misery. The European presence also completely reoriented West Africa's trade routes. Even though the Europeans' insatiable demand for slaves ravaged the inland population, it did not lead to any major European thrust toward the interior or produce colonies overnight. African intermediaries were well organized and strong, and they held off their European competitors for centuries.

Colonization

In the second half of the nineteenth century, the Europeans finally laid claim to virtually all of Africa. In 1884, a conference was convened in Berlin to carve up the African map among the leading European powers. On maps spread across a large table, representatives from these powers drew boundaries, swapped territory, and forged a new map (**Figure 8-7**) that would become a liability in Africa decades later. Amazingly, at no point in the proceedings were African voices heard. While the three-month conference was in progress, most of Africa remained under traditional African rule. Then, in the early 1900s, the European colonial powers moved to take control of all the areas they had marked off and acquired on their maps.

FIGURE 8-6

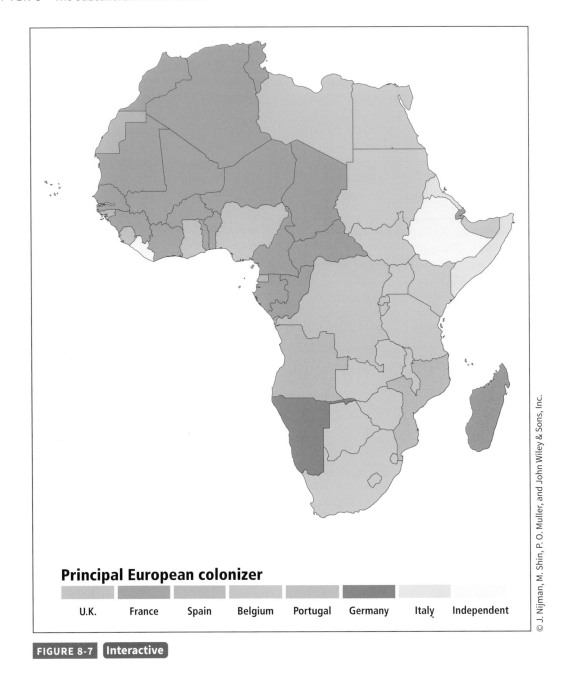

Principal European colonizer

| U.K. | France | Spain | Belgium | Portugal | Germany | Italy | Independent |

FIGURE 8-7 Interactive

It is important to examine Figure 8-7 carefully because the colonial powers governed their new dependencies in very different ways, and their contrasting legacies are still in evidence today. Some colonial powers were democracies at home (Great Britain and France); others were dictatorships (Portugal and Spain). The British established a system of **indirect rule [8]** over much of their domain, leaving indigenous power structures in place and making local rulers representatives of the British Crown. This was unthinkable in the Portuguese colonies, where harsh direct control predominated. The French sought to establish culturally assimilated elites that would disseminate French ideals in the colonies. But King Leopold II of Belgium claimed the Congo Free State as his own personal fiefdom. Subsequently, he unleashed a savagely ruthless campaign of exploitation and genocide. Leopold's reign of terror has become known as Africa's most brutal demographic disaster (after the slave trade): as many as 10 million people were exterminated.

The post-Berlin colonial map entailed a profound political reorientation—yet it lasted barely half a century. Ghana (then known as the Gold Coast) offers a good example. Its colonial ties were formalized despite the constant challenging of the mother country by indigenous ethnic groups. But in 1957, Ghana became the first African country to gain independence—triggering a movement that swept across the realm. Today, only a handful of its countries have not marked 50 years of independence (a list to be completed in 2025), and in retrospect the colonial period should now be viewed as only an interlude rather than a paramount chapter in African history. It did, however, leave a deep and enduring geographic imprint, especially on the realm's cultural landscapes (**Box 8-3**).

Box 8-3 From The Field Notes …

"I teach a one-week university module at the University of Cape Town (UCT) every February and it is interesting to observe the student political activism on campus. Founded in the 1920s, the university's student body was initially almost entirely white but today over half the students are black Africans. UCT today is ranked as the top university in Africa, and its student body has become both increasingly diverse and activist. This activism should resonate with students in the United States who are concerned with the rising cost of higher education, the removal of statues of leaders of the Confederacy, and the *#blacklivesmatter* movement. At UCT, recent successes include the cancellation of the national government's plan to markedly increase tuition fees as well as the removal of the prominent statue of the British colonialist, diamond magnate, and Cape Colony prime minister Cecil Rhodes (see photo). There is a particular irony here in that Rhodes (also of Oxford Rhodes Scholarship fame) bequeathed the very land upon which UCT stands—interestingly, Oxford University in the UK has declared that it intends to keep *its* campus statue of Rhodes.

South Africa's current college students are mostly comprised of the "born-free generation" (individuals born after the 1994 demise of Apartheid): they possess no living memory of Apartheid's racial injustices yet seem united in their deep disillusion concerning the slow pace of change since. Many seek a radical transformation of higher education marked by a complete redirection of the university curriculum, including steps to break down the longstanding colonialist assumption that Western knowledge is superior. A key component in this effort is the students' call for a deeper appreciation and a recentering of African knowledge. Currently, various perspectives on decolonizing the

Roger Sedres/Alamy Stock Photo

university coexist—from supplementing Western knowledge with local research and perspectives from South Africa, the Subsaharan African realm, and other non-Western regions to more extreme views that call for a total break with Western knowledge and ways of knowing. An Africa-centered university might draw more on subjugated histories, voices, languages, and cultures; constantly reconsider for whom its curriculum is designed; and be increasingly sensitive to assumptions made about student backgrounds, cultures, languages, and schooling.

As we enter the 2020s, it remains to be seen how the decolonized university curriculum can be constructed and deployed. As the project evolves, there may be lessons to be learned for the future of global higher education."

Human–Environment Relations

Colonial Legacy

Africa's tropical forests and extensive savannas constitute the world's last refuges for many species of wildlife, from primates and zebras to giraffes to wildebeests. But since colonial times, these natural environments have been under constantly rising threat. Early on, the Europeans introduced hunting as a "sport" (a practice that did not exist within African cultural traditions), thereby triggering the mass destruction of animals. Later, a number of colonial rulers and postcolonial national governments laid out game reserves and other conservation areas, but these were neither sufficiently large nor adequately connected to enable herd animals to adhere to their seasonal migration routes. As we shall see, the same climatic variability that plagues farmers also affects wildlife. When the rangelands wither, the animals seek better pastures; but if the fences of game reserves wall them off, they cannot survive. Where there

are no fences, wildlife invasions of farmland destroy crops, and the farmers retaliate. Clearly, after thousands of years of equilibrium, the competition between humans and animals in Africa has entered a new era of intensifying stress.

It is not solely population pressure that threatens African wildlife. Today, international criminal syndicates are heavily engaged in the illegal exporting of wild animals and animal parts, and this activity has become the world's fourth most profitable illicit trade (after narcotics, human trafficking, and arms smuggling). The horn of the rhinoceros became a valued property for Arabs who fashion them into dagger handles and, in powdered form, an alleged aphrodisiac and cancer treatment for wealthy East Asians. As a result, the northern white rhinoceros is nearing extinction, and the number of all species of rhino has fallen from half a million a century ago to barely 30,000 today. A terrible tragedy is also befalling the African elephant as the value of its ivory tusks has tripled since 2010 (in 2015, a kilogram of ivory in Beijing peaked at U.S. $2100); if this illicit Chinese market is not quickly terminated, this entire species could disappear in the next decade. Note that even though Africans are complicit in international smuggling and that small quantities of ivory are for sale in informal markets in Nigeria and Angola, it is external demand that drives this clandestine trade. Meanwhile, competition for land has more to do with the decline of Africa's lion population (from around 250,000 in 1975 to less than 20,000 today), whereas illegal logging and mining threaten a host of other species, including the great apes.

The Geography of Farming

It would seem that space should be plentiful for both wildlife and humans. **Figure 8-8** does not present a picture of a densely peopled realm—although there are major clusters of population in coastal West Africa, East Africa encircling Lake Victoria, and the African Horn where Ethiopia's highlands sustain a large concentration. Most of the rest of the realm is relatively sparsely populated, even though its overall annual rate of growth is almost two-and-a-half times the global average of 1.1 percent.

Despite an ongoing surge in urbanization, a large majority of Africans still continue to depend on farming for their livelihood. But African environments are difficult for millions of small farmers,

and the realm's highly productive areas (the Great Lakes area, the Western Rift Valley, and the cooler, moister parts of South Africa) are minuscule in comparison to the huge, fertile, alluvial basins of China or India. The great majority of African farmers face daunting challenges that include (1) climatic variability, (2) difficulties in their national policy environments, (3) problems in accessing world markets, and (4) growing competition from agricultural producers in Asia and the Americas—all of which result in unstable prices and diminishing returns for many commodities. Many wealthy governments in the global core subsidize their own farmers, and increasingly environmentally conscious consumers favor more locally sourced and more expensive foods. Not surprisingly, from their point of view, African farmers claim they are having to counter yet another obstacle in the global marketplace.

Although adequate soil and water supplies are critical for farming, a number of political and economic factors are influential as well. Among them are land tenure; commercial (for-profit) versus subsistence (household-level) agriculture; type of farming system (rotational, shifting cultivation); crop prices; government farm policies; and degree of mechanization.

POPULATION DISTRIBUTION: SUBSAHARAN AFRICA

Population per Square Kilometer

- 0 to 5
- 5 to 10
- 10 to 50
- 50 to 100
- 100 to 500
- Over 500

FIGURE 8-8 Interactive

With roughly 60 percent of the realm's livelihoods dependent on farming, the issue of land tenure may be the most important of all. **Land tenure [9]** refers to the way people own, occupy, and use land. African traditions of land tenure are different from those found in Europe or the Americas. In most of Subsaharan Africa, communal, not individual, rights prevail—often held in perpetuity by chiefs, large extended families, and villages. Thus, users of the land have only temporary, custodial rights to it; they cannot sell that piece of land and must follow the rules of their community leaders.

Land Ownership

After the arrival of the Europeans, their colonial administrators prioritized control of the most fertile areas of each colony. In certain cases, this required military conquest; elsewhere, coercion and eviction sufficed. Before the colonial era, many traditional African livelihood systems adhered to sustainable land management practices such as letting fields lie fallow and rotationally grazing their livestock. Viewing this land as unused, however, colonial regimes initiated the process of **land alienation [10]** (confiscation). Much later, after the colonizers withdrew, many newly independent African countries started programs whereby land would revert to traditional forms of ownership and management—a land reform movement that in the end did little to improve conditions for still-marginalized small farmers.

Rapid population growth since the end of the colonial era further intensified pressures on the realm's agricultural land. Traditional systems of land use, involving subsistence farming in various forms ranging from shifting cultivation to pastoralism, work best when populations are fairly stable, and tenure is communal. But the African population explosion of the mid-twentieth century set into motion an accelerated demand cycle in which food producers could no longer rest soils and pasturelands—thereby unleashing new rounds of farmland degradation that steadily drove down crop yields.

Subsistence Farming

Although commercial farming prevails in parts of Africa, most farmers are engaged in subsistence agriculture that entails grain crops (corn, millet, sorghum) in drier areas and root crops (yams, cassava, sweet potatoes) where moisture is adequate. Others herd livestock, mostly cattle and goats, as they deal with the constant challenges of environmental variability (**Box 8-4**). But too many farmers and pastoralists have been unable to secure regular access to markets as well as stable prices for their products because of government policies that often promote a particular export-oriented crop (e.g., peanuts in Gambia; tea and coffee in Kenya).

The decline of African agriculture has been disproportionally hard on women who, according to prevailing estimates, produce about two-thirds of all local food in Subsaharan Africa. Development policies frequently pay too little attention to the role of gender in agricultural production (women account for at least 15 percent of all farmland holders), so that farms operated by women are smaller, more isolated, and endowed with poorer-quality soils than those cultivated by men. Moreover, women always have a harder time than men in obtaining credit.

The **Green Revolution [11]**—the development of more productive, drought-tolerant, pest-resistant, higher-yielding

---

Box 8-4 From the Field Notes …

"Flying a small propeller plane over East Africa's semiarid grasslands, we have a good view of a Maasai village near the border between Kenya and Tanzania. The Maasai are a pastoral people who herd cattle, sheep, and goats. I can clearly see the fenced 'kraals' where they keep their animals at night (during the day they are moved around) and the small dwellings that surround them. The more animals a man (!) owns, the higher his social status. But this way of life is under pressure: many have decided to move to towns and cities because they offer more and better opportunities. This area is also located in the northern extension of Tanzania's famous Serengeti National Park, where wildlife preservation comes into direct conflict with the Maasai way of life. 'It is really impossible,' says the park guide with us on the 10-seater plane, pointing down at the settlement. 'Of course, they want to protect their cattle, but they are killing lions and other animals that should be protected. And we must tell them all the time that they have to stay out of certain areas because the cattle compete for food with the wild animals.' Nodding agreeably, I think to myself: yes, but isn't this their land? As with so many other places around the world where we want to protect nature, we must involve the local inhabitants. That is much easier said than done."

© J. Nijman

varieties of grain—had less impact in Africa than elsewhere during the latter decades of the twentieth century. Belatedly, a renewed effort has now been launched to jump-start Africa's New Green Revolution. Its twin emphases involve the adoption of more effective farming methods and the raising of higher-value crops in order to boost farmer income as well as enhance the realm's food production output. However, the introduction of genetically modified crops is controversial and encounters substantial resistance today. Many African governments, influential media, and individual farmers regard biotechnological solutions with increasing hostility.

Neocolonial Land Grabs

During the postcolonial era, agricultural land in Africa was of little interest to outside investors, but over the past decade an enormous change has taken place. Many African countries now offer large tracts of land for sale or long-term leases to major investors, a number of whom are associated with multinational **agribusiness [12]** corporations and private equity firms. Governments and other large landowners view the sale and leasing of food production resources they control (land, soil, water) as the quickest return on their existing investments to increase agricultural output and maximize foreign revenues. As for the investors, they are motivated by the steady rise in global food prices, brightening prospects for biofuels, and growing shortages of arable land in their own countries. They also see African smallholder agriculture as achieving only a tiny fraction of its potential productivity—which the outsiders believe can be significantly upgraded through the widespread application of contemporary agricultural strategies and technology. Although investors from other geographic realms form the majority of the participants in this land rush, certain African elites (many based in cities) have joined in as well—either directly or in partnership with public and private institutions.

At its peak in the mid-2010s, more than 4 million hectares (9.8 million acres) of African land were under negotiation or managed by foreign investors—an area greater than the combined agricultural land of France, Germany, Belgium, the Netherlands, Denmark, and Switzerland. Most of this land was concentrated in Ethiopia, Congo*, Moçambique, and Sudan. So far, the impact of these land grabs has been mixed and new investment deals have slowed. At the same time, a new phenomenon known as *green-grabbing*—the appropriation of land and resources for environmental purposes—has become more conspicuous. It is driven by the growing need for, among others, biodiversity conservation, carbon sequestration, and improving ecosystem management. Accordingly, new alliances are emerging that involve venture capitalists, scientists, corporations, and nongovernmental organizations in speculative projects. A good example are green-grabs for *biochar* feedstocks (a carbon-rich substance formed when biomass is burned under oxygen-deprived conditions) that have recently occurred in Liberia, Tanzania, and Cameroon.

---

* Two countries in Africa have the same short-form name, *Congo*. In this book, we use **DRCongo** for the much larger Democratic Republic of the Congo, and **Congo** for the smaller Republic of Congo just to its northwest.

For centuries, indigenous farmers here have applied biochar as a conditioner for mineral-poor soils. Today, however, the new prospect is raised of incorporating small farmers into a potential world-scale carbon market, wherein biochar is also used as a tool for reducing the global level of carbon dioxide. Whatever the intention of revaluing African land for a traditional market or an expanding green carbon economy, it appears to be a modern version of neocolonialism. Either way, substantial numbers of small farmers are dispossessed, with adverse economic effects reverberating across affected localities as traditional pathways for accessing land and livelihoods are upended. Not surprisingly, many experts conclude that the appropriation of land by large-scale investors will further exacerbate land scarcity, constrain the potential of smallholder-led development, and put unrealistic pressure on the nonagricultural economy to absorb Africa's rapidly growing labor force.

Population and Health

Medical geography [13] is concerned with the tracking of disease outbreaks to identify their sources, detect their carrier agents, trace their spatial diffusion, and prevent their recurrence. Alliances between medical personnel, epidemiological researchers, and geographers yield significant results. Medical specialists know how a disease ravages the body; geographers know how environmental conditions (such as variations in wind direction and river flow) can affect the spread of disease carriers (*vectors*), and their application of GIS techniques to analyze spatial dimensions of diseases is making myriad contributions to medical research these days.

In Subsaharan Africa today, hundreds of millions of people carry one or more maladies, often without knowing exactly what ails them. A disease that infects many people (the *hosts*) in a kind of equilibrium, without causing rapid and widespread deaths (such as hepatitis or hookworm), is said to be **endemic [14]** to that population. People affected may not die suddenly or dramatically, but their quality of life and productive capacity are hindered as their overall health is weakened and can deteriorate rapidly when a more acute illness strikes.

Epidemics and Pandemics

When a disease outbreak occurs at the local or regional level, it is called an **epidemic [15]**. It may claim thousands, even tens of thousands, of lives, but it remains geographically confined, perhaps limited by the spatial range of its vector. Trypanosomiasis, the disease known as sleeping sickness and vectored by the tsetse fly, is a good example of a regional-scale epidemic (**Fig. 8-9**). The extensive herds of savanna wildlife form the *reservoir* of this disease, and the tsetse fly transmits it both to livestock and people. It is endemic to wildlife, but it also kills cattle, so Africa's herders try to keep their animals in tsetse-free zones. African sleeping sickness appears to have originated in a West African source area during the fifteenth century, and from there

© J. Nijman, P. O. Muller, and John Wiley & Sons, Inc.

WEST AFRICAN
FOCUS
ca. 1400   ca. 1850

1901

1907

ORIGINAL
TSETSE FLY VECTORS
THAT TRANSMIT THE
TRYPANOSOMES

1908

→ Diffusion of
African sleeping sickness

After Willett.

FIGURE 8-9

it spread throughout much of tropical Africa. Its epidemic range is limited by that of the tsetse fly: where there are no tsetse flies, there is no sleeping sickness.

When a disease spreads worldwide, it is known as a **pandemic [16]**. Subsaharan Africa's and the world's most deadly vectored disease is malaria, transmitted by a mosquito and the killer of at least 500,000 people each year. Because people living in the world's poorest (lower-latitude) countries are most vulnerable to malaria, that burden falls overwhelmingly on Africa—which not only accounted for 92 percent of all cases worldwide in 2017, but also 93 percent of the deaths from this disease (mostly children under 5 years of age). Ongoing eradication campaigns against this deadly mosquito vector are scoring some successes, and the UN's World Health Organization reports that better preventive and control measures have reduced malaria mortality in Africa by over 66 percent since 2000. Yet despite these gains, the realm's long war against this scourge is far from being won and is certain to continue well into the future.

Major Advances in the Fight Against HIV/AIDS

AIDS stands for *acquired immune deficiency syndrome*, the body's failure to protect itself against the slow-acting human immunodeficiency virus (HIV). If unchecked, this virus multiplies in the body, causing damage to the immune system and producing AIDS. Worldwide, more than 79 million people have contracted the disease, which killed nearly half of them. Globally, as many as 20 million people do not know they have the virus and need to be reached by HIV-testing services. Africans were most disproportionately affected: AIDS first erupted in this realm during the early 1920s, and a subsequent large-scale outbreak here in the 1970s swiftly became the springboard for a pandemic that

soon reached the United States in 1981. AIDS in Subsaharan Africa peaked around 1997, and the rate of new infections has steadily been reduced since then. Uganda, in particular, has gained wide acclaim in this effort, and its success has inspired a new wave of public health strategies and assistance programs in several AIDS-affected countries.

The still-evolving geography of HIV/AIDS is complex. Even though Subsaharan Africa continues to stand out prominently (with more than 53 percent of the world's cases in 2020), the realm is now experiencing the steepest decline of AIDS-related deaths. The latest World Health Organization (WHO) reports underscore both the accelerating pace of treatment (15.2 million Africans are currently receiving antiretroviral treatment) as well as the upsurge in the number of people who have become aware of their HIV status. Moreover, the upgraded dissemination of public information about the disease has been coupled with the easing of discrimination against AIDS-infected people on religious and other cultural grounds.

Globally, the AIDS outlook is also improving as efforts to mitigate the pandemic intensify. Under the United Nation's Sustainable Development Goal 3, the global community has pledged to conquer AIDS by 2030. Accordingly, UNAIDS "90-90-90" targets aim for (1) 90 percent of all people living with HIV will know their HIV status; (2) 90 percent of all people with diagnosed HIV infection will receive sustained antiretroviral therapy; and (3) 90 percent of all people receiving antiretroviral therapy will have viral suppression. At the end of the 2010s, East and Southern Africa were registering the greatest progress with 81 percent of those living with HIV knowing their status and more than half undergoing treatment.

As for realm's internal regional patterns, most notable are the ten countries of the Southern Africa region that bear the brunt of the global AIDS burden; collectively, they alone account for fully 40 percent of the world total of AIDS cases. When standardized for population size, the health crises of individual countries become clearer: about 25 percent of the adult populations of Eswatini, Lesotho, and Botswana are HIV-positive, and not far behind them at 19 percent is neighboring South Africa. Of these, Eswatini (known as Swaziland before 2018) has the most difficult environment for a civil society to operate in Southern Africa, the result of tight restrictions against gathering in public places, rigid control of information flow by the monarchy, and an entrenched patriarchal society in which men are still in charge of most decisions concerning women's health. (In absolute numbers, the country of South Africa has by far the greatest number of cases: its HIV-infected population in early 2019 stood at 7.2 million.) Comparatively, the HIV prevalence rate is significantly lower in West and Equatorial Africa, ranging from 2 to 5 percent in most countries. That also generally applies in the region of East Africa, where the highest incidence of HIV/AIDS occurs in Uganda (7 percent) and Kenya (6 percent).

Although Africa's HIV/AIDS situation remains serious, the global battle being waged against the disease has recorded some notable successes. Pharmaceutical companies continue to develop and test breakthrough drugs, and by 2019 two patients had been cured, one in London and the other in Berlin.

Throughout the realm today, beneficial public health campaigns are underway to promote better education, wider condom usage, and voluntary testing.

Ebola Crises

Another very dangerous affliction that has ravaged parts of Subsaharan Africa is Ebola virus disease (EVD)—as the world discovered when the disease was first identified in 1976 in the vicinity of the Ebola River in what is now the Democratic Republic of the Congo (DRC). Since then, several outbreaks have occurred, the largest being West Africa's 2014–2016 EVD epidemic. Originating in rural southeastern Guinea, it subsequently spread to urban areas as well as across nearby borders within weeks, and fears arose that it could mushroom into a global pandemic. Yet despite the hyped overreporting of the Western media, this outbreak was almost entirely limited to three adjoining countries: Guinea, Sierra Leone, and Liberia. According to WHO, this 27-month-long epidemic involved 28,652 reported Ebola cases, 11,325 (40%) of which resulted in fatalities, but only 15 deaths occurred outside these three countries. Of the 36 EVD cases beyond Guinea, Sierra Leone, and Liberia, 28 were confined to West Africa's Nigeria and Mali. (In fact, the remaining fatality—a misdiagnosed traveler from Liberia—occurred in Dallas, Texas; the three other EVD patients in the United States all fully recovered.) The outbreak caused serious disruptions to travel, transportation, investment, tourism, and cross-border trade in this part of West Africa, damaging its ties to the outside world and temporarily diminishing agricultural production.

Even though the 2014–2016 EVD epidemic was spatially contained and led to advances in treatment, most notably the development of a promising vaccine, barely two years later another serious outbreak occurred in eastern Equatorial Africa. Originating in August 2018 in the DRC's North Kivu Province (see Fig. 8-18), this soon became the second-largest Ebola outbreak on record. By May 2019, with the number of cases exceeding 1300, the epidemic was still confined to North Kivu and its northern neighbor, Ituri. But these provinces lie in the center of one of Africa's most unstable zones, which has long been plagued by political turmoil, armed conflict, and population displacement. Moreover, the geographic containment of this EVD epidemic is further imperiled by the tens of thousands of Congolese who daily cross international borders to trade and visit with relatives. Indeed, by December 2019—when total Ebola cases reached 3380, with a fatality ratio of 67 percent—cross-border transmission of the virus had become reality as cases surfaced in Uganda and Tanzania (with two other neighbors, Rwanda and South Sudan, also at high risk).

Cultural Geographies

Today, just about 60 percent of Subsaharan Africa's population still inhabit—and work in or near—Africa's hundreds of thousands of villages. They speak one of the 2000-plus languages in use in the realm. Africa's (numerically) largest groups of people constitute major nations, such as the Yoruba of Nigeria and the Zulu of South Africa. Africa's smallest cultural groups number just a few thousand. Overall, as Figure 8-5 amply illustrates, this geographic realm has the most complex and fragmented cultural mosaic on Earth.

African Languages

Africa's linguistic geography (**Fig. 8-10**) is a key component of that cultural-territorial intricacy. It is clear that the Subsaharan

FIGURE 8-10

© J. Nijman, P. O. Muller, and John Wiley & Sons, Inc.

Africa realm begins approximately where northern Africa's Afro-Asiatic language family (mapped in tan) gives way, a changeover at its sharpest in West Africa.

The realm's dominant language family is the Niger-Congo family (mapped in purple), of which the Kordofanian subfamily is a tiny, historic, northeastern outlier (Fig. 8-10), and the Niger-Congo languages extend across the realm from West to East and Southern Africa. The Bantu language forms the largest branch of this subfamily, but Niger-Congo languages in West Africa, such as Yoruba and Akan, also have millions of speakers. Another important language family is the Nilo-Saharan family (mapped here in brown), extending from Maasai in Kenya northwest to Teda in Chad. No other language families are of similar extent or importance: the Khoisan family, of ancient origins, still survives among the dwindling Khoi and San peoples of the Kalahari in the southwest; the small white minority in South Africa speak Indo-European languages; and Malay-Polynesian languages prevail in Madagascar, which was peopled from Southeast Asia long before Africans reached it.

The Most Widely Used Languages

More than 50 African languages are spoken by 1 million people or more, and a dozen by 10 million or more: Hausa (44 million), Yoruba (38 million), Ibo, Swahili, Lingala, and Zulu. Although English and French have become important *linguae francae* in countries such as Nigeria and Ivory Coast (officially *Côte d'Ivoire* in the Francophone manner), African languages also serve this purpose. Hausa is a common language across the West African savanna; Swahili is widely used in East Africa. And pidgin languages—mixtures of African and European tongues—are found along West Africa's coast; millions of Pidgin English (called *Wes Kos*) speakers use this medium in Nigeria and Ghana. However, dozens of African languages are on the verge of extinction, and Nigeria alone has at least 17 endangered, near-extinct languages.

Multilingualism [17] can be a powerful centrifugal force in society, and African governments have tried with varying success to promote national alongside local languages. Nigeria, for instance, made English its official language because none of its 500-odd languages, not even Hausa or Yoruba, had sufficient interregional usage within the country. On one hand, using a European, colonial language as an official medium invites criticism; on the other hand, making a dominant local language official would unleash negative reactions from ethnic minorities. One creative country today is Rwanda, which promotes its Bantu language (Kinyarwanda) in tandem with English for instruction in schools to hasten the separation from its Francophone past as a Belgian dependency.

Religion

Africans had their own belief systems long before Christians and Muslims arrived to convert them. Despite their diversity, all of Subsaharan Africa's cultural groups held a consistent view of their place in nature. Spiritual forces, according to African tradition, are manifest everywhere in the natural environment: in forests, rivers,

mountains, and communal lands—a religious worldview known as *animism*. Gods and spirits are believed to affect people's daily lives, witnessing every move, rewarding the virtuous and punishing (through injury or crop failure) those who misbehave.

As with land tenure, the religious views of Africans clashed fundamentally with those held by the colonizers. Monotheistic *Christianity* first penetrated Africa in the northeast when Nubia and Axum were converted, and Ethiopia has been a Coptic Christian stronghold since the fourth century AD. But the Christian churches' real invasion did not commence until the onset of colonialism after the turn of the sixteenth century. Christianity's various denominations made inroads in different locales: Roman Catholicism in much of Equatorial Africa, mainly at the behest of the Belgians; the Anglican Church in British colonies; and Presbyterians and others elsewhere. And these days, as in South America, Evangelical churches steadily gain adherents.

Islam had a rather different arrival and impact. Today, about one-third of the realm's population is Muslim (see Fig. 7-20). By the time of the colonial invasion, Islam had advanced out of Arabia, across the Sahara, and partway down the Indian Ocean coast of Africa. Muslim clerics converted the rulers of African states and commanded them to convert their subjects. They Islamized the savanna states and penetrated into present-day northern Nigeria, Ghana, and Ivory Coast. They encircled and isolated Ethiopia's Coptic Christians and Islamized the Somali people in Africa's Horn. They established beachheads on the coast of Kenya and took over offshore Zanzibar. Muslim and European-Christian proselytizers competed for African minds, but Islam proved to be a far more pervasive force. Today, tension between Islam and Christianity is especially evident all along

Nigel Dickinson/Alamy Stock Photo

The Muslim faithful kneel during Friday prayers outside a mosque in the city of Kano in northern Nigeria. Whereas most northerners want to practice Islam within the multireligious, multiethnic society of Nigeria, areas governed by strict Islamic (*Sharia*) law are being expanded, bringing official and *Sharia* law into confrontation throughout the north. Nigeria's viability as a unified state is now being threatened by the rise of Boko Haram, the brutally violent jihadist organization that is seeking to expand from its base in the far northeast after declaring itself a caliphate in 2014. Can this country avoid the fate of Syria and thwart the formation of another ISIS-type "Islamic State"? And how will Nigeria's moderate Muslims respond to the terrorist tactics of the extremist Boko Haram cult?

the *Islamic Frontier*, one of the defining regional characteristics—and the dynamic southern border—of the African Transition Zone (discussed on pages 260–262).

Urbanization and Africa's Dual Economies

Subsaharan Africa may still be one of the least urbanized geographic realms, but people everywhere are moving to towns and cities at an accelerating pace. Today, more than 40 percent of the realm's inhabitants reside in urban settlements (up from 28 percent as recently as 2000), and in absolute numbers they total nearly 100 million more than the entire population of the United States. Even though the growing tide of migrants is overwhelming the realm's already inadequate urban infrastructure, rapid urbanization is projected to continue far into the future—a current crisis that will soon be transformed into a social catastrophe because half of all babies born into the world between now and 2050 will be African.

The biggest African cities operate as the anchors of incipient national core areas, housing government offices and the headquarters of the most influential corporations and extractive industries. The **formal economy [18]**, under governmental control that regulates the civil service, business, and industry as well as their workforces, produces a formal urban landscape featuring a modern government district, clusters of newer high-rise buildings, upscale hotels, elite gated residential communities, and even branded fast-food outlets. From a distance, such cityscapes resemble those of urban centers in the global core, but on closer inspection it becomes evident that the formal sector is far from dominant here. In the streets, on sidewalks next to shop windows, in downtown-area slums, and

A typical urban landscape in Subsaharan Africa. This aerial view of the city of Arusha in northern Tanzania (population: about 490,000) shows several features of the settlement layout in urban Africa. The irregular pattern of low-rise, low-density development reflects a general lack of planning, complex land ownership issues, and piecemeal as well as patchwork investment in construction.

on the rough-hewn urban periphery, hustling street hawkers, garment makers, jewelry sellers, and unlicensed taxis signal the functioning of a parallel economy that lies almost entirely outside of government control. This is the **informal economy [19]**, which looms large across every African country (as well as countries throughout the global periphery) and entails an alternative way of living, working, and operating outside of the mainstream national economy. An overwhelming majority of Africans participate in the informal sector by residing in unofficial housing (usually shacks), transacting business by avoiding banks and the payment of taxes, and relying on networks of family, friends, and acquaintances for their survival.

Urban Housing Innovations: The *i-Shack*

Shacks are ubiquitous in this realm's poverty-suffused urban landscapes. Substandard dwellings are always unacceptable, but with some carefully planned and practical design improvements shacks can be creatively enhanced to yield a better quality of life for their disadvantaged inhabitants. The Sustainability Institute of South Africa's Stellenbosch University has produced a new prototype shack design called the *i-shack* ("improved shack"), and a pilot project to develop and apply this concept is underway at Enkanini, an informal settlement in the town of Stellenbosch that lies at the eastern edge of metropolitan Cape Town (**Box 8-5**). I-shacks are rudimentary solar-powered dwellings with the capacity to support two indoor lights, a motion-sensitive outdoor light, and a mobile phone charger. Its windows are strategically positioned for optimal air circulation and sunlight heating, while the roof is sloped to harvest rainwater. Recycled cardboard boxes and "Tetra Pak" containers are used for insulation between the exterior zinc surface and the interior, and flame-retardant paint is employed to minimize the risk of fire. I-shacks can be constructed and fitted for U.S. $660, and represent a significant and affordable upgrade on the everyday self-built shack made out of corrugated iron and discarded wood—which are too damp in winter and especially fire-prone in summer.

Africa Rising?

Despite burdensome colonial legacies, formidable environmental challenges, and a long history of serious adversity, twenty-first-century Africa is steadily emerging on the global scene. Although conditions remain harsh in many places, the achievements of the past two decades are quite encouraging, even stunning in certain cases (such as the lengthening of life expectancy in the realm by a remarkable 11 years since 2000). Gathering economic momentum is the new watchword and for a decade or so "Africa Rising" was a catch-all narrative popularized by the businesses, media, donors, and political leaders. The development outlook remains positive for a select group of countries that have exhibited resilience even in the face of challenges such as the slowdown of the Chinese economy and commodity price declines, both key determinants of the realm's robust growth of the past decade. Entering the 2020s,

Box 8-5 From the Field Notes …

"Shacks are a standard form of housing in the cities of Subsaharan Africa and have proliferated across the realm as the population continues to rapidly urbanize. Because national and local governments have only limited funds to rehouse slum dwellers, shacks are going to remain a signature landscape of the African city well into the future. Given this status quo, various prototypes have been developed to upgrade the experience of living in shacks. The *i-shack* is a good example of just such an improved solar powered dwelling unit. A pioneering effort is being spearheaded by Stellenbosch's Enkanini i-shack pilot project, which I visited on several occasions. Over the last five years the i-shack project has scaled up: in 2019, some 1600 shacks are powered directly by solar energy and closing the energy waiting gap as the community waits to be electrified. The project has also created seven full-time green economy jobs and provided training to another 14 informals in solar power installation, maintenance, marketing and advertising. I-shack experimentation provides a full commitment to maintaining the utility as long as it is needed as opposed to short-term drop and go intervention. Since 2017 the

© J. Nijman/Richard Grant

project has also provided financing to residents in another informal settlement (Siqalo) to purchase their own solar powered systems."

a more realistic assessment is that of a multispeed Africa, as is shown in **Figure 8-11** below (single-view narratives for the realm as a whole have always been too simplistic).

At present, there continues to be a set of countries (e.g., Rwanda, Ivory Coast, Ethiopia, Benin) that exhibit impressive annual GDP growth rates (**Box 8-6**)—while such supposedly

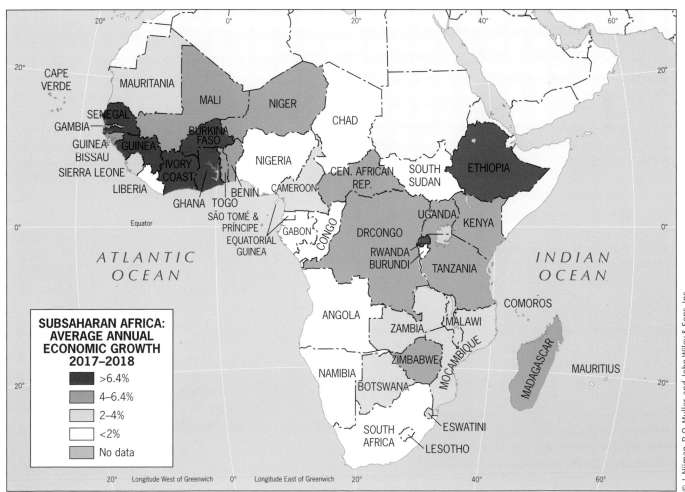

SUBSAHARAN AFRICA: AVERAGE ANNUAL ECONOMIC GROWTH 2017–2018
- \>6.4%
- 4–6.4%
- 2–4%
- \<2%
- No data

© J. Nijman, P. O. Muller, and John Wiley & Sons, Inc.

FIGURE 8-11

Box 8-6 Fastest-Growing Subsaharan African Economies and Leading Exports, 2018

	% Annual GDP Growth	1st and 2nd Biggest Export Products
Rwanda	8.6	Coffee, tea
Ivory Coast	7.4	Oil and natural gas, cocoa
Benin	6.9	Cotton, nuts
Burkina Fasso	6.8	Gold, cotton
Ethiopia	6.8	coffee, animal hides
Senegal	6.8	Oil, phosphate
Niger	6.5	Uranium, oil
Ghana	6.3	Gold, cocoa
Kenya	6.3	Tea, flowers

powerhouse economies as South Africa and Nigeria struggle to deliver jobs, uplift millions out of poverty, tackle corruption, and maintain the transition toward a low-carbon, climate resilient society. In 2019, stirred by economic mismanagement and growing discontent among young citizens over the lack of job opportunities, outbreaks of political unrest engulfed Zimbabwe, Cameroon, Togo, and Guinea-Bissau. Significantly, these concerns were foremost in the minds of voters in the 2019 elections in Nigeria and South Africa, whose consequences now serve as benchmarks for international investors re-evaluating their opportunities in the changing economic and political environment.

Six Subsaharan African states—Ethiopia, Ghana, Rwanda, Ivory Coast, Senegal, and Tanzania—ranked among the world's top 10 countries in GDP growth in 2017 and 2018. Ethiopia, one of Africa's top performers, is led by an authoritarian regime. This suggests that dictatorial rule is associated with strong economic performance; indeed, Ethiopia is following in China's footsteps, serving as a platform for local wage manufacturing activities and massive infrastructural projects (e.g., the Blue Nile's massive Grand Ethiopian Renaissance Dam—see page 242). At the same time, total foreign investment in the realm rose steadily from U.S. $11 billion in 2001 to $42 billion in 2017, much of it focused on the burgeoning resource-extraction sector but with ever more industrial and services projects added to the mix. Nowadays, that financial influx is increasingly channeled into telecommunications, banking, infrastructure, and even manufacturing.

What underlies these far more positive assessments of Subsaharan Africa's development potential? First, the realm is demonstrably closing the gap between its economic performance and that of the global core. Second, new oil and gas discoveries, emerging consumer markets, and hints of the privatization of state-owned companies is commanding the attention of international investors. Third, a new generation of policymakers, activists, and business leaders is emerging across most of Subsaharan Africa—they are Africans to the core, but with an increasingly global mindset derived from their participation in the Internet age, easier international travel and communications, and greater exposure to current worldwide trends and thinking. Fourth, the African debt crisis is winding down, and economic relations with Western donor countries are improving. And fifth, most important of all, China has exploded onto Africa's economic scene, reshaping trade, investment, foreign aid, and diplomatic relations.

China's Role in Emerging Africa

Through much of modern history, Westerners have perceived Africa to be both remote and peripheral to international affairs. Although the subordinate, asymmetrical relationship that long tied this realm to Europe's imperial powers is still visible (in terms of foreign investment, trade, and external oversight), as noted above evidence is mounting that things have begun to change. The recent rise of emerging-market countries such as Brazil, Russia, India, and China—the so-called *BRICs*—as well as the retreat of various Western development models as the only game in town mean that African leaders now have a wider range of options in their international relations.

Many African states are now forging political and economic partnerships of their own choosing. They are casting aside the shackles of colonial thinking and its postcolonial legacies and taking more control of their own international economic policies. Such enhanced agency enables African states to become full-fledged players in international relations rather than the objects of other countries' foreign policies—the dawning of a new era that some are calling a "second independence" for Africa.

Subsaharan Africa is now engaging a host of foreign investors, but nothing marks the elevation of this realm's stature

Billy Miaron/Shutterstock.com

Middle-class youths at a recent midsummer cultural festival in Kampala, Uganda. Demands for education, cultural goods, computer and Internet use, and access to leading entertainment media are skyrocketing, and upscale consumption is steadily on the rise in many of Subsaharan Africa's cities.

more than China's continuing surge of investments in infrastructure, energy-producing facilities, and civic construction. (For instance, in 2017 Chinese construction companies earned no less than 53 percent of their total revenues from projects in just five countries on the African continent.) For the period 2005–2015, Nigeria topped the list with U.S. $32.2 billion of Chinese investments followed by Ethiopia at $17 billion; several other resource-rich African countries, including South Africa, Moçambique, Guinea, and Chad, were awarded over U.S. $10 billion worth of Chinese-funded projects (**Fig. 8-12**). Whereas investment in energy and mining dominated during that ten-year period, a shift into nonresource sectoral investment is clearly evident since the mid-2010s, emphasizing infrastructure (especially transportation), real estate (mostly in South Africa, Kenya, and offshore Mauritius), and manufacturing, which has advanced from textiles and apparel to industries focused on motor vehicles, home appliances, and building materials.

China's *Belt and Road Initiative (BRI)* (discussed at the end of Chapter 6) includes a Subsaharan Africa component (Fig. 8-12). Beijing's BRI strategy (see Fig. 6-20) relies heavily on upgrading infrastructure in several African ports—especially Mombasa (Kenya), Massawa (Eritrea), and Djibouti (Djibouti) as well as a new U.S. $10 billion container port, projected to become Africa's largest, at Bagamoyo, Tanzania, 60 km (37 mi) north of the country's capital, Dar es Salaam. On land, China is investing heavily in building new and/or upgraded railroads which include (1) the new East African Railway (Fig. 8-12) to connect the Kenyan cities of Mombasa and Nairobi (the capital) with lines that serve Kampala (Uganda), Kigali (Rwanda), and Juba (South Sudan); (2) expansion of the Tanzanian rail network that will include long-distance connections to Djibouti on the Red Sea as well as Angola's Atlantic ports clear across the continent; and (3) an upgraded route between the port of Djibouti and Ethiopia's capital, Adis Abeba. Moreover, China Telecoms

CHINESE INVESTMENTS IN SUBSAHARAN AFRICA, 2005–2015

- More than $10 billion
- $5 billion–$10 billion
- $2 billion–$5 billion
- Under $2 billion
- Port project
- East African Railway project

INVESTMENTS BY SECTOR, 2015

Transport 40%
Energy 36%
Real Estate 14%
Metals 7%
Agriculture 3%

Sources: American Enterprise Institute, Mercator Institute for China Studies, *Wall Street Journal*.

FIGURE 8-12

has established a new data center in Djibouti that connects to the company's other regional hubs in Europe, China, and Pacific Asia, facilitating Djibouti's new role as a strategic gateway hub for spreading fiber cable networks throughout eastern Africa. And China's deepening footprint here on the African Horn is also marked by its first-ever overseas military base, again in Djibouti, alongside that small country's U.S. naval base—the only place on Earth where warships of these two countries' navies are moored practically next to each other. As Figure 7-21 and its accompanying discussion underscore, Djibouti occupies a supremely important geostrategic position at the choke-point southern outlet of the Red Sea, through which passes at least one-fourth of the world's oil shipped by supertanker.

Not surprisingly, China has now become the realm's biggest trading partner (see **Box 8-7** and **Fig. 8-13**) as well as

foreign-aid provider and is also a leading source of immigrants (elaborated by Howard French in his monumental recent book, *China's Second Continent: How a Million Migrants Are Building a New Empire in* Africa). Besides being the largest foreign-aid provider, China also is the leading source of oil leases, joint mining ventures, and timber concessions. Perhaps most significant is the difference between the Chinese and Western engagement with contemporary Subsaharan Africa: Beijing does not require African states to emulate its values, only to support its One-China policy (maintaining formal diplomatic relations with Beijing while severing all ties to Taiwan). More generally, Chinese engagement provides wider options for maneuverability and the opportunity for Africans to chart untried development pathways. But skeptics caution that many African countries are increasingly becoming indebted to China, and the term

---

Box 8-7 Geographic Information Analysis

China's Exports to Subsaharan Africa

Since 2001, China has exported an increasing amount of goods and services to Africa. The geography of Chinese exports to Africa was initially rather uneven, with Egypt, Nigeria and South Africa serving as key trade partners. While these three African countries continue to figure prominently in Afro-Chinese trade, recent export patterns include more African countries than ever before.

What is the rationale behind the growing significance of the African continent for Chinese trade? What types of goods and services are being traded between Africa and China? What are the possible benefits and drawbacks to these newly established trading partnerships?

Access your WileyPLUS course to interact with a dynamic version of this map and to engage with online exercises and questions.

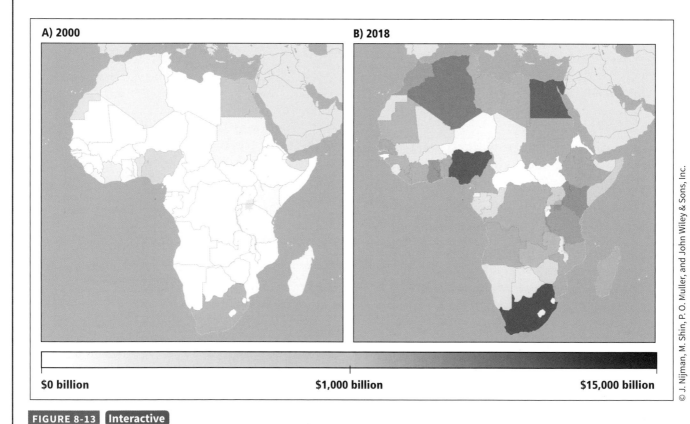

A) 2000

B) 2018

$0 billion $1,000 billion $15,000 billion

© J. Nijman, M. Shin, P. O. Muller, and John Wiley & Sons, Inc.

FIGURE 8-13 Interactive

debt-trap diplomacy has gained currency among economists alarmed by Beijing's new grip on the realm. Should African governments default on their Chinese debts, China's state-owned companies are positioned to cheaply snap up ports, railways, and other facilities.

Africa's ongoing ascent as a vital component of the global economy is largely based on the realm's natural resources. This has brought major new opportunities to develop infrastructure and real estate (e.g., satellite cities, industrial parks, modern CBDs and shopping complexes), provide commercial and producer services for Africa's mushrooming urban population, and accelerate the emergence of a middle class. Furthermore, several of the realm's countries are courted as diplomatic allies in global politics (Eswatini is the sole remaining African state that still has ties with Taiwan). In a world where voting matters more than ever, those of the 54 Subsaharan African states increasingly resonate in international forums that shape trade, environmental, and security-related policies.

Africa's Pressing Need for Supranational Cooperation

Economic interaction among the states of Subsaharan Africa has been stifled for centuries, and geography has played a key role. The realm's huge territorial size facilitated the manipulation of its political geography throughout colonial times in order to isolate individual colonies and connect them to Europe. Accordingly, roads and rail networks were laid out purely to serve the interests of colonial empires: they stressed convergence on colonial capitals, access to resource extraction sites, and, above all, direct linkages to seaports for shipping goods to the mother country. Because so few overland routes penetrated interior border zones, let alone crossing Africa's international boundaries, these colonial arteries reinforced intra-realm fragmentation. And what little cross-border commerce took place was heavily shackled by tariffs, harassment from customs officials, demands for bribes, and other unnecessary protectionist measures. Even today, intra-realm trade accounts for only 18 percent of total trading activity within Subsaharan Africa compared to almost 70 percent inside the European realm.

To overcome such disadvantages, African states need to strengthen their international interaction by committing themselves to the kind of supranational cooperation that is enhancing the internal cohesion of geographic realms from the Americas to Southeast Asia. At the continental scale, the *Organization of African Unity (OAU)* was established for this purpose in 1963 and was superseded by the *African Union (AU)* in 2001. Initially, integration efforts emerged at the regional level in the trading sphere following the launching of the *Economic Community of West African States (ECOWAS)* by 15 countries in 1975. By the early 1990s, 12 additional countries had established the *Southern African Development Community (SADC)* and were soon to be followed by the 20-odd states that founded the *Common Market for Eastern and Southern Africa (COMESA)*. Today, these regional trade agreements are being superseded by a continent-wide free trade pact, the **African Continental Free Trade Agreement (AfCFTA)**.

Signed in 2018 and 2019, AfCFTA is the boldest African-originating initiative to date for fostering economic unity. Coordinated by the African Union, this pact was signed by 54 of the AU's 55 countries (Eritrea is only country that did not (yet) sign on), thereby forging a single African common market to serve a total population of more than a billion, making it the world's largest free trade agreement both in territorial size and number of member-states. Moreover, AfCFTA was conceived as more than a trade pact: it also aims to widen and deepen internal cooperation to promote peace and security, ameliorate environmental and health crises, and upgrade cross-border interactions and related infrastructure. In short, the new agreement is much needed and ambitious. Its outcomes will be followed closely in the coming years.

Regions of Subsaharan Africa

On the face of it, Africa is so massive, compact, and continuous that any attempt to justify a contemporary regional breakdown is fraught with difficulties. Nevertheless, on the basis of environmental distributions, ethnic patterns, cultural landscapes, distance and proximity, historic culture hearths, and other spatial data, we identify a four-region structure (**Fig. 8-14**). By focusing on the contents of that framework, however, we should not overlook a fifth region that straddles the broad divide between Subsaharan Africa and North Africa—the *African Transition Zone*, discussed separately in Chapter 7 as the final region of the North Africa/Southwest Asia realm.

Beginning in the north at the southern margin of the African Transition Zone, we work our way east and then south through the following regions:

1. ***West Africa,*** which includes the countries of the western coast north of the equator and those on the margins of the Sahara in the interior. This populous region of 15 countries is anchored in the southeast by the realm's demographic giant, Nigeria.
2. ***East Africa***, where equatorial natural environments are moderated by elevation and where plateaus, lakes, and mountains, exemplify the countryside. Six countries,

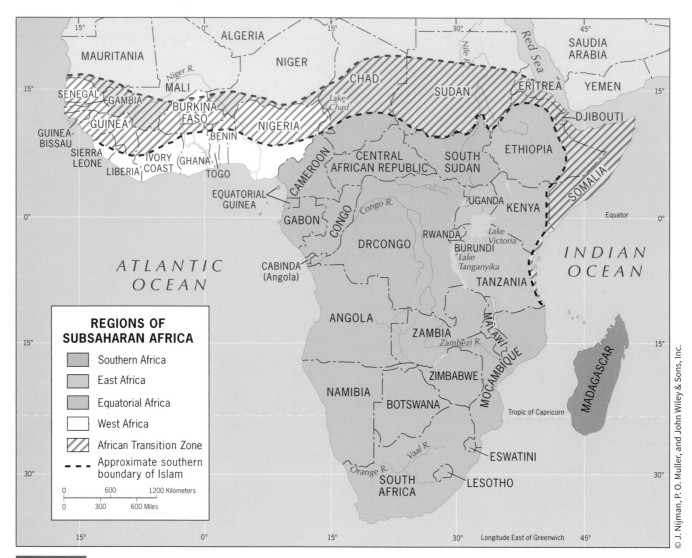

FIGURE 8-14

including the highland part of Ethiopia, comprise this region. The island-state of Madagascar, shaped by Southeast Asian influences, is neither East nor Southern African and has minimal contact with mainland Africa. It is included here only because of its proximate location.

3. *Equatorial Africa,* much of it defined by the basin of the Congo River, where elevations are lower than in East Africa, temperatures are higher and moisture more abundant, and most of Africa's surviving tropical rainforests stand. Among the nine countries that make up this region, which now includes recently formed South Sudan, DRCongo dominates territorially and demographically.

4. *Southern Africa*, extending from the southern tip of the continent to the northern borders of Angola, Zambia, Malawi, and Moçambique. Ten countries constitute this region, which extends southward from the tropics and whose giant is South Africa.

Region West Africa

West Africa occupies most of Africa's northwestern Bulge, extending southward from the margins of the Sahara to the Gulf of Guinea coast and westward from Lake Chad to Senegal (**Fig. 8-15**). A total of 17 states (also counting Chad and offshore Cape Verde) constitute this region. In addition to once-Portuguese Guinea-Bissau and long-independent Liberia, West Africa encompasses four former British and nine former French dependencies. Four countries, comprising an enormous territory dominated by dry environments but sparsely populated, are landlocked: Burkina Faso, Mali, Niger, and Chad.

As Figure 8-15 shows, political boundaries extend perpendicularly from the coast into the interior, so that from Mauritania to Nigeria, the West African habitat is parceled out among parallel, coast-oriented states. The southern half of the region is home to most of the people (Fig. 8-8). Mauritania, Mali,

FIGURE 8-15

and Niger include too much of the *Sahel*—the unproductive belt of Sahara-bordering, semi-arid environment (mapped in Figure 8-15)—to sustain populations as large as those of the most populous countries of Nigeria, Ghana, and Ivory Coast. Another key feature of the region's human geography is West Africa's bifurcation by a religious divide in upheaval—the turmoil-ridden *Islamic Frontier* (see Fig. 7-20) that forms the southern edge of the African Transition Zone (the black dashed line in Fig. 8-14).

Nigeria

Nigeria, this region's cornerstone, is home to 206 million people (**Fig. 8-16**). This is by far the largest population of any African country, and within the region more people live in Nigeria than in all the other countries of West Africa combined. Moreover, in 2014 Nigeria surpassed South Africa to become the largest economy in the realm. But this country is also bisected by one of the most contentious segments of the *Islamic Frontier*, which during the 2010s unleashed the vicious Boko Haram insurgency based in the northeast. And another current security crisis affects the exporting of Nigeria's most valuable resource, oil; pirate attacks persist along the Gulf of Guinea coast, aimed at the main sea lane used by tankers to transport oil to global core countries from the extensive fields that lie beneath the Niger River Delta in the southeastern corner of the country.

Nigeria achieved independence from Britain in 1960, but its new government faced the herculean task of managing a European political creation containing three major nations—the Yoruba, Ibo, and Hausa—and nearly 250 additional peoples. Britain's colonial imprint was always stronger in the two southern subregions than in the north. Christianity became the dominant faith in the south, and southerners, especially the Yoruba nation, took a lead role in the transition from colony to independent state. The choice of Lagos (photos atop page 289),

FIGURE 8-16

the primary port of the Yoruba-dominated southwest, as the capital of a federal Nigeria reflected British aspirations for the country's future. A three-region federation, two of which lay in the south, would ensure the primacy of the non-Islamic part of the state. But this framework did not last long and required frequent modification. Nigeria today has 36 states, and the capital was moved from Lagos to more centrally located Abuja more than a quarter-century ago (Fig. 8-16).

Oil Booms and Busts

Large oilfields were discovered beneath the Niger Delta in the 1950s, and this transformed the agriculture-dominated economy into one based on petroleum exports. But before long, Nigeria's energy riches brought more bust than boom. New misguided development plans focused on ill-considered industrial schemes and the like, while farming, still the mainstay of most Nigerians, fell into neglect. Worse, poor management, corruption, outright theft of oil revenues during military misrule, and excessive borrowing against future oil income led to economic disaster. Much of the country's infrastructure collapsed; in cities, basic services broke down; and in rural areas, clinics, schools, water supplies, and roads to markets crumbled.

Nonetheless, Nigeria has now become the realm's largest economy. Most of that success results from its (2019) position as the world's thirteenth-largest oil producer, with India, the EU, South Africa, and the U.S. as its leading trade partners. But these economic benefits have yet to reach the great majority of the fast-growing Nigerian population; in fact, this oil-rich country recently overtook India as the home of the world's greatest concentration of extreme poverty.

Two faces of the megacity of Lagos, Nigeria, in 2019. To the left, the floating slum of Makoko, near down-town Lagos. More than 2,000 newcomers arrive in Lagos every day and, for many, finding proper housing is a major challenge. An estimated 200,000 people are crammed in this floating slum area. The photo on the right shows construction of "Eko-Atlantic City" on Lagos' Victoria Island, designed to provide an environmentally sustainable "world class" business and residential area, which is expected to cater to the rich and to expat communities.

Islamic Extremism: Boko Haram

In 1999, Nigeria's hopes were heightened as a democratically elected president was sworn into office. But shortly thereafter its challenges magnified when northern States, beginning with Zamfara, began to promulgate strict-Islamic *Sharia* law [20]. When neighboring Kaduna State followed suit, violence between Christians and Muslims was unleashed and devastated that State's venerable capital city, Kaduna. There, and in the 11 other northern States mapped in green in Figure 8-16, the imposition of *Sharia* law led to the departure of thousands of Christians, intensifying the cultural fault line that we now know as the *Islamic Frontier*. That schism continues to deepen, and the kind of Islamic revivalism now gripping the north has reached a level that is imperiling the stability—perhaps even the unity— of West Africa's cornerstone state.

In 2002, a jihadist Islamic organization calling itself **Boko Haram [21]** (meaning "Western education is sinful"), emerged in the northeasternmost State of Borno (Fig. 8-16), an impoverished, underrepresented corner of Nigeria where repressive central government policies had triggered growing militancy. Support for Islamic militants from the State's heavily Muslim population is mainly drawn from young, unemployed, northern Nigerian men, who have lost faith in the politicians and in their own future. Boko Haram's "religious cleansing" campaign was initially confined to Borno, but since 2008 its barbaric activities have geographically expanded across the entire northeastern quadrant of Nigeria, including high-profile terrorist attacks in Kano and even the federal capital, Abuja, well to the west and south. Targeting civilians, Boko Haram attacks in the 2010s killed tens of thousands and forced at least two and a half million to flee their home villages (photo right). The world's attention was seized by the dramatic kidnapping of 276 female adolescents from a school in the village of Chibok in 2014, to be forcibly married to Boko Haram fighters—igniting the global #Bringbackourgirls social media campaign—until most were freed in 2018.

Boko Haram's aggressive terror campaign against Western education has also resulted in the death of more than 600 teachers and the flight of at least 20,000 educators since 2009. Geography teachers, in particular, were singled out as top terrorist targets because of their knowledge of African geopolitics as well as their global perspectives. Moreover, Boko Haram followers hold bizarre worldviews, contending that the Earth is flat and that rainfall patterns are attributable to Allah's divine will and not to climate-physical processes and **anthropogenic [22]** forces. Anthropogenic causes of drought emphasize adverse human influences on the environment, such as overpopulation and land degradation (deforestation, continuous cropping, overgrazing), which lead to more extreme weather, local climate change, and devastating impacts on Sahelian society. Children across the Sahel, particularly in northern Nigeria, face severe malnourishment

The aftermath of a typical nighttime Boko Haram attack in northeastern Nigeria. Here, in the village of Kawuri not far from the city of Maiduguri, 85 residents were murdered and every dwelling was destroyed. A few months later the jihadist killers struck again in a nearby town when 50 gunmen stormed a school at dawn, stabbing 43 students to death to avoid attracting attention by using their firearms and homemade bombs.

and food insecurity. That is one of the drivers of displacement, political instability, and conflict. Thus, an understanding of state security and its relationship to complex human-climate processes is essential to combat the vicious cycle of conflict and food insecurity.

In 2015, following the lead of ISIS, Boko Haram declared itself an independent caliphate (Islamic State's West African Province) consisting of the Nigerian territory it controlled, centered on Borno State's capital, Maiduguri, a city of about one million today (Fig. 8-16). Operating under the banner of its own black flag, Boko Haram at its height dominated an area larger than West Virginia. However, since 2015, Boko Haram has been pushed back, with the Nigerian government not only claiming that at least 40,000 Boko Haram fighters have surrendered but also that well over a million terrorism-displaced people have returned to northern Nigeria. By mid-2019, Boko Haram was under heavy siege by government forces but had not yet been defeated. But even its retreat into splintered factions still poses a dire threat: a regrouping Boko Haram has already demonstrated its capability to launch fierce intermittent attacks, and worse could come if its remaining strongholds were to evolve into safe havens for international terrorists on the run (such as the remnants of ISIS).

Coast and Interior

West Africa's other countries also exhibit distinct regional geographies. Typical of the Gulf of Guinea coastal zone west of Nigeria is neighboring **Benin**, with a population of 12 million and growing cultural and economic ties with northeastern Brazil's Bahía State, where many of its people were taken in bondage and where elements of West African culture survive. Benin's internal geography is comparable to that of neighboring **Togo** (population: 8.2 million): both are markedly elongated, narrow political units with savannas in the north and humid tropical lowlands near the coast, but without usable rivers. In contrast, **Burkina Faso** is representative of West Africa's much drier interior—impoverished and landlocked but containing undeveloped reserves of gold and a commercial economy that relies on exporting cotton. Importantly, with Muslims constituting 59 percent of Burkina Faso's 21 million inhabitants, this country—together with Mali (95 percent), Niger (99 percent), and transitional Chad (58 percent)—lies firmly within the Islamic orbit of northern West Africa.

Among the coastal states, **Ghana** (population: 31 million), known in colonial times as the Gold Coast, was the first modern West African state to achieve independence, with a democratic government and a sound commercial economy based on cocoa exports. Both democracy and the economy soon failed. In the 1990s, a military regime was replaced by a stable representative government, and recovery began. By 2007, Ghana celebrated 50 years of independence, its democracy maturing, its economy forging ahead, corruption persistent but declining, and its international stature rising.

The same year also brought a U.S. $500-million-plus aid package from America's Millennium Challenge Corporation to expand commercial agriculture, further improve infrastructure, and combat poverty—based in no small part on Ghana's solid reputation as an African trailblazer with a growing middle class (**Box 8-8**).

Proceeding westward from Ghana along the southern coast of the African Bulge, **Ivory Coast** (officially *Côte d'Ivoire*) has had a turbulent history. Following independence from France in 1960, it translated the next three decades of autocratic but stable rule into economic progress; however, by the late 1990s the political succession had become badly entangled in the north–south, Muslim–Christian schism that degenerated into two rounds of civil war between 2002 and 2011. Conflict has now largely subsided, but this country of 27 million remains deeply troubled as it strives to rebuild its shattered society and cocoa-based commercial economy. Farther up the coast is another key African Bulge country, **Senegal**. This democratic state demonstrates what stability can facilitate: without oil and other major income-generating resources, and with an overwhelmingly subsistence-farming population of 16.7 million, Senegal nevertheless managed to achieve some of the region's highest GNI levels.

Between Senegal and Ivory Coast lie three fragile coastal countries that have recently emerged from three decades of civil war and horror. **Liberia**, founded in 1822 by freed American slaves who returned to Africa with the help of U.S. colonization societies, was governed by Americo-Liberians for six generations and sold rubber and iron ore abroad. A military coup in 1980 ended that era, and by 1989 full-scale civil war had embroiled virtually every ethnic group in the country. This brutal protracted conflict claimed the lives of about one-tenth of the population (5 million today), and hundreds of thousands of others fled the country. Liberia finally turned the corner in 2006 with the election of (2011 Nobel Peace Prize winner) Ellen Johnson Sirleaf, Africa's first female head of state. She was succeeded in the presidency after 12 years by George Weah, a former FIFA world footballer of the year, who studied for an undergraduate degree in the U.S. upon retirement from the sport; he took over as president in 2018, and promptly launched a 10-year plan to provide a quality education for all Liberians. **Sierra Leone** (population: 8 million) followed an initial similar trajectory from British dependency to republic to one-party state and military dictatorship. In the 1990s, a rebel movement, financed by diamond sales, inflicted dreadful punishment on the local population, but a remarkable turnaround resulted from British intervention during the late 2000s. Diamonds remain the leading export, although now-stabilized Sierra Leone's economy still suffers from the aftereffects of the 2014–2016 Ebola crisis. Unlike the turnarounds in Liberia and Sierra Leone, neighboring **Guinea** continues to be plagued by dictatorial mismanagement and violent power struggles that combine to constrain economic opportunities in agriculture and mining. With a population of 13 million, Guinea remains one of Africa's lowest-income states despite possessing major deposits of bauxite (aluminum ore) and gold, capabilities for

Box 8-8 From the Field Notes …

"Even in the poorer countries of the world, you see things that have become emblematic of twenty-first-century global cities: gated communities, suburban malls, chain coffee shops, and luxury apartment complexes. On a research trip to Accra in 2018, I was struck by the proliferation of upscale high-rise residential towers under construction in the vicinity of the airport. Widening wealth inequality, heightened security concerns, and a surge in traffic gridlock and road accidents have fueled the growth of this niche property market, which stands out in an urban landscape of low-rise residential structures shaped by traditional forms of land ownership. Pictured here is the stunning Villaggio high-rise development, adjacent to Kotoka International Airport's main expressway interchange and a 20-minute drive from downtown Accra. Villaggio combines the amenities of a world-class residential tower found in any global city with a distinctly West African architectural style—inspired by Kente cloth, a traditional silk and cotton fabric made of interwoven cloth strips in the nearby historic Ashanti region of west-central Ghana."

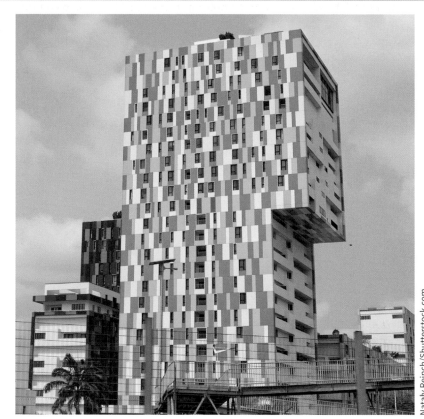

Nataly Reinch/Shutterstock.com

Upscale high-rise apartment buildings are now found in major urban areas throughout the realm and attract upper-income residents, investors from the African Diaspora, and returnee migrants and expatriates who favor congregating in prime locations offering superior access to both the CBD and airport.

producing far more coffee and cotton than it does, and the control of productive offshore fishing grounds.

The numerous conflicts noted in this chapter were and are geographically concentrated in relatively few areas and are not representative of vast, populous West Africa as a whole. Tens of millions of farmers and herders who manage to cope with fast-changing environments in the demanding Sahel live remote from the newsmaking strife along the well-watered coast. Overall, to take fuller advantage of the development opportunities sweeping across the realm, West Africa's foremost challenge is to achieve both economic survival and sustain nation-building through political stability.

Region East Africa

To the east of the chain of Great Lakes that marks the eastern border of the Democratic Republic of Congo, the land rises from the Congo Basin to the East African Plateau. Hills and valleys, fertile soils, and copious rains mark this transition in Rwanda and Burundi. Eastward the rainforest disappears, and the open savanna cloaks the countryside. Great volcanoes tower above a rift-valley-dissected upland. At the heart of the region lies Lake Victoria. Farther north the surface rises above 3300 meters (10,000 ft), and so deep are the trenches cut by geologic faults and rivers there that the land was appropriately called Abyssinia (now Ethiopia). Five countries, in addition to the highland component of Ethiopia, constitute this East African region: Kenya, Tanzania, Uganda, Rwanda, and Burundi (**Fig. 8-17**). Kenya is the most important country from a Western perspective, but Tanzania and Ethiopia have larger populations. And the landlocked countries of Rwanda and Burundi may be small in size, but their fierce ethnic conflicts have spilled over into neighboring countries and contributed to the high regional level of geopolitical instability.

Tanzania

Territorially, Tanzania (its name being a hybrid derived from the 1964 union of **Tan**ganyika and offshore **Zan**zibar) is the largest East African state. Its 60 million people are drawn from more than 100 ethnic groups. No single ethnic group is large

An aerial view, near the Tanzanian–Kenyan border, of an escarpment (right) that lines the edge of East Africa's great Eastern Rift Valley. The lush landscape of this hillside extends clear across the wide valley's well-watered floor, a savanna environment of alternating trees and grasslands, teeming with wildlife.

enough to dominate the state; moreover, Muslims constitute 35 percent of the population, highly concentrated on the coast within the southern, tail end of the African Transition Zone (Fig. 8-17). This is also a country without a core because its clusters of population and zones of productive capacity lie dispersed—mostly on its margins along the Indian Ocean coast, near the shores of Lake Victoria in the northwest, adjoining Lake Tanganyika in the far west, and facing Lake Malawi in the interior south (Fig. 8-8). Experimenting with socialism and farm collectivization in the post-independence period scared off international investors and tourists, and Tanzania soon bottomed out as one of the world's poorest countries. Since 1990, the government has pursued a different course, and prospects have improved. Tanzania today is a leading gold exporter, an emerging oil and natural gas producer, and an increasingly popular safari destination (led by well-managed Serengeti National Park in the central north [and see wildlife photo page 294]), and the growth rate of its economy has hovered around 6 percent a year for the past decade.

Ethiopia

Ethiopia, at 115 million, is Subsaharan Africa's second most populous country after Nigeria. The highland zone of Ethiopia forms the northernmost component of East Africa, but its extreme north and northeast lie within the African Transition Zone behind the *Islamic Frontier* (Fig. 8-17, red-dashed line). Culturally as well as physiographically, Ethiopia is part of East Africa: the two peoples that together dominate its ethnic complexion (the Oromo and the Amhara) are neither Arabized nor Muslim—they are Africans. Uniquely, deep colonial penetration was avoided here. Adis Abeba, the historic capital and headquarters of the Coptic Christian Amharic empire, held its own against Italy's attempted colonial takeover (except for a brief

period in the late 1930s). Indeed, the Ethiopians, based in their mountain fortress (Adis Abeba lies more than 2400 meters/8000 ft above sea level), became colonizers themselves, at times taking control of much of the Islamic part of the African Horn to the east.

During the past decade, Ethiopia has made considerable economic progress and ranks as one of the world's fastest-growing economies. Much of this growth has been attributed to government initiatives, such as selling enormous tracts of publicly owned arable land to foreign agribusiness corporations and welcoming a wide range of Chinese foreign investment. These gains notwithstanding, the country's economy remains focused on such agricultural commodities as coffee, tea, spices, and livestock: no less than 85 percent of the Ethiopian workforce is still composed of subsistence farmers.

The Paradox of Export Manufacturing in Ethiopia

Remarkably for a landlocked country with the lowest road density in Subsaharan Africa, Ethiopia has become a major exporter of shoes and clothing in addition to its traditional export base of livestock and coffee. Ethiopia has overwhelmingly relied (95 percent of its foreign trade today) on Djibouti's port complex since it fought an unsuccessful war to keep neighboring Eritrea from breaking away in 1993 (Ethiopia has longstanding political tensions with its other maritime neighbor, Somalia). Since 2016, Adis Abeba has been connected to Djibouti by an upgraded, 750-kilometer (475-mi) railway link, financed and built by China. By rail, goods can now be transported from Adis to Djibouti in 10 hours versus two days via road. Both countries are benefiting from increased trade, and strategically situated Djibouti has geopolitical ambitions to become the African Horn's main trading hub.

Ethiopia has recorded phenomenal economic growth (annualized near 10 percent) since the late 2000s. The economy is overseen by an authoritarian regime that restricts press freedom, but a generation of returnees and strong financial backing from China have produced greater economic diversification as well as the rise of a major apparel (fast-fashion) and agroprocessing export manufacturing complex. As a next step in advancing its industrialization, Ethiopia is now trying to emulate the experiences of China, Malaysia, and Taiwan in making extensive use of new industrial parks to attract additional foreign investment (30 such facilities are planned to be in operation by 2025).

With the relocation of much of China's footwear and apparel production to the Adis Abeba area, Ethiopia has become a leading exporter, particularly of women's shoes. Granted that Ethiopian workers are not as productive as Chinese workers, but this country's labor rates are among the lowest globally (U.S. $45 per month with no minimum wage laws versus $300 in China). Moreover, a reliable energy grid and the low price of electrical power are added incentives for manufacturers. And even the Chinese work ethic is being introduced into Ethiopian factories—replete with banners transmitting slogans that exhort maximum labor output (lower photo page 294).

FIGURE 8-17

© J. Nijman

The Ngorongoro Crater in northern Tanzania is one of the world's most famous wildlife conservation areas. The background ridge is actually part of the rim of this largest inactive volcano crater on Earth, which formed at least two million years ago when a huge volcano exploded and caved in on itself. This UNESCO World Heritage Site is home to wildebeests and zebras (shown here) as well as dozens of other large mammals, including giraffes, elephants, leopards, lions, and hippos. Ecotourism is an economic mainstay of this immediate area.

Uganda

Uganda is a landlocked, Lake Victoria–fronting country of 46 million, bordered by deeply troubled neighbors. To the north lies newly independent South Sudan, by many measures Africa's

JENNY VAUGHAN/AFP/GettyImages

Workers at the assembly line at the Huajian shoe factory in Dukem, Ethiopia, one of 36 in the Adis Abeba area that collectively export about 5 million pairs of shoes annually. (In the process, Ethiopia has also become a global leader in leather production.) As China's own economy matures, companies such as Huajian are increasingly moving to Africa to find more affordable labor.

poorest state and plagued by incessant civil conflict. To the west lies the ferociously contested eastern frontier of the Democratic Republic of the Congo, where government forces and insurgents fight seemingly endless battles. To the south lies Rwanda, scene of recurrent mass genocides. All this violence has not left Uganda untouched; it is a destination for persistent refugee flows. In addition, Uganda itself had to overcome the legacies of one of Africa's most brutal dictators, Idi Amin, whose economic nationalization drive wreaked havoc on the country's agricultural, mining, and trade sectors throughout most of the 1970s. Another calamity followed in the 1980s at the height of the late-twentieth-century HIV/AIDS pandemic: more than a million Ugandans succumbed, orphaning 1.2 million children. Responding massively in the aftermath, the government launched such an effective public-health campaign that the country became a global poster child in the struggle against this disease.

Rwanda and Burundi

Rwanda and Burundi, each with a population of approximately 12 million, are Africa's most densely inhabited countries. Here, Tutsi pastoralists from the north subjugated Hutu farmers, who had themselves made serfs of the local Twa people. Belgian colonial rulers promoted the minority Tutsi population at the expense of the Hutu majority, and nothing changed following independence in the early 1960s. Not until three decades later did the oppressed majority finally erupt in the horrific genocide of 1994, when almost one million Rwandans were slaughtered in 100 days. Although Rwanda has been stabilized and is one of the fastest-growing African economies (7.5 percent annually over the past decade), it still relies on a narrow export base of coffee and tea—while two-thirds of the population live in grinding poverty. Today, new technology-based entrepreneurship is prized above all. Rwanda now even permits access to its airspace for drone initiatives. One ongoing innovative effort—the urgent delivery of blood supplies to outlying rural clinics—is at the frontier of medical drone applications. Unfortunately, the longstanding conflict continues to simmer in this corner of East Africa, having spilled over the realm's western edge into the Democratic Republic of the Congo in the form of armed combat to control strategic minerals, especially diamonds and coltan (a valuable mineral used in mobile phones and myriad electronic devices).

Kenya

Over the past half-century, Kenya has played a leading role in East Africa. This country of 54 million contains both the region's largest city, its capital of Nairobi, and its principal port, Mombasa. After gaining independence from Britain in 1963, Kenya chose a capitalist path of development, aligning itself with Western interests. Even though it lacked major mineral deposits, Kenya built a relatively prosperous economy based on exports of coffee, tea, and other food products as well as a thriving tourist industry keyed to its magnificent landscapes and wildlife preserves.

But the good times lasted barely a generation as increasingly serious problems arose. Kenya during the 1980s had the highest rate of population increase in the world, and pressure on farmlands and fringes of the wildlife preserves steadily mounted—leading to widespread poaching and a concomitant decline in tourism. Beginning in the mid-1990s, excessive rains triggered massive flooding that was followed by the onset of a long and disastrous drought, resulting in famine across most of the interior. Meanwhile, government corruption siphoned off public funds that should have been invested, and ethnic tensions were allowed to explode into outbreaks of violence, especially after the 2008 fraudulent presidential election. Moreover, the HIV/AIDS pandemic dealt the country a devastating blow, particularly in its western provinces. Kenya's latest misfortune is the growing danger of terrorist attacks since 2010. The Somali-based jihadist group *al-Shabaab* has been in the forefront, killing dozens of shoppers in its high-profile 2013 assault on Nairobi's upscale Westgate Mall—sending shockwaves of insecurity through Kenyan society. Soon thereafter, the jihadists had transformed Kenya's coastal-strip portion of the African Transition Zone (Fig. 8-17)—where most of Kenya's 5 million Muslims reside—into a territorial stronghold overrun by a Boko Haram–style reign of terror, directed at all non-Muslims.

Kenya's greatest challenge is its continuing failure to forge a unified nation out of its more than 30 major constituent ethnic groups. Geography, history, and politics have placed the Kikuyu—who account for just under 20 percent of the population—in a position of dominance. But that position is hotly contested by the other groups, especially the Luhya, Kalenjin, Luo, and Kamba, who collectively constitute nearly 50 percent of the population; and in Kenya's periphery, there are additional major peoples such as the Maasai, Turkana, Boran, and Galla (Fig. 8-17). Adding yet another dimension to this ethnic cauldron, large numbers of Somali refugees from that country's never-ending conflicts have concentrated in Nairobi's Eastleigh neighborhood—triggering worsening ethnic discrimination as well as police harassment, increasingly in response to the threatening stance of the *Islamic Frontier* rooted along Kenya's Indian Ocean seaboard.

Nairobi, like all major African cities, struggles daily with these and many other growing pains of uncontrolled urbanization. But there is also a brighter side to the continuing development of this busy, skyscrapered capital that lies at the heart of a metropolis of five million. Nairobi today is widely regarded throughout the realm as an emerging information technology hub. The centerpiece is Konza Techno City, the new planned satellite city being constructed on the outer suburban frontier southeast of Nairobi (a future technopole already being billed as Africa's *Silicon Savannah*). Here, the Kenyan government envisions the rise of a state-of-the-art, ultramodern sustainable metropolis, driven by a burgeoning IT sector capable of generating as many as 100,000 jobs by 2030. The digital revolution has also made inroads beyond urban Africa (see **Box 8-9**).

Madagascar

Only 400 kilometers (250 mi) off Africa's central east coast lies the world's fourth-largest island, Madagascar (Fig. 8-14). But Madagascar is part of neither East Africa nor Southern Africa. The first human settlers arrived here about 2000 years ago—not from nearby Africa but from distant Southeast Asia. Subsequently, the powerful Malay kingdom of the Merina came to flourish in the highlands, and its language, Malagasy, became the indigenous tongue of the entire island (see Fig. 8-10). Today the Merina remain the largest of nearly 20 ethnic groups in the population of 28 million. Like the African mainland, Madagascar experienced colonial invasion and competition, but the Merina successfully resisted conquest for centuries. Madagascar finally was forced to become part of France's empire in 1897; however, as soon as it could, it split from France in 1960 at the height of Africa's independence movement.

Because of its deep Southeast Asian imprint, Madagascar's staple food is rice, and many of its people also depend on fish. It has some minerals, but the economy is weak and constantly bedeviled by political turmoil. The country also has been burdened for decades by its high population growth rate (it is still almost double the global mean), and its colonial-built infrastructure has crumbled. Madagascar's geographic isolation allowed for the evolution of unique tropical flora and fauna—but the disastrous human onslaught has destroyed no less than three-quarters of the island's rainforest home of this biological treasure.

Region Equatorial Africa

The term *equatorial* has both locational and environmental connotations. The equator bisects Africa, but only the western two-thirds of central Africa displays the conditions associated with the low-elevation tropics: intense heat, extreme humidity, copious rainfall, little seasonal variation, rainforest vegetation, and enormous biodiversity. To the east, where East Africa begins beyond the Western Rift Valley, elevations rise, and cooler, more seasonal

Box 8-9 Technology & Geography

Mobile Money in Subsaharan Africa

Did you know that Subsaharan Africa leads the world in mobile banking? For instance, it is much easier to send and receive money via a text message in Nairobi than it is in Paris. In Rwanda, Ghana, and Uganda, customers can send a text message to a money transfer service to receive a one-time code to use to withdraw money from an ATM—for example, money "sent" to them from a relative or from a small-business partner. For a small fee, mobile-to-mobile payments, withdrawals, and transfers can swiftly be executed via a text message—no bank card or bank account needed! Such services also provide international transfers, payments, and small loans.

According to the World Bank, 15 of the top 20 countries that lead the world in mobile monetary transactions are located in Africa, with Kenya, Uganda, Ghana, and South Africa at the forefront. This system of **mobile money [23]** challenges the more traditional retail banking networks in terms of accessibility and use. The recent rise of mobile money and cashless transactions across Subsaharan Africa can be attributed to a general lack of banking infrastructure, the high costs associated with long-distance financial transactions, as well as the risks associated with all-cash transactions—and to the explosive growth of cellphone usage across this realm. As far back as 2014, it was estimated that more than 80 percent of adults in such countries as Ghana, Kenya, and Tanzania owned cellphones, and smartphone adoption is growing thanks to the arrival of affordable Chinese brands.

In certain African countries, mobile phone penetration and the innovative use of text messaging are replacing local **cash economies [24]** with cashless transactions. Here, mobile money accounts now exceed the number of bank accounts, and more than 10 percent of GDP transactions occur through mobile money. The lack of a retail banking infrastructure also propels the development of mobile money networks, which increasingly rely on local gas stations, bars, and convenience stores—bypassing banks altogether. This is not to say that all of Subsaharan Africa is going cashless or that financial inclusion and banking are now universal across the African continent. The point here is that the future of mobile money is decidedly being shaped in this realm.

The mobile phone revolution has made a dramatic impact on farmers in many parts of the developing world, especially in Subsaharan Africa where distances are lengthy, land lines are absent, and market information is scarce. Mobile phone users in the realm surpassed 500 million in total during the late 2010s, becoming not only the leading communication medium but also a primary platform for a variety of tasks normally performed on computers and/or medical equipment. In rural Africa today, more and more physicians plug hand-held medical devices into smartphones to perform ultrasound imagery. And farmers and market women now employ smartphones to access real-time commodity market prices, transfer money to suppliers and transport companies, and send funds to relatives back in their home villages—all remarkably accomplished without maintaining a traditional bank account.

climatic regimes prevail. Equatorial Africa, lying west of this regional boundary, is physiographically dominated by the gigantic bowl-shaped Congo Basin. This region consists of nine states, of which the Democratic Republic of the Congo—formerly Zaïre, and for convenience labeled *DRCongo*—is by far the largest and most important country (**Fig. 8-18**). Five of the remaining eight states—Gabon, Cameroon, Congo (Republic), Equatorial Guinea, and São Tomé and Príncipe—have Atlantic coastlines. The Central African Republic, the southern part of Chad that belongs to this region, and recently independent South Sudan, are all landlocked and therefore economically constrained. In the limited space we have, most of our focus is on the first and last of these nine countries.

The Democratic Republic of the Congo (DRCongo)

With a territory not much smaller than the United States east of the Mississippi, a population of 90 million, a rich and varied mineral base, much serviceable agricultural land, abundant water supplies, and luxuriant tropical rainforests, DRCongo would seem to have all the ingredients needed to be a leading Subsaharan African economy. But powerful centrifugal forces, arising from its physiography and cultural geography, have always pulled this country apart. The immense, heavily forested heart of basin-shaped DRCongo creates some formidable

FIGURE 8-18

barriers to overland movement between east and west as well as north and south. Many of the country's productive areas lie along its periphery, separated from one another by vast distances. Not surprisingly, these areas tend to look across the border to one or more of DRCongo's nine neighbors for outlets, markets, and ethnic kinship ties.

DRCongo has only a tiny window (37 kilometers/23 mi) facing the Atlantic Ocean, barely enough to accommodate the wide mouth of the Congo River. Oceangoing ships can reach the port of Matadi, inland from which falls and rapids make

it necessary to move goods by road or rail to Kinshasa, the megacity capital of 11.4 million. In the distant other southern corner of DRCongo (3200 kilometers/2000 mi by river and rail from Kinshasa) lies Katanga Province, which contains most of the country's riches of copper and other mineral resources (Fig. 8-18). The narrow protrusion southeast of Lubumbashi into central Zambia is the heart of the Copperbelt, from which these minerals have flowed onto world markets for well over a century via rail lines that connect to ports on the Indian Ocean coast of East and Southern Africa.

DRCongo has been embroiled in conflicts since independence in 1960. Its latest cycle of civil war began in the 1990s when neighboring Rwanda's conflict spilled over into what was then still known as Zaïre. Ever since, competing networks of local rebel groups, insurgents from adjoining countries, and the militaries of various African states have combined to ravage the country. This series of interconnected conflicts has now become known as "Africa's continental war." Among its complex causes are local disputes over land and resources, the expansionist goals of rebel groups and predatory neighboring states, and myriad ethnic political grievances. The latest cumulative death toll estimates stand between 6 and 10 million. As for DRCongo's economic geography, it was being reshaped by rebel groups controlling the export of copper, coltan, and other highly valuable minerals, which increasingly enter the global marketplace through a network of clandestine trade.

Across the River

To the west and north of the Congo and Ubangi rivers lie seven of Equatorial Africa's other eight countries (Fig. 8-18). A pair of them are landlocked. **Chad** (population: 16 million), straddling the African Transition Zone as well as the regional boundary with West Africa, is one of the realm's most remote countries; poverty is rife, although post-2000 oil discoveries in the south and assistance from China in exploiting those reserves are today slowly changing its status. The embattled **Central African Republic** (4.8 million), chronically unstable as well as deeply mired in poverty, never was able to convert its agricultural potential and mineral resources (diamonds, uranium) into significant progress. And one country consists of two small, densely forested volcanic islands: **São Tomé and Príncipe**, a ministate containing a population of just 220,000 whose economy, like several others in the realm, is being transformed by recent oil discoveries.

The four coastal states present a different picture. All possess oil reserves and share the Congo Basin's equatorial forests (Fig. 8-18); accordingly, petroleum and timber figure prominently among their exports. In **Gabon** (2.1 million), which rejoined OPEC in 2016, this combination has produced an upper-middle-income economy. Of the four, coastal Gabon also has the largest proven mineral resources, including manganese, uranium, and iron ore. Its capital, Libreville (the only coastal capital in the region), reflects all this in its high-rise downtown, bustling port, and mushrooming shantytowns.

Cameroon (population: 27 million), not as well endowed with oil or other raw materials, possesses the region's strongest agricultural sector by virtue of its higher-latitude location and higher-relief topography. Western Cameroon is one of the more developed parts of Equatorial Africa but an increasingly violent separatist uprising plagues the English-speaking borderlands along the boundary with Nigeria.

With five neighbors, **Congo** (the "other" [former French] Congo—but much smaller at 5.5 million) could be a major transit hub for this region, especially for DRCongo, if it ever recovers from civil war. Its capital, Brazzaville, lies just across the Congo River from Kinshasa and is linked to the port of Pointe Noire by road and railway. But devastating, unceasing power struggles have thus far neutralized Congo's geographic advantages.

As Figure 8-18 shows, **Equatorial Guinea** consists of a rectangle of mainland territory and the island of Bioko, where the capital of Malabo is located. A former Spanish colony that remained one of the realm's least-developed territories, Equatorial Guinea was dramatically affected by increased oil production and exporting, and in the 2000s briefly became Africa's only high-income economy with GDP growth rates exceeding 10 percent. However, as in so many other oil-rich countries, this bounty had minimal impact on the well-being of most of the people—while ostentatious displays of wealth by the ruling elite elicited international condemnation. Overreliance on these energy resources has since resulted in a heavy toll on the country's economic performance, and by 2018 plummeting world oil prices had produced a negative GDP growth rate of *minus* 7.9 percent.

One other coastal territory would also seem to be a part of Equatorial Africa: **Cabinda**, wedged between the two Congos just to the north of the Congo River's mouth. But Cabinda is one of those odd colonial legacies on the African map—it belonged to the Portuguese and was administered as part of Angola. Today it is an **exclave [25]** of independent Angola, and a most valuable one because it contains sizeable oil reserves.

South Sudan

This newest state to appear on the world political map became independent in 2011, the outcome of a near-unanimous referendum in the southern provinces of former Sudan (the now-truncated country still called *Sudan*, discussed in Chapter 7). South Sudan's birth came in the aftermath of six disastrous decades of postcolonial strife within former Sudan, a brutal conflict in which more than 1.5 million died, magnified by the bifurcation of the country along the religious divide between Islam and Christianity/animism that we know as the *Islamic Frontier* (see Fig. 7-20).

The British effectively ruled former Sudan from the 1890s until independence in 1956. One of the roots of the country's long-running internal conflict, which peaked in the resumption of its post-independence civil war between 1983 and 2005, lay in the decision of the British colonial administration to combine northern Sudan, which was heavily Arabized and Islamized, with a sizeable African/Christian-dominated area to the south. As soon as the British departed in the mid-1950s, the Khartoum-based regime in the north sought to impose its Islamic rule on southern Sudan, immediately triggering the first civil war that lasted until 1972. By the time the bitter second civil war erupted a decade later, former Sudan's strife had impoverished the country to the point that its per capita income ranking was among the lowest on Earth. That slowly began to change in the 1990s when oil was discovered, including major deposits in the embattled south (Fig. 8-18). Eventually, leaders in the south saw oil as a ticket to self-sufficiency.

Animosity between the two halves of former Sudan persists. Indeed, a low-grade, oil-based conflict broke out soon after the political split of 2011—and continues. Grinding poverty, violent internal ethnic conflict, and ineffective governance (South Sudan's population size can still only be estimated—at around 11 million), and intermittent southward flows of displaced people to the ever-dangerous DRCongo have become commonplace. Meanwhile, the overwhelming majority of this country's residents are either subsistence farmers or cattle herders—a deeply traditional livelihood in which individual and family status depends on the number of livestock owned.

Region Southern Africa

As a geographic region, Southern Africa consists of all the countries lying south of Equatorial Africa's DRCongo and East Africa's Tanzania (**Fig. 8-19**). Thus defined, this region extends southward from Angola and Moçambique (on the Atlantic and Indian Ocean coasts, respectively) to South Africa and includes half a dozen landlocked states. Also marking the northern limit of the region are Zambia and Malawi. As noted above, Zambia

FIGURE 8-19

is nearly cut in two by the elongated, mineral-rich land extension from southeastern DRCongo; similarly, Malawi penetrates deeply into Moçambique.

Africa's Resource-Rich Region

Southern Africa constitutes a geographic region in both physiographic and human terms. Its northern periphery marks the southern limit of the Congo Basin in a broad upland that stretches across Angola and into Zambia (the narrow brown lobe extending eastward from the Bihe Plateau in Fig. 8-2). Most of this region is plateau country, and much of the interior upland is framed by the prominent, near-coastal Great Escarpment. There are two consequential river systems: the Zambezi (which forms the border between Zambia and Zimbabwe) and the Orange-Vaal (South African rivers that combine to demarcate southern Namibia from South Africa).

Materially, Southern Africa has long been the continent's richest region. A great north–south corridor of major mineral deposits extends through the eastern heart of the region from the Zambia/DRCongo Copperbelt through Zimbabwe's Great Dyke and South Africa's Bushveld Basin and Witwatersrand to the goldfields and diamond mines scattered around Bloemfontein in the geographic center of South Africa (see Fig. 8-19). Ever since these minerals began to be exploited in colonial times, many migrant laborers have come to work in the mines of these countries. And in Southern Africa's far northwest, huge onshore and offshore oil reserves are propelling the economic development of Angola. The region's agricultural diversity matches its mineral wealth, especially so in South Africa: vineyards drape the slopes of South Africa's Cape Ranges, and the country's relatively high latitudes and its range of altitudes create environments for apple orchards, citrus groves, banana plantations, nut farms, and myriad other crops.

Despite this considerable wealth and potential, only a few of the ten countries of Southern Africa have prospered, and the gap between them is steadily widening (see Fig. 8-11). As Figure 1-11 indicates, three countries—South Africa, Botswana, and Namibia—now rank in the World Bank's upper-middle-income category; in contrast, three remain mired in the low-income category (Malawi, Moçambique, and Zimbabwe). The four remaining states are classified as lower-middle-income countries: Angola, Zambia, and the two mini-states of Lesotho and Eswatini (formerly Swaziland). Among the three most prosperous countries, Botswana and Namibia are both desert-dominated, sparsely settled states, with a combined population of less than 5 million. South Africa, however, stands apart—the region's giant by every measure.

Since we are working our way through the Subsaharan African realm from north to south, we will also approach the region of Southern Africa from that perspective. We begin with the Northern Tier of countries, proceed southward into the Middle Tier, and then put preeminent South Africa in the geographic spotlight.

The Northern Tier

In the four countries that extend across the Northern Tier of Southern Africa—Angola, Moçambique, Zambia, and Malawi—change is occurring everywhere. **Angola**, Southern Africa's most resource-dependent country, was the region's star economic performer during the 2010s. This country of 33 million—which also includes its tiny exclave of *Cabinda* located north of the Congo River's mouth—had a thriving (colonial) economy based on a wide range of mineral and agricultural exports at the time of independence from Portugal in 1975. But the Portuguese did not prepare the country for self-rule, and soon Angola fell victim to the Cold War, with its northern peoples opting to follow a communist course and the southerners falling under the sway of a rebel movement backed by South Africa and the United States. The results included devastated infrastructure, abandoned farms, looting of diamonds, hundreds of thousands of casualties, and millions of landmines that still contaminate large swaths of Angola.

Led by oil, Angola's economic performance was very strong between 2005 and 2015 but sharply contracted as petroleum prices fell during the late 2010s. Yet the country stubbornly scontinued its reliance on (now shrinking) foreign oil income—much of which emanates from exports to China—an accelerating economic burden further aggravated by chronic corruption in this crucial extractive sector. As a result of this overdependence on oil, this fuel now accounts for more than 90 percent of all exports, at least 75 percent of government revenues, and over half of Angola's GDP. Not surprisingly, this heightening vulnerability as a resource economy has exacerbated domestic inequalities: two-thirds of the Angolan population continue to eke out a living on less than U.S. $3 a day.

Angola is Subsaharan Africa's second-largest oil-producing country after Nigeria. This recent view of the center of its burgeoning capital city, Luanda, is reminiscent of the Persian Gulf's showcase cities during their initial building booms. But oil has proven to be double-edged sword: whereas Luanda has become one of the most expensive cities anywhere to live in, the rest of the Angolan population resides in a world of extreme poverty. With 95 percent of the country's foreign income dependent on oil sales, the recent drop in the price of oil has forced government budget cutbacks that only intensify the hardships of life in one of the most unequal societies on Earth.

On the opposite coast, the other major former Portuguese dependency, **Moçambique** (31 million), experienced a comparable trajectory. Upon independence, Moçambique also chose a Marxist course with unfortunate economic and political consequences. Here, too, a rebel movement supported by South Africa caused civil conflict, created famines, and generated refugee flows (mainly toward Malawi). Moçambique possesses neither oil nor as much agricultural land as Angola. But it does have considerable bauxite and coal deposits, and its long Indian Ocean coastline provides an advantageous relative location (as well as access to promising offshore natural gas reserves). In recent years, port traffic has been revived, and the country is working with neighboring South Africa on building the joint Maputo Development Corridor (Fig. 8-19). But this African country still relies too heavily on its mining sector. Unfortunately, the U.S. $2 billion worth of loans acquired in the early 2010s to diversify the economy into fisheries and shipyard projects were squandered, and debt repayments have not been met since 2017. Moreover, the proportion of people living in poverty exceeds 80 percent, making this one of the poorest countries in the world. And to all this must be added the formidable task of rebuilding much of the country following the twin disasters (only six weeks apart) unleashed by cyclones Idai and Kenneth in the nightmarish months of March and April 2019 (see **Box 8-10**).

Box 8-10 Global Climate Change in **Subsaharan Africa**

The United Nations' Intergovernmental Panel on Climate Change has declared Southern Africa a "hotspot" for climate change. For over three decades, the region has experienced rising temperatures as well as more frequent droughts, flooding, and landfalls of cyclones (the term used for hurricanes in the Indian Ocean basin). Peak cyclone activity here occurs from January to March, and since 2000 cyclonic intensity has escalated. In fact, Southern Africa's two deadliest cyclones have both occurred during this century. In 2000, Cyclone Eline struck Moçambique, Madagascar, and Zimbabwe, causing nearly 800 storm-related deaths. In March 2019, Cyclone Idai battered Moçambique, Zimbabwe, and Malawi, resulting in a death toll of 1000+; according to the World Bank, 2.6 million people were severely affected, and the price of recovery would be at least U.S. $2 billion.

Cyclone Idai, a Category-3 storm packing winds in excess of 165 kph (110 mph), quickly became characterized as Southern Africa's Hurricane Katrina—only worse. Hitting just before harvest season, the destructive power of this monstrous tropical storm flattened a large swath of land (850 sq km/330 sq mi), immediately rendering 400,000 people homeless in Moçambique (photo right), uprooting 80,000 people in Malawi and tens of thousands in Zimbabwe. Making landfall near low-lying Beira, a port city of a half-million residents on the central coast of Moçambique, the cyclone devastated most of this urban area as a 4.4-meter (14-ft) storm surge wiped out 80 percent of its residential structures and most of the infrastructure. It then swept inland on a track that passed over eastern Zimbabwe and southeastern Malawi, continuing to destroy communities, farmlands, and their livelihoods, and ultimately leaving fully one million people destitute. Although international humanitarian relief efforts were marshalled to deal with the catastrophe, such temporary relief fails to cope with the ever-present challenges of adapting vulnerable low-lying areas, feeble infrastructure, fragile livelihoods, and makeshift homes to climate change.

Cyclone Idai lays bare the fundamental injustices of climate change. The humanitarian impacts and consequences are plentiful, and in the near-term future include the spread of waterborne diseases, heightening food insecurity and food price hikes, and increased social unrest. A longer-term future shaped by new international migrations may turn out to be the only way that humans can survive the coming climate calamity.

One final note to emphasize is that Idai was no one-off fluke. Only 41 days later, Moçambique was again ravaged by an even more intense (Category-4) storm named Kenneth, which at landfall was declared to be the strongest tropical cyclone to ever strike this country! Luckily, it came ashore in less populated northern Moçambique and rapidly degenerated. Nonetheless, dozens were killed and thousands of homes were destroyed, adding that much more to this country's autumn (while it was spring in the Northern Hemisphere) of utter misery.

Xinhua/Alamy Stock Photo

A week after devastating Tropical Cyclone Idai made landfall on March 15, 2019, Sofala, one of Moçambique's coastal provinces (Fig. 8-19), was still inundated by the floodwaters that had triggered the massive destruction of crops, dwellings, and infrastructure.

Landlocked **Zambia** (18 million), another product of British colonialism, shares the mineral riches of the Copperbelt with Katanga Province in its northern neighbor, DRCongo. However, the stable commodity prices on which the country depends have fluctuated wildly since independence, and the railways leading to Zambia's outlet ports in Angola and Moçambique were rendered inoperative by Cold War conflicts. More recently, the Chinese have taken great interest in Zambia's minerals, and they have invested in railroad repairs as well as the expansion of mining operations—resulting in an annual economic growth rate averaging about 6 percent for the past decade.

Neighboring **Malawi** (19 million), like Zambia, has been able to sustain democracy for more than two decades. Malawi has an almost totally agricultural economic base and is also unremittingly challenged by environmental degradation. This country's dependence on corn as its food staple, its variable climate, and its severely fragmented land-use system inhibit economic growth—which has fluctuated widely over the past decade between 2 and 8 percent annually.

The Middle Tier

Five states lie between the countries of the Northern Tier and South Africa. They form Southern Africa's Middle Tier subregion, and all border the Republic of South Africa: Botswana, Namibia, Zimbabwe, and the two ministates of Lesotho and Eswatini (Fig. 8-19). As the map reveals, four of these five are landlocked. Diamond-exporting (and upper-middle-income) **Botswana** occupies the heart of the Kalahari Desert and surrounding semi-arid steppe. Despite its lucrative diamonds, the great majority of its 2.4 million inhabitants are subsistence farmers, and in recent years Botswana has been the most severely AIDS-afflicted country in all of tropical Africa. **Lesotho** and **Eswatini** (2.3 and 1.2 million, respectively), both traditional kingdoms, depend very heavily on remittances from their workers in South African mines, fields, and factories.

Southern Africa's youngest independent state, **Namibia** (2.6 million), is the former German colony of South West Africa that was administered by South Africa from 1919 to 1990. The country is named after one of the world's driest deserts (the Namib, which lines its coast). Territorially about as large as Texas and Oklahoma combined, only its far north receives enough moisture to enable subsistence farming, which is why most of the Namibians live close to the Angolan border. Mining and ranching constitute the leading commercial activities, but even though orderly land reform is underway, much of the population still lives in poverty.

The Tragedy of Zimbabwe   Zimbabwe (population: 15 million) lies at the heart of Southern Africa, between the Zambezi River in the north and the Limpopo River in the south, and between the Great Escarpment to the east and the desert to the west. Landlocked but endowed with good farmlands, cool uplands, and a wide range of mineral resources, Zimbabwe during the late colonial era had one of Southern Africa's most vibrant economies and seemed to have a bright future. Its core area straddles the mineral-rich Great Dyke that extends across the heart of the country, which contains copper, gold, asbestos, chromium, and platinum. Its farms are capable of producing tobacco, tea, sugar, and cotton in addition to staples for the local market.

But the past five decades have not been fortunate ones for Zimbabwe. During the colonial period, the tiny minority of whites that controlled what was then called Southern Rhodesia (after Cecil Rhodes, the British capitalist of diamond fame and scholarship honors), took the most productive farmlands and organized and ran the agricultural economy. Following a unilateral declaration of independence in 1965 (to keep the white minority in power), their descendants continued to hold huge estates, only parts of which were being farmed. In the absence of comprehensive land reform, let alone progress toward democratic rule, the persistence of this monstrous inequality impeded broad-based development.

In the wake of their successful joint campaign to terminate white-minority rule in 1980, the two peoples that constitute most of Zimbabwe's population—the Shona (70 percent today) and the Ndebele (20 percent)—engaged in acrimonious ethnic conflict. Proper legal reforms did not take effect, and the rule of newly elected Robert Mugabe turned from ineffectiveness to disaster to extreme paranoia. As he encouraged squatters to invade white farms (where a number of owners were killed and many others fled), the agricultural economy began to collapse. Corruption in the Mugabe government resulted in the transfer of land, not to needy squatters but to cronies of top officials. Foreign investment in other Zimbabwean enterprises, including the mining industry, dried up. By the mid-1990s, Zimbabwe was in economic free fall.

Mugabe also turned against the vital informal sector of the economy, which was all most Zimbabweans had left when jobs on farms and in factories and mines disappeared. He ordered his henchmen to destroy the dwellings of some 700,000 "informal" urban slum dwellers, who had little choice other than to join a growing army of persecuted homeless refugees who soon began to stream out of Zimbabwe, many headed for South Africa.

By the early 2010s, heroic Zimbabweans and persistent outsiders were pressuring Mugabe (who turned 90 in 2014) to end his catastrophic dictatorial rule, and his former deputy Emmerson Mnangagwa finally took over the reins in a military-assisted transition in 2017 (he was subsequently elected in 2018, although many claim the voting was rigged). The country remains isolated internationally—except for the embrace of China; not unexpectedly, as Beijing pursues numerous economic opportunities in Zimbabwe, it is almost single-handedly responsible for the country's (unsteady) GDP growth since Mugabe's departure. Otherwise, instability prevails, as when the government in 2019 announced a 130 percent increase in fuel prices, triggering mass protests that were widely met by excessive force. Thus more than ever, its endless internal strife and repression makes Zimbabwe stand out in the sharpest contrast to what is being achieved in the other countries of Southern Africa.

South Africa

The Republic of South Africa (RSA) is the giant of the Southern Africa region, with a population of 60 million and a modern economy second in size only to Nigeria's within this realm. Its historical geography differs somewhat from much of the rest of Subsaharan Africa. The country's lands were fought over by various African nations before the Europeans arrived and the colonial "scramble for Africa" took place. Indigenous peoples had migrated southward—first the Khoisan-speakers and then the Bantu peoples—into the South African *cul-de-sac*. The Zulu and Xhosa nations were fighting over these lands at about the time the first European settlers arrived.

South Africa is one of the most strategic places on Earth, the gateway from the Atlantic to the Indian Ocean, a key source of provisions on the route to Asia's riches. The Dutch East India Company founded Cape Town as early as 1652, and the Hollanders and their descendants, known as *Boers*, have been a part of the South African cultural mosaic ever since. The British took over about 150 years later, and both colonial powers vied for control. The British came to dominate the Cape, while the Boers trekked into the South African interior and, on the high plateau they called the *highveld*, founded their own republics. A war ensued, but by 1910 the Boers and the British had negotiated a power-sharing arrangement, although the Boers eventually achieved dominance that lasted from 1948 to 1994. Having long since shed their European ties, they called themselves *Afrikaners*, their word for Africans.

For more than 40 years, between about 1950 and the mid-1990s, multicultural South Africa was in the grip of the world's most notorious racist policy, **Apartheid [26]**. The word itself means "apartness," but in practice it promotes strict racial segregation and severe discrimination. Out of the concept of Apartheid grew a notion, promulgated by the white (European) minority then in control of the state, known as *separate development*. This would apply Apartheid to the entire country, carving it up into racially based territories whose inhabitants were citizens of those ethnic domains ("homelands")—but not of South Africa as a whole. Predictably, such racist social engineering aroused strong opposition within South Africa and beyond, leading to worldwide condemnation and international sanctions.

By the late 1980s, South Africa seemed headed for a violent revolution, but disaster was averted by a most unexpected turn of events. A leader of the white-minority government, which for decades had ruthlessly pursued its Apartheid policies, and a revered leader of the multicultural majority, who had for 28 years languished in an island prison not far from Cape Town, struck a momentous accord, ushering in a new South Africa virtually overnight. Nelson Mandela walked out of prison a free man on February 11, 1990, and in the RSA's first democratic election in 1994 he became the country's

president—alongside a new, fully representative parliament that became the country's first "rainbow assembly." South Africa's internal political geography was reorganized into a federal arrangement consisting of nine provinces (see **Fig. 8-20**, including the inset map).

Ethnicity and Xenophobia   South Africa has long been the realm's most pluralistic and heterogeneous society. Its multiethnic complexion consists of many indigenous peoples (e.g., Zulu, Xhosa, Tswana, Sotho) plus the communities of various European and Asian settlers. The Dutch brought thousands of Southeast Asians to the Cape to serve as domestics and laborers, as the British did later from their South Asian colonies on the Indian subcontinent. Moreover, a substantial population of mixed ancestry clusters at the Cape, today known as the *Coloured* component of the country's social kaleidoscope. The overall racial composition of South Africa is displayed in **Box 8-11**.

Although heterogeneity marks the urban spatial demography of South Africa, regionalism pervades the national-scale human mosaic. The Zulu nation, almost 12 million strong, remains largely concentrated in today's Kwazulu-Natal Province (Fig. 8-20). The Xhosa (8.5 million) still cluster in the Eastern Cape below the Great Escarpment, and the Tswana (3.5 million) continue to occupy ancestral lands along the northern border with Botswana. Metropolitan Cape Town remains the heartland of the Coloured population; the city of Durban still exhibits the strongest South Asian imprint.

Since the end of the Apartheid era more than 25 years ago, international migrations from all over the realm have streamed into South Africa. These influxes have intensified the RSA's multiculturalism and are most visible in major urban areas, especially in their takeover of much of inner-city Johannesburg. But tension-free coexistence does not always prevail, and violent attacks directed at poor foreign immigrants (mainly those with a high profile in street trade) periodically erupt. Such expressions of **xenophobia [27]** are motivated by the perception that these immigrants are stealing jobs from South Africans—and are further fueled by the dominance of low-end commerce by foreign entrepreneurs whose business skills exceed those of natives (see photo page 305). Nonetheless, efforts to defuse xenophobic sentiments continue: for example, South African schoolchildren learning the alphabet are now taught that the letter "X" stands for xenophobia.

Box 8-11 South Africa's Population by Race, 2020

African nations	48.0 million (80.7%)
Coloured	5.5 million (9.2%)
White	4.5 million (7.6%)
South Asian	1.5 million (2.5%)
TOTAL	59.5 million (100%)

FIGURE 8-20

South Africa's Economic Geography

South Africa's economy, the realm's most modernized, has extensive regional influence. On the global scene, the RSA is regarded as an emerging-market economy as well as a gateway to Subsaharan Africa, particularly concerning corporate command and control. With a land area in excess of 1.2 million square kilometers (470,000 sq mi), South Africa contains the bulk of the region's minerals, most of its good farmlands, its biggest cities, best harbors, most productive factories, and best developed transport network. Mineral exports from the Copperbelt and the Great Dyke move through South African ports. Workers from as far away as Malawi and as close as Lesotho toil in South Africa's mines, industrial plants, and fields. And entrepreneurs from Somalia, Ethiopia, Bangladesh, and China are increasingly entrenched in the commercial transactions of the country's informal economy.

Ever since diamonds were discovered at Kimberley on the *highveld* in 1866, South Africa has been synonymous with minerals. Rail lines were laid from the coast to the diamond complex even as fortune seekers, capitalists, and thousands of African workers, many from as far afield as Moçambique, streamed to the site. Subsequently, prospectors discovered what was long to be the world's greatest goldfield on a ridge named the Witwatersrand (Fig. 8-20). There, Johannesburg soon became the

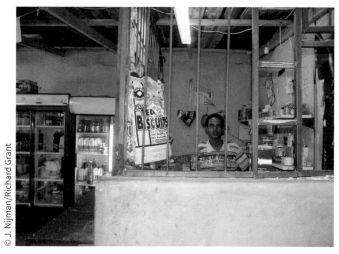

The newest immigrants from Somalia to the RSA often start out by running spaza shops—such as this one in Thokoza Township south of Johannesburg. (*Spaza* is a South African word for a small convenience store in the informal economy.) These Somalis work long hours, live on the premises, and constantly face the hostility of South African spaza competitors as well as unemployed youths. Sometimes they are shot at, and a few have even had their shops burned down. For many desperate immigrants, working in the spazas of dangerous "locations" (their term for the townships) is a rite of passage and the only job that a freshly arrived immigrant can find. Tellingly, the Somali shop-owners who employ them remain in central Johannesburg, choosing to send out only new Somali arrivals to handle the hardships of working the dangerous suburban "locations."

gold capital of the world. During the twentieth century, mining was diversified to include coal, iron ore, nickel, and copper, boosting a sizeable domestic metallurgical and manufacturing sector as well as yielding large revenues on world markets. Before the imposition of severe international sanctions in the 1980s to protest the growing outrages of Apartheid, capital flowed into the country, white immigration expanded, farms and ranches were laid out, and overseas markets proliferated.

Along the way, South Africa's cities grew rapidly. Johannesburg became an industrial agglomeration, major financial center, and immigration hub. To its north, Pretoria became the country's administrative capital, and the two metropolitan areas have coalesced into the even larger Gauteng conurbation that produces over one-third of the national economy's output. In Orange Free State (simply Free State since 1994), substantial industrial development matched the expansion of mining. Durban's port served not only the Witwatersrand but a much wider regional hinterland as well. And Cape Town has become South Africa's second-largest city and a leading international tourist destination; its port, industries, and productive agricultural hinterland gave it primacy over much of the southernmost RSA. (However, as **Box 8-12** shows, it also faces serious environmental challenges.)

But Apartheid frustrated South Africa's potential, and the enduring legacies of injustice have proven formidable and difficult to dislodge. Vast educational and income gaps between blacks and whites persist, making the country's society the most polarized in Subsaharan Africa.

Ongoing Challenges   In many spheres, South Africa is the most important country in Subsaharan Africa, and the entire realm's fortunes are bound up with it. No African country attracts more foreign investment or foreign workers. Its universities, hospitals, and research facilities are the best on the continent. Few have the free press, effective trade unions, independent courts, or financial institutions to match the RSA's. And with a population now surpassing 60 million, South Africa has spawned a substantial, multiracial middle class.

But, alongside its sharp socioeconomic divisions, South Africa also faces daunting political challenges. The African National Congress (ANC) Party—which produced all of the RSA's presidents since 1994—has failed to deliver the jobs and housing progress it has continually promised. Most recently, the presidency of Jacob Zuma (2009–2018) was plagued by serious claims of misconduct, in the eyes of many amounting to nothing less than **state capture [28]** (a term first introduced by the World Bank to indicate a systematic form of political corruption whereby private interests significantly influence government decision-making by bribing public officials). Mounting accusations and evidence of graft forced Zuma to step down from office in 2018.

Efforts to open new economic horizons through black economic empowerment have failed to narrow the economic-opportunity gap and have mostly benefitted the privileged elite. Ironically, even though a black middle class has emerged in urban South Africa, it mainly chooses to wall itself off inside gated residential communities well away from the enormous underclass. As for whites who remained in the country after 1994, they continue to enjoy the rewards of their economic power despite the national political transformation. Thus, for tens of millions of South Africans, very little has changed: they are poorer than ever compared to the newly affluent blacks and the ever-entrenched whites.

It should therefore come as no surprise that in 2019 South Africa recorded the greatest inequality within any of the world's countries. Predictably, there are massive (even widening) gaps in education, personal income, and service provision. South Africa today ranks 137th out of a total of 144 countries in elementary school performance. Overall, progress in righting the wrongs of Apartheid has been far too slow. Even though millions of houses have been built and connected to utilities in the post-Apartheid era, this effort falls drastically short of the number of such dwelling units actually needed. The most recent government reports put national unemployment in the vicinity of 25 percent, a figure easily doubled in the destitute inner-city slums and the sprawling shantytowns of outlying townships. In rural South Africa, farm workers and miners organize protests and strikes to demand fairer wages, and many white landowners have been intimidated by homeless invaders. Land reform

Box 8-12 From the Field Notes ...

"I am in Cape Town in April 2018 as the city is facing the unthinkable *Day Zero*—the date (initially April 12th) when the city's water supply would run dry, forcing Capetonians rich and poor to line up at emergency water distribution points to collect their daily water rations. The possibility of this city of 4 million, a major international tourist destination and economic hub of the southern RSA, running out of water may be a glimpse of what the climatic future holds in store. This 'mother city,' as it is known in South Africa, is not the first global city to run that low on water (Brazil's megacity, São Paulo, nearly ran dry in 2015), but its close scrape may well be a harbinger of a new normal marked by water insecurity.

This water crisis occurred after three consecutive years of cumulative, El Niño–related drought (2015–2018), with rainfall totals running 50–70 percent below average, the lowest level ever recorded here. Cape Town depends almost entirely on rainwater collected in six dammed reservoirs within its metropolitan area. But these facilities are no longer sufficient, because rapid urbanization over the past 20 years has added two million new residents as well as a doubling in water demand.

It was impressive to witness the commitment of locals to conserve water (see photo). Restricted to 50 liters/13.2 gallons per person per day (daily per capita water use in the U.S. averages 350 liters/92.5), people limited toilet flushes to one per day—guided by their new motto, "if it is yellow let it mellow, if it is brown flush it down." Two-minute showers now became the norm, and residents abided by similarly severe restrictions on washing cars, watering gardens, and filling swimming pools. Some people even started showering over buckets in order to reuse this gray water, and many substituted hand-sanitizing fluids for hand-washing. Meanwhile, the Department of Water and Sanitation launched a "war on leaks" campaign, training informal workers to detect and repair leaks in nonrevenue water areas. More controversially, the city launched a water map as a behavior-modification tool to enable neighbors to check on each other to assist in identifying the locations of the heavier users.

The constant presence of *Day Zero* in everyone's mind made people think twice about every drop of water they used. In fact, the campaign was so successful that daily consumption dropped by more than half, and *Day Zero* was continually pushed back until the rains came later in 2018. Following this crisis, which officially ended in early 2019, investment in water-saving technologies is booming,

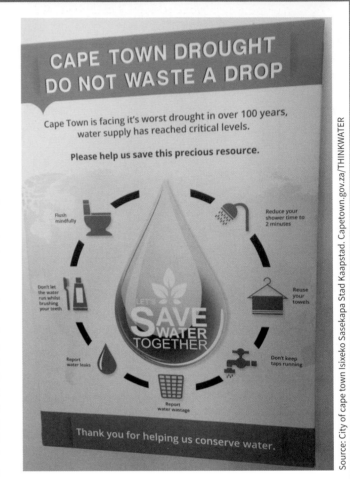

Source: City of cape town Isixeko Sasekapa Stad Kaapstad. Capetown.gov.za/THINKWATER

One of many advertisements by the City of Cape Town leading up to "Day Zero", urging Capetonians to conserve water.

water-efficient appliances are a high priority, three small desalination plants are now on line, and a plethora of new conservation measures is being implemented. Nevertheless, will these actions be enough? The U.N. Intergovernmental Panel on Climate Change predicts a decline in southwestern South Africa's average precipitation, thereby heightening the risk of severe drought for the rest of this century. Therefore, the overriding question in the aftermath of Cape Town's 2018 water crisis is: did it provide a preview of a daunting ecological future?"

and the government's recent decision to expropriate land without paying for it may contribute to a wealth redistribution for some and represent a modest response to the exclusion of most blacks from formal property ownership. But such land reform is risky and, if not well thought out and implemented, could trigger an exodus of white farmers, scare away foreign investors as well as international tourists, and accelerate capital flight.

South Africa has many of the necessary geographic endowments to support a strong economy, but it will take time, greater stability, and better management to achieve that potential. Joining the exclusive *BRICs* (in 2011)—the world's four biggest

emerging-market economies (Brazil, Russia, India, and China)—elevated the country's global stature, even if Pretoria has always been the minor power of the group. However, the strengthened ties with China brought new challenges in the 2010s as the Chinese appetite for South Africa's precious metals, coal, and steel declined sharply. The RSA also appears to be losing its regional advantage in education and skills, and two major new challenges have arisen for the 2020s: adapting to climate change and shifting from fossil fuels to renewable energy. Most of all, South Africa has yet to demonstrate capable leadership that can do for the economy what Mandela did for the society.

As this chapter has made clear, we can no longer accept the "conventional wisdom" conclusion of a Subsaharan Africa described in terms of negative stereotypical judgments. All too often, this realm is still portrayed as a single, monolithic, economic basket-case, where dysfunctional government goes hand in hand with economic underperformance, frequent famines, frightening disease outbreaks, and aimless ethnic violence. On the other hand, neither can the Africa Rising narrative be the broad brush to help us interpret this realm. Entering the 2020s, Subsaharan Africa harbors one of the world's newest emerging-market economies; in many countries, economic forecasts indicate that another decade lies ahead in which growth and development may exceed that of all other geographic realms; and a number of African societies are demonstrating increasing resiliency in combating such monumental challenges as HIV/AIDS and Ebola while continuing to adjust to global economic change.

These positives and potentials notwithstanding, we have also seen that Subsaharan Africa still confronts a formidable set of obstacles. Democracy is taking hold in an increasing number of countries, but elsewhere war and dislocation remain unaffected as authoritarian regimes remain deeply entrenched. Nigeria may have overtaken the RSA to become the realm's largest economy, but its Boko Haram insurgency is far from being defeated. As for South Africa, its foremost internal threat may well be its appalling, still-widening inequalities of opportunities, lifestyles, and personal well-being. And as far as the bigger geographic picture is concerned, energy and mineral resources are the cornerstones of this realm's progress over the past decade—but ever-volatile commodity prices are a crucial reminder that greater diversification in both economic activity and foreign partnering are essential pursuits.

Points to Ponder

- The realm contains more than four dozen countries; the number of languages spoken within Subsaharan Africa is estimated to exceed 2000.

- China's impact on Subsaharan Africa has expanded steadily through activities ranging from trade to investment to education. Turkey and India are also increasing their ties with this realm.

- Shipping a new car from China to a Tanzanian port costs less than hauling it inland from Tanzania to neighboring Uganda.

- The effects of climate change, in the form of severe droughts and the increasing intensity of weather-related events such as cyclones, are wreaking havoc and igniting new humanitarian crises.

- Too many Subsaharan African countries remain resource-dependent economies. Could this realm have a future in manufacturing?

- Although South Africa was liberated from life under Apartheid racism more than a quarter-century ago, the World Bank in 2019 deemed the RSA to be the most unequal society on Earth.

FIGURE 9-1

Two *sadhus* sit at one of the main gates to the old medieval city of Jaisalmer in India's Thar Desert (see Fig. 9-1). Sadhus are "holy men" who dedicate their lives to spirituality. They are Hindus but their beliefs and rituals are hard to categorize. They are not considered priests and generally they are not born into the priestly (Brahminical) caste. They could come from any background and resolve their calling on an individual basis. Many live off alms and the generosity of common people. Some pose for a photograph by a stranger, with the expectation of a small monetary contribution.

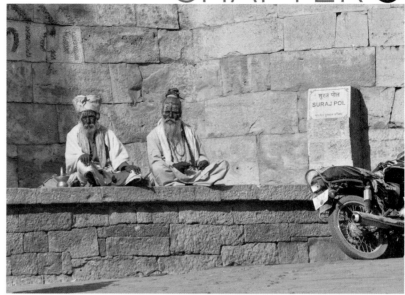

© J.Nijman

The South Asian Realm

IN THIS CHAPTER

- The China–Pakistan Economic Corridor
- South Asia's demographic gender imbalance
- Climate change and extreme weather events in South Asia
- Afghanistan–Pakistan–India: A dangerous triad
- Caste in twenty-first-century India
- Bhutan's happiness index
- Dhaka: The highest-density megacity in the world
- Sri Lanka's fragile ethnic mosaic

CONCEPTS, IDEAS, AND TERMS

309

South Asia is a realm of almost magical geographic names: Mount Everest, the Ganges River, Kashmir, the Khyber Pass. There was a time when this realm was legendary and prized. Remember that it was "India" and its fabled wealth that the European explorers were after, from Vasco da Gama to Columbus to Magellan. Before them, the fourteenth-century North African geographer Ibn Battuta had traveled overland to South Asia, and his writings about its riches were met with astonishment and even disbelief. From the sixteenth century onward, European trading companies derived enormous profits from commerce in this realm.

By the late nineteenth century, however, South Asia had become remote from the affairs of the world—hungry, weak, exploited, the prototype realm of the global periphery. Even after independence in 1947, India as well as the other countries of South Asia long remained among the world's poorest. For decades, population growth outstripped economic expansion.

Today, for a number of reasons, South Asia commands the world's attention once again. It became the most populous geographic realm on Earth in 2011. Two of its states, India and Pakistan, often find themselves in conflict, and both are nuclear powers. In Afghanistan and the remote mountain hideaways of western Pakistan, a terrorist organization's leaders planned attacks that changed the skyline of New York and presaged the battleground of Iraq. In the ports of India, a growing navy reflects the emergence of the Indian Ocean as a new global geopolitical arena in which China, too, is asserting itself. Meanwhile, outsourcing by U.S. companies to India became a hot topic, and India's spectacular rise in information technology has changed that industry. Our daily lives are increasingly affected by what happens in this fascinating part of the world.

Defining the Realm

The Eurasian landmass incorporates all or part of 6 of the world's 12 geographic realms, and few are more clearly defined by nature than the one we call South Asia. **Figure 9-1** shows us why: the huge triangular Indian subcontinent that divides the northern Indian Ocean between the Arabian Sea and the Bay of Bengal is so sharply demarcated by mountain walls and desert wastes that you could take a pen and mark its boundary, from the Naga Hills in the far east through the Great Himalaya and Karakoram in the north to the Hindu Kush and the Iran-bordering wastelands of Balochistan and Afghanistan in the west (**Box 9-1**). Note how short the distances are over which the green of habitable lowlands turns to the dark brown of massive, snow-capped mountain ranges.

This realm extends from Afghanistan in the west to Bangladesh in the east. India, the realm's giant, is flanked by six countries (in clockwise order, Pakistan, Nepal, Bhutan, Bangladesh, Sri Lanka, and the Maldives) as well as Kashmir, a remaining disputed territory in the far north. This realm is home to a kaleidoscope of cultures, suggesting that neither formidable mountains nor forbidding deserts could prevent foreign influences. We will encounter many of these influences in this chapter, but most of South Asia also possessed one unifying historical legacy of sorts: the British Empire, which from its late-nineteenth-century heyday through the late 1940s came to hold sway over almost all of it.

Since Islam prevails in Afghanistan and Pakistan, why are they included with South Asia and not with the North African/Southwest Asian realm? The answer lies in historical geography. First, Pakistan's history is closely tied to that of India—going back to the ancient civilization of the Indus Valley and, more recently, to the role of Islam and of British dominance in the

---

Box 9-1 Major Geographic Features of South Asia

1. Most of South Asia is clearly defined physiographically, and much of the realm's boundary is marked by mountains, deserts, and the Indian Ocean.

2. South Asia's great rivers, the Ganges, Indus, and Brahmaputra, have for thousands of years supported enormous population clusters.

3. South Asia, and especially northern India, was the birthplace of major religions including Hinduism and Buddhism.

4. Because of the realm's natural boundaries, foreign influences in premodern South Asia arrived mainly via a narrow passage in the northwest, the Khyber Pass.

5. Afghanistan is the realm's westernmost country and constitutes a cultural and political transition zone with the North Africa/Southwest Asia (NASWA) realm.

6. The South Asian realm covers less than 4 percent of the Earth's land area but contains almost one-quarter of the world's human population.

7. South Asia's annual monsoon continues to dominate life for hundreds of millions of people. This realm is also particularly vulnerable to the effects of global climate change.

8. Certain disputed territories in South Asia's remote northern mountain perimeter are potentially a source of dangerous friction between India and both Pakistan and China.

9. This realm is still predominantly rural and studded with hundreds of thousands of small villages; but it also contains some of the biggest cities in the world.

subcontinent. The tight integration of Pakistan with the rest of South Asia will not surprise you after you study the realm's physiography in Figure 9-1: the natural boundary in this part of the realm lies west of the Indus River, not east of it. Pakistan today remains part of a realm that begins to change not in the Punjab, but at the Khyber Pass, the highland gateway to Afghanistan. Second, Afghanistan was a staging ground for Islam in the Indian subcontinent, and the cultural boundary between Afghanistan and Pakistan has remained fluid ever since. In political terms, too, there continue to be intricate connections between the two countries that, in turn, impact relations between Pakistan and India. At the same time, Afghanistan today is caught up in developments across Southwest Asia (as well as Central Asia), which is why that country is considered to be a **transition zone [1]** between South Asia and NASWA.

South Asia's Physiography

From snowcapped peaks to tropical forests and from bone-dry deserts to lush farmlands, this part of the world presents a virtually endless array of environments and ecologies, a diversity that is matched by its cultural mosaic. The broad outlines of this realm's physiography are best understood against the backdrop of its fascinating geologic past.

A Tectonic Encounter

As **Figure 9-2** shows, the spectacular relief in the north of this realm is the product of a collision between two of the planet's great *tectonic plates* (see Fig. 1-5). About 10 million years ago, after a lengthy geologic journey following the breakup of the supercontinent Pangaea (diagrammed in Fig. 8-3), the Indian Plate encountered Eurasia. In this huge, slow-motion, accordion-like collision, the Earth's crust surrounding the contact zone was pushed upward, thereby creating the mighty Himalaya mountain range. That process is still going on—at the rate of 5 millimeters (0.2 in) per year—and the massive 2015 earthquake in Nepal proved once again that this is one of the most quake-prone areas in the world. One major outcome of this tectonic collision was that the northern margins of the realm were upwardly thrusted to elevations where permanent snow and ice make the landscape appear polar. The march of the seasons melts enough of this snowpack in spring and summer to sustain the great rivers below, providing water for farmlands that support hundreds of millions of people. The Ganges, Indus, and Brahmaputra all have their origins in the Himalaya. Only south of the Ganges Basin does the enormous plateau begin that marks the much older geologic core of the Indian Plate as it drifted northeastward toward Eurasia.

The Monsoon

Physical geography, therefore, is crucial here in South Asia—but not just on and below the surface. What happens in the

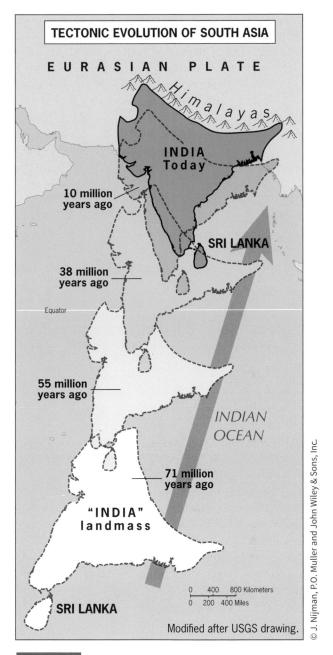

FIGURE 9-2

overlying atmosphere is critical as well. The name "South Asia" is almost synonymous with the term **monsoon [2]** because the annual rains that accompany its onset, usually in June, are indispensable to all forms of agriculture in the realm's key country, India.

Figure 9-3 shows how the monsoon works. As the subcontinental landmass heats up during the spring, a huge low-pressure system is formed above it. This low-pressure system begins to draw in vast volumes of air from over the ocean onto the subcontinent. When the inflow of moist oceanic air reaches critical mass in early June, the *wet monsoon* has arrived. It may rain for 60 days or more. The countryside turns green, the paddies fill, and another dry season's dust and dirt are washed away—the region is reborn (see photos page 313). The moisture-laden

FIGURE 9-3

air flowing onshore from the Arabian Sea is forced upward against the Western Ghats, cooling as it rises and condensing large amounts of rainfall. The other branch of the wet monsoon originates in the Bay of Bengal and gets caught up in the convection (rising hot air) over northeastern India and Bangladesh. Seemingly endless rain now inundates a much larger area, including the entire North Indian Plain. The Himalaya mountain wall blocks the onshore-flowing airstream from spreading into the Asian interior and the rain from dissipating.

Thus the moist airflow is steered westward, drying out as it advances toward Pakistan. After persisting for weeks, this pattern finally breaks down, and the wet monsoon gives way to periodic rains and, eventually, another dry season. Then the anxious wait begins for the next year's wet monsoon because without it India would face disaster. It is estimated that two-thirds of India's agricultural areas do not have irrigation and depend fully on rainfall. For many people, life can hang by a meteorological thread.

Steve McCurry/Magnum Photos, Inc.

The arrival of the annual rains of the wet monsoon transforms the Indian countryside. By the end of May, the paddies lie parched and brown, dust chokes the air, and it seems that nothing will revive the land. Then the rains begin, and blankets of dust turn into layers of mud. Soon the first patches of green appear on the soil, and by the time the monsoon ends all is green. The photo on the left, taken just before the onset of the wet monsoon in the west-coast State of Goa, shows the paddies before the rains begin; three months later when the rice crop is ready to harvest, this same countryside looks as on the right.

Physiographic Regions

Figure 9-3 underscores South Asia's overall subdivision into three physiographic zones: the northern mountains, the southern plateaus, and, in between, a wide crescent of river lowlands.

The *northern mountains* extend from the Hindu Kush and Karakoram ranges in the northwest through the Himalaya in the center (Everest, the world's tallest peak, lies on the crestline that forms the Nepal–China border) to the ranges of Bhutan and the Indian State of Arunachal Pradesh in the east. Dry and barren in Afghanistan and western Pakistan, the ranges become green and tree-studded in Kashmir, forested in the lower-lying elevations of Nepal, and even more densely vegetated in Arunachal Pradesh. Transitional foothills, with many deeply eroded valleys cut by rushing meltwater, lead to the river basins below.

The belt of *river lowlands* extends eastward from Pakistan's lower Indus Valley (the area known as Sindh) through India's wide Gangetic Plain and then on across the great double delta of the Ganges and Brahmaputra in Bangladesh (Fig. 9-3). In the east, this physiographic region is often called the North Indian Plain. To the west lies the lowland of the Indus River, which rises in Tibet, crosses Kashmir, and then bends southward to receive its major tributaries from the area known as *Punjab* ("Land of Five Rivers") to the east.

The *southern plateaus* constitute peninsular India, dominated by the *Deccan*, a massive tableland built of lava sheets that poured out when India separated from Africa during the sundering of Pangaea. The Deccan (meaning "South") tilts toward the east, so that its highest areas are in the west and most of the rivers flow into the Bay of Bengal. North of the Deccan lie two other plateaus,

the Central Indian Plateau to the west and the Chota Nagpur Plateau to the east (Fig. 9-3). On the map, also note the Eastern and Western Ghats: "ghat" means step, and it connotes the descent from Deccan Plateau elevations to the narrow coastal plains below. The onshore winds of the annual wet monsoon bring ample precipitation to the Western Ghats; as a result, here lies one of India's most productive farming areas and one of southern India's largest population concentrations.

Birthplace of Civilizations and Religions

Indus Valley Civilization

A complex and technologically advanced civilization emerged in the Indus Valley by about 2500 BC, simultaneous with other Bronze Age "urban revolutions" in Egypt and Mesopotamia (see Fig. 7-3 for a map of the culture hearths of the Old World). The Indus Valley civilization was centered on a pair of major cities, Harappa and Mohenjo-Daro, which may have been capitals during different periods of its history (**Fig. 9-4**); in addition, there were more than 100 smaller urban settlements. The locals apparently called their state *Sindhu*, and both *Indus* (for the river) and *India* (for the later state) very likely derived from this name. Although the influence of this civilization extended as far east as present-day Delhi, it did not last because of environmental change and, perhaps, because the political center of gravity shifted southeastward into the Ganges Basin.

Aryans and the Origins of Hinduism

Around 1500 BC, northern India was invaded by the *Aryans* (peoples speaking Indo-European languages based in what is now Iran). As the Iron Age dawned in India, acculturated Aryans began the process of integrating the Ganges Basin's isolated tribes and villages into a newly created cultural system, and urbanization made a comeback. The Aryans brought their language, *Sanskrit* (related to Old Persian), as well as a new social order to the vast riverine flatlands of northern India. Their settlement here was also accompanied by the emergence of a religious belief system, *Vedism*. Out of the texts of Vedism and local creeds there arose a new religion—*Hinduism*—and with it a new way of life.

It is thought that the arrival and accommodation of the Aryans in this reorganized society forged a layered system of social stratification that would solidify the powerful position of the Aryans and be legitimized through religion. Starting about 3500 years ago, a combination of regional integration, the consolidation of villages into controlled networks, and the emergence of numerous small city-states produced a hierarchy of power among the people, a ranking from the especially powerful (Brahmins—the highest-order priests) to the weakest. This stratified **caste system [3]** of Hinduism is often criticized in the West (and it is controversial in India, too) because of its rigidity and the ways in which it justifies

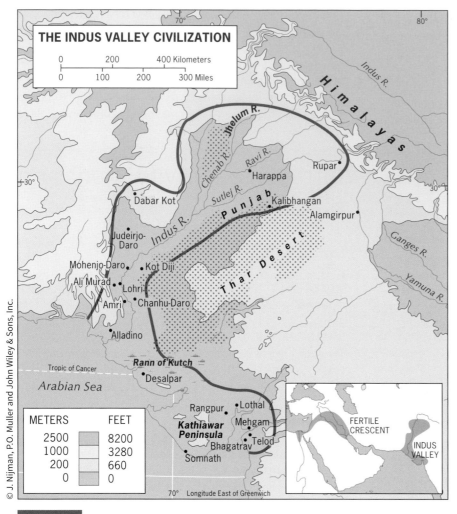

THE INDUS VALLEY CIVILIZATION

FIGURE 9-4

© J. Nijman, P.O. Muller and John Wiley & Sons, Inc.

literary histories. Today, Telugu, Tamil, Kanarese (Kannada), and Malayalam are spoken by nearly 300 million people. In India's northern and northeastern fringes, Sino-Tibetan languages predominate, and smaller pockets of Austro-Asiatic speakers can be found in eastern India and neighboring Bangladesh. To the west, Afghanistan consists of a mosaic of Altaic and Indo-European languages, and Pashto (spoken by two-thirds of the Afghan population) spills across the border into neighboring Pakistan.

Buddhism and Other Indigenous Religions

Hinduism is not the only religion that emerged in this realm. Around 500 BC, *Buddhism* arose in the eastern Ganges Basin in what today is the Indian State of Bihar. The famous story of the "enlightenment" of the Prince Siddhartha (the Buddha) took place in the town of Bodh Gaya, and his adherents soon fanned out in every direction. The appeal of Buddhism was (and is) especially strong among lower-caste Hindus, and large numbers of them have converted through the ages. Interestingly, Buddhism evolved inside what is now India, but its most powerful influences were felt beyond the realm in Southeast and East Asia. Today, only 0.7 percent of the population of India adheres to Buddhism, but it is the state religion in Bhutan, and the great majority (around 70 percent) of Sri Lankans are Buddhist as well.

Another (much smaller) indigenous religion that has evolved alongside Hinduism since ancient times is *Jainism*, often described as a more purist, principled, and deeply spiritual form of Hinduism. It is well known for its uncompromising stand on nonviolence and vegetarianism. Jains today constitute less than one-half of 1 percent of the population in India. Finally, we should take note of *Sikhism* as another of the realm's indigenous religions, a blend of sorts of Islamic and Hindu beliefs. This religion, practiced by 1.7 percent of the population, is of course much younger; it emerged around AD 1500, a few centuries after Islam became a dominant force in much of the South Asian realm.

structural inequality. Those in the lowest castes, deemed to be there because they earned that assignment in their past lives, are worst off, with little hope of advancement and in many ways subordinate to those higher on the social ladder. In recent years, the caste system appears to be eroding from the combined effects of globalization, economic growth, and urbanization; while this is true for India's bigger cities, it is much less so for the large majority of people who still reside in rural areas and small towns and villages. It should be noted that the caste system also implies strong internal communities and social networks, qualities that may have contributed to the longevity of this religious-social organization (it has been more or less intact for more than 3000 years).

Hindus make up about 80 percent of India's population, and the religion also dominates in Nepal (**Box 9-2**). Although Hinduism spread across South Asia and even reached the Southeast Asian realm (especially Cambodia and Indonesia), Indo-European languages never took hold in the southern portion of the subcontinent. As **Figure 9-5** indicates, **Indo-European languages [4]** (several of which are rooted in Sanskrit) predominate in the western and northern parts of South Asia, whereas the southern languages belong to the **Dravidian [5]** family—languages that were indigenous to the realm even before the arrival of the Aryans. But these are not fossil languages: they remain vibrant and have long

Foreign Invaders

The Reach of Islam

In the late tenth century, Islam came rolling like a giant tide across South Asia, spreading across Persia and Afghanistan,

Box 9-2 From the Field Notes . . .

"It is an early February morning, and cremations at Pashupatinath Temple in Kathmandu are already underway. This is one of Nepal's oldest Hindu temples dating back to the fifth century AD (earlier wooden temple structures at this site date to about 400 BC). The main building, its gold-plated roof rising above the rest of this sprawling temple complex, is of the typical Nepalese pagoda style architecture, which betrays China's cultural influence in this region of the South Asian realm. But this a Hindu temple, not Buddhist, and this is where Hindus come to cremate the dead, on the banks of the Baghmati River, a tributary to the Ganges. I am at a distance across the river from the cremations and am told it is permitted to photograph or record the rituals. Hindus cremate their dead as quickly as possible after they pass away

© J. Nijman

to allow peaceful reincarnation or, even better, to help achieve *moksha*, liberation from the cycle of life. I take a moment to wish these souls a good passage."

---

through the Khyber Pass into the Indus Basin, and across Punjab into the Ganges Basin—converting virtually everybody in the Indus Valley and foreshadowing the emergence, several centuries later, of the Islamic Republic of Pakistan. By the early thirteenth century, the Muslims had established the long-surviving and powerful **Delhi Sultanate** that controlled much of the northern tier of peninsular India.

In the early 1500s, a descendant of Genghis Khan named Babur conquered Kabol (Kabul) in Afghanistan, and from that stronghold he penetrated the Indus Basin and Punjab and directly challenged the Delhi Sultanate. In the 1520s, his Islamicized Mongol forces ousted the Delhi rulers and established the **Mughal Empire**. By most accounts, Mughal rule was at times remarkably enlightened, especially under the leadership of Babur's grandson Akbar, who aggressively expanded the Empire but adopted tolerant policies toward the Hindus under his sway; Akbar's grandson, Shah Jahan, made his enduring mark on India's cultural landscape by commissioning magnificent architectural creations, particularly the Taj Mahal in the city of Agra (photo right).

Nonetheless, by the early eighteenth century, the Mughal Empire was in decline. Maratha, a Hindu state in the west, expanded not only into the peninsular south but also northward toward Delhi, capturing the allegiance of local rulers and weakening Islam's hold. Fractured India now lay open to still another foreign intrusion, this time from Europe.

Despite more than seven centuries of Islamic rule in South Asia, it is remarkable that Islam never achieved proportional dominance over the realm as a whole. Whereas today Pakistan is 96 percent Muslim and Bangladesh 90 percent, India—where the Delhi Sultanate and the Mughal Empire were centered—remains only about 15 percent Muslim today. Islam may have arrived like a giant tide, but Hinduism stayed afloat and eventually outlasted the incursion. It also withstood the European

onslaught that culminated in the incorporation of the entire realm into the British Empire.

The European Intrusion

The British took advantage of the weakened and fragmented power of the Mughals and followed a strategy commonly known as "indirect rule." They left local rulers in place as long as the British extracted the desired trading arrangements. Thanks to arrangements with the British, many local maharajas became wealthier than ever before: from northern to southern India,

© J. Nijman

The Taj Mahal, India's supreme iconic seventeenth-century legacy of the Mughals, is one of the world's most admired architectural accomplishments. We are in India, but this structure was built as a Muslim tomb. The Taj Mahal is world famous, but it is also a popular destination for growing numbers of domestic tourists, such as this visiting group from a village in the northern Himalayan State of Uttarakhand.

MAJOR LANGUAGES OF SOUTH ASIA

☐ Indo-European Family (Indo-Aryan Group)	■ Austro-Asiatic Family	
Dravidian Family	Malay-Polynesian Family	
Sino-Tibetan Family	Andamanese	
Altaic Family	☐ Sparsely populated	

LANGUAGE USE IN INDIA

LANGUAGE	SPEAKERS (Percent)
Hindi	27.6
Hindi-associated languages and dialects	12.3
Bengali	8.2
Telugu	7.8
Marathi	7.4
Tamil	6.3
Urdu	5.1
Gujarati	4.8
Kannada	3.9
Malayalam	3.6
Oriya	3.3
Punjabi	2.8
Asomese	1.6
Other	5.3

(Hindi is spoken by nearly 50% and some English is spoken by about 30%.)
Data Source: Census of India.

© J. Nijman, P. O. Muller, and John Wiley & Sons, Inc.

FIGURE 9-5

you can find beautiful palaces (now often converted to museums or hotels) that were built by these rulers, often as recently as the late nineteenth or early twentieth century.

British colonialism in South Asia coincided with the Industrial Revolution in Europe, and the impact of Britain on the realm must be understood in that context. South Asia became, in large part, a supplier of raw materials needed to keep the factories going in Manchester, Birmingham, and other British industrial centers. For instance, when the supply of cotton from the American South came to a halt during the U.S. Civil War in the 1860s, the British quickly encouraged (and indeed enforced) cotton production in what is today western India.

When the British took full power in South Asia during the mid-nineteenth century, this was a realm with already considerable industrial development (notably in metal goods and textiles) and an active trade with both Southeast and Southwest Asia. The colonialists saw this as competition, and soon India was exporting raw materials as well as importing manufactured

products—from Europe, of course. Local industries declined, and Indian merchants lost their markets.

The British colonial legacy included one of the most extensive transport networks of the colonial era, particularly the railway system—even though the network focused on interior-to-seaport linkages rather than fully interconnecting the various parts of the country. British engineers laid out irrigation canals through which millions of hectares of land were brought into cultivation. Coastal settlements that had been founded by Britain developed into major cities and bustling ports, led by Bombay (now renamed Mumbai), Calcutta (now Kolkata), and Madras (now Chennai). These three cities still rank among India's largest urban centers, and parts of their cityscapes bear the unmistakable imprint of colonial architecture.

The British *raj* (the term for Britain's rule between 1857 and 1947) also produced a new elite within the native Indian population. They had access to education and schools that combined English and Indian traditions, and their Westernization was reinforced through university education in Britain. This elite drew from Hindu and Muslim communities, and it was to play a major part in the rising demands for self-rule and independence. Those demands started to gather momentum in the early twentieth century and could no longer be denied after World War II came to an end in 1945.

The Geopolitics of Modern South Asia

Partition and Independence

Even before the British government decided to yield to demands for independence, it was clear that British India would not survive the coming of self-rule as a single political entity. As early as the 1930s, Muslim activists were promoting the idea of a separate state. As the colony moved toward independence, a major political crisis developed that eventually resulted in the separation of India and Pakistan. But **partition [6]** was no simple matter. True, Muslims were in the majority in the western and eastern sectors of British India, but smaller Islamic clusters were scattered throughout the realm. Furthermore, the new boundaries between Hindu and Muslim communities had to be drawn right through areas where both sides coexisted—thereby displacing millions (photo page 318).

The consequences of this migration for the social geography of India were especially far reaching. Comparing the country's 1931 and 1951 distributions of Muslims in **Figure 9-6**, you can see the impact on the Indian Punjab and in what is today the State of Rajasthan. (Since Kashmir was mapped as three entities before partition and as one afterward, the change there represents an administrative, not a major numerical, alteration.) A Muslim exodus occurred even in the east, as is reflected on the map by the State of West Bengal, adjacent to East Pakistan (now Bangladesh), where the Islamic component of the local population declined substantially.

The world has seen many refugee migrations but none involving so many people in so short a time as the one resulting from British India's partition (which occurred on Independence Day, August 15, 1947). As is often the case, it was hard to distinguish between "forced" and "voluntary" migration. Many Muslims, believing they had no choice, feared for their future in the new India and joined the stampede. Others had the means and the ability to make a decision to stay or leave, but even these better-off migrants undoubtedly sensed a threat.

The great majority of Hindus who lived on the "wrong" side of the border moved as well. The Hindu component of

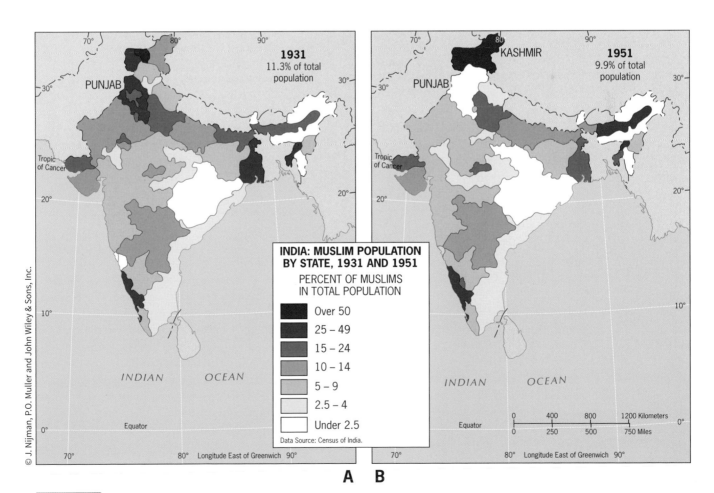

INDIA: MUSLIM POPULATION BY STATE, 1931 AND 1951

PERCENT OF MUSLIMS IN TOTAL POPULATION

- Over 50
- 25 – 49
- 15 – 24
- 10 – 14
- 5 – 9
- 2.5 – 4
- Under 2.5

Data Source: Census of India.

© J. Nijman, P.O. Muller and John Wiley & Sons, Inc.

FIGURE 9-6

Flight was one response to the 1947 partition of what had been British India, resulting in the greatest mass population transfer in human history (involving at least 14 million migrants). Here, two trainloads of eastbound Hindu refugees, fleeing then West Pakistan, arrive at the station in Amritsar, the first city inside India.

present-day Pakistan may have been as high as 16 percent in 1947 but is barely more than 1.4 percent today; in Bangladesh, which was named East Pakistan at the time of partition, it declined from 30 percent to just over 8 percent today. Partition, therefore, created an entirely new cultural and geopolitical landscape in South Asia.

India–Pakistan

From the moment of their separate creation in 1947, India and Pakistan have had a tenuous relationship. Upon independence, present-day Pakistan was united with present-day Bangladesh, and the two countries were respectively called West Pakistan and East Pakistan. As we have noted, the basis for this scheme was religion: in both Pakistan and Bangladesh, Islam is the state religion. Between the two Islamic wings of Pakistan lay 1500 kilometers (930 mi) of Hindu India. But there was little else to unify the Muslim easterners and westerners, and their union lasted less than 25 years. In 1971, a costly war of secession, in which India supported East Pakistan, led to the collapse of this unusual arrangement. East Pakistan, upon its "second independence" in 1971, took the name Bangladesh—and since there was no longer any need for a "West" Pakistan, that qualifier was dropped and the name *Pakistan* remained on the map.

India's encouragement of independence for Bangladesh emphasized the continuing tension between Pakistan and India, which had already led to war in 1965, to further conflict during the 1970s over Jammu and Kashmir, and to periodic flare-ups over other issues. Armed conflict between the two South Asian countries seemed to be a regional matter—until the early 1990s, when their arms race took on ominous nuclear proportions. Since then, the specter of nuclear war has hung over the conflicts that continue to embroil Pakistan and India, a concern not just for the South Asian realm but for the world as a whole. No longer merely a decolonized, divided, and disadvantaged country trying to survive, Pakistan has taken a crucial place in the political geography of a world realm.

The relationship between India and Pakistan is especially sensitive because so many Muslims still live in India. Massive as the 1947 refugee movement was, it left far more Muslims in India than those who had departed. The number of Muslims in India declined sharply, but it remained a huge minority, one that was growing rapidly to boot. Today, it is approaching 200 million (14.2 percent of India's population)—the largest cultural minority in the world and just about equal in size to Pakistan's entire population.

What this means is that a substantial proportion of India's population tends to have more or less natural sympathies toward Pakistan, even though India's Muslims are generally known to be moderate. Their presence in many instances works as a brake on hawkish Indian policies toward the Islamabad government. At the same time, conflict with Pakistan can have detrimental effects on Hindu–Muslim relations inside India, and over the years it has led to communal violence and deadly clashes. Today that issue is further complicated by the alleged involvement of some Indian Muslims in terrorist activities within India that were said to be orchestrated from Pakistan.

Contested Kashmir

When Pakistan became independent following the partition of British India, its capital was Karachi, located on the southern coast near the western end of the Indus Delta. As the map shows, however, the present capital is Islamabad. By moving the capital city, in 1967, from the "safe" coast to the embattled interior, and by placing it on the doorstep of the contested territory of Kashmir, Pakistan announced its intent to stake a claim to its northern frontiers. And by naming the city Islamabad, Pakistan proclaimed its Muslim foundation here in the face of the Hindu challenge. This politico-geographical usage of a national capital can be assertive, and as such Islamabad exemplifies the principle of the **forward capital [7]**.

Kashmir is a territory of high mountains surrounded by Pakistan, India, China, and, along more than 50 kilometers (30 mi) in the far north, Afghanistan (**Fig. 9-7**). Although known simply as Kashmir, the area actually consists of several political divisions, including the Indian State properly referred to as Jammu and Kashmir, a major bone of contention between India and Pakistan.

When partition took place in 1947, the existing States of British India were asked to decide whether they wanted to be incorporated into India or Pakistan. In most of the States, the local ruler made this decision, but Kashmir was an unusual case. It had about 5 million inhabitants at the time, nearly three-quarters of them Muslims, but the maharajah of Kashmir himself was a Hindu. When he decided not to join Pakistan and instead aimed to retain autonomous status, this was answered with a Muslim uprising supported by Pakistan. The maharajah, in turn, called for help from India. After more than a year's fighting and through the intervention of the United Nations, a cease-fire line left most of Jammu and Kashmir (including nearly four-fifths of the territory's population) in Indian hands. Eventually, this line—now known as the *Line of Control*—began to appear on maps as the final boundary settlement, and Indian governments have proposed that it be so recognized.

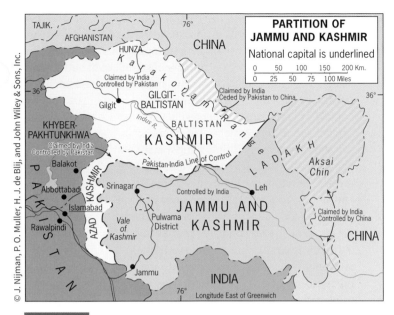

Westernized hotels. Nearly 200 people perished, and hundreds were wounded. Live pictures of smoke billowing from the famous Taj Mahal Hotel in southern Mumbai were broadcast around the world (photo below). The group responsible for these attacks was *Lashkar-e-Taiba* (the Party of the Righteous), a Pakistan-based organization that among other things aims to return Kashmir to Islamic rule.

Here is where the complicated relationships between India, Pakistan, and Afghanistan come into focus. Pakistan's northwestern frontier, bordering Afghanistan, is particularly important in this regard, and we will return to this issue later in the chapter when we consider these regions and subregions in more detail. For now, it is important to understand that this area inside Pakistan is effectively controlled by the **Taliban** [8], the Afghan Islamic extremists who also have a history of collaboration with **Al Qaeda** [9]. This entire border zone with Afghanistan seems to lie well beyond the control of the Pakistani

FIGURE 9-7

With about two-thirds of the Kashmiris adhering to Islam, Pakistan has for decades demanded a referendum in Jammu and Kashmir in which the people can decide for themselves to remain with India or become part of Pakistan. India has refused, arguing that there is a place for Muslims in secular India but not for Hindus in the Islamic Republic of Pakistan.

In 2019, the conflict shifted onto unprecedented terrain when the Indian government took the radical step to revoke the special status of Jammu and Kashmir. Until then, this special status meant that, even though the territory was under the *de facto* control of India, it had its own constitution and administrative autonomy. The revocation changed this fundamentally as the territory became a "Union Territory", i.e., part of the Indian federal state with full official control from Delhi. It also effectively implied the impossibility of any future settlement over the territory with Pakistan. Needless to say, the decision by India's government was heavily criticized by Kashmiri Muslims and Pakistan.

These developments followed in the wake of violent incidents in early 2019, when a Kashmiri Muslim suicide bomber killed 40 Indian policemen in Pulwama District (Fig. 9-7). The bombing was claimed by a jihadist group based in Balakot (in Pakistan, north of Islamabad), and the Indian military responded with air strikes in that area, reporting that hundreds of jihadists had been killed. The Pakistani military, in turn, shot down one Indian fighter jet. By the beginning of 2020, the possibility of a peaceful resolution of the conflict over Kashmir seemed more remote than ever.

Afghanistan-Pakistan-India: A Dangerous Triad

The integration of India's Muslim communities into the fabric of Indian society leaves much to be desired and, even after almost three-quarters of a century of coexistence within the Indian state, the risk of violence is very real. The highest-profile attack to date was in 2008, when terrorists targeted Mumbai's two most upscale,

Smoke billows from the landmark Taj Mahal Hotel in Mumbai on November 29, 2008. The Taj was one of several sites in southern Mumbai that were simultaneously targeted for attack by Islamic militants. The siege lasted four days, and almost 200 people died in the violence. Indian commandos killed all of the terrorists but one, who was captured, tried, sentenced to death, and executed in 2012. It was not the first time that such terrorism had struck India, and several smaller-scale attacks occurred during the 2010s.

government. The U.S. invaded Afghanistan after 9/11 because intelligence indicated the Taliban harbored the Al Qaeda perpetrators of the New York attacks, and along the way the U.S. goal became to defeat the Taliban. But U.S. efforts have in part been thwarted by the Taliban's ability to move back and forth across this border—despite the pressure the United States exerts on Islamabad to confront the Taliban on the Pakistani side of the border in this treacherous mountain refuge. Note that when the U.S. finally found and killed the Al Qaeda leader Osama bin Laden, this was across the border, in Pakistan. The U.S. has no troops on the ground in Pakistan, but it has launched many drone strikes in Pakistan and elsewhere (**Box 9-3**).

Clearly, Pakistan is central to this quandary, but it is an especially delicate geopolitical chess match for the Pakistani regime. It must be careful not to alienate its own Islamic base, even though its military forces engage in frequent clashes with the border-zone extremists, and is also wary of an economically stronger India with growing ties to the United States. India is deeply concerned about Pakistan's role in terrorism on Indian soil and, even worse, the possibility of jihadists taking control of Pakistan's nuclear arsenal; at the same time, it must guard against heightening tensions between Hindus and Muslims inside India. India is also frustrated by American reluctance to choose sides in the Kashmir conflict. The United States, in turn, is sympathetic to the world's largest democracy but needs Pakistan as an ally in the global counterterrorism campaign. Pakistan has for decades received very substantial military aid from the U.S., and Pakistan's governments have been adept in persuading the U.S. to give it support. Each of these parties is walking a tightrope, where the slightest mistake could have tragic consequences.

India–China Relations

Any overview of this realm's geopolitical framework today must also include China's increasingly powerful and sometimes invasive presence. As Figure 9-7 shows, China claims the northeastern extension of India's Jammu and Kashmir State. This issue has been stalemated in recent years, but officially neither China nor India shows any sign of conceding.

China's control over Tibet (called Xizang by the Chinese) and its efforts to influence the lives of Tibetans both within and outside Tibet create additional issues. To the dismay of Beijing, Tibet's exiled Dalai Lama calls India his second home. In recent years, China has been pressuring the government of Nepal, wedged between India and Tibet, to discourage Tibetan immigration and to constrain the activities of Tibetans already in the country. Farther to the east, China claims the bulk of the Indian State of Arunachal Pradesh ("land of the dawn-lit mountains") based on its assertion that the boundary, established in 1914, was never ratified by Beijing—even though it was approved by the then-independent Tibetans. And China even claims rights to the small area of northern India lying between Tibet-adjoining Sikkim and Bhutan, again on the basis that its inhabitants are Tibetans and therefore belong under Chinese jurisdiction.

China's power in South Asia is felt in other ways as well. A few years ago, the Chinese announced plans to construct dams on the Tibetan headwaters of the upper Brahmaputra River, potentially jeopardizing the main water supplies of both northeastern India and Bangladesh. Thus by no means is Kashmir the only place in the realm's lengthy northern frontier zone where trouble can erupt at any time—and within which India directly borders China for 3500 kilometers (2200 mi).

More generally, China needs access to markets for its products and supplies of raw materials to sustain its steadily growing industrial production, and part of that access involves the South and Southeast Asian realms. The Belt and Road Initiative of China includes close collaboration with and massive investments in two countries that border India: Pakistan and Myanmar (we return to these issues later in the chapter). India is monitoring these developments closely and is responding by forging new initiatives or alliances, from Myanmar to Indonesia.

There have also been indications of more conciliatory exchanges between the two Asian giants, including mutual state visits of their presidents. In a notable turnaround, China in 2015 avowed that it would not oppose Indian membership in the U.N. Security Council. More importantly, growing economic interdependence is proving to be advantageous for both countries. China wants to penetrate India's enormous, expanding consumer markets. China–India trade has increased rapidly in recent years, with about two-thirds going from China to India and one-third in the other direction. So far, China has been running a significant surplus, reflecting the fact that the terms of trade are more favorable to China: India mainly supplies raw materials, whereas China largely sells finished, higher-value-added products.

Emerging Markets and Fragmented Modernization

Since the turn of the century, optimistic reports in the media have been proclaiming a new era for South Asia, marked by rising growth rates for the realm's national economies, rewards from globalization and modernization, and tightening integration into the world economy. India, obviously the key to this realm, has even been described as "India Shining" during this enthusiastic wave of optimism.

Nevertheless, well over half of India's 1.4 billion people continue to live in poor rural areas, their villages and lives virtually untouched by what is happening in cities like Mumbai or Bengaluru. Afghanistan and Nepal rank among the poorest countries on Earth; more than one-third of Pakistan's population lives below the offical poverty line, and female literacy still lingers below 50 percent. The World Bank in 2017 estimated that about one-third of the children in South Asia are malnourished and underweight, a majority of them girls. It still remains to be seen whether the benefits of newfound economic growth

Box 9-3 Technology & Geography

Drones of War

Drone warfare [10] today has become a standard element of high-technology military strategy in the arsenal of the United States and other powers. For example, on January 3rd of 2020, the Iranian general Qasem Suleimani was targeted and killed in Iraq (!) by a missile fired from a U.S. military drone. Two years earlier, in May of 2016, the Afghan Taliban leader Mullah Mansoor was assassinated in Pakistan (!) by the U.S. in similar fashion.

Unmanned aerial vehicles (UAVs), or drones, are aircraft that fly without an on-board pilot. Advances in aircraft design as well as imaging, navigational, and communications technology now permit those who control drones to be located halfway around the world from the field of operations. For instance, American military drones in embattled Pakistan, Syria, Yemen, and Iraq are guided by pilots based in the United States who watch live-streaming video feeds via satellite uplinks from cameras mounted on each active drone. Sensors, computers, and on-board instruments enable drones to be flown remotely, similar to a flight simulation game on a computer. When armed with weaponry such as air-to-ground missiles, drones become deadly instruments of war.

Figure 9-8 maps communications relays and satellite links that enable the remote piloting of drones by military personnel stationed on their home territory. Creech Air Force Base near Las Vegas, Nevada, is a major U.S. intelligence and coordination center. From Creech, commands are sent via high-speed transatlantic fiber-optic cable to the U.S. air base in Ramstein, Germany, and from there the command is directly relayed via satellite to drones deployed in Afghanistan and elsewhere. Drone take-offs and landings are controlled locally (see photo).

There are several tactical advantages to using drones in war: they are difficult to detect from the ground, flight times in excess

An airman guides a U.S. Air Force MQ-9 Reaper drone as it taxis toward the runway at Kandahar Airfield, Afghanistan, in March 2016.

of 12 hours are possible without refueling, and advanced imaging, thermal, and night-time-vision capabilities enable drones to "see" what is otherwise invisible to on-board pilots.

Drone technology influences where and how wars are fought, and it also raises issues of international law. The Iranian general was eliminated on Iraqi territory and the Taliban leader was killed while driving in a convoy on a road in southwestern Pakistan. Other drone strikes have targeted Pakistan's northern border with Afghanistan and areas in Somalia. Not surprisingly, local governments often strongly condemn the American incursions: the Pakistani government in 2016 protested the killing of the Taliban leader and the Iraqi government expressed similar condemnation after the killing of the Iranian general inside their borders.

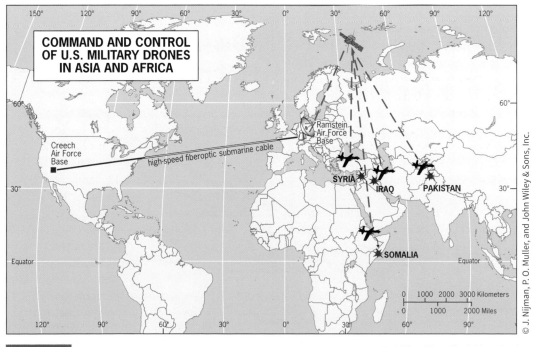

FIGURE 9-8

can be spread around widely enough to improve the living conditions of South Asia's masses.

Economic Liberalization

Most countries in this realm have liberalized their economies since the 1980s as part of a worldwide turn toward **neoliberalism [11]**. This entails privatization of state-run companies, lowering of international trade tariffs, reduction of government subsidies, cutting of corporate taxes, and overall deregulation to stimulate business activity.

The results of these reforms can be observed across South Asia in terms of increased economic growth and higher per capita incomes. Generally, this progress continues in the twenty-first century even though variations between countries are quite pronounced (**Fig. 9-9**), with India, Bhutan, Sri Lanka and Bangladesh showing the strongest growth while Afghanistan, Pakistan, Nepal, and the Maldives have been lagging behind. Most of the growth is in manufacturing, services, and, in India, information technology (**Box 9-4**).

More open economies have attracted increased foreign investment, and during the past two decades a new, urban middle class has emerged. This steadily expanding social stratum may account for only about 25 percent of the realm's population, but in South Asia that translates into over 350 million people—a huge new consumer market for an array of products ranging from smartphones to refrigerators to automobiles. At the same time, it leaves well over a billion South Asians unaffected by this upward social mobility. For them almost nothing has changed;

they remain overwhelmingly dependent on (stagnant) agriculture, and they are not likely to log on to the new information economy anytime soon. As in so many other parts of the world, here too increased overall growth has mostly been accompanied by ever greater economic and social inequality.

One striking feature of this realm is that even though the majority of its people live traditional lives in rural villages, there also are eight **megacities [12]** in which social and economic change is the order of the day. The metropolitan areas centered on Delhi–New Delhi, Mumbai, Karachi, Dhaka, Kolkata, Lahore, Bengaluru, and Chennai all contain populations that exceed 10 million and have grown rapidly over the past two decades (see the box on page 324). Their densities are often overwhelming, environmental conditions are poor, and the widening income gap between rich and poor is staggering. But almost always it is better to be poor in the city than in the countryside, simply because cities offer opportunities that do not exist in the villages—which helps to explain the flow of rural-to-urban migrants that drives the growth of towns and cities all across South Asia. As **Figure 9-10** demonstrates, there is substantial variation in the internal densities and layouts of megacities; those differences also extend to levels of prosperity, and the contrasts between megacities in the Global Core and the Global Periphery are stunning (**Box 9-5**). Dhaka has the highest density of all the world's megacities.

The Significance of Agriculture

More than half of South Asia's workforce is employed in agriculture, ranging from about 40 percent in Pakistan to nearly

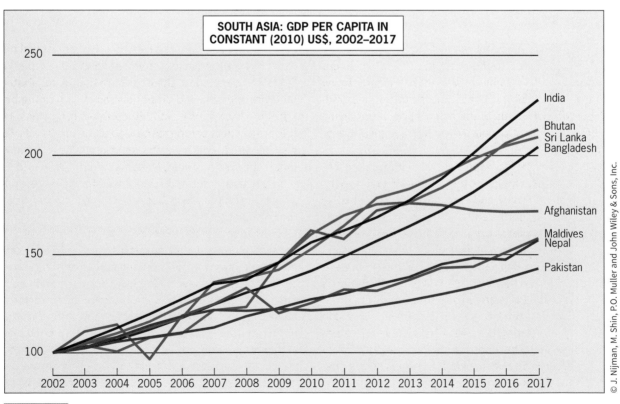

SOUTH ASIA: GDP PER CAPITA IN CONSTANT (2010) US$, 2002–2017

© J. Nijman, M. Shin, P.O. Muller and John Wiley & Sons, Inc.

FIGURE 9-9

Box 9-4 From the Field Notes . . .

"Strolling the grounds of one of India's leading high-tech companies, Wipro, in the southern city of Bengaluru (formerly Bangalore), I am struck by the thought that their use of the word 'campus' was right on target: the great majority of the 33,000 employees at this facility (120,000 in total) are in their mid-20s, and almost all have a college degree; a large number of them are involved in research and product development; and the layout of the premises could easily be mistaken for a (well-funded) university. This Silicon Valley–like campus is the heart of the company's booming outsourcing and consulting business. Wipro is active in most major Indian cities and has a sizeable global presence. I am shown around by Rohit, a 26-year-old with a bachelor's degree in engineering from an Indian college and an MBA from Singapore National University. He revels in the opportunity: 'I joined the company just four months ago in their marketing division and feel extremely fortunate. Everybody wants this job! It's a great professional opportunity, and the salary is very good.' And Bengaluru is a pleasant place to live: at 900 meters (3000 ft) above sea level, even the summer weather is tolerable. Bengaluru, however, is growing so rapidly that its infrastructure cannot keep up: thus congestion, traffic jams, and commuting times are all growing as well."

© J. Nijman

70 percent in Nepal. But overall productivity is low, and the contribution of agriculture to the national economy averages only around 20 percent. Incomes in rural areas are much lower than in the towns and cities, and the same is true for the standard of living. With two-thirds of the realm's population living outside urban areas, most people either work in agriculture or rely on it indirectly.

Millions of lives every year depend on a bountiful harvest. As Figure 9-3 shows, the wet monsoon brings life-giving rains to the southwestern (Malabar) coast, and a second branch from the Bay of Bengal sweeps across north-central India toward Pakistan, losing strength (and moisture) as it marches westward. This means that in amply watered eastern India, States like Asom (Assam) (**Box 9-6**) that surround Bangladesh as well as the southwestern coastal strip, rice can be grown as the staple crop. But drier northwestern India and Pakistan can only support wheat.

Farmers' fortunes tend to vary geographically, as can be illustrated by comparing the two sides of India's Western Ghats upland (Fig. 9-3). The monsoon rains are generally plentiful along the Arabian Sea–facing western slopes of this linear highland, from the southern tip of the subcontinent as far north as the vicinity of Mumbai. Here you can see the hillside vegetation assume all shades of fresh green after the rains

begin in June, a promising sign that the harvest will be abundant. But on the eastern (rain shadow) side of the Western Ghats and deeper into the interior of the Deccan Plateau, the rains are sporadic and do not last as long; farming becomes a gamble with nature, and life becomes precarious. Many of the farmers here belong to the lowest-ranking castes, landless and indebted, and have the most difficulty in making ends meet. Almost every year, Maharashtra State's inland districts report several thousand farmer suicides as desperately poor peasants end their lives because they can no longer support their families.

It is clear that the majority of people in this realm depend on agriculture and that governments must target economic policies at improving farm productivity to raise rural standards of living. But they have a long way to go. Demands on governments are voluminous, and they often seem distracted by economic sectors that can make a faster and greater contribution to the tax base—such as manufacturing, financial services, and information technology—economic activities that almost always are centered on the bigger cities far from the impoverished countryside. Furthermore, it is the direct dependence of well over half the population of South Asia on agriculture that makes it particularly vulnerable to climate change (**Box 9-7**).

Box 9-5 Geographic Information Analysis

Comparing the Megacities of Dhaka and Los Angeles

South Asia is the most heavily populated geographic realm on Earth. But its overall level of urbanization is quite low at 34 percent because the great majority of people inhabit rural areas. In absolute terms, however, South Asian cities are growing rapidly, and the realm now contains eight megacities. In comparison, the North American realm has only two: New York and Los Angeles.

Significant contrasts mark the world's megacities, as shown in the pair of maps below that differentiate the metropolitan areas of Dhaka and Los Angeles. Analyze the two maps and provide a concise description of the fundamental differences. Can you think of some reasons that explain these differences?

Access your WileyPLUS course to interact with a dynamic version of this map and to engage with online exercises and questions.

© J. Nijman, M. Shin, P.O. Muller and John Wiley & Sons, Inc.

FIGURE 9-10 Interactive

South Asia's Population Geography

Given its enormous human population, South Asia's areal size is relatively small and totals less than 40 percent of the territory of similarly populous East Asia. Comparing the world's two giants shows that China's territorial extent is almost three times as large as India's. The total population of Subsaharan Africa is less than half of South Asia's, in an area almost five times as large. Adjectives such as *teeming*, *overcrowded*, and *crammed* are often used to describe the realm's habitable living space, and with good reason. South Asia's intricate cultural mosaic is tightly packed, with only the deserts in the west and the mountain fringe in the north displaying extensive empty spaces. The outlines of the densely populated river basins are clearly visible in the clustering patterns on the population distribution map (**Fig. 9-11**).

The field of **population geography [14]** focuses on the characteristics, distribution, growth, and other aspects of spatial demography in a country, region, or realm as they relate to soils, climates, land ownership, social conditions, economic development, and other factors. In the South Asian context, it is useful to concentrate on four demographic dimensions: the role of density; the demographic transition; demographic burdens; and the gender bias in birth rates. As we shall see, population issues are often more complicated than they first appear.

Population Density and the Question of Overpopulation

Population density [15] measures the number of people per unit area (such as a square kilometer or square mile) in a country, province, or an entire realm. We distinguish between two types of measures. *Arithmetic density* is simply the number of people per areal unit, usually a country. **Physiologic density [16]** is a more meaningful measure because it takes into account only land that is arable and can be used for food production. In Pakistan, for example, the two measures are quite different because of that country's large deserts and inhospitable mountainous zones: its physiologic density is more than three times its arithmetic density.

Until recently, South Asia's persistent poverty was often related to its enormous, rapidly growing population and its high population densities. The idea was that there were simply "too

Box 9-6 From the Field Notes . . .

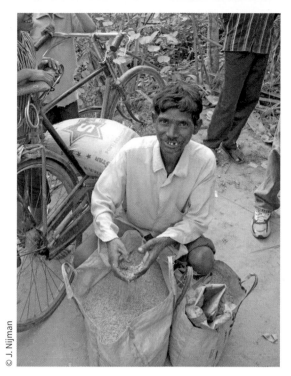

© J. Nijman

"I am traveling in northeastern India and visiting some agricultural areas in the Brahmaputra Valley that benefit from plentiful irrigation. Tea and rice are two major crops. Because big companies run the large tea plantations, many of the rice farmers have to make do with small plots of land. Here a farmer in Asom shows off part of his rice harvest. I encounter him and several others at a small, machine-operated mill shop. Note that the rice is brown—milling it removes the bran layer and makes the rice white. 'We have the best rice in all of India,' he proudly says, as he slowly poured it from his hand back into the bag."

The Demographic Transition

The relevance of population issues to development goes far beyond density, which is really just a snapshot of the population pattern at a moment in time. It gets more interesting—and more complicated—when we relate population change to economic trends. For instance, for a considerable time South Asia's population grew faster than the realm's economy. Clearly, that was a problem because more and more people had to survive on less. Today, fortunately, it is the other way around: in most of the realm, the economy is growing faster than the population.

The term **demographic transition [17]** refers to a structural change in birth and death rates resulting, first, in rapid population increase and, subsequently, in decreasing growth rates and then a stagnant or even a slowly declining population (**Fig. 9-12**). Note that Stage 2 and part of Stage 3, with high birth rates and low death rates (due to healthcare advances), entail a sizeable population expansion. The key issue, of course, is for birth rates to come down so that overall growth rates will diminish and the population will stabilize. Different regions and countries around the world find themselves at different stages in the transition. Japan, Russia, and Italy are in the fifth stage; their populations are currently shrinking. The United States, Canada, and Australia are in the fourth (stagnant) stage. South Asian countries range from the latter part of Stage 2 (Afghanistan) to Stage 4 (Sri Lanka), meaning their populations are generally still increasing. Note that the demographic transition pertains to natural growth only (i.e., death and birth rates); migration is another piece of the puzzle and can be especially critical to countries in Stage 5 (see Chapter 5 on Europe).

Fertility rates [18] (the total number of births per woman of childbearing age) have dropped across South Asia over the past quarter-century, today ranging from a still very high 4.6 in Afghanistan to 3.5 in Pakistan, 2.3 in India, and 2.1 in Sri Lanka. Considering that a fertility rate of 2.1 reflects a stable population, it is clear that South Asia, overall, is still growing.

Demographic Burdens

The immediate significance of demography to economics lies in what is called the **demographic burden [19]**. This term refers to the proportion of the population that is either too old or too young to be productive and that must be cared for by the productive population. Typically, the most productive population in developing countries is represented in the age cohorts between 20 and 50 years. A country with low death rates and high birth rates will have a relatively large share of old and young people, and thus a large demographic burden. Obviously, the way to reduce this burden is to lower birth rates.

Let us now examine **Figure 9-13** and compare today's **population pyramids [20]** (diagrams showing the age–sex structure) for India and China. China has more vigorously (and controversially) curtailed births since 1980, so it now faces a lower demographic burden than India. But, interestingly, what is advantageous today can soon turn into a disadvantage. Figure 9-13 also displays the population profiles

many mouths to feed"—the realm was "overpopulated." The notion of *overpopulation* can be compelling and seems to make sense at an intuitive level because every country or region can be thought of as having a limited "carrying capacity."

But things are more complex than that. Some countries with high densities, such as the Netherlands or Japan, are highly developed, and not necessarily because they possess an impressive natural resource base (neither does). The point is that high density in itself is not always a problem and that, in certain circumstances, population can be considered as a human resource. If productivity is high, there does not appear to be a problem; but if productivity is low, then large populations can be a drain on the economy. Countries with high education levels, institutional effectiveness, and technological know-how are able to use their resources more efficiently. Thus in South Asia, with its substantial number of undereducated and illiterate inhabitants, population functions as an impediment rather than a resource—the problem being that too many people are not sufficiently productive.

Box 9-7 Global Climate Change in South Asia

Exposure to global climate change in this realm is part of its physical-geographic package: the effects of global warming on the Himalaya Mountains, whose glaciers and powerful rivers are the main water source for billions of people; the low-lying terrain of areas such as the Ganges and Brahmaputra deltas, and archipelagos such as the Maldives, that make them so susceptible to inundation from rising sea level; and the high ocean-surface temperatures, moisture conditions, and air circulation patterns that are conducive to the formation of tropical cyclones (hurricanes). The World Bank in 2018 estimated that climate change, if it continues at the present speed, is likely to sharply erode the living conditions of 800 million (!) people in South Asia.

Climate change vulnerability, however, is not simply about atmospheric warming trends. According to the latest meteorological research in India, the South Asian wet monsoon is becoming increasingly unpredictable, with the number of **extreme weather events [13]** in South Asia having tripled over the past 35 years (see photo pair). These events take the form of:

1. Severe heat waves, such as northern India's successive record-breaking summer temperatures in 2015, 2016, and 2019.

2. Excessive torrential downpours outside the summer monsoon season, unleashing major flooding and/or landslides. Examples include the massive flooding of Pakistan's lower Indus Valley in 2010; the deadly floods and landslides in northern India's Uttarakhand State in 2013; the record-breaking unseasonal rains in the southern Indian State of Tamil Nadu in 2015; and yet another record-breaking monsoon storm that killed more than 1200 people across South Asia in August 2018.

3. Intensified droughts, such as the one that accompanied the heat wave of 2016 across central India, from Maharashtra State in the west to Odisha in the east. Across much of this belt, the monsoons of 2014 and 2015 had been inadequate to fulfill the needs of farmers.

In this densely populated realm, it is the exposure of so many people to the impacts of extreme weather events that necessitates more comprehensive and coordinated climate change mitigation. That involves infrastructural and disaster planning to adapt to these growing threats (e.g., better water storage, levies, dikes). But myriad institutional and bureaucratic hurdles stand in the way, and financial resources in South Asia are limited.

Recent extreme weather events in India may be related to global climate change. On the left, people wade along a water-logged roadway in the southern megacity of Chennai during December 2015 (well beyond the monsoon season). Torrential rains caused the city's airport to close, and thousands of residents were temporarily stranded in their homes or in traffic. On the right, villagers in the northern State of Uttar Pradesh (India's most populous), near the city of Prayagraj (Allahabad), walk on parched soil in the bed of the shrunken Mansaita River in May 2016. A third of a billion people in northern India were suffering from drought at this time, desperately awaiting the arrival of the tardy monsoon (it finally arrived about four weeks later, and by early August copious "normal" rains were pelting this crucial part of the Ganges Basin).

that are projected to exist one generation from now, in 2044. Assuming that India will be able to further reduce its birth rate in the coming years, its demographic burden will be less than China's one generation from now. In China, today's productive cohort will have moved on to old age, thereby adding to the national demographic burden. This is yet another reason for India boosters to be optimistic about the future. However, other major challenges remain, and certain forms of birth control across South Asia are increasingly under fire.

Gender Bias: The Missing Girls

Issues of family planning and birth control shed some fascinating light on the "fragmented modernization" of South Asia. As we have seen, the realm is in an advanced stage of the demographic transition wherein birth rates have begun to decline. But take a good look at India's population pyramid for 2019 (Fig. 9-13) and note that among young children boys far outnumber girls. In fact, males outnumber females well into middle age.

POPULATION DISTRIBUTION: SOUTH ASIA

Population per Square Kilometer

0 to 5
5 to 10
10 to 50
50 to 100
100 to 500
Over 500

© J. Nijman, M. Shin, P.O. Muller and John Wiley & Sons, Inc.

FIGURE 9-11 **Interactive**

Traditionally, boys are valued more than girls because they are thought to be more productive income earners, because they are entitled to land and inheritance, and because they do not require a dowry at the time of marriage. When a couple gets married (often arranged and at a young age), the bride comes into the care of the groom's family, where she also contributes her work in and around the house. For this, the bride's family must provide a dowry, which may impose a major expense for her parents. For these reasons, the birth of a boy is a greater cause for celebration than that of a girl. "Raising a daughter," as one saying goes, "is like watering your neighbor's garden."

One reason for the high fertility rates of the past was that families would continue to have children until there were enough sons to take care of the parents in their old age (the girls, after all, would be taking care of their future husband's parents). Hence, this gender tilt is in itself a major factor in South Asia's population growth—but that is not the only problem. When a poor couple repeatedly produces girls and not boys, in some instances the family decides to end the life of the newborn daughter. It is this *female infanticide* or *gendercide* that causes the unnatural gender bias in South Asia's population profiles (the same applies to parts of China as well).

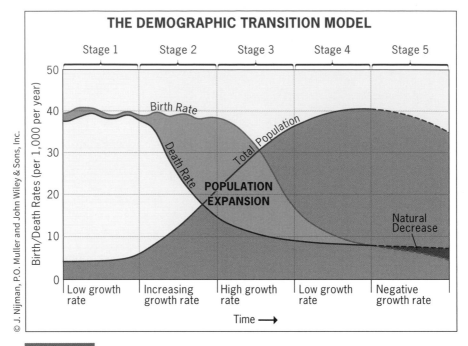

© J. Nijman, P.O. Muller and John Wiley & Sons, Inc.

FIGURE 9-12

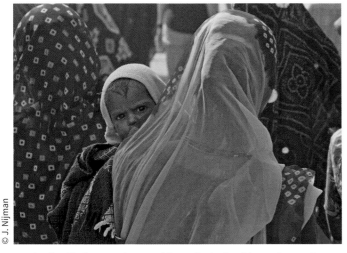

© J. Nijman

A Rajasthani infant peering over his mother's shoulder at the market in the northwestern Indian town of Pushkar. Across the subcontinent, women tend to marry and bear children at an early age. Yet fertility rates have been declining, and in India the average number of children per woman has dropped from 3.9 in 1990 to 2.3 today—a substantial decrease but still generating a national population growth rate well above the global average.

But why—as the economic situation has improved, as birth rates have receded, and as modernization has begun to set in—do we still observe this skewed **sex ratio [21]**? The answer is that with fewer children, the importance of having at least one boy has for many families become even more pressing. And here is where "modernization" throws another curve ball: newly available technologies of ultrasound scanning plus rising incomes (i.e., the growing affordability of a scan) have induced many families to determine the gender of the unborn child and decide on abortion if the child is female. Thus in recent years the sex ratio has become more, not less, skewed, and the most extreme ratios

are now found in some of the most developed parts of South Asia, such as the Indian States of Punjab and Haryana (**Box 9-8**).

In the long run, of course, this leads to a shortage of females, which becomes particularly apparent at marriage age. In some areas, families now face a problem in finding brides for their sons, and this "bachelor angst," in turn, is leading to a change in attitudes. Look at India's population pyramid for 2044 and note that over the next quarter-century the sex ratio is projected to become less skewed.

More broadly, discrimination against women and girls in South Asia is expressed in a variety of ways in daily life. One matter that is now receiving more attention is the lack of sanitation facilities for females. In many workplaces, schools, slum

© J. Nijman

A group of girls in a park in Delhi showing off their henna tattoos. "Mehndi" is a centuries-old festive tradition involving decorative painting on the hands and wrists. These designs usually wear off within a few weeks.

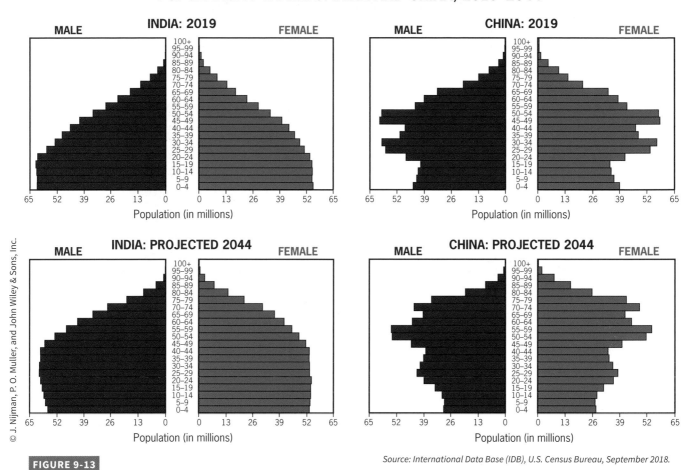

POPULATION PYRAMIDS: INDIA AND CHINA, 2019–2044

INDIA: 2019

MALE   FEMALE

Population (in millions)

CHINA: 2019

MALE   FEMALE

Population (in millions)

INDIA: PROJECTED 2044

MALE   FEMALE

Population (in millions)

CHINA: PROJECTED 2044

MALE   FEMALE

Population (in millions)

© J. Nijman, P. O. Muller, and John Wiley & Sons, Inc.

FIGURE 9-13

Source: International Data Base (IDB), U.S. Census Bureau, September 2018.

neighborhoods, and public spaces, women still have no (or insufficient) access to toilets. For safety and personal reasons women need more privacy, but in India at least 300 million females lack proper access to toilets, with major implications for health and social functioning. Imagine a small rural school with one toilet (if indeed there is one): it will be used by boys to the exclusion of girls. Girls will have no available access all day. This simple fact leads to girls missing school and even dropping out. Recent efforts by development organizations to install toilets for girls have so far produced significant reductions in female-student dropout rates.

The recent increase of highly publicized incidents of rape in India has expanded the debate surrounding the country's gender relations. It is an alarming trend, especially where it signals a reaction by men against liberalized behavior among young urban women. But it also reflects the increasingly open disposition of

India's mass media as well as a new assertiveness among women (though it is hard to know to what extent this increase is due to a higher rate of reporting and not of rape itself). And, relatively speaking, the incidence of (reported) rape in India is still far below the level of most Western countries, including the United States.

It is not easy to make generalizations about gender relations across this most populous of all geographic realms, in part because of religious and regional diversity as well as rural–urban differences. To be sure, these are in many ways male-dominated societies, especially in the younger age cohorts, and more so in Afghanistan and Pakistan than in India, Sri Lanka, and Bhutan. Nonetheless, keep in mind that Pakistan, India, Sri Lanka, and Bangladesh have all had female prime ministers who held their countries' most powerful political office. This has yet to happen in the United States.

Regions of South Asia

The South Asian geographic realm can be subdivided into five regions (**Fig. 9-15**):

1. **The West** region consists of a pair of Islamic-dominated countries. War-ravaged *Afghanistan* currently forms a transition zone between the South Asian and North African/ Southwest Asian realms. Beleaguered *Pakistan* revolves

around a core area formed by the Indus River and its tributaries in east-central Punjab.

2. **India** is the realm's giant and historic heart. It is anchored by its core area focused on the broad Ganges Basin. It also contains key subregions based on cultural and other criteria.

Box 9-8 Geographic Information Analysis

Female/Male Sex Ratios Across India

The imbalance of sex ratios in India is traditionally attributed to better care for boys than girls, selective abortions, even female infanticide. Improving economic conditions and better medical care are expected to redress this imbalance. **Figure 9-14** shows child sex ratios (females per 1000 males) across India's States in 2001 and 2011. A lower number indicates a greater imbalance.

How have things changed over time? Are changes uniform spatially? Are there States with a high imbalance where conditions deteriorated further? Can you identify a State with a low imbalance where things became better still? What are possible explanations for these changes?

Access your WileyPLUS course to interact with a dynamic version of this map and to engage with online exercises and questions.

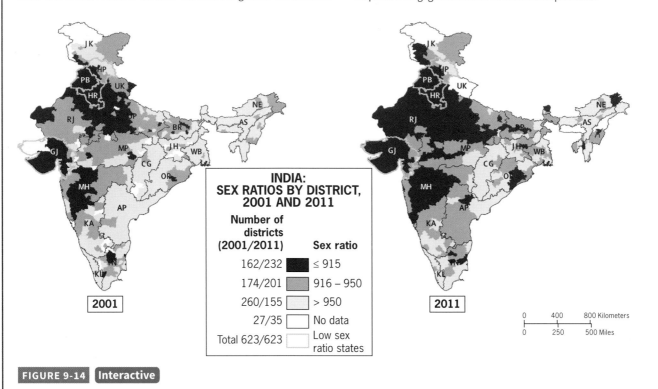

FIGURE 9-14 Interactive

Adapted from P. Jha et al, "Trends in selective abortions of girls in India: Analysis of nationally representative birth histories from 1990 to 2005 and census data from 1991 to 2011." *The Lancet*, Vol. 377 (2011), page 8.

Camel traders gearing up at sunrise in northwestern India's Thar Desert, not far from the border with Pakistan.

3. **Bangladesh** centers the densely populated, low-lying double delta formed by the Ganges and Brahmaputra rivers, which creates a unique and challenging environment.

4. **The Northern Mountain States** region comprises two noncontiguous Himalayan states: *Nepal* and *Bhutan*. Nepal is plagued by a continuing political struggle following the failure of longtime royal rule; Bhutan remains a peaceful traditional kingdom.

5. **The Southern Islands** region consists of two very different countries. Dominantly Buddhist, ethnically divided *Sri Lanka* is located a short distance to the southeast of India's southern tip. Farther offshore, on the southwestern side of the subcontinent's tip, lies the island chain known as *The Maldives,* an Islamic mini-state imperiled by rising sea levels.

© J. Nijman, P. O. Muller, and John Wiley & Sons, Inc.

REGIONS OF SOUTH ASIA

Afghanistan Transition Zone
West Region

POPULATION

● Under 250,000
● 250,000–1,000,000
● 1,000,000–5,000,000
● 5,000,000–10,000,000
● Over 10,000,000

National capitals are underlined

0   200   400   600   800 Kilometers
0   100   200   300   400 Miles

FIGURE 9-15

Region The West: Afghanistan's Transition Zone

Geography and history appear to have conspired to divide Afghanistan. As Figure 9-1 reveals, the towering Hindu Kush range dominates the center of the country, creating three broad environmental zones: the comparatively well-watered, fertile northern plains and basins; the rugged, earthquake-prone central highlands; and the desert-dominated southern plateaus. Kabol (Kabul), the capital, lies on the southeastern slopes of the Hindu Kush, linked by narrow passes to the northern plains and by the Khyber Pass to Pakistan and the rest of South Asia (**Fig. 9-16**).

Across this rugged, variegated landscape moved countless peoples: Greeks, Turks, Arabs, Mongols, and others. Some settled here, their descendants today speaking Persian, Turkic, and other languages (see Fig. 9-5). Others left archeological remains or no trace at all. The present-day population of Afghanistan (39 million) contains no ethnic majority. This is

FIGURE 9-16

a country of minorities in which the Pushtuns of the east are the most numerous but make up only 42 percent of the total. The second-largest minority are the Tajiks (27 percent), a world away across the formidable Hindu Kush, concentrated in the far north near Afghanistan's border with Tajikistan. Other groups include Hazaras, Uzbeks, Turkmen, and Balochs (Fig. 9-5).

Afghanistan has a long history of conflict, complicated by the persistent involvement of outside powers vying for influence in this strategically located territory. The protracted nineteenth-century competition between the Russian and British empires over control of Afghanistan—with the wider purpose of controlling adjacent Central Asia—went down in history as "the Great Game." In fact, modern Afghanistan owes its existence to this contest: the Russians and British agreed to

tolerate it as a geopolitically neutral cushion—or **buffer state [22]**—between them. During the second half of the twentieth century, Afghanistan was pulled into the Cold War, then suffered a Soviet invasion and ten years of occupation, the emergence of violent Islamic militants (particularly Al Qaeda), the presence of U.S. troops since 2002, and most recently a resurgence of the vicious Taliban jihadist forces. And yet, despite this turmoil, Afghanistan remains much as it was before the first inning of "the Great Game"—a feudally balkanized country with a weak and ineffectual government in Kabol. It has also now become a **transition zone**, an arena of spatial change in which the peripheries of adjacent realms overlap.

As Afghanistan has found itself time and again drawn into geopolitical contests between foreign powers and disturbed by conflict and war, its economic development has been compromised throughout. Even by South Asian standards, Afghanistan is desperately poor, its per capita income at U.S. $590 per year barely one-third of India's, while fertility rates are the highest in the realm. Afghanistan has been urbanizing very slowly: just over a quarter of the population lives in cities. It is also marked by **urban primacy [23]**: Kabol is by far the biggest city with 4 million people, but the second biggest, Kandahar, contains only about half a million.

The Emergence of the Taliban and Al Qaeda

In 1994, what at first seemed to be just another one of fractious Afghanistan's warring factions appeared on the scene: the *Taliban* ("seekers of religion"), who originated at religious schools (*madrassas*) in Pakistan. Their avowed aim was to end Afghanistan's chronic factionalism and endemic corruption by imposing *Sharia* (strict Islamic) law. Popular support across the war-weary country, especially among the Pushtuns, led to a series of successes, and by 1996 the Taliban had taken Kabol. The harsh Taliban version of *Sharia*, however, was uncompromising: restrictions on the activities of women ended their professional education, employment, and freedom of movement; the lives of children were even more devastatingly impacted; public stonings and amputations rigidly enforced the Taliban's code.

Afghanistan became a haven for Islamic revolutionary groups whose agendas and goals went far beyond those of the Taliban: they plotted attacks on Western interests throughout southwestern Asia and threatened regimes they deemed to be compliant with Western priorities. Taliban-ruled, cave-riddled, remote, and isolated Afghanistan turned out to be an ideal refuge for these outlaws. Already in possession of arms and ammunition (Soviet as well as American) left over from the Cold War, they also took advantage of Afghanistan's huge, thriving, illicit opium trade. Afghanistan has long served as the world's foremost producer of opium, accounting for more than three-quarters of the global heroin supply, and much of this drug revenue was quickly diverted into the coffers of the conspirators (photo right).

In 1996, a terrorist organization named *Al Qaeda* took root in the country, an expanding global network that worked to further the aims of the once loosely allied revolutionaries. Afghanistan now became Al Qaeda's headquarters, and Osama bin Laden, a notorious Saudi renegade, its director. Al Qaeda's most fateful operation came with the suicide attacks of September 11, 2001, on New York and Washington, D.C., which destroyed the World Trade Center's twin towers, damaged the Pentagon, killed over 3000 people, and inflicted massive psychological damage. Several weeks later, U.S. and British forces, with the acquiescence of Pakistan, attacked both the Taliban regime and Al Qaeda in Afghanistan.

The U.S.-Led Invasion of Afghanistan and Its Aftermath

The American-led invasion following 9/11 initially appeared to be successful. The Taliban were swiftly driven off, a more representative government was taking shape, local warlords were co-opted or sidelined, Pushtun and other refugees were streaming back into Afghanistan from Pakistan and elsewhere, and life in Kabol and other urban centers returned to a semblance of normal. By 2004, Afghanistan had even held an election that produced a representative government. It took nearly a decade of U.S. intelligence work, but in 2011 Bin Laden was finally discovered and killed in an American military raid—not in Afghanistan but in the city of Abbottabad in neighboring northern Pakistan (Fig. 9-16). With Al Qaeda markedly weakened and having allowed for a transitional period in which to fully train Afghan security forces, the United States at the end of 2014 withdrew the last of its combat troops from Afghanistan—a residual U.S. (advisory) force of 14,000 remained.

During the following years, American and NATO advisors found it extremely challenging to foster a workable consensus among various Afghan factions in order to build a more democratic state. One serious problem was the country's economy, which had been all but shattered by decades of unremitting conflict and war, forcing the government to depend heavily on foreign aid. Soon, Taliban fighting units had regrouped into smaller militias and were again attacking security forces, government facilities, and Western compounds. Much of this activity first took place in the far north adjacent to Tajikistan (Fig. 9-16), but by 2016 was spreading across the Hindu Kush to resume hostilities in and around Kabol. Although Al Qaeda in Afghanistan had been dealt a stunning blow at the core, the organization has since proven resilient. As discussed in Chapter 7, by 2013 it had branched out beyond Afghanistan to

Local farm workers harvesting opium sap from a poppy field in southern Afghanistan's Kandahar Province in spring 2015. This country is by far the world's largest producer of opium (the raw material of heroin). Cultivating these poppies is considerably more profitable than raising legal crops and is believed to involve various networks of organized crime, including Al Qaeda.

NurPhoto/Getty Images

become a major player in unstable countries, such as Somalia, Yemen, Libya—and Pakistan.

By 2019, after 18 arduous years and spending around $45 billion per year, the U.S. had failed to defeat the Taliban, and it had failed to install a stable democratic government. This painful truth resulted in renewed calls for the U.S. to fully withdraw from Afghanistan.

Region The West: Pakistan

In the Indus Valley of Pakistan lay South Asia's earliest urban civilizations, whose innovations diffused southeastward into the massive triangular peninsula. Today, this elongated river basin runs a like spine through Pakistan, from north to south, contiguous to the great Islamic realm to the west and irrevocably linked to India's enormous Muslim minority to its east. Pakistan's cultural landscapes bear witness to its transitional location. Teeming Karachi is the typical South Asian megacity. Historic, architecturally Islamic Lahore is reminiscent of the scholarly centers of Muslim Southwest Asia. In Pakistan's east, the boundary of the 1947 partition divides a Punjab subregion that continues well into India—a land of villages, wheat fields, and irrigation canals. In the northwest, Pakistan closely resembles the Afghanistan Transition Zone in its huge migrant populations and mountainous borderland.

The Indus River and its principal tributary, the Sutlej, nourish the ribbons of life that form the heart of this populous country (Fig. 9-16). Territorially, Pakistan is not large by Asian standards; its area is about the same as that of Texas plus Louisiana. But Pakistan's population of 221 million makes it the world's fifth most populous state. Among Muslim countries (its official name is the Islamic Republic of Pakistan), only Southeast Asia's Indonesia is larger.

A Hard Place to Govern

At independence in 1947, (West) Pakistan had a bounded national territory, a capital, a cultural core, and a population—but few centripetal forces to bind state and nation. The disparate subregions of Pakistan shared the Islamic faith but little else. Karachi and the coastal south, the southwestern desert of Balochistan, the city of Lahore and surrounding Punjab, the rugged northwest along Afghanistan's border, and the mountainous far north remain worlds apart. Urdu is the official language, yet English remains the *lingua franca* of the elite. Several other major languages, however, prevail in different areas (see Fig. 9-5), and ways of life vary enormously as well.

Successive Pakistani governments, civilian as well as military, turned to Islam to provide the common bond that history and geography had denied this state. In the process, Pakistan became one of the world's most theocratic countries. But even Islam itself is not unified in restive Pakistan. Roughly 80 percent of the people are Sunni Muslims, and the Shia minority accounts for most of the remaining 20 percent. Sunni fanatics

intermittently attack Shi'ites, leading to retaliation and establishing grounds for subsequent counterattacks.

To govern so diverse and fractious a country would challenge any political system, and so far Pakistan has failed the test. Democratically elected governments have time and again been overthrown by military coups. Pakistan's modest economic growth in the last 15 years has not filtered down to the poor; literacy rates have risen only modestly; health conditions are not improving significantly; national institutions are weak (for instance, there are only about 2.5 million registered taxpayers); and one consequence of the global antiterrorism campaign is that Pakistanis, who used to be overwhelmingly secular in their political choices, are now increasingly joining Islamic parties.

Meanwhile, too little is being done to confront a growing water-supply crisis, an insurgency festers in Balochistan, the army is incapable of establishing control over mountainous Waziristan (where Al Qaeda and Taliban militants reign), the issue of Kashmir costs Pakistan dearly, and relations with neighboring India remain conflicted, as discussed earlier in this chapter.

The Four Subregions of Pakistan

Punjab constitutes Pakistan's core area, the Muslim heartland across which the post-independence boundary between Pakistan and India was superimposed (Fig. 9-16). (As a result, India also has a State named Punjab on the other side of the border.) Pakistan's Punjab is home to just over half of the country's population; in the triangle formed by the Indus and Sutlej rivers live more than 100 million people. Punjabi is the language here, and wheat farming is the mainstay.

Three cities anchor this core-area subregion: Lahore, the foremost center of Islamic culture in the realm; Faisalabad; and Multan. Lahore, now home to more than 11 million people, lies close to the border with India. Founded about 2000 years ago, Lahore was situated favorably to become a great Muslim center during the Mughal period when Punjab was the main corridor into India. Punjab's relationship with Pakistan's other three provinces is one of the country's weaknesses. Both the governments and residents of those provinces feel uneasy about the dominance of Punjab, the populous, powerful core of the country from which most of the military is drawn.

The lower Indus River is the key to life in **Sindh**, but Punjab controls the waters upstream, which is one of the issues that divide Pakistan (Fig. 9-16). When the Punjab-dominated regime proposes to build dams across the Indus and its major tributaries, Sindhis (who make up almost one-fifth of the national population) are reminded of their underrepresentation in government and talk of greater autonomy. The ribbon of fertile, irrigated, alluvial land along the lower Indus, where the British laid out irrigation systems, makes Sindh a Pakistani breadbasket for wheat and rice. Commercially, cotton is king here, supplying textile factories in the cities and towns (textiles account for more than half of Pakistan's exports by value).

But the dominant presence in southern Sindh is the chaotic megacity of Karachi. With little effective law enforcement, Karachi has become a hotbed of terrorist activity, but somehow the city still functions as Pakistan's (and Afghanistan's) major maritime outlet and the seat of Sindh's provincial capital.

As Figure 9-16 indicates, the province of **Khyber Pakhtunkhwa** lies wedged between powerful Punjab to the east and troubled Afghanistan to the west, with the territory long known as the *Tribal Areas* intervening in the south. The name "Pakhtunkhwa" translates as "belonging to the Pushtuns," the Afghan-associated tribes that inhabit this subregion.

The Pakistani government's reach into these parts has always been quite limited. The province's mountainous physical geography reflects its remoteness and the isolation of many of its people. Mountain passes lead to Afghanistan; the Khyber Pass, already noted as the historic route of invaders, is legendary (photo next page). Coming from Afghanistan, the Khyber's road leads directly to the provincial capital, Peshawar, which lies in a broad fertile valley where wheat and corn drape the countryside. Khyber Pakhtunkhwa is a conservative, deeply religious, militant province, where Islamic political parties and movements are stronger than in any other part of the country.

The fourth subregion mapped in Figure 9-16 is **Balochistan**. By far the biggest of Pakistan's four provinces, it accounts for nearly half of the national territory (not including Kashmir) but is inhabited by only about 14 million people, barely 5 percent of the country's population. For a sense of its terrain, take a look at Figure 9-1: much of this vast territory is desert, with mostly barren mountains that wrest some moisture from the air only in the northeast. Sheep-raising is the leading livelihood here, and wool is the primary export. In its northern extremity, Balochistan abuts the Tribal Areas and Afghanistan. This is where Quetta lies, capital of a province that could easily be called the "South West Frontier Province."

Yet Balochistan is of considerable economic-geographic importance. Beneath its parched surface lie possibly substantial reserves of oil as well as coal, and already this province produces most of Pakistan's natural gas. The new seaport of Gwadar on the southwestern coast not only handles raw materials from Balochistan but is also a transshipment point for oil and gas from Iran and the Caspian Basin, destined for markets in East Asia. China has a key interest in Gwadar and in a much larger infrastructural project to connect it all the way to the China-Pakistan border in the north, often referred to as the **China-Pakistan Economic Corridor [24]** (see **Fig. 9-17**). But the Baloch people are treated poorly by Pakistan's central government—the majority have no access to electricity or clean water. Short-lived local rebellions have flared up here since

FIGURE 9-17

travelib prime/Alamy Stock Photo

The pivotal Khyber Pass that links Afghanistan and Pakistan across the Hindu Kush (see Fig. 9-16). One of the most strategic passes in the world, it was no easy passage for invading armies in times past. Nowadays, the road and its tunnels facilitate common trade along with the movement of refugees, drugs, and weapons, which are used by militant separatists on both sides of the border. Pakistan's Khyber Pakhtunkhwa Province, especially its provincial capital of Peshawar, is a hotbed of such activities.

Pakistan became independent, and during the 2000s they escalated into a serious insurgency led by the Balochistan Liberation Army (BLA), whose attacks on security forces continue.

Region India: Giant of the Realm

A Federation of States and Peoples

That India has endured as a unified state is a politico-geographical miracle. The country contains a cultural mosaic of immense ethnic, religious, linguistic, and economic diversity and contrast; it is truly a state of many nations. Upon independence in 1947, India adopted a democratic, secular, federal system of government, giving regions and peoples some autonomy and identity, and allowing others to aspire to such status.

The map of India's political geography shows a federation of 29 States, eight Union Territories (UTs), and one National Capital Territory (NCT) (**Fig. 9-18**). The federal government maintains direct authority over the UTs, all of which are small in both territory and population. The NCT, however, includes most of the Delhi/New Delhi urban region and now contains almost 28 million inhabitants (photo below and satellite image p. 338). The country's State boundaries reflect the broad outlines of the country's cultural mosaic: insofar as possible, the system recognizes languages, religions, and cultural traditions. Indians speak 14 major and numerous minor languages, and although Hindi is the official language, it is by no means universally used (see Fig. 9-5).

As Figure 9-18 shows, the territorially largest States lie in the heart of the country as well as on the massive southward-pointing landmass. Uttar Pradesh (208 million) and Bihar (119 million) together cover much of the Ganges Basin and also form the core area of modern India. Maharashtra (120 million), anchored by the great coastal city of Mumbai (called Bombay before 1996), also has a population larger than that of most countries. West Bengal, the State that adjoins Bangladesh, contains 98 million residents, 15 million of whom live in its urban focus, Kolkata (known as Calcutta before 2000).

These are staggering numbers, and they do not decline much toward the south. Southern India consists of four States linked by a discrete history and by their distinct Dravidian languages. Facing the Bay of Bengal are Tamil Nadu (76 million) and Andhra Pradesh (53 million), both part of the hinterland of the city of Chennai (formerly Madras) that is located on the coast near their joint border. Facing the Arabian Sea are Karnataka (66 million) and Kerala (35 million).

India's smaller States lie mainly in the northeast, on the far side of Bangladesh, and in the northwest, toward Jammu and Kashmir (Fig. 9-18). The map becomes even more complicated in the remote northeast, beyond the narrow corridor between Bhutan and Bangladesh. The dominant State here is Asom (the old spelling is Assam), home to just over 35 million, famed for its tea plantations, and important because its petroleum and gas production amounts to more than 40 percent of India's total. In the Brahmaputra Valley, the axis of Asom, this State

© J. Nijman

The Delhi urban region consists of Old Delhi (from the 17th century and before), New Delhi (dating to British colonial times), and massive newly developed edge cities like Gurgaon and Noida. The photo was taken in 2019 on an early morning at Chawri Bazar, Old Delhi, where businesses are opening their doors, couriers and carriers are delivering their goods, and many working men (there are few women on the streets here) begin the day with a cup of chai (Indian tea). This bazaar is a wholesale market for brass, copper, paper products, and more.

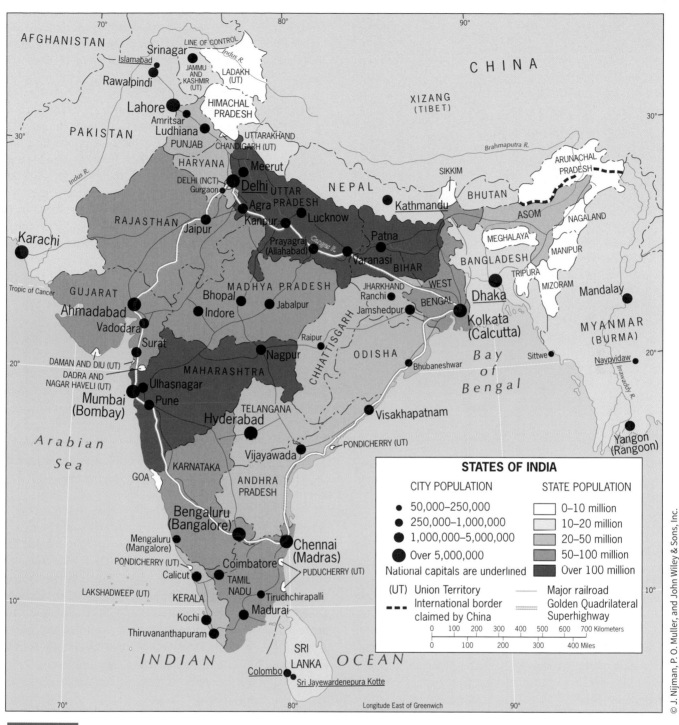

FIGURE 9-18

resembles the India of the Ganges. But in almost every direction from Asom, things change.

North of Asom, in sparsely populated Arunachal Pradesh (1.5 million), we are in the Himalayan offshoots once again. To the east, in Nagaland (2.1 million), Manipur (3 million), and Mizoram (1.2 million), lie the forested and terraced hills that separate India from Myanmar (Burma). This is an area of myriad ethnic groups (more than a dozen in Nagaland alone) and of frequent rebellion against the central government. To the south, the States of Meghalaya (3.3 million) and Tripura (4 million),

hilly and still wooded, border the highly populated floodplains of Bangladesh. Here in the country's northeast, India confronts one of its strongest regional challenges.

India's Changing Map

Devolutionary pressures have continued throughout India's existence as an independent state. In some of the cases, the federal government and the military have come down hard on

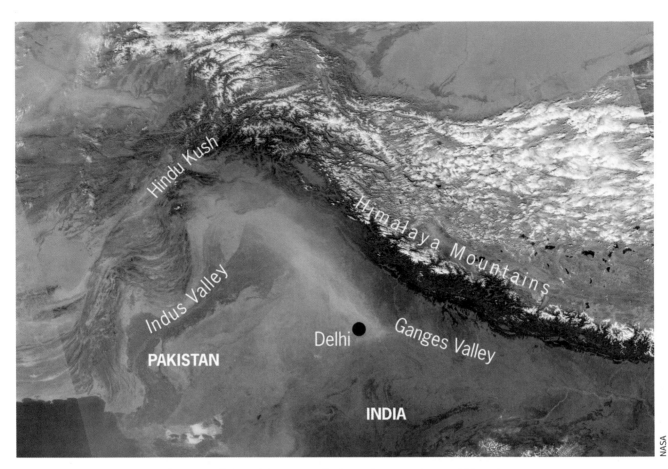

NASA

The Delhi metropolis is notorious for its air pollution. This NASA satellite image from 2016 shows a thick coat of smog and smoke hovering over northwestern India and Pakistan, caused by industrial and traffic emissions along with crop stubble burning in the agricultural fields of Punjab. The problem is also geographical: coastal winds from the Arabian Sea "push" the polluted air north toward the low-lying basins of the Ganges and the Indus, where it gets trapped against the mighty Himalayas and the Hindu Kush (see Fig. 9-1).

the insurgents, but in others the government has been more inclined to negotiate. In 2000, for instance, three new States were recognized: Jharkhand, carved out of southern Bihar State on behalf of 18 poverty-stricken districts there; Chhattisgarh, where tribal peoples had been agitating since the 1930s for separation from the State of Madhya Pradesh; and Uttarakhand (originally named Uttaranchal), which split from Uttar Pradesh. Most recently, in 2014, the new State of Telangana (India's 29th) was created out of the northwestern sector of Andhra Pradesh. Telangana is an inland area that long saw itself marginalized by coastal elites. The new State contains more than 38 million people, covers roughly one-third of pre-2014 Andhra Pradesh, and its capital is Hyderabad—one of the country's leading technopoles and its sixth-largest city—which will serve jointly as the capital of the new State of Andhra Pradesh until 2024.

Yet another challenge to India's federal government is posed by communist-inspired rebellions in some of the more deprived regions of the country. It is known in India as the *Naxalite* movement, named after a village in the State of West Bengal where it was founded in the 1960s. Mainly active in India's poorest and most disaffected States—such as

Bihar, Jharkhand, and what is now Telangana—the Naxalites appeal to the poor and other minorities (especially tribal people), whose plight is blamed on India's elites and their neoliberal economic policies. They blow up railroad tracks, attack police stations, and intimidate villagers. The Indian government has embarked on a major counteroffensive which appears to have had some success as Naxalite violence has declined in recent years.

Communal Tensions

The term **communal tension [25]** (communal disharmony) refers to the several different categories of conflicts that recur among India's highly diverse sociocultural groups. Most commonly, these conflicts have a base in (politicized) religion, but they can also be caste-based, as we shall see.

The *Sikhs* (the word means "disciples") adhere to a religion that was created about five centuries ago to unite warring Hindus and Muslims into a single faith. The Sikhs rejected what they considered to be the negative aspects of Hinduism and Islam, and Sikhism gained millions of followers in Punjab and surrounding areas. During the colonial period, many Sikhs supported the

© J. Nijman

One of India's most famous landmarks, the Golden Temple in the city of Amritsar in Punjab State, which dates back to the late sixteenth century. Sikh places of worship are called *gurdwaras,* and this is the holiest one of all, known to the faithful as *Harmandir Sahib* ("The Temple of God"). More than 100,000 Sikhs pray here every day of the year. Adherents of all religions are welcome, but visitors must dress appropriately, remove their footwear, and cover their heads.

British administration of India, and by doing so they won the respect and trust of the colonialists, who employed tens of thousands of Sikhs as soldiers and police officers. By 1947, Punjab was home to a large Sikh middle class. Today they still exert a strong influence over Indian affairs, far in excess of the 1.7 percent of the population (about 23 million) they constitute.

For a period after India's independence, Sikh extremists created India's most serious separatist problem, demanding the formation of an independent state they wanted to call *Khalistan.* The conflict intensified in the 1970s and culminated in an attack by federal military forces on Sikh rebels barricaded in the Golden Temple in the city of Amritsar, Sikhdom's holiest site (photo above). Shortly after this 1984 attack, then Prime Minister Indira Gandhi was assassinated by her own Sikh bodyguards, unleashing a lengthy period of anti-Sikh violence across much of India. Since these dramatic events of the 1980s, both sides have worked to gain better understanding.

Its *Muslim* minority may constitute less than 15 percent of India's population, but they number 200 million, nearly the total population of Pakistan. Muslims are in the majority in the States of Jammu and Kashmir, Asom, and West Bengal, and their absolute numbers are substantial in the big States of Uttar Pradesh,

Bihar, and Maharashtra. To India's great advantage, the Muslim minority is not geographically concentrated, avoiding what could otherwise lead to a regional secessionist movement.

The status of Muslims within Indian society is not that easy to separate from India's relations with Pakistan, from the confrontation in Kashmir, and from acts of Islamic terrorism that have bedeviled India in recent years. As a result, tensions between Muslims and Hindus have been rising. The most serious impediment to a better integration of Muslims into Indian society, aside from discrimination and the rise of *Hindutva* (see below), arises from their comparatively low level of education and socioeconomic standing. While Muslims make up 15 percent of the overall population, they represent only 3 percent of college students. And according to one recent government report, Muslims tend to be as as poor as *Dalits,* the members of India's lowest-ranked Hindu caste (**Box 9-9**).

Hindu extremism or fundamentalism may seem to be a contradiction in terms, but a movement has come to the fore in this century that seeks to remake India as a society in which Hindu principles prevail. *Hindutva* [26], or Hinduness, is variously expressed as Hindu nationalism, Hindu heritage, or Hindu patriotism. It has been the guiding agenda for the

Box 9-9 From the Field Notes . . .

"It is a Friday afternoon as we make our way through a main thoroughfare in one of Mumbai's biggest slums. Dharavi 'houses' approximately half a million people on less than 2 square kilometers! About 20 percent here are Muslims, and the overwhelming majority of all others are Dalits. Increasingly, the Muslims live spatially segregated in their own tightly knit neighborhoods. There are several mosques but little space inside them. So the men, taking a break from work nearby, place their mats alongside the road and prepare for prayer outdoors. Having a faith is important to people living and working in Mumbai's slums, especially for the Muslim minority."

© J. Nijman, Photo by Zach Woodward

Bharatiya Janata Party (BJP), a powerful force in national politics as well as in large States like Maharashtra, Gujarat, and Madhya Pradesh. *Hindutva* fanatics want to impose a Hindu curriculum on schools, change the flexible family law in ways that would make it unacceptable to Muslims, constrain the activities of non-Hindu religious proselytizers, and forge a new India in which non-Hindus are essentially outsiders. This movement is also reponsible for a number of name changes, intent on erasing the Islamic (Mughal) past of India: for example, the city of Allahabad, one of the most holy cities for Hindus, situated at the confluence of the Ganges and Yamuna rivers, was recently renamed Prayagraj, the Hindu name it carried before Muslim rulers gave it the name Allahabad ('abode of God').

It is easy to see how Hindu hardliners and Islamic revivalists can fuel each other's agendas and maintain a vicious cycle of conflict. This naturally worries Muslims as well as other minorities, but it also concerns those who understand that India's secularism—its separation of religion and state—is indispensable to the survival of democracy. Moderate Hindus and non-Hindus in India oppose this strengthening of *Hindutva*, whose notions are as divisive as any India has faced.

The Caste System

Hinduism's sprawling and accommodating properties (there are few orthodoxies in Hinduism, and there is no central authority) combine with a system of social stratification that is often derided in the West and by some Indians as well. The *caste system* has its origins in the early social divisions into priests, warriors, merchants, and farmers, and it is also thought to have had a racial basis (the Sanskrit term for caste, *varna*, means color). More specifically, caste became associated with specific occupations, and over the centuries its complexity expanded until India had thousands of castes, some of them containing a few hundred members and others consisting of millions.

Hindus believe in reincarnation, and a person is believed to be born into a particular caste based on his or her actions in a previous life. Hence, it would not be appropriate to counter such ordained caste assignments by permitting mobility (or even contact) from a lower caste to a higher one. Persons of a particular caste could perform only certain jobs and worship only in prescribed ways at particular places. They and their children could not eat, play, or even walk with people of a higher social status.

The *untouchables* occupying the lowest tier of all were the most debased, wretched members of this rigidly structured social system. Indeed, the term *untouchable* acquired such negative connotations that it was replaced on several occasions. Mahatma Gandhi, a powerful critic of the system, introduced the term *harijans*, meaning children of God. But more recently this label was perceived to have a condescending connotation, and it gave way to the term **Dalits [27]** (the oppressed), indicating a greater sense of awareness and assertiveness among them. Today, Dalits are estimated to comprise about one-sixth

of all Hindus; Brahmins, the highest-ranking caste, account for a similar proportion. The rest of the population are members of a wide range of intermediate castes.

Untouchability was officially abolished at independence, but the caste system has proven difficult to dismantle. Many Dalits have chosen to convert to other religions, most notably Buddhism. Successive Indian governments have now introduced an elaborate system of affirmative action on behalf of *Scheduled Castes* (the official government label for Dalits). This effort has had more effect in urban than in rural areas of India. In any case, Dalits now have reserved for them places in the schools, a fixed percentage of State and federal government jobs, and a quota of seats in national and State legislatures. These jobs are often highly desirable because they tend to be white-collar and much better paid than those generally available to Dalits.

Just how far India's political pendulum can swing was demonstrated in the 2007 provincial legislative election in Uttar Pradesh State, where a Dalit party won an absolute majority and where a woman named Mayawati Kumari was the first Dalit to become the chief minister of India's most populous State. It was a stunning victory that made headlines throughout the country and shook up the political establishment, revealing the growing power of the lowest castes and marking a turning point for India's representative government.

Most lower-caste Indians are still faced with limited opportunities, widespread discrimination, and extreme forms of poverty, particularly in less accessible rural areas. In traditional India, caste provided stability and continuity; in contemporary India, it constitutes an often painful and difficult legacy.

Economic Geography

The most commonly cited, and most clearly evident, regional division of India is between north and south. The north, and especially the Ganges Valley, is India's heartland, where Hinduism was born (as was Buddhism), and where Indo-European languages (mostly variations of Hindi) are spoken. In the Dravidian south, an entirely different language family prevails, and as a second language, southerners prefer English over Hindi. In the south, too, people tend to have darker skin, the food gets spicier, and the climate is more tropical.

But there is another, potentially more significant, divide across India. In **Figure 9-19**, draw a line from Lucknow on the Ganges River south to Chennai on the southeastern coast. To the west of this line, India's economy is stronger and more modern, and the map shows that incomes are higher. To the east, India has more in common with the less promising economies of Bangladesh and Myanmar (Burma) that also face the Bay of Bengal. India's heavy industries built by the state near Kolkata in the 1950s are increasingly outdated, uncompetitive, and in steady decline. The State of Bihar represents the stagnation that afflicts much of India east of our line: by several measures, it ranks among the poorest of the 29 States.

Now compare this situation to western India's. The State of Maharashtra, the hinterland of Mumbai, leads India in many categories, and Mumbai leads Maharashtra. Many smaller private industries have taken root here, manufacturing goods that range from umbrellas to satellite dishes and from toys to textiles. Across the Arabian Sea lie the oil-rich economies of the Persian Gulf countries. Hundreds of thousands of workers from western India have found jobs there, sending remittances back to families from Punjab to Kerala. More important, several have used their foreign incomes to establish service industries back home. Outward-looking western India, in ever sharper contrast to the inward-looking east, is well in the lead in strengthening linkages to the global economy. For instance, satellite and fiber-optic cable links have driven Bengaluru (formerly Bangalore) to become the center of a burgeoning software-producing complex that reaches world markets.

Manufacturing and IT Industries

The industrial geography of India is in part a legacy of colonial times, with coastal Mumbai, Kolkata, and Chennai anchoring major industrial zones, and textiles—the entry-level industry of disadvantaged countries—dominating the manufacturing scene (Fig. 9-19). A number of indus-

FIGURE 9-19

© J. Nijman, P. O. Muller, and John Wiley & Sons, Inc.

tries, such as steel, machinery, and building materials, have become much stronger, and some leading companies have already become major global players. For example, ArcelorMittal is the biggest steel company in the world and effectively run and owned by Indians (the CEO is Lakshmi Mittal); the Tata Steel Group, also Indian, ranks eighth in the world. It should be noted, though, that these are now very much multinational companies with much of their production and employment located outside India.

India's information, communication, and technology (ICT) services, clustered in metropolitan Bengaluru, Delhi, Hyderabad, and Mumbai, draw much international attention. Leading Indian software companies such as Infosys, Wipro (see Box 9-4), and Cognizant are now household names all across urban India. The growth of software and ICT services has been spectacular and now accounts for about 8 percent of GDP and no less than one-quarter of merchandise exports. But this sector employs only about 2 percent of the workforce.

What India needs far more, in terms of employment growth, is manufacturing industries that competitively sell goods in the global marketplace, putting tens of millions to work and transforming the economy. Clearly, India is very different from China. The two may have comparably sized populations, but the Chinese economy employs about ten times as many workers in manufacturing and has a much larger urban middle class. India may have experienced spectacular recent

growth, but too much of it is concentrated in the IT sector and benefits only a small, highly educated urban elite.

Life Is in the Harvest   Agriculture provides more jobs in India than any other employment sector, and India's fortunes (and misfortunes) remain strongly tied to farming. It is estimated that nearly two-thirds of the population still lives on (and from) the land, spread in and around the country's hundreds of thousands of villages. There, traditional farming methods persist, yields per unit area remain among the world's lowest, and malnutrition still continues to afflict millions even as grain surpluses accumulate. It should be noted that agricultural productivity can vary substantially from region to region; for example, the rich wheat lands of the Punjab stand in contrast to the drought-prone millet farms in eastern Maharashtra.

As everywhere, food production is closely tied to climate and physiography, and farm output varies significantly across the country. In India, the monsoon is absolutely critical, its abundance and timing a reliable predictor of the harvest. **Figure 9-20** maps India's agricultural geography and reflects the rainfall patterns and monsoonal cycle depicted in Figure 9-3. Rice dominates all along the Arabian Sea-facing southwestern coast (on the rainy side of the Western Ghats) and in the monsoon-drenched peninsular northeast; where drier conditions develop, wheat and other grains prevail.

INDIA: AGRICULTURE
DOMINANT CROP ZONE

Rice   Millet

Wheat   Plantation

Cotton   Chickpea

Groundnut   Shifting Cultivation

Corn   Coconut

0 100 200 300 400 500 600 700 Kilometers

0 100 200 300 400 Miles

© J. Nijman, P.O. Muller and John Wiley & Sons, Inc.

FIGURE 9-20

Urbanization

India is very well known for its enormous cities, but the country is not yet an urbanized society. Only about 34 percent of the population currently lives in cities and towns—compared to an average of around 80 percent across the developed world. In absolute terms, however, urbanization in India is a massive phenomenon. People by the hundreds of thousands are arriving in the already overcrowded cities, swelling urban India by about 3 percent annually—twice as fast as the country's overall population growth. Not only do the cities attract as they do everywhere, but many villagers are also driven off the land by desperate conditions in the countryside. As these migrants manage to establish themselves in bigger and smaller cities, they help their relatives and friends to join them in squatter settlements that often are populated by newcomers from the same locality, bringing their language and customs with them and cushioning the stress of relocation.

Figure 9-18 displays the distribution of major urban centers in India. Except for Delhi, the biggest cities have coastal locations: Kolkata dominates the east, Mumbai the west, and Chennai the south. The overriding influence of these coastal cities is a colonial legacy. But urbanization also has expanded in the interior, notably within the core area (symbolized by Delhi overtaking Mumbai in 2011 to become India's largest city). In 2016, the country had more than 50 metropolitan areas containing populations of at least one million—and an urban population exceeding 400 million (well above the U.S. population total).

When you arrive in any Indian city, you are struck by the abundance of small shops everywhere—tiny businesses wedged into every available space in virtually every nonpublic building along every street. Even the upper, walk-up floors are occupied by shops, their advertising signs suspended from windows and balconies. What keeps all these small stores in business? Most of them earn very little and can afford to stay open only because they are part of the **informal sector [28]**—they are essentially unregistered, pay no taxes, utilize family labor, have been handed down through generations, and keep going because of the strong social and communal networks that characterize India's societal fabric. Moreover, high densities and the relatively low level of spatial mobility of many urbanites combine to create substantial demand for the local availability of daily goods and services.

This bustling retail scene notwithstanding, change is now under way throughout urban India. The country's growing consumerism, propelled by a rapidly expanding middle class

Even though the country produces an ample variety of crops, yields usually are disappointingly low as a result of inefficient land ownership and small farmers' lack of access to inputs such as fertilizer, irrigation equipment, and machinery. If there is any to sell, getting produce to market is yet another struggle for millions of farmers. Many of India's 600,000-plus villages cannot be accessed by truck and in this era of modern transportation, animal-drawn carts are still estimated to outnumber cars nationwide.

To meet these challenges, India is now engaged in an unprecedented series of infrastructural improvement projects, ranging from highways to state-of-the-art airports. The most ambitious of these mega-projects is the construction of a 5900-kilometer (3700-mi)-long superhighway known as the Golden Quadrilateral. Completed in 2012, this fifth-longest highway on Earth interconnects the four anchors of the Indian urban system (Delhi, Mumbai, Chennai, and Kolkata) in a gigantic circuit that also passes through 15 other major cities (Figs. 9-18, 9-19). The impacts of this all-important artery are multiple: it is expanding urban hinterlands; commuters are using it to travel farther to work than ever before; several once-remote rural areas now have a link to markets; and it is accelerating the rural-to-urban migration flow that will continue to transform India well into the future.

"As I set foot into this spacious (and partially air-conditioned) mall from the steamy and crowded Andheri Link Road in Mumbai, I wonder: is this the future in a land of small family-owned shops where 'retail' has a different and historic meaning? Infinity Mall, an upscale shopping center opened in 2005 in suburban Mumbai, caters to the megacity's new middle class. My Indian friend Pankaj thinks this is all for the better. 'On the weekends my wife and I often come here. There is a nice food court and the shopping is great. These are the things you Americans want, and these are the things we Indians want.' Malls have sprung up across India's major cities in the 2010s, transforming both urban landscapes and consumer behavior."

estimated to number almost 400 million in 2018, has led to the introduction of a new feature on the landscape of major cities—the shopping mall (**Box 9-10**). As recently as 2000, India did not have a single shopping center. But only five years later the 100th had opened its doors, and by 2019 the total exceeded 700. You will see American fast-food restaurants among the establishments in these malls as well as the brand names of numerous other foreign companies, proving that the tide of globalization has already washed up on India's shores.

Nonetheless, if the 28 percent of the Indian population that makes up the middle class can afford to shop at a mall, the remaining one billion people cannot. Can India's ongoing economic transformation run its course with such inequalities?

Region Bangladesh

On the map of South Asia, Bangladesh looks like another State of India: the country occupies the area of the **double delta [29]** of India's great Ganges and Brahmaputra rivers, and India almost completely surrounds it on its landward side (**Fig. 9-21**). But Bangladesh is an independent country, born in 1971 after its brief war for independence against Pakistan, with a territory about the size of Wisconsin. Today, this country of 165 million is still largely a nation of subsistence farmers and among the poorest in the world. Only a third of the population lives in urban settlements, and more than half of the workforce is engaged in agriculture. Dhaka, the megacity capital, is home to over 17 million; the cities of

FIGURE 9-21

Chittagong, Rangpur, Khulna, and Rajshahi are the only other urban centers of consequence. Bangladesh has one of the highest physiologic densities (people per unit area of arable land) in the world.

The country's relations with neighboring India have at times been strained over water resources (India controls the Ganges River, which is the lifeline of Bangladesh), cross-border migration (about 8 percent of the population is Hindu), and transit between parts of India across Bangladesh's north.

A Vulnerable Territory

Not only is Bangladesh economically challenged; it also is highly susceptible to damage from natural hazards. During the twentieth century, two of the deadliest weather disasters in the world struck this small country, both caused by cyclones (as hurricanes are called in this part of the world). The greatest disaster in modern times was triggered by Cyclone Bhola in 1970, which killed half a million (!) of the country's inhabitants. The other of the two occurred in 1991 when a cyclone killed more than 150,000 people. Smaller cyclones strike several times every year and almost routinely kill anywhere between dozens and thousands of people.

The reasons for Bangladesh's vulnerability can be deduced from Figure 9-21. Southern Bangladesh consists of the vast deltaic plain of the combined Ganges–Brahmaputra river system, integrating extremely fertile alluvial (river-deposited silt) soils that attract farmers with the low elevations that endanger them when the water rises. Moreover, the shape of the Bay of Bengal forms a funnel that steers cyclones and their storm surges of wind-driven water to barrel into the double delta's coast. Without money to build seawalls, floodgates, elevated shelters in sufficient numbers, or even adequate escape routes, hundreds of thousands of people are at risk—sometimes with deadly consequences.

It is not just the exposure to cyclones that threatens Bangladesh: with so much of the country located in this enormous double delta, it is prone to *river* flooding during the summer monsoon. Heavy monsoon rains in 2017 swelled the rivers, causing them to overflow and flood a huge part of the country. At least 100 people died, millions of people were affected, and there was major damage to agricultural lands and infrastructure.

Limited Opportunities, Creative Development Strategies

Even though geography seems to offer Bangladesh limited options, some notable successes have been achieved here in recent years. First, since 1990, life expectancy in Bangladesh has increased by ten years and now exceeds that of India, despite India's considerably higher incomes. Second, primary school enrollment among girls today is double that of 2010, and the female literacy rate is steadily improving. And third, since 1990 infant (< 1 year) mortality rates have been more than halved, and child (< 5 years) mortality rates as well as maternal mortality rates have dropped by about three-fourths. According to these indicators, Bangladesh outperforms India and also fares much better than Pakistan.

Interestingly, these improvements were achieved with a relatively small economy and low incomes—even though, as Figure 9-9 showed, the country's economic growth in the last 15 years has outstripped that of Pakistan and is around the same level as Sri Lanka and Bhutan. Bangladesh is an overwhelmingly Muslim society, but in some ways it has also been a very progressive one, particularly when it comes to the role of women. For example, 30 seats in the national legislature are reserved for women. Much of this success is attributed to the role of **nongovernmental organizations (NGOs) [30]** that, independently of (but supported by) the national government, have promoted programs to improve the quality of life. An important example concerns family planning programs in which women play the lead role (free contraceptives and education). The fertility rate in Bangladesh fell rapidly from 7.0 in 1970 to 3.4 in 1994 to 2.1 in 2018. Consider this: when Pakistan and Bangladesh split in 1971, each had a population of around 65 million; today, their populations are, respectively, 221 and 165 million.

The NGOs also have been the key providers of so-called **micro-credits [31]**, small loans at favorable terms to the poor, along with guidance and counseling, allowing them to invest in the means to secure a proper and sustainable livelihood. Frequently, such loans have gone to small farmers to buy a piece of land or construct a farmhouse, or to starter entrepreneurs to buy machinery or transport equipment. The success of micro-credits has spread from Bangladesh around the global periphery and is now widely considered to be a key element in formulating economic development strategies.

Over the past several decades, Bangladesh has become known as the "garment capital of the world," today accounting for nearly three-quarters of all export earnings. Almost half of all industrial workers are employed in this industry, and most are women. Many of the cheaper items of clothing that are sold worldwide under popular labels from big-name marketers are made in factories in Bangladesh that are unsafe and where the minimum wage is little more than one U.S. dollar a day. Western companies impose strict production quotas and deadlines, and factory managers force their employees to work long hours, sometimes under hazardous conditions. There are no unions to protect workers; disastrous fires and building failures have made the international news (e.g., the 2013 collapse of a factory building in the Dhaka suburb of Savar that killed more than 1300 laborers). Globalization has made Bangladesh's garment industry a valuable contributor to the nation's commercial economy, but those low prices you see advertised in Western stores raise painful questions about what constitutes "fair trade."

While Bangladesh is praised by many for its progressive social and economic programs, it is at the same time confronted by the rising militancy of Islamic fundamentalists. In the past few years, there have been dozens of assassinations of liberal politicians, gay activists, academicians, and journalists who had been vocal about democratic freedoms and what could be perceived as criticism of fundamentalist Islamic organizations. It is not clear whether the perpetrators are associated with Al Qaeda or ISIS, or whether they belong to isolated local terrorist cells. Some observers, and many liberal Bangladeshis along with them, worry that the country's progressive course is at risk and that Bangladesh may be going down the same violent path as Pakistan.

Region The Northern Mountain States

As Figure 9-15 and **Figure 9-22** show, two landlocked countries lie in the central sector of the mountainous zone that walls India off from China: *Nepal* and *Bhutan*. Together they constitute a distinct region of South Asia, but both are relatively small (Nepal is roughly the size of Illinois, and Bhutan only one-quarter of that). Importantly, these countries function geopolitically as buffer states between the two Asian giants.

Nepal

Culturally and (until recently) economically, Nepal lies within India's sphere of influence, but China's presence here is being felt ever more strongly. Maoist-communist groups are supported by China but loathed by India, resulting in fractious and

© J. Nijman

Bhaktapur is a small town a few miles east of Kathmandu, known for its fantastic medieval architecture, temples, and royal buildings from the time this was the capital of the Malla Kingdom from the fifteenth to the eighteenth century. The picture, taken in February 2019, shows Durbar Square (the central town square), with the palace to the left; to the right is the ongoing reconstruction of Dattatraya Temple after it was partially destroyed in the 2015 earthquake; and in the middle, on top of the pillar, a statue of the historic King Bhupatindra Malle, facing the palace. Many more buildings, old and new, were damaged in the earthquake, and recovery has been painfully slow.

contentious governance. China's economic clout, as in other parts of the world, has been growing in Nepal.

Located just to the northeast of India's Hindu core area, Nepal has a population of 30 million, of whom 82 percent are Hindu. Nepal's Hinduism, however, is a unique blend of Hindu and Buddhist ideals. Thousands of temples and pagodas,

FIGURE 9-22

ranging from the simple to the ornate, grace the cultural landscape, especially in the Valley of Kathmandu, the country's core area. The Nepalese are a people of many origins, including India, Tibet, and interior Asia. Even though well over a dozen languages are spoken, most of the people also speak Nepali, a language related to Indian Hindi.

Closer examination of Figure 9-22 shows that Nepal contains three geographic zones: a southern, subtropical, fertile lowland called the Terai; a central belt of Himalayan foothills with swiftly flowing streams and deep canyons named the Mahabharata Range (or Lesser Himalayas); and the spectacular high Himalaya itself, topped by Mount Everest in the north. The capital, Kathmandu, lies in the east-central part of the country in an open valley of the central hill zone. Most of the population is concentrated in the foothills and the Terai.

The country's soaring Himalayan peaks are a world-renowned attraction for tourists and especially for mountain climbers and hikers, but the revenues from tourism are modest. This is a poor country with the realm's second-lowest GDP per capita after Afghanistan. Many young people go abroad to find work, and Nepal depends heavily on remittances. It is also faced with environmental degradation, overused farmlands, soil erosion, and deforestation.

It does not help to be located in a geologically volatile part of the world (Fig. 9-2). In 2015, Nepal suffered a massive earthquake that killed nearly 9000 people, injured thousands more, caused major material destruction, and left nearly three million people homeless (photo page 345). The government proved poorly equipped in the rescue and rebuilding efforts, which were largely taken over by foreign relief agencies.

Bhutan

Mountainous Bhutan, with a population of 775,000, lies wedged, fortress-like, between India and China's Tibet (Figs. 9-1 and 9-22). Bhutan has a long history as a constitutional monarchy, ruled by a king whose absolute power was unquestioned by his subjects. But in 2007, the newly crowned monarch, who had just succeeded his father—and perhaps with an eye on the recent overthrow of the royal rulers in nearby Nepal—decided to order his subjects to vote for a political party in an abruptly created democracy. Thus, Bhutan overnight went from an absolute monarchy to a multiparty democracy, and Thimphu, the capital, became the seat of a newly elected National Assembly. Throughout the mountainous countryside, the symbols of Buddhism, the state religion, dominate the cultural landscape, and the government's development policies emphasize the importance of spiritual fulfillment alongside the satisfaction of material needs.

Forestry, hydroelectric power, and tourism all have much potential here, and Bhutan also possesses ample mineral resources. The country did not open up to foreign visitors until the 1970s, and even today tourism is kept within strict limits. Anyone wishing to travel to Bhutan is required to spend a minimum amount of money per day/night in designated accommodations. Although isolation and inaccessibility continue to preserve traditional ways of life, Bhutan is not immune to the turbulent social forces that permeate this realm. Tensions between the Buddhist Bhutanese majority and a Nepalese Hindu minority have led many Nepalese Hindus to flee across the sliver of India that separates the two mountain states (Fig. 9-22). To date, the governments of Bhutan and Nepal have been unable to work out a solution to this problem.

Bhutan's **Gross National Happiness Index (GNHI) [32]** was first created by the King of Bhutan in 1972. It is conceived and designed to measure well-being in ways that go beyond economic measures such as GDP per capita and to give equal importance to noneconomic aspects such as living standards, health, education, time use, psychological well-being, environmental conditions, and governance. Ever since, the Happiness Index has been replicated or imitated in other parts of the world, and the UN came up with a Human Development Index that also combines economic and noneconomic dimensions of well-being. Bhutan's GNHI was not meant to compare Bhutan to other countries or for worldwide comparisons—its purpose was and is to assess the development and well-being of Bhutan over time and across its populations and provinces. Environmental conditions and natural amenities are an important part of it—as you might expect in a country so richly endowed with natural beauty (**Box 9-11**). While it is clear that happiness will always be difficult to measure, the introduction of the GNHI constitutes an important contribution and reminder of the multifaceted nature of human well-being.

Region The Southern Islands

As Figure 9-15 shows, South Asia's subcontinental landmass is flanked by several sets of islands. The largest, *Sri Lanka*, lies just to the southeast of the southern tip of India. *The Maldives* form an archipelago farther out in the Indian Ocean to the southwest of southernmost India. And the Andaman and Nicobar Islands (which will not be discussed here), a Union Territory of India, mark the southeastern edge of the Bay of Bengal.

The Maldives

The Maldives consists of some 1200 tiny islands whose combined area is just 300 square kilometers (115 sq mi) and whose highest elevation is less than 2 meters (6 ft) above sea level. The Maldives' low elevation has been singled out repeatedly in assessments of the future impact of rising sea levels caused by global warming (see **Box 9-12**). The population of 540,000 drawn from Dravidian and Sri Lankan sources is now virtually 100 percent Muslim, with one-quarter of it concentrated on the island of the capital, named Maale.

This country actually has the realm's highest GNI per capita, thanks to the Maldivians having translated their palm-studded,

Box 9-11 From the Field Notes . . .

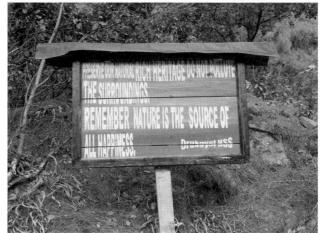

© J. Nijman

"I am on a day hike from the town of Paro (about 15 miles west of Thimphu; see Fig. 9-22) to 'Tigers Nest,' one of Bhutan's famous Buddhist monasteries up in the mountains. I am about halfway there when the monastery comes into view. It looks spectacular, unreal. To the Bhutanese, it is known as Paro Taktsang, and it dates from the seventeenth century—it must have been incredibly hard to build, precariously perched on steep cliffs about 900 meters (3000 ft) above the Paro Valley. The monastery was built there because of the nearby caves, where the founder of Buddhism in Bhutan, Guru Padmasambhava, was said to have meditated in the eighth century. The legend has it that he flew to the caves on the back of a tiger, from Tibet, hence the name Tigers Nest. On the way to the monastery, I come across various signs reminding bypassers of the importance of cleanliness and respect for nature—important ingredients of the Bhutanese notion of happiness."

beach-fringed island chain into a tourist mecca that annually attracts tens of thousands of mainly European visitors. The Maldives, however, also faces social and political challenges—its nascent democracy was dealt a blow in 2012 when the first freely elected president, Mohamed Nasheed, was deposed in a *coup* engineered by the country's former dictator and radical Islamists. In the spring of 2019, after several years of exile, Nasheed had been able to make a comeback with a landslide win in new elections, which put his Maldives Democratic Party firmly back in power (for the time being, at least).

Sri Lanka: A Fragile Ethnic Mosaic

Sri Lanka (known as Ceylon before 1972), the compact, pear-shaped island located just 35 kilometers (22 mi) across the Palk

Box 9-12 From the Field Notes . . .

"We have just taken off from the Gan airstrip on Addu Atoll, in the southern Maldives. Within seconds, the neighboring island of Villingi comes into full view. It is a resort island, one of many that are vital to the economy of this 'tropical paradise.' Consisting of about 1200 islands with an average elevation of less than 1.5 meters (4 ft 11 in), the Maldives have also been labeled 'ground zero' in debates about global warming and rising sea levels. At the moment it is low tide, but you can see how far inland the water reaches at high tide. With stormy weather, flooding is guaranteed. I just spent three days here, and people told me that they face growing problems of beach erosion and soil salinity, and that fish are steadily migrating to greater depths farther from the islands. Given the ominous forecasts of global sea-level rise, this may all be gone (or at least depopulated) by the end of the century."

© J. Nijman

Strait from southernmost India (see Fig. 9-15 and **Fig. 9-23**), became independent from Britain in 1948. There were good reasons to create a separate sovereignty for Sri Lanka. This is neither a Hindu nor a Muslim country: just over 70 percent of its population of 21 million are adherents of Buddhism.

The majority of the Sri Lankan population is descended from migrants who arrived at the island from northwestern India beginning about 2500 years ago. Most of them probably walked across to the island, which was connected to southernmost India by a land bridge (now submerged beneath the Palk Strait) known as *Rama's Bridge*. If you look closely at Figure 9-1, you can see the Sri Lanka–end remains of that land bridge, which was further eroded by a powerful cyclone in the fifteenth century. The migrants brought with them the advanced culture of their source area, building towns and irrigation systems, and introducing Buddhism. Today, their descendants are known as the Sinhalese and speak Sinhala, which belongs to the Indo-European language family of northern India (Fig. 9-5).

The Tamil-speaking Dravidians, who lived on the mainland side of the Palk Strait, arrived much later. During the nineteenth century, the British colonizers brought hundreds of thousands of Tamils to the island to labor on their tea plantations, and they soon became a substantial component of then-Ceylonese society. Not only did the Tamils bring their Dravidian language to the island—they also introduced their Hindu faith. At the time of independence, one year after the India–Pakistan partition, this minority constituted more than 15 percent of the island's population; today, they comprise just over 11 percent.

When Ceylon became independent, it was one of the great hopes of the postcolonial world. The country had a sound economy as well as a democratic government, and it was renowned for its tropical beauty. Its reputation soared when a massive effort succeeded in eradicating malaria, which was followed by a family-planning campaign that reduced the birth rate while the rest of South Asia was experiencing a substantial population explosion. Rivers from the cooler, forested,

© J. Nijman, P. O. Muller, and John Wiley & Sons, Inc.

FIGURE 9-23

interior highlands fed the paddies that provided ample rice; commercial crops from the moist southwest paid most of the bills, and the capital, Colombo, grew to reflect the optimism that prevailed. Economically, Sri Lanka is better off than most of the rest of South Asia, with its GDP per capita second only to that of The Maldives.

But the island's political multicultural challenges have been very substantial. Sri Lanka's Tamil minority soon began proclaiming its sense of exclusion, demanding better treatment from the Sinhalese majority. Tensions steadily escalated until a full-scale civil war broke out in 1983. Militant activist leaders in the Tamil community demanded a separate Tamil state encompassing the north and east of the country, coinciding with the Tamil-speaking areas mapped in stripes in the inset map of Fig. 9-23. This violent, devastating conflict endured for 26 years (with a death toll estimated as high as 100,000), finally ending in 2009 when the Sinhalese-dominated government declared victory. In the still bitter aftermath, northernmost Sri Lanka was permitted to have a provincial government dominated by the leading Tamil party—but the central government in Colombo maintains very tight sovereign control.

In recent years, Sri Lanka's ethnic mosaic was put under further strain. First, in 2018, hardliners in the Sinhalese Buddhist community fueled a surge of anti-Muslim violence, not unlike what happened in Myanmar in recent years (see Chapter 11). The government had to declare a state of emergency following mob attacks against Muslim businesses and neighborhoods. Second, on Easter Sunday 2019 the country experienced a wave of horrendous attacks against Christian churches and international hotels, killing more than 300 worshippers and wounding hundreds more. This happened in Colombo, the capital city, which has a mix of ethnicities and religions and is not dominated by any one group in particular (Fig. 9-23). The attacks were attributed to a Muslim jihadist group, and within a few days it was applauded internationally by ISIS leaders. Sri Lanka, this naturally beautiful and bountiful island, has its political challenges spelled out.

Sinhalese Buddhist monks visit the famous ruins of the twelfth-century Hatadag Temple in the ancient city of Polonnaruwa in north-central Sri Lanka. The temple that once stood here was also known as the "tooth relic temple" because it was said to have treasured an actual tooth of Prince Siddhartha—the Buddha.

Points to Ponder

- The most skewed sex ratios in the population tend to occur in the most prosperous parts of South Asia.
- According to population growth projections, by 2030 India will be home to the largest national population of Muslims in the world.
- India's ICT sector accounts for one-quarter of all exports but employs barely 2 percent of the national workforce.
- Does the United States have the right to conduct drone strikes on other countries' territories, such as Pakistan or Iraq, without consent of those governments?
- South Asia's population is likely to be more severely affected by global climate change than that of any other world realm.

FIGURE 10-1

Variations of urbanization in the East Asian realm: to the left, the sprawling outskirts of the fast-expanding city of Ulaanbaatar, Mongolia; to the right, a view of Shenzhen, China. In Mongolia, urbanization often implies a shift from nomadic to sedentary lifestyles but many continue to live in their *gers* (the white round tents visible on the photo), ready to move elsewhere if the need arises. At the other end of the spectrum is Shenzhen, which grew from a handful of small villages in 1980 to a high-rise metropolis of 13 million people in 2019.

The East Asian Realm

IN THIS CHAPTER

- The Belt and Road Initiative
- Japan's aging population
- China's megacity regions
- At Korea's Demilitarized Zone
- Sea-level rise in the Pearl River Delta
- Mongolia's urbanization challenges
- The Chinese makeover of Lhasa, Tibet

CONCEPTS, IDEAS, AND TERMS

Tsunami **[1]**	Gender imbalance **[10]**	Special Economic Zone (SEZ) **[20]**
South-to-North Water Diversion Project **[2]**	Gender pay gap **[11]**	Overseas Chinese **[21]**
	Intercontinental ballistic missile **[12]**	Regional growth pole **[22]**
Dynasty **[3]**	Buffer state **[13]**	Foreign direct investment **[23]**
Sinicization **[4]**	Belt and Road Initiative **[14]**	China–Pakistan Economic Corridor **[24]**
Hanification **[5]**	Floating population **[15]**	
Asian Tiger **[6]**	*Hukou* system **[16]**	Regional complementarity **[25]**
High-value-added goods **[7]**	Megacity region **[17]**	State capitalism **[26]**
Dependency ratio **[8]**	Urban system **[18]**	Conurbation **[27]**
One-Child Policy **[9]**	High-speed rail (HSR) **[19]**	Technopole **[28]**

East Asia is a geographic realm like no other. At its heart lies the world's most populous country, the product of what may be the world's oldest continuous civilization. On its Pacific mainland shores an economic transformation has taken shape with no parallel in world history. Its offshore islands witnessed the first use of atomic weapons on civilian populations and the postwar emergence of one of the world's most powerful economies. Few lives in this world were left unaffected, directly or indirectly, by the momentous events that have occurred in East Asia over the past two generations. Just look around you. Chinese-assembled smartphones (and so much more), Japanese-made automobiles, South Korean televisions, Taiwanese computers—from toys to textiles and from semiconductors to software, East Asian products fill streets and stores, homes and hotels.

Defining the Realm

It has all happened with astonishing speed. About half a century ago, you were no more likely to find anything useful made in East Asia than you were to buy anything from Russia. By the early 1970s, things began to change rapidly. Japan led the way, turning its 1945 World War II defeat into an economic triumph in less than three decades. By the mid-1970s, that country's remarkable economic growth— compared to China's seemingly total stagnation—appeared to justify Japan's recognition as a discrete geographic realm, an economic engine for the world, and a strong competitor for the top-ranked U.S. economy. People began referring to the Yen Bloc, the East and Southeast Asian part of the world dominated by Japanese trade. On a smaller scale, Hong Kong, South Korea, and autonomous Taiwan soon followed Japan's example, showing the world what other peoples in East Asia were capable of.

All this was prologue to the main event—the unprecedented, remarkably swift, and massive rise of China. Less than five years after the 1978 implementation of its game-changing open-door and market reform policies, this gigantic country had taken center stage in the realm's economy. Since the 1990s, led by burgeoning China, East Asia has emerged as the most dynamic realm in the global economy. Today, it continues to reshape the world—and in the process is itself undergoing a profound transformation.

Geographic Dimensions

As **Figure 10-1** indicates, the East Asian geographic realm forms a roughly triangular wedge between the vast expanses of eastern Russia to the north and the populous countries of South and Southeast Asia to the south, its edges often marked by high mountain ranges or remote deserts. The darker brown on the map designates the highest mountains and plateaus, which create a vast arc north of the Himalaya before bending southward and becoming lower (tan shading) toward Myanmar, Laos, and Vietnam in Southeast Asia. Here in the southwest, where Tibet is located, mountains and plateaus alike are covered by permanent ice and snow, the soaring ranges crumpled up like the folds of an accordion. Three major rivers, their valleys parallel for hundreds of kilometers, disclose the orientation of this high-relief topography. Northward, note how rapidly the mountains give way to broad, flat deserts whose names appear prominently: the Takla Makan in the far west, the Gobi where China meets Mongolia, and the Ordos in the embrace of what looks like a huge bend of a river we will learn more about shortly, the Huang He (Yellow River).

The green areas on the map, which have the lowest relief and (often) the most fertile soils, are home to the vast majority of this realm's population. Here the great rivers that come from the melting ice and snow in the interior highlands have been depositing their sediment load for eons, and when humans domesticated plants and started to grow crops, this was the place to be. That was thousands of years ago—perhaps as long as 10,000 years—and ever since, this has been the largest human cluster on Earth.

The East Asian realm is not confined to the mainland of mountains and river basins. You can imagine how its offshore islands became populated: the Korean Peninsula seems to form part of a bridge pointing toward the southernmost large island (Kyushu) of what is today Japan, and from there it seems likely that the early migrants moved farther north until they reached Hokkaido. In warmer times, they may even have ventured beyond, onto the Kurile Islands. And in the south, Taiwan lies even closer to mainland China than Japan does to Korea, whereas near-tropical Hainan Island, the realm's southernmost extremity, is almost—but not quite—connected to the small peninsula that stretches toward it.

In terms of total land area, East Asia is mostly mainland— but the islands and their peoples have played leading roles in forging this realm's regional geography. The waters between mainland and islands (the Taiwan Strait, the South China Sea, the East China Sea, the Yellow Sea, and the Korea Strait, among others) also figure prominently in the geographic evolution of this realm. Today, the Japanese and the Chinese are arguing over the ownership of small islands in these waters, specks of land surrounded by possible oil reserves and fishing grounds claimed by both sides. So this realm's map is considerably more complicated than that wedge-shaped triangle in Figure 10-1 initially suggests—including even the spellings of many of its parts (**Box 10-1**).

China's Dominance

It is all too easy to refer to China when you mean East Asia because China is East Asia's dominant country, contains more than 85 percent of the realm's population, and has commanded

Box 10-1   Major Geographic Features of East Asia

1. East Asia is encircled by snowcapped mountains, vast deserts, and Pacific waters.

2. East Asia was one of the world's earliest culture hearths, and China is one of the world's oldest continuous civilizations.

3. This is the second most populous geographic realm after South Asia; its population remains heavily concentrated in its eastern regions.

4. The People's Republic of China (PRC), the world's largest state demographically, is the current rendition of an empire that has expanded and contracted, fragmented and unified many times during its long existence.

5. Key components of China's sparsely peopled western regions are now being rapidly developed—not only because they are strategically important to the state but also because they lie exposed to minority pressures and Islamic influences.

6. The post-Mao economic transformation launched four decades ago along China's eastern seaboard has in places

expanded westward, accompanied by a major in-migration of ethnic (Han) Chinese.

7. Widening regional disparities—exacerbated by breakneck, massive-scale urbanization—are straining Chinese society and challenging the government which officially still espouses a communist ideology.

8. Japan's economy is ranked among the world's wealthiest, even though its economic performance in recent decades pales in comparison to that of China.

9. Geopolitically, this realm is home to the world's newest superpower as China's economic and political influence is increasingly projected beyond East Asia.

10. Mongolia has long been a "buffer state" situated between China and Russia, but in recent years Chinese investments and engagements have pulled it more into China's orbit.

11. East Asia contains a number of geopolitical flashpoints capable of generating conflict, including North Korea, the PRC–Taiwan relationship, and several contested island groups.

an increasingly prominent role on the global stage. But there are five other political entities on East Asia's map: Japan, South Korea, North Korea, Mongolia, and Taiwan. Note that we refer here to *political entities* rather than *states*. In this realm, the distinction is important. Taiwan refers to itself as the Republic of China (ROC), but it is not recognized as a sovereign state by most members of the international community; the communist administration in Beijing, capital of the People's Republic of China (PRC), regards Taiwan as part of China and as a temporarily wayward province. And North Korea is widely viewed as a rogue state, as an archaic dictatorship that has profoundly failed its people. Having compiled one of the world's most dreadful human rights records, North Korea is not even a fully functional member of the United Nations.

Nevertheless, China is the realm's predominant entity—demographically, economically, and politically. It is important to keep in mind, as you read this chapter, that portions of what we map today as regional components of China were not part of the country in the past, and that other areas now lying outside China are regarded by many Chinese as China's property (for example, a large sector of the Indian State of Arunachal Pradesh and portions of the Russian Far East). China's imperial past saw the state expand, contract, and then expand again, accumulating unsettled border issues on land as well as at sea. On issues like these, Chinese national sentiments can run quite deeply.

Environment and Population

To understand the complex physical geography of East Asia mapped in Figure 10-1, it is useful to refer back to Figure 1-5 in the introductory chapter. The high, snowcapped mountain

ranges of the realm's southwestern interior result from the gigantic collision of the Indian and Eurasian tectonic plates (also see Fig. 9-2), pushing the Earth's crust upward and creating not only the mountain ranges of which the Himalaya is the most famous, but also popping up the enormous, domelike

Satellite image of the heavily damaged Fukushima Daiichi nuclear power complex, three days after the catastrophic Tohoku earthquake and tsunamis struck northeastern Japan in 2011. This strongest earthquake ever recorded in the country, together with the tsunami flooding, triggered the meltdown of three of the facility's six nuclear reactors and a series of explosions and fires, in turn causing major radioactive contamination. 140,000 people were evacuated. Eight years later, the clean-up had progressed but still was not finished. Some wary residents have returned, but others permanently left. Today, Japan still relies heavily on nuclear energy and, following a major program of safety upgrades across the country, most facilities had resumed operations by 2018.

DigitalGlobe/Getty Images

Qinghai–Xizang (Tibetan) Plateau. In Figure 1-5, note the high incidence of earthquakes associated with this collision, converging on a narrow zone of instability that crosses southwestern China and stretches into Southeast Asia.

From Figure 1-5 it is also obvious that the Pacific Ring of Fire, with its lethal combination of volcanism and earthquakes, endangers Japan far more than it does China. This vulnerability became a reality on March 11, 2011, when a monstrous 9.0-magnitude earthquake struck off Japan's northeastern coast near the city of Sendai. **Figure 10-2** displays the configuration of the underlying tectonic plates that converge on Japan. The Pacific Plate moves westward at an average of 7–10 centimeters (3–4 in) per year, and the only way it can do so is by pressing forward (or subducting) *beneath* the North American Plate (much of Japan sits atop the western tip of the former). It was the most powerful Japanese earthquake ever recorded and one of the five biggest in the world since 1900. Even worse, this massive temblor and its torrent of violent aftershocks triggered a series of devastating **tsunamis [1]** (seismic sea waves) that swept across the narrow, densely populated coastal plains that hug the shoreline of Honshu (Japan's largest island) north of Tokyo. And as if the destruction and death left behind was not enough, the leakage of radioactivity from a heavily damaged nuclear power complex near Fukushima south of Sendai (photo previous page and Fig. 10-2) may have caused serious future health problems in the local population. The final death toll of this event, now officially known as the Tohoku earthquake, has been reckoned to be just under 20,000.

Japan's vulnerability to such disasters results from a dangerous combination of circumstances: it is located in a particularly active tectonic-plate collision zone, and many of the country's habitable (and most densely populated) areas are confined, low-lying plains on the islands' eastern coasts open to flooding by Pacific tsunamis. Not surprisingly, the location of most Japanese nuclear power plants along these susceptible shorelines remains a hotly debated issue.

Looking at the layout of East Asian climates (**Fig. 10-3**, left map), we should not be surprised that the western and northern sectors of this realm are dominated by conditions that do not favor substantial population clusters. Permanent snow and ice cover much of the area mapped as **H** (highland climates), including Tibet (Xizang) and Qinghai. Northward, the dry **B** climates (desert and steppe) prevail because this vast expanse lies about as far from maritime influences (and moist air masses) as you can get in Asia. Mongolia is one of the driest countries in the world, but even here—and also in frosty Tibet—there are places where people manage to eke out a living. The map leaves no doubt as to why most East Asians congregate in the eastern segment of this realm.

When we compare the climates prevailing in the East Asian realm to those familiar to us in North America (Fig. 10-3, right map), it is immediately obvious that the **Cf** or humid-temperate climates are more extensive in the United States than in East Asia. Note especially the comparative location of the milder **Cfa** climate, which in the United States extends beyond 40° North latitude up to New England, but which in China yields to colder **D** climates at a latitude equivalent to Virginia's. Thus the capital, Beijing, has the warm summers but long, bitterly cold winters characteristic of **D** climates. Take a closer look at Figure 10-3 and it is clear that both the Korean Peninsula and the Japanese islands lie astride this transition from **C** to **D** climates. South Korea is significantly milder and moister than North Korea, and Japan's largest island (Honshu) is temperate in the south but cold in the north. Not unexpectedly, the least densely populated major island of Japan is northernmost Hokkaido, where the climate is quite similar to that of northern Wisconsin.

Comparing just the United States and China, observe that whereas **C** or **D** climates prevail over more than half of the United States, these climates predominate over less than one-third of China—even though the United States has only 330 million people compared to China's 1.5 billion. That is what makes the population distribution map (**Fig. 10-4**) so noteworthy: the overwhelming majority of East Asia's inhabitants are located in the easternmost one-third of the realm's territory, creating the largest and most densely settled population cluster in the world, which is mainly associated with the limited green-colored area in Figure 10-1.

The fertile river basins and coastal plains of East Asia supported ever-larger farming populations, whose descendants still live on that same land: for all its ongoing industrialization and urbanization, fully 40 percent of the population of China remains rural to this day. Environment, in the form of terrain, water supply, soil fertility, and climate, played the crucial role in the evolution of the population distribution displayed in Figure 10-4—and will continue to do so for centuries to come.

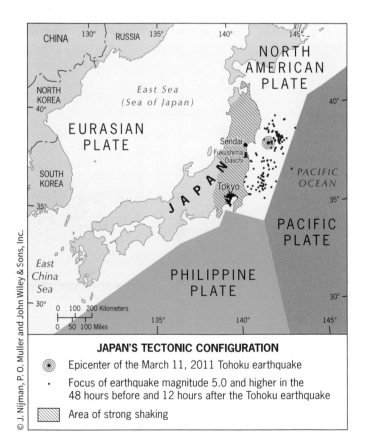

JAPAN'S TECTONIC CONFIGURATION

◎ Epicenter of the March 11, 2011 Tohoku earthquake

· Focus of earthquake magnitude 5.0 and higher in the 48 hours before and 12 hours after the Tohoku earthquake

▨ Area of strong shaking

© J. Nijman, P. O. Muller and John Wiley & Sons, Inc.

FIGURE 10-2

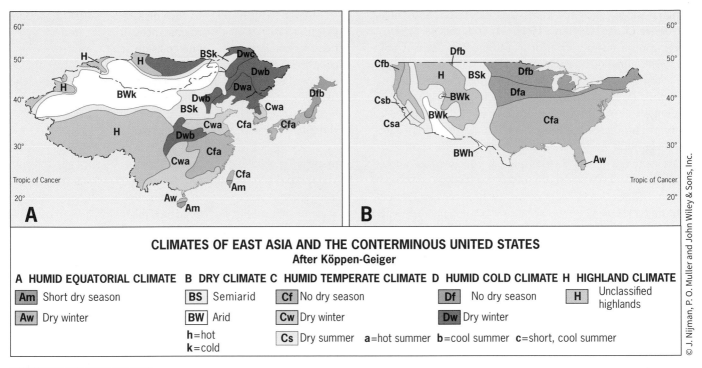

CLIMATES OF EAST ASIA AND THE CONTERMINOUS UNITED STATES
After Köppen-Geiger

A HUMID EQUATORIAL CLIMATE   B DRY CLIMATE   C HUMID TEMPERATE CLIMATE   D HUMID COLD CLIMATE   H HIGHLAND CLIMATE

| Am | Short dry season | BS | Semiarid | Cf | No dry season | Df | No dry season | H | Unclassified highlands |

| Aw | Dry winter | BW | Arid | Cw | Dry winter | Dw | Dry winter |

h=hot
k=cold

Cs | Dry summer   **a**=hot summer   **b**=cool summer   **c**=short, cool summer

FIGURE 10-3 **Interactive**

© J. Nijman, P. O. Muller and John Wiley & Sons, Inc.

The Great Rivers

China is in many respects the product of four great river systems and their basins, valleys, deltas, and estuaries. These rivers and their tributaries are visible but do not stand out as clearly in Figure 10-1 as they do on the physiographic map (**Fig. 10-5**). So important are they as shapers of the realm's regional geography that we should acquaint ourselves with them here. Of the four, the two in the middle are in many ways the most important of all: the **Huang He** (Yellow River, named for the yellow sediments carried by the river) that makes a huge loop around the Ordos Desert and then flows across the North China Plain into the Bohai Gulf, and the **Yangzi** (often still spelled Yangtze), probably the most famous river in China's historical geography, known as the *Chang Jiang* (Long River) upstream.

As the map illustrates, the Yellow River and its tributaries form the sources of water essential to the historic core area of China, the North China Plain, where the capital, Beijing, is located. The Yangzi River is the primary artery of the Lower Chang Basin; at its mouth lies China's largest city, Shanghai, and in its middle course the water flow is controlled by the world's biggest dam (see photo). Both the Huang and the Yangzi originate in the snowy highlands of the Qinghai–Xizang (Tibetan) Plateau, a reminder that these remote environments are critical to hundreds of millions of people who live thousands of kilometers away.

The other two rivers have much shorter courses, but the *Pearl River* outlet of the one in the south, the **Xi** Jiang (West River), forms an estuary that has become China's (and East Asia's) foremost hub of globalization. This is where you find

Hong Kong and, right next to it, the fastest-growing major city in the history of humankind—Shenzhen.

Finally, the northernmost of China's four major rivers, the **Liao** River, originates near the margin of the Gobi Desert and then forms an elbow as it crosses the Northeast China Plain to reach the Bohai Gulf flowing southward. Here the climate is much colder, lowlands scarcer, agriculture lagging, and population smaller than in the more southerly river basins, but mineral resources create opportunities for mining and industry. Each of East Asia's river-based population clusters has its own combination of potentials and problems.

VCG/Getty Images

Three Gorges Dam is emblematic of China's contemporary "era of the mega-project." The dam wall rises 180 meters (600 ft) above the inundated valley floor of the Chang/Yangzi River. It is 2 kilometers (1.2 mi) wide and creates a reservoir—largest of its kind in the world—that extends more than 600 kilometers (375 mi) upstream.

POPULATION DISTRIBUTION: EAST ASIA

Population per Square Kilometer

0 to 5	
5 to 10	
10 to 50	
50 to 100	
100 to 500	
Over 500	

© J. Nijman, M. Shin, P.O. Muller and John Wiley & Sons, Inc.

FIGURE 10-4 Interactive

But even these major river systems, together with China's lesser ones, are increasingly pressed to satisfy the country's skyrocketing demands for water. In order to meet these urgent needs, China is now proceeding with the massive *South-to-North Water Diversion Project* to bring new supplies to its thirsty northeastern core area, especially metropolitan Beijing and Tianjin (**Box 10-2** and Fig. 10-5).

Along the Coast

East Asia's Pacific margin is a jumble of peninsulas and islands. The Korean Peninsula looks like a near-bridge linking Asia to Japan, and indeed it has served as such in the past. The Liaodong and Shandong peninsulas protrude into the Yellow Sea, which continues to silt up from the sediments of the Huang and Liao rivers. Off the mainland lie the islands that have played such a crucial role in the modern human geography of Asia and, indeed, the world: Japan, Taiwan, and China's Hainan. Japan's environmental range is expressed by cold northern Hokkaido and warm southern Kyushu, but Japan's core area lies on its main island, Honshu. As Figure 10-1 shows, myriad smaller islands flank the mainland and dot the East China and South China seas. As we will discover, some of these smaller islands increasingly affect the realm's political geography.

Natural Resources

Given that the East Asian realm contains nearly one-fourth of the world's population, it is not difficult to imagine the magnitude of the demand for natural resources here. The world first received notice of this about 125 years ago when Japan's imperial expansion was in part driven by the need for raw materials to feed its rapidly expanding industrial base. Today, East Asia's demand for natural resources is expressed not in imperialism but in the global marketplace: it is a hugely important driver of natural resource exploitation around the world, from Russia to Subsaharan Africa to Brazil to Australia.

When China took off in the 1980s, economic geographers cast a wary eye on the country's natural resources map. Until then, China's biggest natural asset had been its fertile, river-deposited alluvial (silt) soils: despite the communist regime's best efforts, the plans of state-run industries to satisfy domestic needs were no match for globalizing Japan. Most Chinese were farmers, and staving off famine was an unrelenting preoccupation. However, when China opened its doors to the world, and its industries and cities mushroomed, its needs—for oil, natural gas, metals, electrical power, cement, wood, and water—multiplied. Before long, China had replaced Japan as Australia's leading customer. Moreover, Chinese manufacturers and suppliers were now searching for commodities from

**CHINA AND MONGOLIA:
PHYSIOGRAPHY AND
NATURAL RESOURCES**

South-to-North Water Diversion Project:

⊔⊔⊔⊔⊔ Diversion route

Western route under construction

ENERGY AND MINERAL RESOURCES		
Gasfield	●━━━●	Gas pipeline
Oilfield	◇━━━◇	Oil pipeline
Coalfield		

Sb	Antimony	**Mo**	Molybdenum
Cu	Copper	**P**	Phosphates
Au	Gold	**Ag**	Silver
Fe	Iron	**Sn**	Tin
Pb	Lead	**W**	Tungsten
Mn	Manganese	**U**	Uranium
Hg	Mercury	**Zn**	Zinc

FIGURE 10-5

© J. Nijman, P. O. Muller and John Wiley & Sons, Inc.

Indonesia to Iraq and from DRCongo to Chile. If China holds an advantage, it is in the so-called *rare earth elements* that are not commonly known, such as thulium (used in lasers), praseodymium (aircraft parts), lanthanum (electric automobiles), and promethium (X-ray equipment). Nowadays, China's deposits account for roughly 90 percent of the global production of these minerals, which also are increasingly used in missile technology and "green" energy applications.

But China needs the world because East Asia's storehouse of other known resources is not particularly favorable (Fig. 10-5). Widely dispersed coal reserves can satisfy the country's expanding coal-fired energy system—but only at the cost of thousands of miners' lives every year. Oil reserves are also widespread, but they tend to be rather modest and diminishing. Deposits in northeastern and far western China are the largest, and exploration is proceeding offshore. Yet nothing in China compares with the abundant iron ore deposits and plentiful ferroalloys available to Russia when it industrialized in the early twentieth century. Similarly, China cannot begin to match the immense gas and oil reserves of contemporary Russia, let alone the gargantuan deposits that surround Southwest Asia's Persian Gulf. Hence, China competes with Japan for energy shipments from Russia and with

other sources for its industrial raw material imports. So, in just a quarter-century, China has become both the world's biggest consumer of raw materials and the world's leading exporter of finished goods.

Peoples of the East Asian Realm

Among the many peoples encircling the Chinese are not only the Koreans and the Japanese in the northeast, but also the Mongols and Tatars to the north and northwest; the Kazakhs, Kyrgyz, Tajiks, and Uyghurs to the west; the Tibetans and Nepalese to the southwest; and peoples too numerous to identify individually to the south, including both majorities (Burmans, Thais, Vietnamese) and minorities in states that were still forming in South and Southeast Asia.

China ranks among the world's oldest continuous civilizations, and the Chinese imperial state can be traced back to at least 1766 BC when the first ruling **dynasty [3]** came to power. Most important of all was the formative Han Dynasty

Box 10-2 Global Climate Change in China

China's South-to-North Water Diversion Project

China's **South-to-North Water Diversion Project [2]** (SNWDP) is the biggest inter-basin water transfer scheme in the world. While diversion of water from the south to China's dry northeast has a long history, the needs have accelerated in recent decades due to urbanization and climate change. The project was about two-thirds complete in 2019, and the cost at that benchmark has been estimated between U.S. $50 and $80 billion. Its purpose is to deliver approximately 40 billion cubic meters of fresh water annually to northern Chinese cities facing chronic and increasingly severe shortages, exacerbated by surface water pollution, steady depletion of underground supplies, and increasingly frequent droughts. (By comparison, New York City consumes about 1.5 billion cubic meters of water each year.)

Using the three diversion routes mapped in Figure 10-5, these aqueducts originate from the Huang He (Yellow) and Chang Jiang (Yangzi) rivers that emanate from the snow packs of the icy highlands of the Tibetan Plateau. The *Eastern Route*, which opened in 2013, follows the watercourse of China's ancient Grand Canal, which dates back to around 500 BC. It has the capacity to supply up to 14.8 billion cubic meters of water per year to the coastal-area provinces of Jiangsu, Anhui, Shandong, Hebei, and the megacity of Tianjin. The *Central Route* was completed in 2014 and can provide up to 9.5 billion cubic meters of water per year from the Danjiangkou Reservoir on the Han River (a major tributary of the Yangzi). Water delivery to residents of Beijing via this route is now well underway (see photo). The far more challenging (in part due to the mountainous terrain) *Western Route* is just entering its 30-year (!) construction stage. By mid-century, it is expected to begin diverting about 20 billion cubic meters annually from three tributaries of the upper Chang Jiang around Qinghai's Bayankala Mountains and into the Huang He not far downstream from its source.

Although the SNWDP represents an amazing feat of regional planning, calls are rising for more sustainable local water management solutions in northern China. It is feared that the artificial provision of massive water supplies from beyond the North China Plain will obviate the need to address the underlying causes of water shortages. These include widespread pollution, lack of recycling, and inefficient water usage for agricultural, industrial, and urban activities. Moreover, critics argue that in the longer run climate change and recurrent droughts of growing intensity may undermine the entire scheme, even if demand for water in the northeast will only increase. The problem is that there may not be enough water to divert, especially because of the growing vulnerability of the crucial Han River (Fig. 10-5).

Xinhua News Agency/Getty Images

An open canal segment of the South-to-North Water Diversion Project's Central Route flowing past Zhengzhou, capital of Henan Province, on its way north to metropolitan Beijing. This smog-choked industrial city of more than 4 million lies at the midpoint of the SNWDP's 1400-kilometer (865-mi)-long middle route.

(206 BC–AD 220), whose legacy was the molding of the Chinese nation and the consolidation of its territorial base. In fact, to this day the Chinese still define themselves ethnically as the *People of Han*. Like all empires, China expanded and contracted over time. But during its final dynasty, that of the Qing (1644–1911), many of the states and peoples mentioned in the previous paragraph fell under Chinese rule, and the effects of **Sinicization [4]**—also known as **Hanification [5]**—were considerable even if Chinese domination was resented. This expansion, however, would prove to be the emperors' last hurrah: during the dynasty's final decades, European and Japanese imperialists challenged Qing rule, took control over most of the state's core area, ousted the Chinese from much of the periphery, and left the country in chaos, bringing about the end of nearly 3700 years of dynastic rule in 1911.

By that time, the ethnic geography of the East Asian realm had become highly complicated. Even after losing control over peoples from Korea to Vietnam and from Mongolia to Burma (today called Myanmar), the successor Chinese state of the past century still continues to govern numerous minorities. Thus East Asia remains an extremely complex mosaic of ethnicities and languages.

Figure 10-6 highlights the cultural diversity of a realm in which China dominates numerically but infusions from elsewhere are evident. We already know of the Mongols, the Muslim Uyghurs, and the Buddhist Tibetans; but the most varied and most numerous minority groups inhabit the southeastern corner of this realm, from Hainan Island northward to the mouth of the Yangzi River. For example, the Yue language (colored teal on the map—it used to be called Cantonese) is the common language in the pivotal Pearl River Estuary. Many of the other minority tongues shown here have links to Southeast Asian languages.

In Figure 10-6, note the approximate location of the Great Wall: this is a simplified map of an amazing barrier built by the Han Chinese over many centuries to keep the Mongols at bay. During the thirteenth and fourteenth centuries, the Mongolian empires ruled in these parts, and it was under their control that Beijing rose to prominence. Most parts of the Great Wall were built after the Han Chinese empire had been restored and they moved their capital to Xian, at a safer distance. Beijing is only about an hour's drive from the nearest portion of the wall, a stretch that was built in the fifteenth century and that has been restored (see photo next page).

FIGURE 10-6

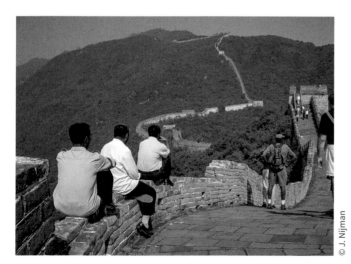

Chinese visitors, along with foreign tourists, admire the Great Wall of China. This portion, about an hour from Beijing, was originally built in the fifteenth century and has been restored. It served to keep the Mongols, who had ruled parts of China in earlier times, at bay.

War and Revolution in the Twentieth Century

The 1911 fall of the 267-year-old Qing Dynasty was followed by four decades of revolutionary upheaval. Much of this chaotic period was marked by the ultimately futile effort of the Nationalists to gain control over all of China. They successfully expelled the Europeans by 1927, and in 1934 drove their chief domestic enemy—the fighters of the strengthening communist movement—into the remote western interior. But the Nationalists were unable to resist Japan's expanding imperial thrust into China during the 1930s, and they bore the brunt of the devastating full-scale war that broke out in 1937—soon to be absorbed into the much wider conflict of the Asia–Pacific theater of World War II. Although the Japanese lost that war in 1945, much of China lay in ruins. The well-armed communists chose that moment to burst out of their far western stronghold and launched widespread attacks on the exhausted Nationalist forces.

Communist China

With the Japanese finally out of the way, civil war resumed with a vengeance between the Nationalists (commanded by Chiang Kai-shek) and the Red Army (led by the chair of the Communist Party, Mao Zedong). After a four-year losing struggle, Chiang's forces were driven off the mainland, with no other option than to regroup on nearby Taiwan as the newly formed Republic of China. Meanwhile, on October 1, 1949, standing in front of the assembled masses at the Gate of Heavenly Peace in Beijing's Tiananmen Square, Mao proclaimed the birth of the People's Republic of China.

Under communism, Chinese society was completely overhauled. The communist regime, dictatorial and brutal though it was, attacked China's weaknesses on several fronts, mobilizing virtually every able-bodied citizen in the process. Land was confiscated from the wealthy; farms were collectivized; dams and levees were built with the hands of thousands; general health conditions improved; child labor was reduced; literacy was broadened.

At the same time, the regime committed colossal errors and systematically repressed its citizens at an unheard-of scale. The so-called *Great Leap Forward* (the propaganda term for what this aimed to accomplish) was a failed attempt at the end of the 1950s to accelerate the growth of industrial and agricultural productivity; it ended as perhaps the worst human-engineered catastrophe in world history, causing around 30 million deaths, mostly from starvation.

Another calamitous episode of Mao's rule was the so-called *Great Proletarian Cultural Revolution*, launched during his final decade in power (1966–1976). Fearful that Maoist communism was becoming contaminated by Soviet "deviationism" and concerned about his own stature as its revolutionary architect, Mao unleashed a vicious campaign against what he viewed as emerging elitism in society (most likely, it was a strategy of power consolidation in the name of culture). He mobilized young people living in cities and towns into cadres known as Red Guards and commanded them to attack "bourgeois" and "revisionist" elements throughout China, criticize Party officials, and violently root out "opponents" of the system. As a result, thousands of China's leading intellectuals were murdered; moderate leaders and older revolutionaries, including some of Mao's closest allies, were purged and imprisoned; the economy remained stagnant, especially agriculture, and an estimated 3 million people were killed in the violence of this period. Although the full economic impact of the Cultural Revolution is still debated, the repetitive political campaigns, purges, and violence created a culture of interpersonal distrust that would take much time to repair. One of those who survived the political upheaval was a Communist Party leader who himself had been purged and thereafter reinstated—Deng Xiaoping. Deng was destined to lead the country into the new era of economic reform following Mao's death in 1976.

East Asia's Economic Transformation

Japan: From Military Defeat to Economic Prowess

After Japan's defeat in 1945, the United States forced it to accept a new constitution as well as minor territorial adjustments. The new constitution stipulated that the country could not spend more than 1 percent of its GDP on the military, and it had to accept the permanent stationing of American troops on its soil. Most importantly, this changed postwar condition induced Japan to reset its focus from military might to economic prowess.

Japan's accelerated economic recovery during the postwar era, which led to its rise as a world economic superpower, was one of the greatest success stories of the second half of the twentieth century. Even though Japan had lost a war, only four decades later it had completely reinvented itself as an industrial behemoth, a technological pacesetter, a highly urbanized society, a political power in the new global arena, and one of the most affluent nations on Earth.

By 1980, the U.S. automobile industry had effectively been competed out of the market by the likes of Toyota and Honda. The same thing happened in the domain of consumer electronics and other high-technology products. Japan did it better and cheaper than anyone else. Now cities everywhere had reliable Japanese cars on their streets; people across the globe adopted Japanese video and portable audio products; laboratories the world over utilized Japanese precision equipment. Japan today continues to be one of the world's richest countries, but its trading dominance—particularly in the sphere of consumer electronics—has moderated since the mid-1990s. Part of the reason was the steady rise of competitors from within the realm.

The Asian Tigers

Japan's sensational development success did not go unnoticed, particularly in the Asia Pacific's smaller-scale, dynamic, upwardly bound economies that were soon being labeled the four **Asian Tigers [6]**: Hong Kong, South Korea, Taiwan, and Singapore. (The first three belong to the East Asian realm and the last to adjoining Southeast Asia.) In the 1960s and 1970s, all four embarked on similar strategies that resulted in rapid industrialization propelled by the attraction of foreign direct investment and the creation of export processing zones for manufacturing **high-value-added goods [7]**, including computers, mobile phones, kitchen appliances, and a plethora of electronic devices. The highly competitive Tigers (first called "The Four New Japans") quickly became formidable exporters to the most affluent markets of North America and Europe, taking full advantage of their large, ultramodern, fortuitously located ports.

China is the world's biggest exporter and home to three of the five busiest container ports in the world: Shanghai, Shenzhen, and Hong Kong (shown here). The other two—Singapore and South Korea's Busan—are also located in the Asian Pacific Rim.

Japan and the Tigers in tow shared an important attribute in their export-driven trajectories of industrial upgrading—the strong intervention of the state. National political leaders worked closely with industrial corporations, targeting such strategic sectors as automobiles, microelectronics, and ship-building; nationalized financial institutions were tapped to supply the necessary funding that was soon augmented by the infusion of foreign investment capital. These national governments played an important part in the stimulation of new urban industrial regions that became engines of the national economies—in some ways a model that would be replicated, at a much larger scale, by China (**Fig. 10-7**).

EAST ASIA: ECONOMIC GEOGRAPHY

- China's original Special Economic Zones
- Major urban-economic region
- Nuclear power plant
- Major highway
- Railroad

National capitals are underlined

0   300   600 Kilometers

0 100 200 300 Miles

FIGURE 10-7

© J. Nijman, P. O. Muller, and John Wiley & Sons, Inc.

China's Economic Miracle

After Mao's death in 1976, China began another historic metamorphosis that was to have the widest global impact. The essence of China's transformation—at a much grander

Employees on the assembly line at the massive Foxconn plant in Shenzhen. If the United States has long represented "big" to the rest of the world—whether in terms of houses, automobiles, skyscrapers, or the size of soft drinks—China today increasingly stands for "huge." Foxconn, also known as Hon Hai, is a giant electronics contract manufacturer of products for Apple, Dell, Hewlett-Packard, PlayStation, and more. It was established in 1978 in Taiwan and opened its first gigantic factory on the mainland ten years later in Shenzhen. In 2019, Foxconn was operating plants in nine cities across the PRC. This Shenzhen plant employs around 250,000 workers on a sprawling "campus" complete with dormitories, stores, banks, and sports facilities as well as a food court that daily dispenses three tons of meat. Worldwide, Foxconn now employs about 1.5 million workers. You are quite likely to own more than one item made by Foxconn, the leading maker of iPhones, iPads, Kindles, video games, and much more—even if its name is not on the product.

Qilai Shen/Corbis Historical/Getty Images

scale—was comparable to what the Asian Tigers had achieved earlier: the creation of a favorable environment for foreign investment to support the growth of a state-of-the-art manufacturing sector geared mostly toward exports. Chinese wages, at least at the outset, were kept low, and technology transfers in the form of training programs were aimed at constantly upgrading the skills of the local workforce. At the same time, political conditions remained stable because in China, more than anywhere else, the (Party-dominated) government maintains very tight control. This was, and remains, a communist state—but one that proved to be extremely adept at understanding how global capitalism operates and how best to put it to use. These days, pragmatism has overwhelmingly become the hallmark of Chinese policies, and their model is sometimes labeled "state capitalism" (**Box 10-3**).

Over the past quarter-century, China has emerged as the most dynamic and fastest-growing component of the global economy. To be sure, Japan and the Tigers had also experienced double-digit growth rates, but it is a far different story when the developmental centerpiece is a country of one and a half billion. As far back as 2010, China surpassed Japan to become the world's second-largest national economy.

East Asia's Population Dynamics

Economic Growth and Demography

The spectacular economic rise of the East Asian realm was primarily driven by carefully designed economic policies: low wages, well-trained workers, an attractive environment for

---

Box 10-3 From the Field Notes …

"I could not quite believe my eyes when I came upon this Starbucks in the middle of Beijing's Forbidden City: this icon of American consumerism, wrapped in ancient Chinese architecture, in the heart of this revered complex, in a country that calls itself communist. It was back in 2004 that the Chinese authorities had agreed to allow the Starbucks café on the premises in return for a hefty financial contribution to ongoing restoration efforts. But not everybody agreed this was a good idea. Several Chinese groups protested against this 'erosion' of Chinese culture, and by 2007 the government felt compelled to reverse its decision. The deal was rescinded, and Starbucks was evicted from the Forbidden City. It was a splendid illustration of the pragmatism that marks the culture of the Chinese and permeates their economic policies. Call it capitalism or authoritarianism, but the Chinese prefer to do whatever works given the circumstances, and they seem to have little time for inconvenient ideological principles."

© J. Nijman

foreign investment, export-oriented and value-added production, and so forth. But in some ways it was also supported by advantageous demographics. We discussed some basic concepts of demography in Chapter 9 and also compared India and China, but it is useful to reiterate a few points here, and then discuss how demography is likely to play a key role in the economic future of this realm.

The rise of Japan and the Asian Tigers, and subsequently of China, happened at a time when the working-age population was growing and birth rates were declining. This meant that the so-called **dependency ratio [8]** was relatively low (the share of the population that is not working and must depend on the productive population to care for them). In China, the *One-Child Policy* was introduced in 1979 specifically for this purpose, though it should be noted that across the realm many families decided for themselves to have fewer children. It had to do with urbanization, decreasing needs for children to work on family farms, more secure jobs, and more women joining the workforce. Thus these East Asian countries were simultaneously experiencing major demographic change during this period.

Changing Demographics: A Moving Target

The thing about demographic shifts—changes in birth rates and dependency ratios—is that they can also have major implications for the future: i.e., a low dependency ratio yesterday could mean a high ratio tomorrow. More precisely, the low birth rates of the past few decades mean that in the near future there will be fewer people in the workforce as well as more older people, and thus the dependency ratio could rise quickly. It is against the backdrop of these previous (and continuing) low birth rates that the East Asian realm today is aging faster than any other (Europe is a close second, with North America following suit). Take the Demographic Transition Model (see Fig. 9-12): East Asia is about to transition into Stage 5, where the overall population is about to start shrinking. Japan is the worldwide trendsetter, with the largest share of population over the age of 65: about one-quarter today and projected to be one-third by 2030. Taiwan in 2019 had the lowest birth rate in the world: 1.2 children per woman (recall that 2.1 signifies the replacement level of reproduction to keep a population at its current size). In South Korea, which also exhibits very low birth rates, ongoing public debates identify demographics as the country's "most dangerous enemy." And then, of course, there is China, where government involvement in family planning has recently taken a drastic turn.

The Rise and Demise of China's One-Child Policy

China has long been the world's most populous country, and during the population explosion of the twentieth century it was also one of the fastest growing. Throughout Mao's rule (1949–1976), when China still was a dominantly agricultural society, China grew at a high annual rate of about 3 percent. When the reformers took over after Mao's death, the new leadership resolved that curtailing China's rate of growth was critical to its economic future. Accordingly, the new regime in 1979 embarked on a severe population-control program that imposed on Han Chinese families (but not minorities) a one-child limit. By the mid-1980s, China's annual population growth had been reduced to 1.2 percent, and by 2019 the rate was estimated to be down to around 0.5 percent (less than half the worldwide average of 1.1 percent and two-thirds the U.S. rate of 0.7 percent).

For many years, the **One-Child Policy [9],** while controversial from an ethical point of view, was considered a key instrument in China's development strategy and one that had been successful. But a generation later, the policy was increasingly challenged and not just on ethical grounds. It gradually became clear that in the longer run, the policy could have devastating effects. China might grow old before it grew rich! In the coming years, China's only-child cohort will constitute the mainstay of the country's workforce, and they will have to provide for a rapidly growing elderly population. By the mid-2010s, debates over the unpopular One-Child Policy spilled out into the open, and the government decided to officially abandon it in 2016. Indeed, by 2019 the Chinese government was introducing measures to stimulate families to have *more* than one child. But that came after 36 years (one and a half generations) of enforcement of the One-Child Policy, and many demographers wonder if it is too late to avoid a population crisis that could derail China's economic progress. As one more sign of the seriousness of the problem, the government in 2017 relaxed some of its draconian anti-immigration regulations to allow Vietnamese immigrant laborers to cross the border between China and Vietnam for seasonal agricultural work.

Evolving Gender Relations in East Asia

One of the side effects of the One-Child Policy was that, as male children were generally preferred, rates of female fetus abortion, infanticide, and abandonment increased, resulting in a gender imbalance. Apart from the moral issues involved, this kind of demographic will also cause problems down the road. China has the most serious and prolonged **gender imbalance [10]** (sex ratio) imbalance on Earth: data for 2018 showed that only 100 girls were born for every 115 boys. Demographers calculate that if the policy is not modified, China as soon as 2030 would be short some 30 million brides.

This gender imbalance is not unique to China. In less extreme form, it can be observed across many male-dominated societies in Asia and beyond (for example, see Chapter 9 for a discussion about India's "missing girls"). This reflects gender relations in a broader sense, where women are in many ways subordinate to men—sometimes by law, but more often on the basis of cultural norms and institutions.

Since the 1960s and 1970s, women have entered the realm's workforce as never before in Japan, the Asian Tigers, and China. In certain ways, this has affected birth rates, as noted above, but it also influences gender relations and the social position of women. As women entered the workforce, many became more independent and less reliant on men for their economic well-being, particularly among the upper-income and higher-educated social strata.

But major inequalities remain. In Japan, the **gender pay gap [11]** is 25 percent, that is, women earn on average 25 percent less than men in the same job (for comparison: in the U.S. the gap is 19 percent, and in countries like Denmark and New Zealand it is less than 10 percent). In terms of women's political representation, Japan ranks a poor 158th among 193 countries—only 10 percent of the seats in their House of Representatives are held by women (in the U.S., women in 2019 held 23 percent of the seats in the House of Representatives, and in Denmark and Costa Rica these numbers were 37 and 46 percent).

The gender pay gap in South Korea is even greater (at 37 percent), and women hold just 17 percent of the seats in Parliament. Interestingly, the South Korean government is now trying to counteract the demographic crisis, and encourages citizens to have more babies via new legislation that is more female-friendly. For example, it introduced laws allowing both parents to take parental leave, thereby allowing more women to continue their careers; babies can now have the mother's last name, giving women a sense of empowerment; and there are stronger incentives to appoint equal numbers of women and men to corporate boards. All of this, it appears, is primarily about stimulating higher birth rates, yet greater gender equality could be a much-needed side benefit (see photo pair).

Geopolitics in East Asia Today

Sino-Japanese Relations

Understandably, relations between China and Japan have been complex and plagued by problems during the century following the end of Chinese imperial rule. In the years leading up to World War II, Japan proclaimed to stand for "pan-Asian" ideals, and that is how it legitimized its invasion (1931) and occupation of China (through 1945). But the Japanese invasion and the atrocities that were committed against the Chinese along the way left deep scars.

Decades later, when China's economic reforms of the 1980s and 1990s led to a renewed Japanese presence in China, the Chinese public and its leaders called for Japan to acknowledge and apologize for these wartime crimes against humanity. The unqualified apology that the Chinese desire, in word and deed, has not been forthcoming. Some Japanese history textbooks still avoid acknowledging what happened to the Chinese, and surveys indicate strong public sentiment on this still-sensitive issue. Now that China has surpassed Japan in terms of economic prowess, it is China that is on its front foot and brimming with confidence. The two countries

Kyodo News/Getty Images

Shin Woong-jae/The Washington Post/Getty Images

In South Korea, gender roles and the position of women are very much part of the current public debate. The photos show just two examples of the ways in which women's roles seem to be changing in South Korea's big cities. The left photo, taken at Seoul's Shinsa rail station, shows advertisements for cosmetic facial surgery, a booming industry that also attracts large numbers of customers (mainly women) from China. The right photo shows a demonstration in Seoul on International Women's Day in 2018 that was part of the worldwide #MeToo campaign.

have close economic ties, but their diplomatic relations are strained by historical memory, cultural friction, and clashing interests.

One of the flashpoints involves the Sino-Japanese (and Taiwanese) dispute over the Senkaku Islands in the East China Sea (**Fig. 10-8**). These tiny, uninhabited islands were seized by Japan in 1895, yet are claimed not just by China, which calls them the Diaoyu Islands, but also by Taiwan. It is not so much the intrinsic value of the islands that is at stake here: it is far more a matter of national pride and entitlement. Moreover, some oil and gas deposits have recently been discovered in this sector, and ownership of the islands grants rights to their surrounding territorial waters and what lies beneath them. China's claim to these islands is in part based on evidence from eighteenth- and nineteenth-century maps (emanating from both China and Japan) showing that they were in Chinese hands before the Japanese takeover.

The fact that the U.S. is a close Japanese ally and has had troops and military bases in Japan since World War II weighs heavily in China–Japan relations. While it provides important security cover to Japan, it can also be said to heighten Chinese wariness and distrust of the Japanese.

As noted above, under the new constitution in 1945, Japan could only have a modest "Self Defense Force," and military expenditures were not supposed to exceed 1 percent of its GDP. However, since China's rapid ascent in this realm (and the world), Japan has been strengthening its military posture. First, military expenditures have been going up since around 2010 and, even as they do not (yet) exceed the 1 percent limit, it now has the eighth largest military budget in the world (remember Japan has a huge GDP and still is the third-biggest economy in the world). As far back as the mid-2000s, Japan's government indicated that it was their prerogative to raise spending beyond 1 percent. Moreover, the U.S., which played a decisive role in drafting the Japanese constitution in 1945, has recently emphasized the need for its allies (Japan among them) to increase their spending. Second, Japan is shifting its Self Defense Force toward more offensive capabilities, including longer-range missiles that can reach China. And third, Japan has expanded its military base in Okinawa (where U.S. troops are stationed as well).

The Korea Factor

Throughout history, the Korean Peninsula has repeatedly been divided, partitioned, colonized, and occupied. After Japan—which had annexed Korea as its colony in 1910—was defeated in World War II, the Allied powers divided the peninsula for administrative purposes. In 1945, the territory north of the 38th parallel was placed under the control of the forces of the Soviet Union; south of this latitude, the United States was in control. But only five years later, communist armed forces from North Korea invaded the South in a forced-unification drive, unleashing a devastating three-year-long conflict that first swept southward across the peninsula, then back northward across the 38th parallel, and finally drew in the Chinese military, who pushed the front southward again. By 1953, a military stalemate halted the hostilities, and the Korean War ended at the cease-fire line not far from where the 38th parallel had marked the original 1945 boundary. Ever since, a heavily armed demilitarized zone (DMZ) has more or less hermetically sealed North from South—with the two Koreas, having grown apart, still in danger of renewed conflict (**Box 10-4**).

But this is not just a bilateral issue. The Korean conflict has long had realmwide and even global implications. A major reason is that North Korea's nuclear capability remains in the hands of a tightly closed, unpredictable, and repressive regime: its nuclear missiles may be aimed at South Korea, but (theoretically at least) they can also reach China, Russia, and Japan—and perhaps even North America. The North Korea issue has also deepened divisions within the realm. South Korea and Japan are diametrically opposed to this regime based in Pyongyang, whereas China takes a decidedly more neutral position. The Chinese also appear to be using North Korea in their dealings with Japan and the United States, since it is widely thought that China is crucial to containing the North Korean threat.

One result of the international community's failure to constrain North Korea's nuclear aspirations may be Japan's military revival. As just noted, the country has gradually increased its military posture, and vast resources have been invested in a state-of-the-art missile defense system to protect against potential North Korean (or Chinese) attacks.

EAST ASIA: GEOPOLITICAL FLASHPOINTS

★ Recent incidents of armed conflict

■■■ Naval bases (China, U.S., Taiwan)

© J. Nijman, P. O. Muller, and John Wiley & Sons, Inc.

FIGURE 10-8

Box 10-4 From The Field Notes …

"I am spending a day on the south side of the Korean Demilitarized Zone, a strip of land running east to west across the peninsula, dividing it roughly in half, between North Korea and South Korea. It is an agreed-upon (in 1953) border barrier zone, 250 kilometers (160 mi) long and 4 kilometers (2.5 mi) wide. It is like a no-man's land where military activities or installations are prohibited—but the front lines abutting the DMZ are among the most weaponized in the world. The westernmost segment of the DMZ lies a few miles north of South Korea's capital, Seoul, well within reach of North Korean missiles on the other side. There is a small Joint Security Area within the DMZ, used for occasional meetings between North and South, as well as relics from the Korean War that are kept for educational purposes. The most interesting site, to me, is the Dorasan Train Station, about 700 yards south of the DMZ. Built in 2002, it looks brand new—because it has never been used, its purpose being mainly symbolic. The photo shows the next train leaving for Pyongyang (North Korea's capital) departing from Track 1 at 16:27 hours—but there is no such train, and nobody is selling tickets. Dorasan Station was built to show the aspirations of (South) Koreans to normalize relations between south and north, perhaps even unify the peninsula. Construction was funded by private donations from South Korean citizens, and there is

a memorial outside that honors them. Yes, North and South Korea are enemies, but they are also one people, and this is what makes the conflict so intractable. As I am about to leave, I spot the map on the wall (right side of the photo), and its message is clear: a railroad between north and south would also connect South Korea to China overland, and thereby to China's ambitious Belt and Road Initiative. South Koreans are wary of the North but hopeful about the future, both for ideological and practical reasons. But it could be a while before that train to Pyongyang leaves the station."

© J. Nijman

The 2011 demise of the North's leader, Kim Jong-il, and the accession of his son, Kim Jong-un, coincided with one of the many critical food shortages that have afflicted the North Korean people since 1945. With the country desperately in need of outside aid, in 2012 it appeared that an agreement had been reached between the new regime and the United States: the U.S. would supply emergency food shipments, and in return the American government extracted a promise to end the testing of nuclear missiles. However, the deal collapsed only a few months later, following North Korea's launching of a rocket capable of delivering a ballistic missile. North Korean threats of war against South Korea and even the United States swiftly became part of the standard playbook, and usually peaked at the time of the annual joint military exercises of South Korean and U.S. forces. Given the unpredictability of the North Korean leadership, these threats continue to raise alarms throughout the realm and well beyond (see **Box 10-5**).

Figure 10-8 displays some of these international geopolitical flashpoints in East Asia. They include Chinese–Japanese conflict in the East China Sea as well as military encounters between North and South Korea on- and

offshore. The *Cheonan* was a South Korean naval vessel patrolling off the country's west coast when it was torpedoed and sunk by a North Korean submarine in 2010. That same year, North Korean forces fired artillery shells and rockets at South Korea's Yeonpyeong Island, hitting both military and civilian targets. In the last few years, the North Koreans have employed overland drones sent south across the border for intelligence purposes (several have been shot down by the South Korean military). North Korea has also tested nuclear devices since 2017, both intercontinental ballistic missiles (that could reach the U.S.), and short-range missiles (mainly aimed at South Korea). U.S. efforts to reach a deal with North Korea about demilitarization (especially denuclearization) had yet to bear fruit as 2020 began.

Taiwan: The Other China

Mention the island of Taiwan in mainland China, and you are likely to be greeted with a frown and a headshake. Taiwan, your host may tell you, is a problem foreigners do not understand.

Box 10-5 Geographic Information Analysis

Where Can North Korea's Missiles Reach?

North Korea is one of the few countries of the world with nuclear weapons. It also possesses missile technology that can easily deliver a nuclear warhead to countries within the East Asian realm, particularly Japan and South Korea. North Korea aspires to develop an intercontinental ballistic missile (ICBM) that is capable of reaching U.S. territory in North America. What do you think is the geopolitical

rationale behind North Korea's nuclear missile ambitions? How far would a North Korean ICBM have to travel to reach North America, and which major U.S. cities are the most vulnerable?

Access your WileyPLUS course to interact with a dynamic version of this map and to engage with online exercises and questions.

NORTH KOREA'S INTERCONTINENTAL MISSILE TECHNOLOGY

AREA WITHIN REACH OF HWASONG-14 MISSILE,
WITH ESTIMATED MINIMUM RANGE OF **6,700 KM**

© J. Nijman, M. Shin, P.O. Muller and John Wiley & Sons, Inc.

FIGURE 10-9 **Interactive**

Virtually all of the 23.8 million people of Taiwan are Chinese. Taiwan was part of China during the Qing Dynasty. Taiwan was stolen from China by Japanese imperialists in 1895, when it was known as Formosa. Then, when communists and Nationalists were fighting each other for control of mainland China right after World War II, and the communists were about to

win, the Nationalists in 1949 fled by plane and boat to Taiwan, where they overpowered the locals. Even as Mao Zedong was proclaiming the birth of the People's Republic of China (PRC) in Beijing, the loser, Chiang Kai-shek, named his regime in Taiwan the Republic of China (ROC)—and told the world that he headed China's "legitimate" government.

The PRC ridiculed this assertion, but the ROC had powerful friends, especially the United States. Chiang's regime was soon installed at the United Nations in China's seat. Washington sent massive aid to support the island's economic recovery and weapons to ensure its security. While the PRC languished under communist rule, Taiwan—the name commonly employed for the ROC—advanced economically, and over time its political system matured into a functioning (if turbulent) democracy.

To the PRC, Taiwan is a "wayward province" that must be reunited with the motherland. When U.S. president Richard Nixon arrived in Beijing in 1972 for a historic visit, Taiwan was a bargaining chip. Soon, the ROC's United Nations delegation was dismissed, and representatives of the Beijing government were seated in its place. Many countries all over the world that had recognized Taiwan as the legitimate heir to China's leadership now swiftly changed sides (including the United States). So much for the ideological convictions of the West! Meanwhile, Beijing's leaders set about trying to isolate the ROC, and to a considerable extent they succeeded.

Geography, however, was to intervene. With billions of U.S. dollars in reserves, fruitful connections to Overseas Chinese in Southeast Asia, and its emergence as an Asian Tiger economic powerhouse, Taiwan had some significant cards to play—and in the late 1970s the Beijing regime discovered it could not afford to deny Taiwanese companies permission to exploit opportunities in the PRC's new development zones. And so, via the "back door" of Hong Kong, Taiwanese entrepreneurs built thousands of factories in mainland China, many of them located directly across the Taiwan Strait. Taiwanese businesspeople now pumped hundreds of millions of dollars into China's development boom, and the economies of Taiwan and the PRC found themselves on a path toward ever tighter integration. In 2018, 29 percent of Taiwan's exports went to China along with substantial Taiwanese foreign direct investment. And in 2015 almost 1.5 million Taiwanese were living and working (temporarily) on the mainland.

Today, per capita annual income in Taiwan exceeds U.S. $25,000—more than two-and-a-half times that of China. And even though a sizeable Taiwanese majority opposes reunification with the PRC, few would want to see their economic well-being imperiled by political adventures.

Mongolia's Place in East Asia Today

Mongolia has little in common with the other countries in the East Asian realm, even if there are regional continuities across the southern border into China's Inner Mongolia province. First, unlike all other countries in the realm, it is landlocked. As such, it has not developed a similar kind of economy benefitting from oceanic trade and shipping. Second, Mongolia is positioned as a **buffer state [13]** between Russia and China, and in some ways it can be considered a transition zone between the two realms. Mongolia became

independent from China in 1921, its autonomy very much a matter of compromise between Russia and China. Third, Mongolia's standard of living is lower and its economy is not nearly as developed, relying heavily on agriculture and mining. Over the past two decades it has received growing foreign investment in the mining sector. China is the biggest investor, and is by far Mongolia's most important trading partner. With Russia a much less significant economic player, it is hard to see how this cannot affect Mongolia's geopolitical outlook. Surprisingly, Japan is quite actively increasing its economic involvement with Mongolia, particularly in the spheres of investment and trade. Rather pointedly, Japan has presented itself as Mongolia's "Third Neighbor," motivated by securing its own needs for raw materials—but also to balance China's dominance in northeastern Asia. Mongolia, in turn, is faced with new choices. Economically, it will want to take all the help it can get; politically, it is likely to move cautiously, as it has done since independence just a century ago.

China's Belt and Road Initiative (BRI)

It is hard to think of an issue in the international media that, over the past few years, has received more attention than China's **Belt and Road Initiative [14]**. BRI involves massive investments by China in the infrastructure of more than a hundred countries around the world. Some of these projects are modest whereas others are huge. All are part of a single design to provide multiple infrastructural connections from China to the world. The Chinese leader Xi Jinping has called it "the project of the century," and in terms of its aggregate costs and investments it probably is.

The BRI project is sometimes referred to as the "New Silk Road," a remake of the ancient trading routes from China across much of Asia toward Europe. But some would say it is not really a single infrastructural project but rather a wide-ranging series of investments by China, trying to secure access to global markets—markets for buying raw materials, markets in which to sell finished goods, and, where possible, markets for making a profit through the construction of roads, railroads, seaports, airports, bridges, tunnels, and buildings.

The name of the overall project, Belt and Road Initiative, is rather clumsy because it is not all that clear as to what is Road and what is Belt. **Figure 10-10** shows the general framework of BRI in the Eastern Hemisphere (additional projects are located in South America as well). This map suggests that both overland and maritime routes are linearly connected to China, yet in reality many infrastructural projects stand by themselves.

The Chinese position on BRI is that this is a win-win situation both for them and the other countries involved: China

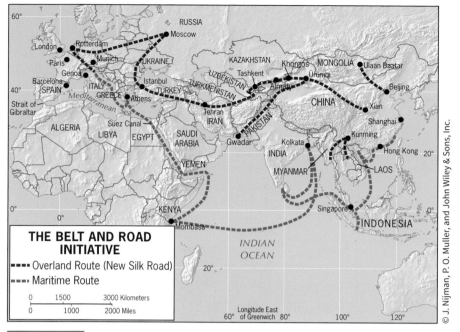

FIGURE 10-10

provides aid to those countries, which often do not have their own capital and/or expertise for such projects, to develop and move forward. Some U.S. critics believe that BRI is a ploy by the Chinese to dominate Asia, and subsequently the rest of the world. They cite the "debt-trap diplomacy" that underlies the Initiative: countries must borrow money from China to finance these projects, but often they cannot repay, at which time the Chinese take possession of (or rights to control) the project, be it a port, a railway, or a similarly valuable new asset.

More objective observers argue that a number of countries are in a position to benefit from this kind of Chinese investment, and for them it could indeed be a win-win proposition. But the poorer countries, especially, would have to be wary about getting involved because they may the ones surrendering major economic or even political concessions to China. Those observers often also point out that, rather than criticizing China for this bold initiative, the U.S. itself should make sure to stay involved around the world and especially in the Asian geographic realms.

Regions of East Asia

East Asia presents us with a good opportunity to illustrate the changeable nature of regional geographies. Our regional ordering is based on current circumstances and may change over time. At the present, we identify seven geographic regions in the East Asian realm (**Fig. 10-11**), the first three of which comprise the People's Republic of China:

1. **China's Coastal Core.** Anchored by the country's eastern coastal provinces, this is not only China's core area but also East Asia's most influential region, as well as the most dynamic spatial component of the global economy. The development success of this region since 1980 is the biggest miracle of all—the conversion of a repressed economic backwater of more than a billion people into an ultramodern urban-industrial colossus. By the end of

the 2010s, the seaboard component of the Coastal Core had fully consolidated, and its transformative energy is now being redirected westward; the leading edge of this inland penetration of China's core area follows upgraded transport corridors that lead to the key interior cities of Chengdu, Lanzhou, and Kunming.

2. **China's Interior.** This vast crescent of territory centered on Lanzhou, for centuries the gateway to far western China, is the country's fastest-growing region today. Propelled by new government policies to reverse the massive and deepening economic divide between the booming Pacific seaboard and the stagnant, poverty-plagued, still-rural interior, huge investments are pouring into this region to implement a high-priority "Go West" program to

FIGURE 10-11

counter this uneven development before it triggers protest movements that could threaten the political stability of the PRC.

3. **China's Western Periphery.** Two large Autonomous Regions, remnants of Han and Qing Dynasty imperialism, constitute China's western tier. One is Xizang—better known as Tibet—whose small Buddhist population is dispersed across the enormous, high-altitude Tibetan Plateau that is framed on its south and west by the world's mightiest mountain range. The other is even larger Xinjiang, whose vast desert basins and encircling mountains comprise the homeland of the Muslim Uyghurs, a primary ethnic component in the human geography of the vital frontier borderland where China meets Islamic Central Asia.

4. **Mongolia.** The landlocked desert state of Mongolia is vast, sparsely peopled, and relatively poor—but in the past decade it experienced a development boom

thanks to rich mineral deposits and substantial Chinese investment.

5. **The Korean Peninsula.** The Korean people form a single nation, but they have been partitioned by ideology since the end of World War II in 1945. Hermetically sealed, China-supported North Korea is a communist dictatorship. In stark contrast, South Korea has evolved into a vibrant democracy as well as an economic powerhouse.

6. **Japan.** The postwar economic transformation of this longtime rival of China is one of modern history's most successful development stories. Today, Japanese society still ranks among the world's most affluent. The Japanese have a strong national identity, and their relationship with China is strenuous.

7. **Taiwan.** This thriving Asian Tiger continues to maintain its own regional identity within the East Asian realm, even as the PRC views the island as a "wayward province" that must be reunited with the mainland state.

Before embarking on our survey of East Asia's regions, we first need to set the stage with a general overview of China. This includes a profile of its political-administrative structure, which is critical to the way this huge country is governed.

The People's Republic of China (PRC)

Citizens of the PRC have no doubt: not only does their country still have the largest population of any state in the world ("still" because India is closing in), but theirs is the oldest continuous civilization on Earth. Yet China's territorial size does not match its demography: even with its 1.5 billion people, the PRC is only slightly larger than the conterminous United States (**Fig. 10-12**). Distances in China are comparable to those in the Lower-48: it is about the same distance from Beijing to Shanghai as it is from New York to Chicago. And as the map shows, east–west distances are similar as well. Latitudinally, however, China extends farther north as well as south of the conterminous United States, creating a greater range of natural environments. In its far northeast, China comes very close to Siberian cold. The extreme south exhibits Caribbean-like tropical warmth—with typhoons (as hurricanes are called in Pacific Asia) to match.

Political-Administrative Divisions

China and the United States may be about the same size, but they exhibit very different governance structures: the United States is a federation, whereas China is a highly centralized unitary state that is ruled from the capital, Beijing. For administrative purposes, China is divided into the following units (shown in **Fig. 10-13**):

4 Central-government-controlled Municipalities (each is known as a *zhi-xia-shi,* or *shi* for short)

22 Provinces

2 Special Administrative Regions (SARs)

5 Autonomous Regions (ARs)

The four central-government-controlled *Municipalities* (*shi's*) are the capital, Beijing; the nearby port city of Tianjin; China's largest metropolis, Shanghai; and the upper Chang River port of Chongqing in the Sichuan Basin. Most notably, the *shi* status of these megacities allows the central government to exert direct control over China's largest internal population agglomerations (see Fig. 10-4).

We should keep in mind that the administrative map of China continues to change. The city of Chongqing was made a *shi* in 1996, and its municipal territory was enlarged to incorporate not only the central urban area but an enormous hinterland covering all of eastern Sichuan Province. As a result, the urban population of Chongqing is now officially 30 million, making this the world's "second metropolis"—but, in truth, the central urban area contains "only" 8.9 million inhabitants. And because Chongqing's population is officially not part of the province that borders it to the west (Fig. 10-13), the official population of Sichuan dropped by 30 million when the Chongqing *shi* was created.

China's 22 *Provinces*, like U.S. States, tend to be smallest in the east and largest toward the west. The territorially smallest are the three easternmost provinces on China's coastal bulge: Zhejiang, Jiangsu, and Fujian. The two largest are Qinghai, flanked by Xizang (Tibet), and Sichuan, China's Midwest. As with all large countries, some provinces are more important than others. Hebei Province nearly surrounds Beijing. Shaanxi Province is centered on the great ancient city of Xian. In the southeast, momentous economic development is occurring in Guangdong Province, whose urban focus is Guangzhou (now China's third-largest city).

In 1997, the British dependency of Hong Kong was taken over by the PRC and became the country's inaugural *Special Administrative Region (SAR)* (Macau followed in 1999, when it was handed over to China by Portugal). Hong Kong is now supposedly in a transitional administrative phase: China has sovereignty, yet this SAR retains an "independent" government and even its own currency (the Hong Kong dollar), and it is exempt from China's notorious "great firewall" that blocks major Western Internet websites. After this transitional stage of 50 years (until 2047), Hong Kong is supposed to be fully integrated into the political and administrative structures of the PRC.

That seems a long way off, especially considering the steadily rising anxieties of Hong Kong residents regarding the PRC's long shadow. It is telling that two new infrastructural projects of the late 2010s received high construction priority to more efficiently connect Hong Kong to the homeland—a 45-kilometer (28-mi) bridge to Zhuhai and a high-speed rail line to Shenzhen and Guangzhou (see Fig. 10-15). Now

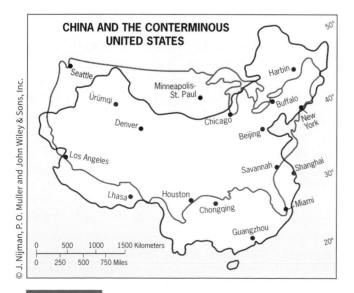

CHINA AND THE CONTERMINOUS UNITED STATES

Seattle · Ürümqi · Minneapolis-St. Paul · Harbin · Buffalo · Denver · Chicago · New York · Los Angeles · Beijing · Savannah · Shanghai · Lhasa · Houston · Chongqing · Miami · Guangzhou

0　500　1000　1500 Kilometers
0　250　500　750 Miles

FIGURE 10-12

FIGURE 10-13

completed, and put in the context of wider issues that sparked protests throughout the SAR in 2019, they are overwhelmingly perceived as part of a comprehensive PRC plan to increasingly tighten its grip on Hong Kong (see upper photo next page).

In 1999, Portugal similarly transferred Macau to Chinese control, and this former colony, situated opposite Hong Kong on the Pearl River Estuary, became the second SAR under Beijing's administration.

The five *Autonomous Regions (ARs)* were created in order to recognize the non-Han ethnic minorities living there. Some laws that apply to the Han Chinese do not apply to certain minorities. As in the case of the former Soviet Union, however, demographic changes and population movements affect such regions, and the policies of the 1940s may not work in the twenty-first century. Han Chinese immigrants now outnumber several minorities in their own ARs. The label "autonomous" should not be taken literally here, and more than anything else it is a political gesture because Beijing maintains very tight control throughout the PRC. The five Autonomous Regions are (1) Nei Mongol AR (Inner Mongolia), (2) Ningxia Hui AR (adjacent to Inner Mongolia), (3) Xinjiang Uyghur AR (China's broad northwestern corner),

(4) Guangxi Zhuang AR (bordering Vietnam in the far south), and (5) Xizang AR (Tibet, cornerstone of the southwest).

China's "Capitalist" Turn

Although China made considerable progress in developing heavy industry (particularly defense industries) during the 27 years of Mao's regime, this advance cannot begin to be compared with the spectacular achievements of Japan and the Asian Tigers. As the post-Maoist era dawned in the late 1970s, it was clear that the old-fashioned socialist production system had to change because now China was falling ever farther behind, especially in terms of technology.

Taking their cues from the accelerated development of Asia's offshore economic powerhouses, the new leadership saw opportunities that could not even be whispered about under Mao's rule. For example, the Lower Chang Basin—focused on the port of Shanghai (China's largest city) and interconnected by the Yangzi and its tributaries—still retained some of its historical identity and energy. Drab, teeming, and

Hong Kong protesters march amidst teargas fired by riot police on October 6, 2019. For much of this year, residents of Hong Kong came out to agitate against Beijing's tightening grip of this Special Administrative Region whose autonomy is increasingly under threat. This particular round of protests followed the declaration by the authorities of the "Anti-Mask Law," banning the use of face masks in public gatherings (note that China is a world leader in facial recognition technologies). Earlier protests reacted against new laws that would allow Hong Kong to extradite political refugees from mainland China back to the mainland. The Hong Kong crisis reflects the dimming prospects of democracy in China and its outcome is likely to be a harbinger of China's political future.

decrepit though it had become, Shanghai had not fully lost its intellectual vigor, artistic individuality, risk-taking entrepreneurs, or even its opponents of communist dogma. And the city's geographical assets continued to offer immense possibilities. Thus from day one, the new regime in Beijing, led by Deng Xiaoping and his pragmatists, saw in Shanghai what they could not yet foresee in their own backyard. Here was a vibrant hub on the Pacific coast right at the mouth of the realm's greatest river, loaded with talent and accustomed to taking chances. Moreover, Shanghai's vast hinterland contained much more than a farmscape of flood-prone paddies and wheat fields: the Sichuan Basin alone had a population of more than 100 million, growing everything from rice to tea and from fruits to spices.

If China's planners needed any further encouragement, all they had to do was look to southern China's Pearl River Estuary, the main outlet of the Xi River Basin. There, situated at the estuary's mouth, was the burgeoning market economy of Hong Kong, the Asian Tiger still ruled by the British. When Deng took charge, Hong Kong was not only a successful port city importing raw materials by the shipload, but also disgorged manufactured products that sold on markets all around the world. Hong Kong may have been a British crown colony, but it was Chinese managers and Chinese workers who propelled its beehive economy. If you were a visitor to Hong Kong in the 1970s, you would be taken to a place on a nearby hillside from where you could peer across the fortified border into "Red China"—and observe some villages with duck ponds and rice-producing paddies as well as wooden fishing boats. The contrast was crystal clear.

China's newly formulated policies of the late 1970s, however, did not entail an official departure from communism. This was (and is) still the People's Republic of China: the Communist Party remains firmly in control, the Politburo still calls the shots, Mao continues to be officially revered like a deity, and freedom of speech remains anything but. Yet the party leadership fundamentally reorganized the way the economy operates in China—beginning with certain vital parts of it. China created highly favorable conditions for foreign investors from around the world with a docile, skilled, hardworking labor force; increasingly efficient facilities; a steadily improving infrastructure; fiscal advantages; and outstanding accessibility vis-à-vis the rest of East Asia, the most rapidly developing regional-scale component in the global economy.

China's Urban Transformation

As recently as 1980, barely one-fifth of the Chinese people were urban dwellers; today, 60 percent live in towns and cities. China's urban transformation is on a scale the world has never seen before. Much of this growth has been planned and controlled by the government, and that alone is an incredible feat. Most significantly, it has brought profound changes to Chinese society, uprooting tens of millions, infusing awareness even in remote rural areas (where a quarter of China's villages have been abandoned since 2000), creating inequality both within cities and between city and countryside, and spawning an enormous **floating population** [15] consisting of temporary urban dwellers with restricted residency rights.

The so-called *hukou* **system** [16] is based on residency permits that indicate where individuals are from and where they may exercise such rights as education, healthcare, and housing. The *hukou* tradition dates back to ancient China and is not uncommon in Japan, Vietnam, and other parts of Asia. During Mao's rule, it became far more rigid and was widely used as a tool to control migration, manipulate labor supplies in the urban-industrial as well as the rural-agricultural sector, and minimize government investment in urban services.

As the market reform movement has intensified over the past three decades, so have the liabilities of the *hukou*

A 2019 view of Kowloon, the most populous ultra-urban area of Hong Kong that dominates its mainland sector (see Fig. 10-17).

system. In Shanghai, for instance, millions of migrants continue to pour into China's largest metropolis, their numbers having tripled in the last ten years alone. In 2018, they accounted for over 10 million of the 24 million people who inhabit this mushrooming urban region. They also accounted for nearly 70 percent of Shanghai's 20–34 age cohort, the critical life-cycle stage during which many couples start a family. Not surprisingly, the metropolis contains several hundreds of thousands of children below the age of 6 who were officially categorized as migrants lacking a Shanghai *hukou*. With the total of migrant workers in China now exceeding 300 million, the *hukou* tradition has become an ever greater social problem throughout the PRC. Even though these workers are major contributors to the expanding economies of the metropolitan centers in which they are employed, they are treated as second-class citizens who receive minimal recognition and are denied access to basic local urban services.

China now counts over 100 (some say 160) cities containing at least one million people, heavily dominating the metropolitan landscape of this realm. This is more than three times as many as the United States.

Toward Megacity Regions   What is perhaps most impressive about China's urbanization is the formation of enormous polycentric urban regions (**Fig. 10-14**). The three most famous are *Jingjinji* (with Beijing at the core), the *Yangzi Delta* (centered on Shanghai), and the *Pearl River Delta (or Estuary)* (that includes Hong Kong, Shenzhen, and Guangzhou). The population totals of these urban formations are staggering. Consider this: Jingjinji, with 112 million people, has more residents than all but ten of the world's countries.

China is now aggressively pursuing policies to reduce further growth of several individual cities such as Beijing and Shanghai, and it is trying to get people to move to smaller (but still big) cities within relative proximity. At the same time, it is rapidly developing the infrastructure needed to integrate these **megacity regions [17]**. One way to do this is with high-speed rail (**Box 10-6**), allowing for fast connections and commutes among the multiple urban centers of a megacity region.

China is pursuing this planning strategy because it will take pressure off highly congested cities like Beijing and Shanghai; because it allows newly planned urban development in relatively proximity; and because it is likely to provide China's government more flexibility in creating productive and more

FIGURE 10-14

© J. Nijman, P. O. Muller and John Wiley & Sons, Inc.

Box 10-6 Technology & Geography

High-Speed Rail in China

The introduction of railroads to North America and Europe in the nineteenth century sparked a revolution in the circulation of goods and people. Railroads were critical both in the integration of regional economies and propelling the expansion of **urban systems** [18]. In the United States, further improvements in long-distance transportation during the twentieth century saw the railroads supplemented and then surpassed by the rise of new interstate-highway and air-travel networks.

East Asia's economic development over the past half-century evolved during a different historical-technological context, one that offered a new and more efficient means of overland mobility: **high-speed rail (HSR)** [19]. Today, China, Japan, and South Korea account for well over half of the world's 17,000 kilometers (10,500 mi) of HSR operations. Although there is no global standard as to what exactly constitutes HSR, its trains typically average speeds of 200–300 kilometers per hour (125–185 mph)—at least twice as fast as conventional express trains. Dedicated HSR routes also utilize specially welded rails to minimize track vibrations, and trains are powered electrically via overhead cables. Moreover, aerodynamic designs, engine technologies, and advanced braking systems set them apart from conventional trains. Their dedicated tracks, particularly within built-up areas, are separated from all other rail traffic, including their access to railroad terminals.

Most HSR networks across the East Asian realm are designed for passenger travel between major cities—because speed matters more for moving people than freight. At medium distances of up to about five hours, HSR offers important time–distance advantages over air travel: main stations are located in the city center, whereas airports are almost always located in the urban periphery. Furthermore, rail travel takes less check-in time and dispenses with most airport-type security procedures, delays (especially weather-related) are less common, seating is both faster and more comfortable, and ticket prices tend to be cheaper. Finally, HSR is more environment-friendly than cars and planes; in HSR-rich Europe, this has become a major reason for those concerned about their carbon footprint to fly less and "train" more.

China is particularly well suited for HSR because it is the best way to connect its emerging megacity regions and provide efficient transport for longer-distance commuters, businesspeople, and migrant workers. As **Figure 10-15** shows, many of the intercity corridors in China's core area are well served by HSR, and its network here constantly expands. The prevalence of HSR in China, however, stands in sharp contrast to its near absence in the United States, whose extensive (but conventional) rail infrastructure is almost exclusively used for freight (see Chapter 2). Americans, of course, are wedded to their cars—a luxury that most Chinese cannot yet afford. But one has to wonder whether HSR could also benefit the U.S., an issue that has been debated for decades. In the meantime, China continues to grow its HSR network at a feverish pace, especially within and between its megacity regions.

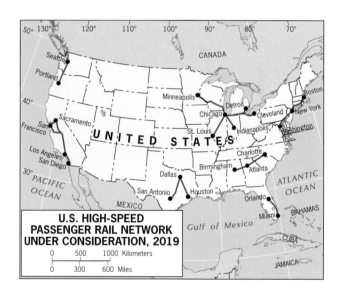

CHINA'S HIGH-SPEED PASSENGER RAIL NETWORK, 2019

U.S. HIGH-SPEED PASSENGER RAIL NETWORK UNDER CONSIDERATION, 2019

© J. Nijman, P. O. Muller, and John Wiley & Sons, Inc.

FIGURE 10-15

livable cities in economically diversified regions. To be sure, other parts of the world, including the U.S., have also been experiencing the emergence of polycentric urban regions; in China, it is the enormous scale of this development, together with the planning controls, that sets it apart.

Rapid urbanization has been closely tied to equally rapid economic growth, as will be explained in the next section (see also **Box 10-7** and **Fig. 10-16**). But the future is uncertain. Will

there continue to be enough jobs to employ all these rural–urban migrants? What will be the consequences of China's accelerated program to expand its urban housing stock by millions of new units every year? Can cities that grow at such a frenzied pace be managed in an environmentally sustainable way? Will the Communist Party be able to maintain control over the new concentrations of urbanites with their changing lifestyles, identities, and expectations?

Box 10-7 Geographic Information Analysis

Urbanization and Economic Growth in China

Underscoring the prominence of China's Coastal Core region, this map shows the distribution of Chinese cities whose populations exceed 5 million as well as the size of the economy in each province and autonomous region. China's rate of urbanization since 1980 has been so phenomenal that more than 100 cities now contain at least one million people (more than twice as many in the United States). Simultaneously, the Chinese economy steadily expanded, soon to dominate global trade like no other. Generally speaking, what is the geographic relationship between urban growth and economic development in China? How would you explain that relationship?

Access your WileyPLUS course to interact with a dynamic version of this map and to engage with online exercises and questions.

PROVINCIAL GROSS REGIONAL PRODUCT (GRP), 2018 (billions of US dollars)

| No Data | Below $100b | $100b - $400b | $400b - $600b | $600b - $1,000b | $1,000b and over |

URBAN POPULATION CHANGE, 1970 to 2010

| No Data | Below 100% | 100% - 250% | 250% - 1,000% |

© J. Nijman, M. Shin, P. O. Muller and John Wiley & Sons, Inc.

FIGURE 10-16 Interactive

Even though China has made impressive progress and now holds a position of considerable power on the global stage, it faces formidable and growing internal challenges. First, rising (regional) inequality will be difficult to rationalize by a regime that still pays homage—at least on paper—to socialist principles. "Let some people get rich first," were the famous words of the lead reformer Deng Xiaoping four decades ago. Yet now it seems that getting rich has become an

Imaginechina Limited/Alamy Stock Photo

A crowd of Chinese passengers lined up at an entrance to Beijing Railway Station, one of the capital's two main terminals, to return home for the Spring Festival (Chinese New Year). For many migrants, this official holiday is the only one long enough for such a visit; in recent years, this event has become internationally known as the biggest single movement of people on Earth. In 2019, the festival ran from January 21 to March 1, and it produced an estimated three billion trips, by automobile, rail, and air. In no small way, this army of annual migrants serves to maintain what is left of the ties between China's rapidly developing cities and the widely dispersed countryside.

obsession for many but is attainable for relatively few. That trend is evident in measures of income disparity: China's world ranking is comparatively high—behind South Africa and Brazil, but well ahead of the United States. Second, there has been a major shift in the constituencies of the Communist Party: peasants and workers, the traditional pillars of support for the regime, have grown increasingly critical as they feel left ever farther behind. With the Party today most strongly supported by the urban middle class and the *nouveaux riches*, how much longer can it maintain the ideological foundation of its policies? And third, the proliferation of the Internet and social media are providing new channels of information flow and opinion expression. Although most online forums are non-political, and the "Great Firewalls" were imposed to limit and police Internet communication (you cannot access Google or Facebook in the PRC), this powerful information-transmission technology is bound to become a new platform of political contestation in China.

Region China's Coastal Core

Turning now to our survey of the seven East Asian regions mapped in Figure 10-11, we begin with the realm's cornerstone—China's Coastal Core, the crucible of the PRC's post-Mao economic transformation. This revolutionary shift began quietly and was officially launched in 1980 at a few carefully chosen places. By establishing new economic rules that would initially apply only to certain urban areas on the Pacific coast, most of the rest of the country and the majority of the population would remain comparatively unaffected—at first.

Special Economic Zones

The government introduced a three-tier system of **Special Economic Zones (SEZs) [20]**, which would attract technologies and investments from abroad and reshape the economic geography of China's seaboard (see Fig. 10-7). Within these economic zones, investors were offered numerous incentives that included low taxes, eased import–export regulations, and simplified land leases. The hiring of labor under contract was permitted. Products made in the economic zones could be sold on foreign markets and, under some restrictions, in China as well. Even Taiwanese enterprises could operate here. And profits earned were allowed to be sent back to the investors' home countries.

When the government made the decisions that would reorient China's economic geography, location was a primary consideration. Beijing wanted China to participate in the global market economy, but it also wanted to produce as little impact on interior China as possible—at least in the beginning stages. The obvious answer was to position the SEZs along the coast. Accordingly, four SEZs were established in 1980, all with particular geographical properties (Fig. 10-7): *Shenzhen* (adjacent to booming, then-British Hong Kong on the Pearl River Estuary in Guangdong Province), *Zhuhai* (next to Macau), *Shantou* (opposite southern Taiwan), and *Xiamen* (on the Taiwan Strait). Interestingly, the last three of these areas were former colonial port cities from which many Chinese had fled the country during the Chinese takeover, to settle elsewhere in East and Southeast Asia. Whether or not this was part of the plan, these **Overseas Chinese [21]** were going to play a major role in China's remarkable economic revolution (elaborated in Chapter 11).

Three more SEZs were proclaimed, in 1988, 1990, and 2006; respectively, these were: *Hainan Island*, with a forward location on the doorstep of Southeast Asia; *Pudong*, on the east bank of the Huangpu River directly facing central Shanghai; and *Binhai New Area*, the coastal zone of the northern megacity port of Tianjin (about 100 kilometers [70 mi] southeast of Beijing). It should be noted that the distinctions between the SEZs and other areas and regions in China have decreased over time and become blurred, because so much of the country has been opened up, to various degrees, to market activity and foreign investment, even though government design and control of the economy has remained intense.

The grand design of China's economic planners, therefore, was to stimulate economic development in the coastal provinces and to capitalize on the exchange opportunities created by location: (1) the availability of funding; (2) the proximity of foreign investors in Southeast Asia, Taiwan, Japan, and—importantly—still-British Hong Kong; (3) the presence of abundant cheap labor; and (4) the promise of world markets eager for lower-cost Chinese products.

To date, the poster child of China's economic planners' design is Shenzhen (**Fig. 10-17**). In the 1970s, the name "Shenzhen" was hardly known, but within a decade this small fishing village had become the fastest-growing city in human history—from about 30,000 (mainly fishing folk and farmers) in 1980 to 3 million in the early 1990s to more than 13 million

FIGURE 10-17

today. Virtually everyone who lives in Shenzhen comes from somewhere else, and because all are really outsiders they speak to each other in Standard ("Mandarin") Chinese rather than the Cantonese dialect that prevails all around them.

Accompanying this unprecedented growth was the headlong conversion of the surrounding Pearl River Delta (PRD) conurbation into a world-class industrial powerhouse. What caused that to happen? When China opened the Shenzhen SEZ, hundreds of companies moved their factories here from neighboring Hong Kong and offshore Taiwan, capitalizing on lower wages and taxes, looser environmental regulation, and weaker official oversight. At the same time, the Chinese government built business-friendly infrastructure in the form of state-of-the-art seaport facilities, airports, highways, and railroads. Practically overnight, Shenzhen became the icon of Deng's Open China policy, to be emulated, inasmuch as possible, by the other SEZs.

Think about it: Shenzhen, an urban center barely 40 years old, now contains more residents than New York City, and Guangdong Province has an industrial workforce larger than that of the entire United States! As for the wider Pearl River Delta conurbation, it has become nothing less than the most productive regional economy of its size in the world. Today it accounts for as much as half of all of China's exports, so if you possess a car, TV, laptop, smartphone, and/or any similar device, it is certain to contain components that originated in this massive manufacturing complex. Although Shenzhen looks like an ultramodern city from above, its population is largely composed of first-generation rural–urban migrants. Moreover, the rapid expansion of cities like Shenzhen and Guangzhou (Fig. 10-14) has swallowed up many small villages, and their former residents now often find themselves in a radically transformed environment.

Regional Engines of Growth

None of this accelerated progress would harm Hong Kong's capitalist economy. In fact, it ushered in a new and even more prosperous era for the still-British colony that would rejoin China as its first SAR in 1997. By the 1990s, Hong Kong had become less of a manufacturing center and far more specialized in international banking, finance, fiscal management, and business services, taking full advantage of the growing skills of its workforce in these sectors. Such occupational transitions are the concern of the field of *economic geography*, which focuses on raw-material distributions, historical-cultural factors (such as the after-effects of colonialism), environmental issues, and particularly the role of spatial economic networks in regional development.

China's development strategy was predicated on the understanding that rapid economic growth can be achieved and managed if it is spatially concentrated, because local conditions can be manipulated and controlled. Once growth takes off, it can have a stimulating effect on surrounding places, and the benefits can be spread further by a powerful central government. This approach—built on a remarkable blend of capitalist principles and socialist-government control—includes competition among these **regional growth poles [22]** and cities, using tax breaks and other incentives to lure domestic and foreign investors. Thus urban municipalities are allowed important freedoms to stimulate local economic growth. At the same time, the central government encouraged millions of *Overseas Chinese* (most of whom had left China from those very coastal provinces to settle—and often thrive—in other countries) to invest their financial resources in their ancestral homeland, thereby attracting ever more substantial **foreign direct investment [23]** from Japan, the United States, and other key countries of the Global Core.

The overall result has been an incredible concentration of economic activity and people in China's emerging megacity regions on the coast (**Box 10-8; Figure 10-18**).

Coastal Core Expansion

As noted above, China's economic miracle began with the Special Economic Zones, and the PRC government continues to use that strategy to launch new regional engines of growth through enormous, concentrated investment plans. But economic development has also spilled over into adjoining areas and diffused more widely to other provinces, so an increasing number and variety of localities now carry a "special" designation of one type or another.

We should not get too fixated on the SEZs because some have received far less attention and investment than others. Hainan, for instance, has only experienced minimal increases in manufacturing, mostly in the processing of agricultural products.

Furthermore, it does not necessarily require the official status of SEZ to obtain inflows of investment (especially from domestic sources) and to achieve economic growth. Thus the entire tier of seaboard provinces, from Guangdong in the south to Liaoning in the north (Fig. 10-11), now forms the heartland of a highly productive, Pacific Rim–based economy; its relentless growth in recent years has generated nothing less than the inland expansion of the Coastal Core region as far west as the central Sichuan Basin. **Figure 10-19** shows that the coastal provinces dominate the Chinese economy, but also that some of the fastest growth is now in the lesser developed areas, such as Tibet (with the highest GDP growth of all provinces), while some established provinces such as Shandong now show growth that is below the national average.

Although export industries remain largely clustered along the coast, economic development is decidedly being steered toward the interior. Much of this growth (resulting from the government's "Go West" program referred to earlier) is occurring around such thriving inland manufacturing centers as Wuhan, Chongqing, Zhengzhou, and Xian, all now engulfed by the expanding Core (Fig. 10-11). As for the leading edge of this inland-moving region, by 2019 three extended lobes had reached the major gateway cities to the western half of China: *Lanzhou* on the Huang River in the north, the jumping-off point for accessing the crucial westernmost ARs of Tibet and Xinjiang; Sichuan's capital, *Chengdu*, in the center; and *Kunming* in the south, a hub for travel into mainland Southeast Asia. This westward expansion of China's Coastal Core has posed a challenge in terms of infrastructure and connectivity between the country's cities and regions. High-speed rail has been hugely important in this regard, and China has become the preeminent global leader in this technology (Box 10-6; Fig. 10-15).

Figure 10-19 affirms the continuing dominance of the coastal provinces, but the bottom map also shows that the highest growth rates are now in the southwestern interior. Yet China

The megacity of Chengdu (population: 11.4 million), capital of Sichuan Province, has grown and modernized at a staggering pace since 2000, riding the crest of China's urbanization wave. Not long ago, this was part of the Interior; now the city has been drawn into China's westward-advancing Core region and forms its spearhead. Traditionally known as one of the country's most laid-back and livable cities, Chengdu has quickly acquired a new reputation as a pacesetting economic center that specializes in information technology, finance, and biopharmaceutical industries.

Box 10-8 Geographic Information Analysis

Population Change in the Pearl River Delta

The Pearl River Delta is home to one of the world's most dynamic megacity regions. It is also one of the key economic drivers of the Chinese economy, a global leader in manufacturing (e.g., computer hardware, smartphones, consumer electronics, textiles, toys), and is one of the fastest-growing urban areas on Earth. However, growth

within the PRD conurbation is somewhat uneven. Explore the interactive map and identify the largest cities in this region, and those projected to grow the most by 2020 (based on China's 2010 census). Alternatively, where is growth projected to slow or decline?

Access your WileyPLUS course to interact with a dynamic version of this map and to engage with online exercises and questions.

POPULATION CHANGE IN THE PEARL RIVER DELTA, 2000 to 2020 (projected)

-50,000   No Change   +250,000

FIGURE 10-18 **Interactive** © J. Nijman, M. Shin, P. O. Muller and John Wiley & Sons, Inc.

has only begun to face the formidable challenges of uneven regional development—and enormous chasms persist not only between the affluent cities and the deprivation-studded rural areas, but also at the national scale between the enlarging Coastal Core and the rest of China.

Region China's Interior

China's Interior region offers a stark contrast to the affluent environs of Shanghai or Beijing. Situated between the Coastal Core and Western Periphery, it stretches in a vast

crescent from the Russian border in the far northeast to the city of Yumen (midpoint of the Hexi Corridor leading to Xinjiang) in the northwest to the Vietnamese border in the farthest reaches of the south (Fig. 10-11). The Interior can be divided into three subregions: the Northeast, Central China, and the South.

The Northeast

Most of this subregion, once known as Manchuria, consists of Liaoning, Jilin, and Heilongjiang provinces as well as the northern prong of the Nei Mongol AR (Inner Mongolia) (Fig. 10-11). It is relatively well endowed with coal and metallic

FIGURE 10-19

mineral reserves (**Fig. 10-20**), and Mao's regime made the industrial redevelopment of the war-ravaged Northeast a priority. Large, state-owned heavy industrial enterprises were established here, particularly steel mills. By the late 1950s, it produced more than one-quarter of China's manufacturing output, but its state-run factories never achieved real efficiency. Most of the Northeast's industrial plant was abandoned soon after the reformers came to power, and the 1980s and 1990s were marked by layoffs, factory closures, worker protests, and out-migration.

Such decline notwithstanding, in the 2010s the Northeast again became a planning priority, and parts of this former "rustbelt" have changed for the better. Liaoning Province, in particular, has undergone significant development and is now fully integrated into the Coastal Core region—linking Core and Northeast as never before. Indeed, since 2010 the Northeast has grown faster than any other regional economy in China (with the exception of the rest of Inner Mongolia, part of Central China to be discussed next), much of it due to accelerating investment in electronics and the auto parts industry

FIGURE 10-20

© J. Nijman, P. O. Muller, and John Wiley & Sons, Inc.

from nearby South Korea and Japan. Undoubtedly, proximity to Beijing and superior coastal access were major factors in this transformation. Liaoning's leading city, Shenyang, today showcases a diversified economy that, besides heavy industry, includes aerospace, electronics, and banking. The Northeast's primary outlet port—Dalian, located where the Bohai Gulf meets the Yellow Sea—has become a leading export focus for trade with Japan and the Koreas.

Central China

The Central subregion of China's Interior borders Mongolia in the north, Xinjiang and Tibet in the west, and the Chengdu protrusion of the Coastal Core in the south (Fig. 10-11). The very sparsely populated western margins of this subregion are dominated by Tibetan pastoralists. Lanzhou, focal point of the upper basin of the Huang He, is Central China's leading urban center. This key crossroads city for overland travel to the Western Periphery is also known for its oil refineries and petrochemical industries. Much of the central Interior is arid country that includes parts of the Gobi and Ordos deserts and, farther west, the salt lakes of the Qaidam Basin (Fig. 10-5). But this subregion also includes the middle course of the Huang with its agriculturally productive Loess Plateau (*loess* consists of fertile, windblown deposits of glacier-pulverized rock). Add the water of this great river plus an adequate growing season, and a sizeable population can be supported.

To the south, Central China includes the northern half of the *Sichuan Basin*, another highly fertile area, crossed by China's other great waterway, the mighty Chang Jiang. Despite its vulnerability to devastating earthquakes (most recently in 2013), this huge natural amphitheater has supported human communities for millennia, and you can readily discern its population cluster in Figure 10-4. The Sichuan Basin, encircled as it is by mountains, is also one of the world's most sharply delineated physiographic regions; the settlement pattern of its roughly 130 million inhabitants reflects that configuration.

Although there are ethnic minorities in the Interior, the overwhelming majority of the people are Han Chinese. Serious regional disparities have long existed between the central Interior and coastal China, and the economic gap has widened during the era of market reform. China's Pacific Rim transformation drew tens of millions to the coastal factories and myriad construction sites in what turned out to be the largest internal national migration in human history. Those who stayed behind were too remote from the opportunities introduced by the reform, and now they faced major new problems in the form of water shortages and the negative effects of rapid farmland conversion to nonagricultural uses. Not surprisingly, protests and social conflicts have steadily risen throughout Central China.

In conjunction with its "Go West" crusade, Beijing's leadership is responding to the development imbalance by extending the privileges enjoyed by the coastal SEZs to the urban centers of the Interior. As workers' wages steadily rise in the seaboard provinces, a growing number of companies are finding it profitable to move their production facilities into this Central sector of the Interior region, abetted by the new economic zones that operate in cities such as Lanzhou and Kunming. A further incentive is the large-scale construction of public works in the Central subregion, especially the ongoing diversion of waters from the Chang to the increasingly thirsty northern cities of the Core (see Fig. 10-5). Yet another lure involves the rapidly improving linkages to the Coastal Core in the form of thousands of kilometers of new expressways, a steadily expanding high-speed rail network (Fig. 10-15), and dozens of new or expanded airports.

The South

The South subregion of the Interior, like its two counterparts, is dominated by Han Chinese, but it also includes sizeable concentrations of minorities, especially near the borders with Myanmar, Laos, and Vietnam (Fig. 10-6). Indeed, adjoining Vietnam in the southeast, we find the Guangxi Zhuang AR, which was specially created with these minorities in mind (Fig. 10-13). This is where China is now allowing Vietnamese cross-border migrants to do seasonal agricultural work, as discussed in the section on population dynamics earlier in this chapter.

The South is largely covered by the Yunnan Plateau, source of the streams that feed the Xi Basin and its Pearl Estuary outlet. Much of southeastern China is quite hilly and in places even mountainous. Among other things, this challenging terrain has inhibited overland contact between China and Southeast Asia—an obstacle soon to be overcome by new routes extending outward from Kunming (see Chapter 11). Yunnan Province itself, in addition to its geopolitical significance, possesses important mineral deposits as well.

Region China's Western Periphery

Historically, empires have expanded and contracted, acquiring and losing territory as well as subjects, and leaving their imprints where they once ruled. Russians in Central Asia, British in East Africa, French in Indochina, and Japanese in Taiwan all came and went, leaving behind languages, religions, infrastructures, and traditions.

And so it was with China. The Qing Dynasty had forged a sprawling empire covering eastern Asia but lost most of its periphery after 1850. Nonetheless, Figure 10-11 reveals that China is still an empire today. The PRC's beyond-the-Han domain includes a pair of large Autonomous Regions (Xizang and Xinjiang) as well as a third one lining the Mongolian border (Nei Mongol) that is undergoing rapid integration into Han

China itself. Even though they are called autonomous regions, the governance structures of China's far western minority entities are fully integrated into the nationwide Party and government hierarchies. Ethnically, the bureaucracies of these ARs may be a joint force of minority and Han administrators, but almost always under the top leadership of a Han.

The Western Periphery has been gaining in importance for China not only because of its natural resources but also its strategic access to Central Asia (especially Kazakhstan; see Chapter 6) and South Asia (especially Pakistan; see Chapter 9).

Xizang (Tibet)

Tibet (called *Xizang* by Han Chinese) is the icebound heartland of Tibetan Buddhist culture that extends northeastward into Qinghai Province and, most importantly, into a corner of far northeastern India in the Indian State of Arunachal Pradesh. Tibet shook off Chinese domination at the close of the nineteenth century, but in 1950, almost immediately after the communist regime took control in Beijing, the Red Army was ordered to go into Xizang to recapture the territory (see **Box 10-9**).

Tibetan society had been organized around the fortress-like monasteries of Buddhist monks who paid allegiance to their supreme leader, the Dalai Lama. The Chinese government wanted to modernize this feudal system, but the resistant Tibetans clung to their traditions, and in 1959 the army had to crush an uprising. During the Cultural Revolution, the Chinese destroyed much of Tibet's cultural heritage, looting its religious treasures and works of art.

Under Deng's rule, part of this heritage was returned to Xizang, but Beijing also encouraged Han Chinese to move to Tibet and built "the world's highest railroad" (reaching an altitude of 5072 meters/16,640 ft) across the Tibetan Plateau to the capital of Lhasa (Fig. 10-11). While the Dalai Lama travels

the world making the 3-million-plus Tibetans' case and asking not for independence but for genuine autonomy, Hanification continues. A complex set of cultural policies has come to prevail in Tibet: separate school systems, one taught in Chinese Mandarin, the other in Tibetan; dual healthcare systems, one specializing in traditional Tibetan therapeutics, the other in Han (Western) medical practices; and a quota system mandating that a certain percentage of government employees be Tibetan. Nevertheless, driven by the twinned processes of Hanification and modernization, Tibetan culture is existentially threatened. Keep in mind that most Tibetan citizens were born after the PRC seized control, and for them the Chinese presence has become a fact of life.

Tibet is fiercely controlled by the Chinese government, not only for geopolitical reasons (e.g., the border with India) but also because so many critical parts of China depend on the water provided by the two great eastward-flowing rivers that rise in this soaring highland. The Chinese government refers to the Tibetan Plateau as "China's water tower." Its resources are so abundant that planners are even considering the diversion of some of these waters to serve the expanding populated areas of the Western Periphery's arid northern interior.

Forward economic momentum in Tibet is strong, and incomes are climbing. But popular resistance continues, intermittently accompanied by violence and drama. Between 2009 and 2019, there have been reports of more than 155 self-immolation suicides, mainly by Buddhist monks, to protest the Chinese occupation, the marginalization of Tibetan culture, and the alleged destruction of the Tibetan environment through mining.

Xinjiang

Xinjiang constitutes the westernmost margin of China's modern empire and is even larger than Xizang (Fig. 10-11 and **Fig. 10-21**). With more than 25 million inhabitants, roughly

---

Box 10-9 From the Field Notes . . .

"This is my fourth visit to Lhasa in the last ten years, and each time I see more construction. This is the capital of Tibet, but it is being remade in China's image. China's urban transformation began over 40 years ago in its eastern coastal provinces. Today it has spread not only to the surrounding provinces of the interior but also to the remote autonomous regions of the far west. This photo shows the large-scale construction of apartment buildings on the outskirts of Lhasa. The official Chinese presence in Tibet is symbolized by the red flags near the right margin of the photo. The developers and construction companies are all said to be Chinese, and so too will be the future owners of these apartments. The Han Chinese influx into this urban area continues, and during the 2020s Lhasa's Tibetan population is likely to fall below the 50-percent level. The city grew from about 220,000 in 2000 to 330,000 in 2018."

© J. Nijman

40 percent of them Han Chinese, the Xinjiang–Uyghur AR is even more important than its Buddhist, Tibetan neighbor. Here China meets the peoples of Central Asia and the faith of Islam; here China has highly significant energy reserves; and here, in the remote, cloudless, desert-dominated far west, China built its original space program. Moreover, across the border to the west lies Kazakhstan, itself endowed with abundant natural resources and providing transit to huge oil and gas reserves in the Caspian Sea Basin. And Kazakhstan is also becoming a key segment of China's Belt and Road Initiative, connecting western China through Central Asia to Turkey and Europe (see Fig. 10-10).

The modern, tightly controlled, Han-dominated, urbanized component of Xinjiang is located in the northern Junggar Basin, focusing on the revitalized capital of Ürümqi and the nearby model city of Shihezi. The traditional, predominantly Islamic, rural component is anchored by the historic oasis city of Kashgar in the extreme western corner of the southern Tarim Basin (Fig. 10-21).

The Xinjiang–Uyghur AR is no more autonomous than Tibet. During the Qing Dynasty, the Muslim peoples here—including Uyghurs, Kazakhs, Tajiks, Kyrgyz, and others—fell under Chinese imperial control and continued their numerical dominance for a half-century after the communist regime came to power in

FIGURE 10-21

© J. Nijman, P. O. Muller, and John Wiley & Sons, Inc.

Beijing. But in the twenty-first century, Hanification has shifted into high gear: today around 40 percent of the population are Han, and Xinjiang has become a two-tiered society. Led by the Uyghurs, who still constitute just under half of the AR's population, the peoples of the lower, Islamic tier disdain "Xinjiang," preferring the label "East Turkestan" to signify their inter-realm connection to the Turkic nations of neighboring Central Asia.

When the Uyghurs confront the Han, as often occurs in the factories of Ürümqi and the streets of Kashgar, the results can be tragic. In 2009, a skirmish between Uyghur workers and Han managers at a plant outside Ürümqi spun out of control and led to rioting in which some 200 Han Chinese citizens were killed and hundreds more wounded. Despite the growing risk of central-government retaliation, Uyghur militants continue to pursue a sporadic—and fruitless—campaign of violence against Han domination.

The Chinese government's grip on Xinjiang was markedly tightened at the end of 2014, with the opening of the Lanxin (Lanzhou–Xinjiang) high-speed rail line. This 1776-kilometer (1100-mi) line, China's longest and the first high-speed railway built across a high-altitude zone, operates at 250 kph (155 mph) and reduces the travel time between Lanzhou and Ürümqi from about 24 hours to under 12. The top photo shows the new bullet train at Hami station near Ürümqi a few days before its maiden eastbound trip. This significant improvement in the connectivity of the Western Periphery is part of a massive effort to modernize China's most remote region. However, the lower photo, which was taken about a year before the inaugural Lanxin run, is a reminder that the Han–Uyghur ethnic conflict continues to percolate and hamper Xinjiang's progress. This assemblage of armed military police in central Ürümqi was a deliberate show of force in response to a surge of Uyghur unrest (partially inflamed by Muslim extremists) in western Xinjiang.

In the meantime, Beijing is tightening its grip on the Western Periphery for economic as well as geopolitical reasons. In 2010, Kashgar was designated an "economic development zone," the first in far western China. Consider its geographic significance (Fig. 10-21): as the Belt and Road Initiative takes shape, Kashgar has a long history as a key stop on the ancient Silk Road; it lies close to the borders with Pakistan, Afghanistan, Kyrgyzstan, Uzbekistan, and Tajikistan; and it is already the largest marketplace in this part of Eurasia. Moreover, Kashgar is also the northern terminus of the **China–Pakistan Economic Corridor [24]** that connects to the Pakistani port of Gwadar (see Box 9-12). Thus Kashgar is advantageously located to be a potent link between China and this portion of Eurasia, and plans call for its development into a primary trading and logistical hub.

Both Xizang and Xinjiang exhibit the properties of peripheries, where indigenous cultures and traditional economies are increasingly intertwined with expanding national interests and global systems. As a result, the widening disparities between the Western Periphery and China's Coastal Core at the national scale are being reproduced at the autonomous-region level, fueled by a deepening core–periphery-type schism between the successful Han and the struggling peoples who constitute the local social mosaic.

News reports from Xinjiang are scarce, and the media are tightly controlled by the Chinese government, but since 2018 a series of indications have pointed to severe repression of the Uyghur population. According to some estimates, close to a million Uyghurs (mostly men) are being held in so-called re-education camps, where they receive vocational training of various sorts but are also required to learn the Chinese language and official Chinese history, and disavow their Islamic faith. It is also reported that many are subsequently moved to forced-labor factories. Some observers underscore that Uyghur Islamic beliefs differ markedly from those held in the Arab world, and that this form of repression may well drive Uyghurs into the arms of international militant groups, potentially worsening the problems that plague China in its far western borderlands.

Region Mongolia

We now turn to the four East Asian regions beyond China and begin by looking to the north, to Mongolia, where the effects of the realm's economic transformation have only lately arrived. This landlocked, isolated country wedged between China and Russia, with an area larger than Alaska but a sparse population of just 3.2 million, suggests a steppe-and-desert-dominated vacuum between two of the world's most powerful countries (Fig. 10-11).

Mongolia used to be the domain of a powerful people who, many centuries ago, swept westward to challenge the Russians and southeastward to rule China. But in more modern times, Mongolia became a vulnerable country whose 500,000-plus herders and their millions of sheep today follow nomadic tracks along the fenceless fringes of the immense Gobi Desert, where Siberian cold periodically causes severe human and livestock losses. During the Soviet era, the location of the capital of Ulaanbaatar symbolized the country's security against Chinese encroachment, and Mongolia functioned as the typical **buffer state** that separated two political adversaries.

In the 2010s, Chinese involvement and investment have been soaring here, and much of this revolves around Mongolia's storehouse of raw materials. Major deposits of gold, copper, and coal have been found, and exploration by large foreign companies ensued. Mongolia was the fastest-growing national economy in the world in 2011 (nearly 18 percent), but it has since declined due to allegedly deficient governmental management and declining world commodity prices. The country has become overly dependent on a small number of gigantic foreign mining companies, and negotiations concerning taxes and regulations have proven to be a cause of economic volatility. The main consumer of Mongolia's commodities is China, and Chinese investments continue to pour in, largely for infrastructural projects. It is not clear whether ordinary Mongolians, among the poorest people in Asia, will benefit from this bonanza anytime soon; that uncertainty has not restrained some economists, who speculate that Mongolia is evolving into another Qatar—where a small population grew rich from its enormous natural endowments.

The capital, Ulaanbaatar, contains more than 40 percent of the country's population, reflecting a desire of many people to leave the hardships of the nomadic life behind. On the rapidly growing outskirts of the city, slums have emerged in recent years, consisting of makeshift houses and *gers*, the traditional nomadic tent-homes (these areas are sometimes referred to as "*ger* districts"). Mongolia faces a stiff challenge in providing livelihoods for such new urban dwellers. Poverty is rampant and local environmental conditions are atrocious, with Ulaanbaatar's air pollution ranking among the worst in the world. Part of the problem is that all electricity comes from coal-fired power plants, and another is that many new migrants in the cities, living in their *gers*, heat their homes with coal or firewood. The city lies in a valley surrounded by mountains, and, especially in winter, the smog is awful. A recent study from the University of British Columbia found that one in ten deaths in the city are related to chronic air pollution (see photo).

Region The Korean Peninsula

Take another look at Figure 10-11 and you will see one prominent (bicolored) peninsula that seems to reach out from the East Asian mainland toward the islands of Japan. As Koreans will tell you, this isn't just another peninsula. This is a place where human geography took momentous turns as a corridor of migration, an incubator of culture, a cauldron of warfare, and a cradle of economic miracles. Today, this Idaho-sized tongue of land has a population of 77 million (Idaho: 1.8 million) that is

One of Ulaanbaatar's *ger* districts, where newly-arrived rural-urban migrants have settled. There is minimal infrastructure in these areas, and settlers are on their own when it comes to heating their homes in the brutally cold winters. They do so with coal or wood, thereby contributing to the always-severe air pollution that bedevils the city.

divided between two states steeped in the same linguistic and cultural traditions: 51 million live in the south and 26 million live in the north.

The Korean Peninsula has been a most turbulent stage. Thousands of years ago, even before there was a China, lower sea levels made the crossing from Korea to Japan possible, and migrants from the mainland began to challenge the even earlier inhabitants of the islands. Later, long-stable states, always influenced by rising China, forged a durable cultural geography marked by distinctly Korean attributes including ethnicity, language, and customs. Then, more than a century ago, disaster struck as Japan embarked on a campaign of colonial conquest that eventually overran not only Korea but also large parts of neighboring China. Subsequently, in the aftermath of Japan's World War II defeat, the Korean Peninsula became the scene of a terrible great-power conflict between China-supported communist armies and American-led anticommunist forces. For more than three years (1950–1953) this devastating conflict swept back and forth across the peninsula, claiming more than 3 million lives and ending in an armistice whose ceasefire line can be seen in **Figure 10-22**. Ever since, that line has separated—almost hermetically sealed—a vicious communist dictatorship in North Korea from a South Korea that originated as a brutal capitalist dictatorship but then transformed itself into both a free-wheeling democracy and an Asian Tiger. The partition of this region remains at the heart of a politico-geographical tragedy.

North Korea

As the map shows, North Korea is territorially larger than South Korea, but its population of 26 million is half the size of South Korea's. Seven decades of the most barbaric variety of communist rule have turned North Korea into one of the poorest,

FIGURE 10-22

hungriest, and most regimented countries on Earth. But it also has one of the largest standing armies as well as advanced nuclear weapons and missile technologies. A regime that incarcerates its citizens for the slightest offenses operates a gulag of prison camps reputedly worse than that of the Stalinist-era Soviets, starves its people as punishment, and isolates its subjects from the rest of the world—yet invests heavily in weaponry with which to blackmail its neighbors and infect the world. Refugees' and escapees' stories tell of the most extreme forms of poverty and misery. But China, North Korea's ideological ally,

supports the regime based in the capital of Pyongyang in the interest of regional "stability" rather than pressuring it to ameliorate its policies.

North Korean dictators rule with powers unimagined even by the likes of Fidel Castro, Saddam Hussein, and Bashar al-Assad. Under this extreme form of tyranny, the North Koreans have become a society of informers—on each other. Any sign of disaffection risks imprisonment or worse, and mass displays of regimented solidarity reminiscent of Soviet public travesties attend dynastic transfers of power. So it was when the ostensibly affable Kim Jong-un, son of the previous dictator Kim Jong-il, took control of the state in 2011. Shortly thereafter he threatened to resume nuclear testing and followed through, with himself and his ruling elite kept comfortably in power by China's provision of food and energy (North Korea borders China to the north).

South Korea

The tragedy of the Korean Peninsula is especially painful when put in geographic perspective. As environmental maps reveal, North Korea and South Korea actually need each other. The North has raw materials demanded by South Korean industries. The South produces food the North could urgently use. This condition is known as **regional complementarity [25]**, but in Korea it is precluded by political segregation.

As noted above, South Korea emerged from the Korean War in 1953 as an unstable, dictatorial, politically and economically corrupt nation where, as in the North, dissent could cost you your life. But South Korea's rulers also encouraged industrial development, driven by powerful industrial conglomerates. This kind of **state capitalism [26]** triggered rapid growth, and the politicians knew that workers would demand better treatment and that the economic and political landscape would have to change. And so it did: dictatorial rule ended, representative government took hold, corruption was confronted, and the economy took off like a rocket. South Korea in short order became a boisterous democracy, an economic giant, a nation in universal pursuit of education, the world's leading shipbuilder, and a manufacturer of iron and steel as well as chemicals and electronics plus countless other industrial products (in addition to producing ample food supplies). South Korean social institutions became ever stronger, and the media were now as free as any on Earth. Freedom of religion prevails, Korean art and culture find a global market, and Korean athletes excel in the Olympic Games and other multinational sports events. As a key member of the international community, South Korea participates in peacekeeping and security operations worldwide.

The map of South Korea (Fig. 10-22) reveals several of these achievements, most notably the massive urban agglomeration formed by Seoul and neighboring Incheon, which today constitutes East Asia's second-largest metropolitan complex and contains almost half of the country's population. Even before democracy took hold, South Korea's rulers decided to further enhance the country's infrastructure by introducing a high-speed rail network to connect Seoul to the traditionally separated subregions of Geyongsang and Jeolla. For a long time, workers in parts of the country not favored by politicians and manufacturers had trouble finding employment in fast-growing industrial areas, an economic-geographic pattern that has still not completely been erased. As regional integration progressed, the southeastern subregion of Geyongsang, centered on the city of Busan, was always ahead of the southwest, where Jeolla was historically marginalized by the powers in Seoul. Today Geyongsang remains in the lead: here

300,000-ton tankers are built along with smaller freighters and warships; here are the famous Hyundai and Kia automobile plants; and here is the home base of POSCO, headquartered in Pohang, one of the world's largest steelmaking corporations.

Region Japan

Japan consists of four main islands—Honshu, the largest; Hokkaido to the north; and Shikoku and Kyushu to the south—in addition to numerous small islands and islets (**Fig. 10-23**). Most of the country is mountainous and steep-sloped, geologically young, earthquake-prone, and studded with volcanoes. A mere 18 percent of the country is considered to be habitable. Japan's high-relief topography has compressed its economic development. Except for the ancient capital of Kyoto, all of Japan's major cities are perched along the coast, and virtually all lie partly on artificial land reclaimed from the sea. Note that all major cities and industrial regions are situated on Honshu's south coast, which offers fine natural bays and harbors; the north coast, in contrast, is dominated by agriculture and fishing.

Sailing into Kobe harbor near Osaka (the country's second-largest city), one passes artificial islands designed for high-volume shipping and connected to the mainland by automatic, space-age trains. Enter Tokyo Bay, and the refineries and factories to the east and west stand on huge expanses of landfill that have pushed the bay's shoreline outward.

With nearly 124 million people, seven-eighths of whom reside in towns and cities, Japan uses its habitable living space very intensively—and expands it wherever possible. As Figure 10-23 indicates, farmland in Japan is both limited and regionally fragmented. Urban sprawl has invaded much of the cultivable land. In the hinterland of Tokyo, around Osaka, and surrounding Nagoya, major farming zones are under relentless urban pressure. All three of these lowlands lie within Japan's fragmented but well-defined core area (delimited by the red line on the map), the heart of the country's prodigious manufacturing complex.

Coastal Development

Figure 10-23 also tells us a great deal about the nature of Japan's external orientation and its dependence on foreign trade. All of the country's primary and secondary regions are found on the coast. Dominant among them is the *Kanto Plain*, the heart of Japan's core area focused on metropolitan Tokyo, which contains about 30 percent of the country's population (see satellite photo). Among its advantages are an unusually extensive area of low relief, a fine natural harbor at Yokohama, a relatively mild and moist climate, and centrality with respect to the country as a whole. Its principal disadvantage lies in its acute vulnerability to major earthquakes (see Fig. 10-2).

The second-ranking primary economic region within Japan is named the *Kansai District* (Fig. 10-23); it contains the Osaka–Kobe–Kyoto triangle and is located at the eastern end

FIGURE 10-23

of the Seto Inland Sea between Honshu and Shikoku. Osaka and Kobe are leading industrial centers and busy ports, but the Kansai District also yields large harvests of rice, Japan's staple food. Kyoto is often viewed as Japan's cultural hearth; it is also the nation's former imperial capital (photo at right). Between the Kanto Plain and the Kansai District lies the *Nobi Plain*, centered on the megacity of Nagoya. And, as the map shows, the Japanese core area is anchored in the west by the **conurbation [27]** focused on *Kitakyushu*, situated not on Honshu Island but in the northwestern corner of Kyushu, Japan's southernmost major island. This still-expanding edge of the core area is particularly favored by its location relative to South Korea and China.

A Trading Nation

Japan is a remarkable country that despite its poor natural endowments (except for its advantageous location) has done extremely well. It became the dominant power of East and Southeast Asia through military force prior to World War II, and its postwar economic miracle astounded the world. In the 1970s and 1980s, people referred to East and Southeast Asia as the "Yen Bloc," one that was dominated by the Japanese economy.

© J. Nijman

Kinkaku-ji (Temple of the Golden Pavilion), one of Kyoto's most famous Zen Buddhist temples, is placed in a beautiful, serene garden that is meticulously maintained. Japan's Buddhist heritage is infused with both Taoist influences and Zen traditions.

The country defeated the odds by calling on organizational efficiency, massive productivity, dedication to quality, and tight adherence to common goals. Keep in mind that this has always been an exceptionally homogeneous, collectively oriented, and consensus-minded nation with a strong work ethic and a deep-seated conviction that common interests override those of the individual. It is hard to overestimate this cultural context in Japan's outstanding achievements, even if it can become a liability—as when collective rigidity impedes adaptation to new circumstances.

More concretely, Japan dealt with its own limited geographic opportunities by engaging the alternatives. Japan had no choice but to engage other countries, near and far, in order to compensate for what it does not possess itself. Japan, therefore, became wealthy through trade, imported raw materials, accomplished production and processing of high-value-added goods, and developed the ability to turn these goods into lucrative exports. Look at Figure 10-7 as well as Figure 10-23 and note Japan's substantial reliance on nuclear energy—a strategy not without major risks in this earthquake-menaced region, as became so painfully clear in the aftermath of the catastrophic Tohoku quake/tsunami in 2011 (see Fig. 10-2). Before the 2011 disaster, about one-third of all energy consumption directly relied upon the country's 22 nuclear power plants. In the wake of the disaster, almost all were closed under severe political pressure from the Japanese population; by 2019, after years of inspections of reactors around the country, most were operating again.

Planet Observer/Universal Images Group/Getty Images

A satellite image of central Honshu's Tokyo Bay, with Japan's capital located just to the west of the bay's northernmost corner. Even this Landsat-7 satellite image—measuring roughly 80 by 115 kilometers (50 by 70 mi)—cannot capture the entirety of this massive conurbation.

Japan's Slowdown

At the end of the 1980s, it seemed that Japan's meteoric rise had been so prodigious that it would only be a matter of time before it would surpass even the United States' economy. But by the mid-1990s, Japan had entered a stall, and for the past quarter-century the country's economy has been moving much more slowly than in its heyday.

One often-cited cause of the slowdown was Japan's lack of adaptability: its culturally embedded economic system seemingly could not cope with rapidly changing conditions. For example, with bankruptcy a near-taboo due to the tight internal networks of firms and their traditions of social responsibility, weak companies were often kept alive. This kind of economic sclerosis inhibits the creation of new businesses (the United States in any given year has three times the number of start-ups as Japan and twice the number of bankruptcies).

But perhaps a more basic explanation rests with the rapid rise of so much competition from the Asian Tigers and China. Suddenly, Japan was forced to compete with automakers in South Korea and electronics manufacturers in Taiwan. All the same, thanks to much lower wages in China, Japanese companies began to shift much of their production out of Japan and into China. The brand names would still read Kawasaki, Sony, or Honda, but the labels would proclaim "Made in China." This meant that Japan, as also happened to the United States decades earlier, started to lose employment because producers moved abroad. The results were not as dramatic in Japan as in the U.S. (where the after-effects of the deindustrialization crisis of the late twentieth century still linger), but it is clear that Japan's economy lost some of its overall momentum. Nonetheless, Japan today remains a formidable economic power, and solidly retains its status as the world's third-largest economy (after the U.S. and China).

Region Taiwan

When we see where Taiwan is located, a mere 200 kilometers (125 mi) from the mainland, we would expect it to be inextricably bound up with China. Not surprisingly, Taiwan has a history of immigration from coastal Chinese provinces. As noted earlier, China's rulers view Taiwan as a "wayward province" that must one day return to the People's Republic. Taiwan lost its seat in the United Nations in 1971 and was replaced by the PRC. The United States at the time also transferred its official recognition of China from Taiwan to the PRC in order to improve relations with the mainland government.

Taiwan's Island Geography

Taiwan, as **Figure 10-24** shows, is not a large island. It is smaller than Switzerland but has a much larger population (23.8 million), most of it concentrated in a crescent lining the western and northern coasts. The Chungyang Mountains, an elongated zone of high elevations (some exceeding 3000 meters [10,000 ft]),

© J. Nijman, P. O. Muller, and John Wiley & Sons, Inc.

FIGURE 10-24

Taiwan's high-tech manufacturers constitute a force to be reckoned with in the global science industries. Domestically, one of the leading contestants for the title of "Taiwanese Silicon Valley" is the burgeoning technopole centered on the city of Hsinchu—a metropolis of about half a million located on the Taiwan Strait 45 minutes southwest of the capital, Taipei. R&D predominates here, and the top photo shows the Synchrotron Radiation Research Center, focused on its new-generation nuclear particle accelerator. Hsinchu is not without rivals, however, and an hour further south lies the smaller city of Taichung, whose high-tech centerpiece is its thriving Science Park. The lower photo shows one of that complex's state-of-the-art factories, whose owner is one of the world's foremost manufacturers of flat-screen panels. Demand for flat screens has surged in recent years, prompting the AU Optronics Corporation to expand production onto the nearby Chinese mainland.

steep slopes, and dense forests, dominate the eastern half of the island. Westward, these mountains yield to a zone of hilly topography and, facing the Taiwan Strait, a wide substantial coastal plain. Streams from the mountains irrigate the paddy fields, with farm production more than doubling between 1950 and 1980, even though hundreds of thousands of farmers have since left the land for jobs in Taiwan's booming industrial sector.

Today, the lowland urban-industrial corridor lining western Taiwan is anchored by the capital, Taipei, at the island's northern end and by the rapidly growing port city of Kaohsiung in the far south (the latter is also home to Taiwan's biggest naval base). The Japanese developed Chilung, Taipei's outport, to export nearby coal, but now the raw materials flow the other way. Taiwan imports raw cotton for its textile industry, bauxite

(aluminum ore) from Indonesia, oil from Brunei, and iron ore from Africa. Taiwan has a steel industry, nuclear power plants, shipyards, a substantial chemical industry, and modern transport networks. Most of all, it has become a leading producer and exporter of high-technology products, especially personal computers, telecommunications equipment, and precision electronic instruments. Taiwan offers a wide range of brainpower resources, and many foreign firms join in the research and development effort. Assisted by the government, a world-class **technopole** [28], focusing on microelectronics and PC development, has been established at Hsinchu on the northwestern coast (see photo pair). Farther south, the budding "science city" of Tainan specializes in microsystems and information-technology innovations.

The creation of China's Special Economic Zones went a long way toward mitigating longstanding problems between Beijing and Taipei. China needed investment capital and Taiwan had it, and in the almost-anything-goes environment of the SEZs, Taiwanese entrepreneurs could buy or build factories just as "real" foreigners were able to do. In fact, the Xiamen SEZ (see Fig. 10-7) was laid out directly across from Taiwan for this purpose (people there will tell you that almost all of the locals have relatives across the Taiwan Strait). Consequently, thousands of Taiwanese-owned factories began operating in the PRC, their owners and investors having a strong interest in avoiding major confrontations over political issues.

Taiwan's unresolved status entails risks, but there are signs that both parties want to find a solution. Given that the PRC will never allow Taiwan to attain independence, and the growing prospect that the United States, Taiwan's chief guarantor, could not secure Taiwan against Chinese military intervention, many Taiwanese as well as their allies and adversaries are seeking a long-range, negotiated solution. Exactly what such an agreement would look like remains to be seen.

Points to Ponder

- The fortunes of China and Japan have always seemed to be inversely related: when one was rising, the other was in decline.

- China in 2019 was engaged with over a hundred countries around the world through its Belt and Road Initiative; in the meantime, the U.S. appeared to be withdrawing from the world stage.

- Widening social inequalities resulting from China's rapid economic growth are becoming ever more difficult for China's self-proclaimed communist government to justify.

- In South Korea, there are strong sentiments about the possibility of a future reunion with the North; after more than half a century of indoctrination in the North, do the North Korean people feel the same?

- Mongolia has long been a buffer state between Russia and China; in a move apparently to balance China's growing influence in the region, Japan has stepped up its economic involvement in Mongolia, officially presenting itself as Mongolia's "Third Neighbor."

- In Mongolia, people are leaving their nomadic ways of life behind and moving to the city—but many continue to live in their traditional *gers*.

© J. Nijman, P. O. Muller, and John Wiley & Sons, Inc.

FIGURE 11-1

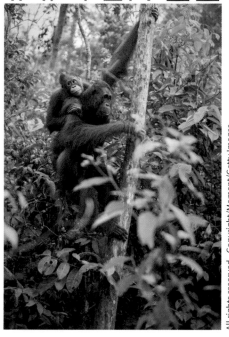

CHAPTER 11

Orangutans, widely viewed as the most intelligent of primates, are found only in the rainforests of the Indonesian islands of Sumatera and Kalimantan (the Indonesian portion of Borneo). They belong to the taxonomic family of Great Apes that also includes gorillas, chimpanzees, and bonobos. There are three different species of Orangutans, two of which are native to Sumatera and the other from Borneo. All three species are considered critically endangered, with the proliferation of palm oil plantations and accompanying deforestation the leading threat (see Box 11-3).

The Southeast Asian Realm

IN THIS CHAPTER

- The mighty Mekong under threat
- China's growing presence in Southeast Asia
- Palm oil's environmental price tag
- Hybrid warfare in the South China Sea
- Global climate change in Myanmar
- Communist Vietnam allied with the U.S.
- Islamic communities under pressure in Myanmar, Thailand, and the Philippines

CONCEPTS, IDEAS, AND TERMS

Indochina **[1]**
Archipelago **[2]**
Tsunami **[3]**
Biofuels **[4]**
Biodiversity **[5]**
Ground subsidence **[6]**
Overseas Chinese **[7]**
Emerging market **[8]**
Node **[9]**
World city **[10]**
ASEAN **[11]**

AFTA **[12]**
Cold War **[13]**
Containment **[14]**
Domino theory **[15]**
State territorial configuration **[16]**
Compact states **[17]**
Protruded states **[18]**
Elongated states **[19]**
Fragmented states **[20]**
Nine-Dashed-Line map **[21]**
Hybrid warfare **[22]**

Remote sensing **[23]**
Primate city **[24]**
Golden Triangle **[25]**
Rain shadow **[26]**
Subsistence farming **[27]**
Entrepôt **[28]**
Growth Triangle **[29]**
Exclave **[30]**
Business process outsourcing **[31]**

Southeast Asia is a realm of peninsulas and islands, a corner of Asia bounded by India to the northwest and China to the northeast (**Fig. 11-1**). Its western coasts are washed by the Indian Ocean, and to the east stretches the gigantic Pacific. From all these directions, Southeast Asia has been penetrated by outside forces. From India came traders; from China, settlers; from across the Indian Ocean, Arabs to engage in commerce and Europeans to build empires; and from across the Pacific, Americans. Southeast Asia has been the scene of countless contests for power and primacy—the competitors have come from near and far.

Defining the Realm

Southeast Asia's geography—summarized in **Box 11-1**—comprises a mosaic of relatively small countries adjacent to two of the world's largest states: India and China. The French, during colonial times, referred to the eastern mainland of this realm as **Indochina** [1], reflecting the importance of this geographic situation. Throughout history, both of these giant neighbors have projected their cultural (and sometimes political) influence into this realm. From the introduction of wet rice cultivation (originally from China) to the adoption of Hinduism or Buddhism (both from India), the ancient imprint on Southeast Asian culture is apparent. This realm has long been a cultural, political, and economic crossroads, and not only vis-à-vis China and India. The impact of Western (and Japanese) imperialism was felt here, too, and later the realm became embroiled in the Cold War. The U.S., of course, was a main party to the Vietnam War at that time. To present-day visitors to the realm, it is the great cultural diversity that stands out, in terms of ethnicity, language, religion, food, and more. Behind the everyday scenes, though, is a restless political geography in which China is an ever-increasing presence.

A Geographic Overview

Figure 11-1 displays the relative location and dimensions of the Southeast Asian realm, an assemblage of 11 countries situated on the Asian mainland as well as on thousands of islands, large and small, extending from Sumatera in the west to New Guinea in the east, and from Luzon in the north to Timor in the south. We will become familiar with only the largest and most populous of these islands, but if you were sailing the seas of this part of the world, you would pass dozens of smaller ones every day. And if you could stop, you would find even the tiniest populated islands possess an individual character resulting from the cultural sources of their inhabitants, the environmental challenges they face, the opportunities they exploit, their modes of dress, the layout of their dwellings, and even the vivid colors they use to decorate their boats.

The giant of this realm, both territorially and demographically, is the far-flung **archipelago** [2] (island chain) of Indonesia, labeled appropriately in Figure 11-1 by the largest lettering. In its easternmost sector, the Indonesian state extends beyond the Southeast Asian realm into the Pacific realm because it controls the western half of a large island—New Guinea—whose indigenous peoples are not Southeast Asian. We discuss this unusual situation later (and focus on New Guinea as a whole in Chapter 12), but note it here because unusual borders and divided islands are a hallmark of this realm.

On the Asian mainland, as noted above, Southeast Asia is bordered by China and India, both sources of immigrants, cultural infusions, economic initiatives, and other relationships evident in the realm's cultural landscapes. Also originating

---

Box 11-1 Major Geographic Features of Southeast Asia

1. Southeast Asia extends from the peninsular mainland to the archipelagos offshore. Because Indonesia controls part of New Guinea, its functional region reaches from the Southeast Asian geographic realm into the neighboring Pacific realm.

2. The realm consists of two main parts: the Mainland and the Insular region.

3. Southeast Asia is a global crossroads with strong Indian and Chinese cultural influences, but it is also shaped by Western intervention.

4. The realm's physiography is dominated by high relief, crustal instability marked by volcanic activity and earthquakes, and tropical climates.

5. A majority of Southeast Asia's people live on the islands of just two countries: Indonesia, with the world's fourth-largest population, and the Philippines. The rate of population increase in the Insular region exceeds that of the Mainland region.

6. Although the great majority of Southeast Asians have the same ancestry, cultural divisions and local traditions abound.

7. The Mekong River rises in China and borders or traverses five Southeast Asian countries, sustaining tens of millions of farmers and fishing people.

8. Singapore is the leading world city in Southeast Asia and lies at the realm's center of trade and business networks.

9. The realm contains a number of rapidly emerging markets and fast-growing economies that in certain respects imitate development in the neighboring East Asian realm.

10. China's influence in this realm has increased markedly in recent years, sometimes triggering local backlash.

there are rivers that play a pivotal role in the lives of many millions of people residing to the south of the Chinese border. As we shall discover, the core areas of many of Southeast Asia's most populous countries are located in the basins of major rivers whose sources lie outside the realm.

Before we get started, it is helpful to become acquainted with the key states of this realm, some of which will already be familiar. North of Indonesia lies the Philippines, well known to Americans because this was once an American colony and is still a source of many immigrants to the United States. Also lying north of Indonesia is Malaysia, easy to find on the map because its core area is located on the lengthy Malay Peninsula that almost connects mainland Asia with Indonesia. At the tip of that peninsula lies another famous geographic locality: Singapore, the spectacular economic success story of Southeast Asia and a world city by every measure.

In the realm's mainland component, the name Vietnam still resonates in the United States, which fought a bitter and costly war there in the 1960s and 1970s—today, half a century later, the U.S.–Vietnam relationship is being fundamentally recast, as we shall see. The map shows that Vietnam is shaped as an elongated strip of land extending from its border with China to the delta of the greatest of all Southeast Asian rivers, the Mekong. Looking westward, neighboring Laos and Cambodia may not be all that familiar, but next comes centrally positioned Thailand, the dream (and reality) of millions of tourists and one of the world's most fascinating countries. And then, on the realm's western margin, we come to Myanmar (formerly Burma): here human potential, natural endowment, and opportunity were thwarted by a half-century of extremely harsh military rule that plunged the country into devastating poverty—but that quite unexpectedly relaxed its vise-like grip in 2012 and reopened Myanmar to the world.

Physical Geography

Physiographically, Southeast Asia is in some ways reminiscent of Middle America, a fractured realm of islands and peninsulas flanking a populous mainland studded with high mountains, deep valleys, and stunning landforms (**Box 11-2**). It is useful to look back at the world plate tectonics map in Figure 1-5 to see why: both Southeast Asia and Middle America contain dangerous terrain, where the Earth's crust is unstable as tectonic plates collide, earthquakes are a constant threat, volcanic eruptions take their toll, tropical cyclones lash sea and land, and floods, landslides, and other natural hazards make life riskier than in most other parts of the world.

More broadly, the realm forms part of the Pacific Ring of Fire—and all the geologic hazards this designation carries with it (Fig. 1-5). In 2004, an undersea earthquake off westernmost Indonesia produced a **tsunami [3]** (seismic sea wave) in the Indian Ocean that killed more than 230,000 people along coastlines from Indonesia's Sumatera to eastern Africa's Somalia. Yet this was only one of the most recent in a string of geologic disasters originating in Southeast Asia whose effects were felt far beyond the realm's borders. In 1883, the Krakatau volcano lying between Sumatera and Jawa exploded, resulting in a death toll estimated at more than 36,000. In 1815, the Tambora volcano in the chain of islands east of Jawa known as the Lesser Sunda Islands blew up, darkening skies throughout the world and affecting climates around the planet (the year that followed is still known as the "year without a summer" when crops failed, economies faltered, and people went hungry as far away as Egypt, France, and New England). Research on what may well have been the most calamitous of all such eruptions suggests that, about 73,000 years ago, the Toba volcano

---

Box 11-2 From the Field Notes ...

"After a three-hour drive by bus southward from Hanoi, we reached the Vanlong Nature Reserve in northeastern Vietnam's Red River Delta. It lies in a limestone terrain known as karst, part of the same geology that continues to the north through Halong Bay and into southern China beyond. Karst landscapes are formed through the slow but steady dissolution of limestone by (naturally) acidic water; the harder rock formations are more resistant to erosion and remain as towers, resulting in unique and sometimes whimsical landforms (left photo). This natural reserve is a biologically rich ecotourism site that was established in 2001 and involves local communities. Women from nearby villages guide visitors around in their basket boats, and they welcome the extra income in an area barely touched by Vietnam's emerging economy (right photo)."

on Sumatera exploded with such force that its ash and soot not only darkened skies and affected weather, but altered the global climate for perhaps as long as 20 years and threatened the very survival of the human population, which then was still small in number and widely scattered. In fact, a number of scientists postulate that the widespread impact of this eruption gave rise to so many casualties that human genetic diversity was significantly diminished.

Thus, high relief dominates Southeast Asia, from the Arakan Mountains in western Myanmar to the glaciers (yes, glaciers!) of New Guinea. Take a close look at Figure 11-1 and you can see how many elevations approach or exceed 3000 meters (10,000 ft), and, using the elevation guide in the map legend, how mountainous and hilly much of the realm's topography is. Lengthy ranges form the backbones not only of islands such as Sulawesi and Sumatera but also of the Malay Peninsula, most of Vietnam, and the border zone between Thailand and Myanmar. In Figure 1-5, it is possible to trace these volcano-studded mountain ranges by their earthquake epicenters and volcanic records.

Exceptional Borneo

Among the islands, however, there is one major exception. The bulky island named Borneo in Figure 11-1 exhibits high elevations (Mount Kinabalu in the north reaches 4101 meters [13,455 ft]) yet has no volcanoes and negligible Earth tremors. This island has been called a stable "minicontinent" amid a cauldron of volcanic activity, a slab of ancient crust that long ago was uplifted high above sea level by tectonic forces and then subsequently eroded into its present landscapes. Borneo's soils are not nearly as fertile as those of the volcanic islands, so that an equatorial rainforest developed here that long survived the human onslaught, providing sanctuary for countless plant and animal species that include Southeast Asia's great ape, the orangutan (photo p. 395). That era is ending today as human encroachment on Borneo's tropical habitat accelerates: logging is destroying vast swaths of the remaining forest, and roads and farms are penetrating the island's interior. Borneo, along with other parts of Indonesia and Malaysia, has also experienced the rapid expansion of palm oil plantations at the expense of tropical woodlands (see **Box 11-3**).

Borneo's tropical forests are a remnant of a much larger stand of equatorial rainforest that once covered most of this realm. Besides Borneo, eastern Sumatera still contains limited expanses of rainforest, including a few orangutan sanctuaries. So does less populous and more remote New Guinea, which was never reached by these great apes. As we noted in Chapter 1, equatorial rainforests can still be found today in three low-latitude zones: the Amazon Basin of South America, west-equatorial Africa, and here in Southeast Asia. The climatic conditions that sustain these forests—persistently warm temperatures and year-round rainfall—produce the biologically richest and ecologically most complex vegetation regions on Earth. Enormous numbers and varieties of trees and

other plants grow in very close proximity, vying for space and sunlight both horizontally and vertically. Yet despite all this luxuriant growth, the soils beneath rainforests are nutrient poor. Most of the surface nutrients are derived from the decaying vegetation on the forest floor, which nurture the growth of the next generation of plants. Long-term indigenous inhabitants have developed methods of shifting cultivation to use these soils, plants, and animals to eke out a living, but when migrant farmers from elsewhere remove the trees to utilize the soil, mistakenly assuming it will be able to support their crops, failure swiftly results. This reality notwithstanding, intensifying population pressure throughout the tropics continues to shrink what remains of the world's rainforests—an issue also discussed in the chapters on Middle and South America (Chapters 3 and 4).

Relative Location and Biodiversity

Look again at Figure 11-1, but in a more general way. The Malay Peninsula, adjacent Sumatera, Jawa, and the Lesser Sunda Islands east of Jawa all seem to form a series of stepping stones toward New Guinea, and in the extreme lower-right corner of the map you can see Australia's Cape York Peninsula also appearing to reach toward New Guinea. Throughout Southeast Asia's geologic history, sea levels have repeatedly dropped, periodically drying up those narrow bodies of water lying between the islands as well as certain peninsulas. During such prehistoric intervals, the orangutans, whose descendants survive in Indonesia today, were able to migrate from the mainland into even warmer equatorial latitudes because islands were temporarily connected by land bridges. Later, early human arrivals here also received assistance from nature: Australia's Aborigines probably managed to cross what remained of the deeper trenches between islands by building rafts, but their epic journey was facilitated by land bridges as well.

Thus time and again, Southeast Asia was a receptacle for migrating species. Combine this with the realm's tropical environments, and it is no surprise that it is known for its **biodiversity** [5]. Scientists calculate that 10 percent of the Earth's plant and animal species are found in this comparatively small realm. Biogeographers can trace the progress of many of these species from the Asian mainland across the Indonesian archipelago toward Australia, but some were halted by the deeper trenches that remained filled with water even when sea levels dropped. An especially deep trench lies in the Lombok Strait between the islands of Bali (next to Jawa) and Lombok (Fig. 11-1), a key biogeographical boundary (discussed in Chapter 12) first recognized by the naturalist Alfred Russel Wallace (1823–1913), a contemporary of Charles Darwin (1809–1882). As we shall presently observe, Southeast Asia's biodiversity had a fateful impact on its historical geography. Among the realm's specialized plants, it was the spices that attracted outsiders from India, China, and Europe, with consequences still visible on the map today.

Box 11-3 Deforestation and Palm Oil Plantations

Palm oil is used worldwide in processed foods and myriad other products. This form of commercial agriculture is quite lucrative because the yield of oil palms per hectare (which equals 2.5 acres) is much higher than the yield of such competing edible oil crops as soybeans and sunflowers. Moreover, the price of this commodity on global markets has risen steadily over the past several years, triggering successive rounds of production increases. Malaysia and Indonesia have benefited the most, together accounting for 90 percent of the world's palm oil. But because these countries take advantage of economies of scale and produce palm oil on huge, ever-expanding plantations, this crop has become the single biggest threat to tropical forest preservation (**Fig. 11-2**). Most at risk for survival are such key wildlife species as orangutans, pygmy elephants, Sumatran rhinos, and tigers.

Ironically, back in 2008 there was strong support for palm oil from across the globe because it was viewed as a key ingredient of environment-friendly **biofuels [4].** Governments around the world (including the U.S.) initiated laws that encouraged the use of vegetable oil in fuels (especially for diesel engines) to reduce CO_2 emissions. Biodiesel imports in the U.S. went from zero in 2006 to 100 million gallons a month by 2016. While CO_2 emissions are indeed much less compared to those of fossil fuels, and oil palms are renewable, this policy has proven to be a major miscalculation: the deforestation involved in the creation of palm oil plantations has been an absolute disaster. It is contributing massive amounts of CO_2 emissions through the burning and clearing of forests for new plantations—which is estimated to exceed the benefits. Indonesia is now estimated to be the world's fourth-largest source of emissions (see photo p. 428).

A section of a vast palm oil plantation in Sarawak State in the southern part of Malaysian Borneo. These continually enlarging plantations have proliferated in Malaysia and Indonesia (see Fig. 11-2), earning sizeable export revenues but at ever-increasing environmental cost.

PALM OIL PRODUCTION IN INDONESIA AND MALAYSIA, 2019

FIGURE 11-2

© J. Nijman, P. O. Muller, and John Wiley & Sons, Inc.

Four Major Rivers

Water is the essence of life, and Southeast Asia is comparatively well endowed with moisture (Fig. 1-6, Chapter 1). Ample, occasionally even excessive, rainfall fills the rice-growing paddies of Indonesia and the Philippines. On the realm's mainland, where annual rainfall averages are somewhat lower and the precipitation is more seasonal, major rivers and their tributaries fill irrigation channels and form fertile deltas. The map of the mainland's population distribution is dominated by this spatial relationship between rivers and people, expressed most clearly in distinct coastal clustering atop the biggest deltas (see Fig. 11-4).

As Figure 11-1 reminds us, rivers that are crucial to life in one realm may sometimes flow from sources in neighboring realms, a geographic issue that can lead to serious regional discord. What right does an upstream state in one realm (or even the same realm) have to dam or otherwise siphon water away from a river whose flow is vital to a state (or states) downstream? In Southeast Asia's case, three of its four principal rivers originate in China: (1) the *Mekong*, which crosses the heart of the mainland from north to south; (2) the *Red River*, which reaches the sea via its course across northern Vietnam; and (3) the *Irrawaddy*, which forms the lifeline of Myanmar. The fourth is an intra-realm river, the *Chao Phraya*, which functions as Thailand's leading artery.

The Endangered Mighty Mekong   From its headwaters, high in the frozen uplands of China's Qinghai–Xizang (Tibetan) Plateau, the Mekong River rushes and flows some 4200 kilometers (2600 mi) to its delta in southernmost Vietnam. This "Mother of Rivers," as its name connotes, traverses or borders five of the realm's countries, supporting rice farmers and fishing people, forming a vital transport route where roads are few, and providing electricity from dams upstream. Tens of millions of people depend on the waters of the Mekong, from apartment

dwellers in China in the north to subsistence farmers in Cambodia in the south. The Mekong Delta in southern Vietnam is one of the realm's most densely populated areas and produces enormous harvests of rice and, closer to the sea in more brackish water, has lots of shrimp farms. The Mekong Basin as a whole (mapped in green in **Fig. 11-3**) is reputed to be the world's largest and most productive area for freshwater fishing.

But the Mekong is increasingly under threat, and so are the livelihoods of millions in its basin, as demand for hydroelectric power accelerates all across mainland Southeast Asia. To the north, China has already built seven dams for generating hydropower, and many more are either under construction or planned (Fig. 11-3). In the Mekong's middle basin, impoverished Laos considers hydropower its most strategic commodity and promising route to development, and it is making big plans to sell much of that energy to China, Thailand, and Vietnam. Favored by its hilly terrain, Laos is planning seven new dams, and its downstream neighbor, Cambodia, wants to build another three.

Among these projects, it is the ongoing construction of the Xayaburi Dam in Laos that has drawn special attention and criticism from environmental activists because it sits squarely in the Mekong's main stream (Fig. 11-3; see photo). As such, it will substantially reduce the river's downstream transportation of

Construction of the controversial Xayaburi Dam in northern Laos, located on the Mekong River upstream from the capital, Vientiane. For security reasons, this site is zealously guarded, and current photos are hard to find (this one dates from late 2012). Opposition groups say this dam will displace and adversely affect countless communities along the Mekong as well as produce significant ecological harm. Meanwhile, the Laotian government is primarily focused on developing the dam's hydropower potential, which is planned to be a main driver of national economic growth through the rising sale of electricity to China as well as other countries. Xayaburi Dam was completed and began operating in late 2019.

MEKONG RIVER DAMS, 2018
- Existing
- Under construction
- Planned
- Mekong River Basin

National capitals are underlined.

0   300   600 Kilometers
0   200   400 Miles

FIGURE 11-3

silt, the alluvial sediment that is highly crucial to the formation of fertile soils on the floodplains of central Cambodia and particularly the massive delta in southern Vietnam. And the sheer number of dams planned for the Mekong Basin is certain to damage ecosystems all along the Southeast Asian courses of the river and its tributaries (starting with the disruption of the migratory routes of freshwater fish). In the long run, therefore, prioritizing energy over environment and even food production may well prove to be a disastrous choice.

The risks are compounded by the fact that the Mekong delta is slowly sinking. This **ground subsidence [6]** is caused by groundwater extraction for rice cultivation—yes, there is plenty of water in this delta, but cultivation requires measured irrigation, and especially during the dry season (November to April), there is widespread use of wells. Most of the delta is only about 1–2 meters (4–6 feet) above sea level, but in recent years it has been subsiding a few centimeters (about an inch) per year. This, in turn, caused salination in the areas nearest the mouth of the delta, and that impedes rice growing. And it increases exposure to cyclones and associated flooding. Altogether, these are precarious times for the Mekong Delta.

Population Geography

Examine the map of Southeast Asia's population distribution (**Fig. 11-4**), and you are immediately struck by the huge concentration of people on a relatively small island in Indonesia—a cluster larger than any other in the realm, the four mainland river

POPULATION DISTRIBUTION: SOUTHEAST ASIA

**Population
per Square Kilometer**

0 to 5
5 to 10
10 to 50
50 to 100
100 to 500
Over 500

© J. Nijman, M. Shin, P. O. Muller, and John Wiley & Sons, Inc.

FIGURE 11-4 **Interactive**

deltas included. This is Jawa, and its population of approximately 150 million not only accounts for well over half of Indonesia's national total but also exceeds that of every other country in the realm.

This population concentration is particularly noteworthy because Indonesia is not yet a highly urbanized country. Today nearly half of all Indonesians still live in rural areas, and although Jawa is the country's most urbanized island, more than 50 million people still live off the land. What makes all this possible is a combination of highly fertile volcanic soil, abundant water, and extremely warm temperatures that enable Jawa's farmers to raise three crops of rice in a single paddy during a single year, helping feed a national population still growing faster than the global average. But we should also be aware that the very high density on Jawa denotes fast-growing urban areas, cornerstones in the building of a new economy with increasingly global linkages.

Within Indonesia, the contrasts between Jawa and the four other major islands—Sumatera, Kalimantan (Indonesia's portion of Borneo), Sulawesi, and most especially Indonesian New Guinea—reflect the core–periphery relationship between these two sectors of the country. As the map suggests, similar contrasts, less sharply defined, also mark other Southeast Asian countries, and the primate cities of this realm (such as Bangkok, Manila, Jakarta, and Kuala Lumpur) are particularly dominant. In fact, Vietnam even contains two such anchors, which respectively represent its historic northern (Hanoi) and southern (Ho Chi Minh City) core areas (Fig. 11-4).

The states of Southeast Asia are not, by world standards, especially populous. Indonesia today is the world's fourth-ranking country in population size (274 million), but no other Southeast Asian country contains even half that total. Three mainland countries—Vietnam, Thailand, and Myanmar—contain between 50 and 100 million people. But take Laos, quite a large country territorially (about the size of the United Kingdom), and note that its population is just over 7 million; similarly, Cambodia, half the size of Germany, is home to only 16.5 million. In part, such modest numbers on the mainland reflect natural conditions less favorable to farming than those prevailing on volcanic soils or in fertile river valleys; but more generally, this realm did not grow as explosively as neighboring realms did during the past century. Indeed, the populations of three large mainland states of Southeast Asia—Vietnam, Thailand, and Myanmar—are now growing at a rate below the global average of 1.05 percent annually.

Cultural Geographies

Religions of Southeast Asia

Southeast Asia, more than any other realm, is a historic crossroads of religions (see Fig. 1-9). With the migrants from the Indian subcontinent came their faiths—first Hinduism and Buddhism, later Islam. The Islamic religion, promoted by the increasing number of Arab traders who appeared on the scene, soon became the predominant religion in Indonesia (where 88 percent of the population adheres to Islam today). But in Myanmar, Thailand, and Cambodia, Buddhism remained supreme, and in all three of these mainland countries at least 90 percent of the people are currently adherents. In culturally diverse Malaysia, the Malays are Muslims (to be a Malay *is* to be a Muslim), and almost all Chinese are Buddhists; but most Malaysians of Indian ancestry remain Hindus.

Although Southeast Asia has generated its own local cultural expressions, most of what remains in tangible form has resulted from the infusion of foreign elements. For instance, take Angkor Wat, the enormous complex of religious structures built in Cambodia during the twelfth century AD and today one of the world's most famous monuments (see photo). It was originally constructed as a Hindu temple, dedicated to the god Vishnu. The carvings on the walls of Angkor Wat tell the stories of the Hindu epics, and the temple's designs are closely associated with Hindu cosmology. But during the fourteenth century, Buddhism took over this part of the realm, and the temple complex was transformed into a major place of worship for Buddhists.

A group of young Buddhist monks is working the grounds around Cambodia's Angkor Wat, one of the world's most famous historic religious sites and a blend of Hindu and Buddhist influences.

Vietnam represents what is surely the ultimate fusion of diverse religious influences in Southeast Asia. In this country, for centuries there has been an almost casual blending of early Hinduism with Buddhism, Daoism, and Confucianism—all mixing comfortably with age-old traditions of ancestor worship.

The Ethnic Mosaic

Southeast Asia's peoples come from a common stock just as (Caucasian) Europeans do, but this has not prevented the emergence of regionally or locally discrete ethnic and cultural groups. **Figure 11-5** displays the broad distribution of ethnolinguistic groups in the realm, but be aware that this is a broad generalization. At the scale of this map, myriad smaller groups cannot be represented (for example, Myanmar alone has some 135 different ethnic groups).

The map shows the rough spatial coincidence on the mainland of major ethnic groups and the territories of contemporary political states. The Burman dominate in the country formerly called Burma (now Myanmar); the Thai inhabit the state once known as Siam (now Thailand); the Khmer form the nation of Cambodia and extend northward into Laos; and the Vietnamese occupy the long strip of territory facing the South China Sea—again, though, these larger ethnic groups can be divided into many smaller ones.

Territorially, by far the largest population mapped in Figure 11-5 is classified as Indonesian, the inhabitants of the great island chain that extends from Sumatera west of the Malay Peninsula to the Malukus (Moluccas) in the east and from the Lesser Sunda Islands in the south to the Philippines in the north. Collectively, all of these peoples shown on the map—the Filipinos, Malays, and Indonesians—are known as Indonesians, but they have been divided by history and politics. And note as well that the Indonesians in Indonesia itself include Javanese, Madurese, Sundanese, Balinese, and other major groups; moreover, hundreds of smaller groups again cannot be mapped at this scale. In the Philippines, too, island isolation and contrasting ways of life are reflected in the cultural mosaic. Also part of this Indonesian ethnic-cultural complex are the Malays, whose heartland lies on the Malay Peninsula but who form minorities in other areas as well. Like most Indonesians, the Malays are Muslims, although Islam is a more powerful force within Malay society than it generally is in Indonesian culture.

In the northern margins of mainland Southeast Asia, numerous minorities inhabit remote corners of the countries in which the Burman (Burmese), Thai, and Vietnamese dominate. Those minorities tend to occupy such peripheral areas because the terrain is mountainous and the forest is dense, thereby hindering the governments of their national states from exerting complete control. This remoteness and sense of detachment give rise to aspirations of secession, or at least resistance to government efforts to establish full authority, often resulting in bitter ethnic conflict.

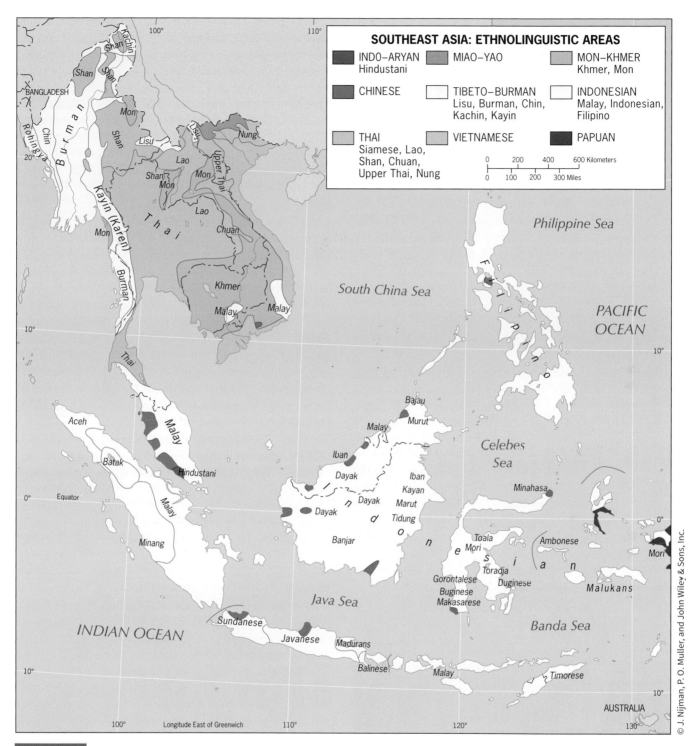

FIGURE 11-5

Immigrant Communities Figure 11-5 reminds us that Southeast Asia also is home to substantial ethnic minorities from outside the realm. On the Malay Peninsula, note the South Asian (Hindustani) cluster. Such Hindu communities with Indian ancestries exist in many parts of the peninsula, but in the southwest they form the majority in a small area. In Singapore, too, South Asians form a significant minority. These communities emerged during the European colonial period. However, South Asians had arrived in this realm many centuries earlier,

propagating Buddhism and leaving their architectural and cultural imprints on landscapes as far removed as Jawa and Bali. Most South Asian communities blended in reasonably well, though in recent years there has been growing tension between national governments and Muslim minorities in peripheral locations in the Philippines and in Thailand (see the separate sections on these countries, later in this chapter). And, in the far northwestern corner of the realm, along Myanmar's Bay of Bengal coastline, similar tensions escalated into outright violence and *ethnic cleansing* of a Muslim population known as Rohingyas. Originally from what is now adjacent Bangladesh, this long-downtrodden ethnic group has in recent years been singled out for intensified persecution—ironically at the hands of Myanmar's dominant Buddhist majority.

The Rohingya Refugee Crisis

Since 2012, the relaxation of harsh military dictatorship in Myanmar, accompanied by tentative democratic openings, has been widely celebrated. But this transition has also unleashed previously suppressed ethnic tensions: in 2013, the world was stunned by the wave of savage violence perpetrated by Buddhists against a small Islamic minority residing in Myanmar's northwestern Rakhine State. Known as the Rohingyas (see Fig. 11-5), they have lived in Myanmar for generations, and in 2013 numbered about 2 million. These Muslims, who were never fully accepted into Myanmar's society, had been denied citizenship and were subject to discrimination and intermittent acts of violence. The Buddhist majority has long referred to them as "Bengalis," a reference to their origins in neighboring Bangladesh that borders Rakhine on the north (Fig. 11-5).

Since this upsurge in ethnic violence began in 2013, at least 200,000 Rohingyas have been interned "under protection" in (or fled to) refugee camps operated by the government. Many more fled the country, and it is estimated that only about half a million (of the two million) are left in Myanmar. About one million are in refugee camps in Bangladesh, and the rest took flight to Malaysia, Saudi Arabia, Pakistan, Thailand, and India. Rohingyas by the thousands desperately took to the Indian Ocean after Myanmar military blocked off overland routes (photos below). By 2015, this refugee crisis had finally attracted international attention and condemnation of the Myanmar government.

Two important countries, however, kept a low profile and did not openly criticize Myanmar: India and China. We return to this matter in the section on Southeast Asian geopolitics later in the chapter.

The Overseas Chinese in Southeast Asia

By far, the largest immigrant minority in Southeast Asia is of Chinese origin. The Chinese began arriving here during the Ming and early Qing (Manchu) dynasties, with the largest exodus occurring in the late colonial period (1870–1940), when as many as 20 million immigrated. The European powers at first encouraged this influx, employing the Chinese in administration and trade. These **overseas Chinese [7]** soon began to congregate in the realm's major cities, where they established Chinatowns and gained control over much of the local commerce. World War II was about to break out, and the colonial era would end soon thereafter.

Figure 11-6 shows the migration routes and current concentrations of Chinese in Southeast Asia. Most migrants originated in southern China's Fujian and Guangdong provinces, and a large number invested much of their wealth back in China when the country opened up to foreign businesses in the 1980s. Clearly, the overseas Chinese of Southeast Asia have played a significant role in shaping the People's Republic of China's (PRC's) economic miracle.

Today Southeast Asia is home to almost 30 million overseas Chinese, more than half of this Sinic diaspora's worldwide total. Their lives here have often been challenging. The conquering Japanese relentlessly persecuted Chinese living in Malaya during World War II. Later, in the 1960s, Chinese in Indonesia were accused of communist sympathies, and hundreds of thousands were killed. In the late 1990s, Indonesian

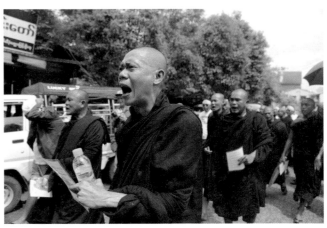

AP Images/S. Yulinnas

Aubrey Belf/REUTERS

The ethnic conflict between the Buddhist majority and the Rohingya Muslim minority in Myanmar's Rakhine State became one of Southeast Asia's major news stories of the 2010s. It first came to light in 2013, and then prominently resurfaced in the spring of 2015 when the accelerating flight of Rohingyas created an international refugee crisis. In the left photo, dated May 20, 2015, Rohingya migrants await rescue from an overloaded boat in the Andaman Sea just offshore from the northernmost portion of Sumatera, one of Indonesia's main islands (Fig. 11-1). The right photo, taken just one week later in Myanmar's largest city (Yangon), captures the protest march of several hundred irate Buddhist monks to denounce foreign criticism of the country's mistreatment of its Rohingya population. Today, about three-quarters of the Rohingya have fled abroad.

CHINESE IN SOUTHEAST ASIA

→ Migration route

Major Concentration

Secondary Concentration

National capitals are underlined

FIGURE 11-6

because "old" Chinese and "new" Chinese today refer to distinctly different identities within Singapore's multicultural mosaic.

Interestingly, the Chinese government has taken notice and has been responding with a series of gestures aimed at improving relations with Singapore (and elsewhere in Southeast Asia), and at nurturing the Chinese identity of these overseas migrants who left their home country often generations ago. They organize subsidized exchanges, conferences, and visits of the migrants to their ancestral villages in China. To some observers, though, it only heightens the suspicion that China is looking to increase its influence even more.

Economic Geography of Southeast Asia

Southeast Asia's Emerging Markets

If Japan and the Asian Tigers set the early example in the 1960s and 1970s, with post-Mao China following suit since the mid-1980s (Chapter 10), several Southeast Asian countries have since the turn of the century joined the ranks of the world's rapidly **emerging markets [8]**. Vietnam, Indonesia, and Malaysia, particularly, have in recent years attracted substantial foreign investment and exhibited decidedly robust economic growth rates.

But this realm is marked by decidedly extreme economic disparities. It is highly developed Singapore that records Southeast Asia's highest income—at a level about halfway between those of wealthy Luxembourg and Switzerland. The tiny, oil-rich state of Brunei is second, and Malaysia comes in a distant third, followed by Thailand. Populous Indonesia and the Philippines exhibit per-capita incomes less than one-tenth that of Singapore. Finally, Vietnam, Laos, Cambodia, Myanmar, and East Timor average around half that of Indonesia. Indeed, economically this realm is so internally dissimilar that its countries fall into all four World Bank income categories (see Fig. 1-10, Chapter 1).

Singapore's Leadership

There can be little doubt that Singapore is the economic heart of Southeast Asia. With a mere 5.9 million citizens and only 619 square

mobs again attacked Chinese and their property because they resented their relative wealth, and because many Chinese had converted to Christianity during the colonial era and were now targeted by Islamic throngs. Such hostility continues to smolder in various parts of Southeast Asia, and episodic flare-ups against the Chinese are likely to continue.

In recent years, Singapore has experienced its own unique challenges with the latest wave of overseas Chinese. Unique, because the overwhelming majority of native Singaporeans are themselves of Chinese descent—even though they are third-, fourth-, or fifth-generation and have to varying degrees blended into the surrounding Malay culture (Singapore and Malaya were both part of the same British colonial entity for more than 130 years before independence in the 1960s). However, that shared ethnicity has not prevented Singaporeans from becoming ever more critical of the ongoing influx of "newly rich" mainland Chinese. Resentment continues to rise concerning the role of purportedly arrogant, wealthy immigrants who are driving up real-estate prices, commandeering the best jobs, and disrespecting the city-state's strict codes of public behavior. Overall, this clash increasingly erodes social relations kilometers (240 sq mi) of territory, we are obviously not referring to size—this is all about the connections and centrality that enable Singapore to function as the predominant **node [9]** in a realm-wide economic network. Singapore ranks as a top-tier **world city [10]** because it has major international linkages and exerts global influence. Singapore's container port is not only the largest in the realm but also the second-biggest in the world, underscoring its key role within and far beyond Southeast Asia (photo below). Moreover, ultramodern Changi Airport Singapore (SIN) is consistently ranked as one of the world's best international airports.

Singapore's commanding regional position results from its exceptional relative location. In the seventeenth century, it was Malacca farther up the Malay Peninsula that was the leading hub for trade and shipping in Southeast Asia. The Strait of Malacca (Fig. 11-8), named after that town, was already the most important sea route, providing access to the realm's waters for ships coming from the west. When the British sought to displace Dutch dominance here during the eighteenth century, they discovered an even better local base of operations: Singapore Island. Because it possessed a larger and deeper natural harbor than Malacca to accommodate the bigger steamships of the time, Singapore swiftly rose to prominence as the British consolidated their power over this key corner of the world.

Most important, Singapore today is a symbol of modernity, a model for Southeast Asia's future. Throughout the realm, those who can afford it go there to shop, to connect to intercontinental flights, to transact business, to invest in real estate, or to send their children to one of the city-state's highly ranked universities.

Singapore, Southeast Asia's unmatched world city, has a state-of-the-art container port whose annual cargo-handling volume is second only to Shanghai's. This thriving city-state simultaneously functions as the nucleus of the realm's spatial economic system as well as its vitally important gateway to the global economy.

Prospects of Realm-Wide Integration: ASEAN

The overall development of Southeast Asia still has a long way to go. Political stability and increased regional integration will facilitate the process, and that is the long-term goal of **ASEAN [11]**, the Association of Southeast Asian Nations. Founded during the late 1960s, this supranational organization has primarily been concerned with security. But that has been a constantly challenging effort because a wide range of conditions mark its ten member-states. With the lone exception of minuscule East Timor (which has long sought to join), ASEAN encompasses all of the realm's countries. These include one influential city-state (Singapore), an Islamic oil state (Brunei), two impoverished communist regimes (Vietnam and Laos), and a reforming military dictatorship whose population is rising from the ranks of the world's most deprived (Myanmar).

Another problem that ASEAN has failed to resolve, one that literally affects health across much of this realm, is the recurrent air pollution caused by Indonesia's massive, human-ignited forest fires. Depending upon weather conditions and prevailing winds, thick plumes of smoke emanating from Indonesia's Sumatera (where most of the burning occurs, often to enlarge oil palm plantations) stream out toward Jawa, Singapore, Malaysia, and countries farther afield (see photo on page 428). Repeatedly, Singapore's government has had to advise residents to stay indoors (a new record-high pollution level was set in 2013). Indonesia has been especially slow to address this environmental crisis and even refused to ratify the 2002 ASEAN Agreement on Transboundary Haze Pollution.

In 1992, 25 years after its founding, ASEAN was able to expand into the economic domain through **AFTA [12]**, the ASEAN Free Trade Agreement, and here the payoff has been more substantial. AFTA has both engendered the lowering of tariffs and encouraged the expansion of trade within Southeast Asia. With lower wages than China, certain foreign-investment flows (e.g., in the garment industry) have shifted to Southeast Asia. And intra-realm trade has indeed surged in recent years—a critical forward step to avoid being completely overshadowed by China.

China in Southeast Asia Today

Since the turn of this century, Southeast Asia has increasingly been drawn into the economic (and political) orbit of China. In 2010, ASEAN and China concluded a landmark free-trade agreement, with the Chinese most interested in acquiring raw materials from mainland Southeast Asia as well as accessing the realm's growing export markets. As **Figure 11-7** in **Box 11-4** shows, trade shifts resulting from this agreement have been significant.

East Asia and Southeast Asia are now steadily growing closer economically. This heightened interaction is not only about trade but also about investment flows, industrial development, divisions of labor, and infrastructural integration. With wages in Southeast Asia currently about one-third those of China's, this differential is fueling the relocation of industrial activity from China to Southeast Asian countries. Those manufactures are then exported directly from those countries to China and/or other parts of the world. Increasingly,

Box 11-4 Geographic Information Analysis

Trade Patterns in Southeast Asia

The maps show the leading trade partners for countries across Southeast Asia in 2000 and 2015. In 2000, China was the leading import partner for how many Southeast Asian countries? What about in 2015? Which countries were displaced as leading (a) import, and

(b) export, partners between 2000 and 2015 across Southeast Asia? What can changing trade partnerships across the realm tell us about the global economy?

Access your WileyPLUS course to interact with a dynamic version of this map and to engage with online exercises and questions.

I: Leading import partner
E: Leading export partner

China · Hong Kong · Indonesia · Japan · Malaysia · Singapore · Thailand · United States

FIGURE 11-7 Interactive

it is low-skill jobs that move to Southeast Asia, whereas the better-paying, higher-skilled jobs remain in China.

In order to facilitate the increased production and transportation of goods within as well as to and from China, the Chinese established the Asian Infrastructure Investment Bank (AIIB) in 2015. The AIIB's members include most of the realm's states, and several European countries plus New Zealand have also joined (the United States has not). With the Chinese now investing heavily to exploit Southeast Asia's raw materials, infrastructural improvements to upgrade connections to China are a leading priority. Hence, it is no surprise that this realm occupies an important place in China's *Belt and Road Initiative* (see Chapter 10).

One important project is a high-speed rail network originating in Kunming, interior southern China's routeway hub for continental Southeast Asia and capital of Yunnan Province. The main rail corridor leads south from Kunming across northern

Laos to Vientiane, the Laotian capital on the Mekong River that also forms the border with Thailand (**Box 11-5**). From there, state-of-the-art rail lines will fan out in stages to Thailand, Cambodia, Vietnam, Malaysia, and eventually Singapore. Another high-priority project is the recently completed pair of oil and gas pipelines from Myanmar's Indian Ocean coast to Kunming (discussed later in the section on Myanmar).

In part as a result of this Chinese regional economic stimulation, Southeast Asia is growing faster than any other developing realm. And since the building of this inter-realm relationship has been fast-tracked, East and Southeast Asia are already more economically integrated than either North America with Middle America or Europe with Africa. China is the biggest investor in most Southeast Asian countries; the United States is not the biggest investor in any country in the realm, and in terms of trade, too, its presence has been fading rapidly over the past decade.

Box 11-5 From the Field Notes . . .

"It is the afternoon of February 21, 2019, and we are slowly cruising up the Mekong River towards the old royal city of Luang Prabang in north-central Laos (see Fig. 11-12). A mighty river it is, I am guessing about 200 meters (660 ft) wide most of the time. On the banks of the river, among the lush tropical vegetation, we see some small villages and croplands, and small fishing boats tied up to wooden poles or docks. Laos has a relatively small population, with only a third living in urban areas, and this area is quite underdeveloped. It makes for a quiet, peaceful boat ride. But change is in the air. Upstream, the construction of a huge new bridge is underway, part of the China-led road and rail corridor from Kunming to Laos's capital of Vientiane, south of here. Laos is key to Chinese ventures in this realm, and given its limited options it has been a willing participant in the *Belt and Road Initiative*. I wonder what this area will feel like ten years from now."

© J. Nijman

Even though China's boost to the realm's economic development has been quite successful so far, Chinese domination is creating unease in some quarters. Some of the anxiety stems from China's sheer economic prowess, which produces asymmetrical relationships; for instance, many infrastructure projects are formidable joint ventures that require comparatively larger investments from these much smaller Southeast Asian states. Moreover, the issue is not just about economics: for countries such as Vietnam, historic sensitivities and/or long-standing rivalries inevitably come into play.

Geopolitics in Southeast Asia

The Impact of Colonialism Before 1950

When the European colonizers arrived in Southeast Asia, they encountered a patchwork of kingdoms, sultanates, principalities, and other traditional political entities whose leaders they tried to co-opt, overpower, or otherwise fold into their imperial schemes. These colonizers forged empires here, often by playing one state off against another; the Europeans divided and ruled. Out of this foreign intervention, and subsequent decolonization and independence, emerged the modern map of Southeast Asia, and only Thailand (formerly Siam) survived the colonial era as an independent entity—because it conveniently served as a buffer between the French sphere to the east and the British to the west.

Figure 11-8 shows the colonial framework in the late nineteenth century (before the United States assumed control over the Philippines in 1898). The colonial powers divided their possessions into administrative units as they did in Africa and elsewhere. Some of these political entities became independent states when the colonial powers withdrew or were ousted by force. As noted earlier, the French named their empire *Indochina*. The *Indo* part refers to cultural imprints received from South Asia: the Hindu presence; the importance of Buddhism, which came to Southeast Asia via Ceylon (Sri Lanka) and its seafaring merchants; and the influences of Indian architecture and art (especially sculpture), writing, and literature as well as social structures. The *China* in the name Indochina signifies the role of the Chinese here: its emperors coveted Southeast Asian lands, and China's power penetrated deeply into this realm. French Indochina consisted of present-day Vietnam, Cambodia, and Laos (photo pair p. 410).

The British ruled a pair of major entities in Southeast Asia—Burma and Malaya—in addition to a large part of northern Borneo and many small islands in the South China Sea. Burma (now Myanmar) became independent in 1948, and the Federation of Malaysia was created in 1963 by the political unification of recently independent mainland Malaya, Singapore, and the former British dependencies on the largely Indonesian island of Borneo. Singapore, however, left this federation in 1965 to become a sovereign city-state, and the remaining components were later restructured into peninsular Malaysia and, on Borneo, Sarawak and Sabah. Thus the term *Malaya* properly refers to the geographic territory of the Malay Peninsula, including Singapore and other nearby islands; the term *Malaysia* identifies the politico-geographical entity of which Kuala Lumpur is the capital.

The Dutch had come to Southeast Asia in the late sixteenth century, primarily attracted by the so-called spice trade, and they ended up controlling most of what is now Indonesia. Among the plants domesticated by the local people on these islands of the "East Indies" were black pepper, cloves, cinnamon, nutmeg, ginger, turmeric, and other condiments essential to flavorful meals. In Figure 11-1 you can detect, in eastern Indonesia between Sulawesi and New Guinea, a group of small islands called the Maluku Islands (formerly the Moluccas). The Dutch colonizers called these

FIGURE 11-8

the *Spice Islands* because of the lucrative commerce in spices long carried on by Arab, Indian, and Chinese traders. Java (Jawa), the most populous and productive island, became the focus of Dutch administration; from its capital at Batavia (now Jakarta), Dutch colonialism threw a girdle around Indonesia's more than 17,000 islands, paving the way for the creation of the realm's largest and most populous state that is now home to more than a quarter-billion people.

Finally, the Philippines had been controlled by Spain since 1571. But these colonizers were confronted with a major uprising here just as the Spanish–American War broke out over the Caribbean island of Cuba in 1898 (there, as well, the locals revolted against Spanish domination). As part of the settlement of that brief war, the victorious United States replaced Spain in Manila as colonial proprietor and eventually guided the Philippines to independence in 1946. Today all 11 of Southeast Asia's states are independent, but centuries of imperial rule have left deeply embedded cultural imprints. In its urban landscapes, education systems, languages, and many other spheres, Southeast Asia will long carry the legacies of its colonial past.

Examples of the French colonial imprint in Ho Chi Minh City (formerly Saigon), Vietnam. To the left, the Notre Dame Cathedral that was built in 1880; to the right, the nearby post office from 1891, presenting a blend of European architectural styles and incorporating some Asian features.

In the years leading up to and during World War II, Japan occupied most of this realm and drove out the Western colonial powers. But in 1945, Japan was defeated, the power of former Western rulers was deflated, and this set the stage for independence: the Philippines in 1946; Burma (now Myanmar) in 1948; Indonesia, 1949; Laos, 1953; Cambodia, 1954; Malaysia, 1957; and Singapore, 1965. In 1954, the French colonizers were finally driven out of Indochina, but Vietnam's independence was a lot less straightforward. That country had become divided in a conflict between the communist north and French-supported south; and with the Cold War gearing up and the French departing, the United States entered the fray. It was not until 1975 that Vietnam achieved independence, and became a unified communist state one year later.

Southeast Asia in the Cold War

The confrontation between the United States and the Soviet Union during the **Cold War [13]** (1945-1991) took place for the most part on the margins of the Eurasian continent: Europe, the Middle East, South Asia, and Southeast Asia. These areas were accessible by sea for the United States, and they surrounded, sometimes from a distance, the Soviet Union. They conformed to the U.S. strategy of **containment [14]**, meant to prevent the Soviets from expanding their sphere of influence through a wide-ranging series of alliances and agreements with third countries.

The U.S. viewed Vietnam very much as a proxy of the Soviets and as a staging ground for Soviet expansion, and, given its location (a long coastline on the South China Sea and also bordering China, Laos, and Cambodia), Vietnam was considered especially important. When the French pulled out in 1954, the

U.S. moved in. By this time, the Cold War had gathered momentum, and the U.S. formally entered the war in 1964. The strategy of containment was supplemented with the so-called **domino theory [15]**, which held that destabilization and conflict from any cause in one country can result in the collapse of order in one or more neighboring countries, triggering a falling-domino chain of events that can affect a series of contiguous states in a region. In other words, the U.S. was concerned that the conflict might spill over from Vietnam into Laos and Cambodia, and from there into Thailand, Malaysia, and even then-Burma.

The Vietnam War (it is called the American War in Vietnam) ended in 1975 when the U.S. finally cut its losses. For Vietnam, it had been horrific and left deep scars. For the U.S. it had been traumatic, not only for the loss of lives but also because the U.S. could not win this war, and because it had caused so much suffering in Vietnam and so much division at home. The dominos never fell. After hostilities terminated in 1975, as many as 2 million Vietnamese refugees set out in often-flimsy watercraft onto the South China Sea; of those "boat people" who survived, a majority settled in the United States.

Political–Territorial Configurations

The present-day political geography of Southeast Asia is a function of physiography (natural boundaries such as coastlines, mountain ranges, rivers) and of the historical evolution of the regions of the realm, including the colonial impact. Look again at Figure 11-5 to see the correspondence of relatively long-settled ethnic groups and today's state boundaries to understand that the formation of these boundaries has been a long historical process that started well before nineteenth-century colonialism.

Political boundaries define and delimit states. They also create the mosaic of interlocking territories that give individual countries their shape or **state territorial configuration** [16]. A country's shape can affect its condition, and in certain cases even its survival. Vietnam's extreme elongation has influenced its existence since time immemorial. And, as noted later in this chapter, Indonesia has tried to redress its fragmented layout—consisting of thousands of far-flung islands—by promoting unity through the "transmigration" of residents from its most populous, core-area island (Jawa) to many of the others.

Political geographers have identified four types of state territorial configuration. Not all states will neatly fit in one of these categories, and clearly there are many that combine some of their different features. But all of these prototypical shapes can be seen on the political map of Southeast Asia, and **Figure 11-9** displays the terminology and examples.

Compact states [17] have territories shaped somewhere between round and rectangular, without major indentations. This encloses a maximum amount of territory within a minimum length of boundary. Southeast Asian example: Cambodia.

Protruded states [18] have a substantial, usually compact territory from which extends a peninsular or other corridor that may be landlocked or coastal. Southeast Asian examples: Thailand and Myanmar.

Elongated states [19] have territorial dimensions in which the length is at least six times the average width, creating a state that lies astride environmental and/or cultural transitions. Vietnam is a good example. Its pronounced elongation played an important part during the war. The country is widest in the north and south, with a slender "waist" in its middle zone—ensuring that North and South Vietnam were worlds apart in a number of ways.

Fragmented states [20] consist of two or more territorial units separated by foreign territory or a substantial body of water. Subtypes include mainland–mainland, mainland–island, and island–island. Southeast Asian examples: Malaysia, Indonesia, the Philippines, and East Timor.

For so comparatively small a realm, the 11 countries of Southeast Asia display a considerable variety of shapes. But one note of caution: states' territorial configurations do not determine their viability, cohesion, unity, or lack thereof; they can, however, influence these characteristics. Cambodia's compactness, for instance, has not ameliorated its divisive political geography.

© J. Nijman, P. O. Muller, and John Wiley & Sons, Inc.

FIGURE 11-9

Hybrid Warfare in the South China Sea

If China's economic role in Southeast Asia is frequently viewed with mixed feelings, its recent geopolitical and military forays are increasingly met with indignation and opposition. Much of this revolves around China's maritime ambitions and its claims to territorial waters in the South China Sea. Since 2009, the PRC government has circulated its so-called **Nine-Dashed-Line map [21]** displaying Chinese claims in these waters—a map said to date back to 1949, the year the PRC was founded. The "nine dashes" refer to the delimitation of Chinese maritime claims that effectively cover most of the South China Sea, and are mapped in red in **Figure 11-10** in **Box 11-6**. In the words of one high-ranking Chinese official in 2012, "China does not want all of the South China Sea, it just wants 80 percent." But that 80 percent includes some 40 islands (which China says are illegally occupied by other countries) as well as vital international shipping lanes and seafloor zones potentially rich in deposits of petroleum and natural gas.

Box 11-6 Geographic Information Analysis

© J. Nijman, M. Shin, P. O. Muller, and John Wiley & Sons, Inc.

MARITIME CLAIMS IN THE SOUTH CHINA SEA
- Brunei
- China
- Malaysia
- Philippines
- Vietnam

(Note: Taiwanese claims not shown)

FIGURE 11-10 Interactive

Maritime Claims in the South China Sea

The South China Sea is now the world's most contested maritime geopolitical space. The name South *China* Sea reflects the dominance of China, but other countries, such as the Philippines, Malaysia, and Vietnam, have staked their claims as well, mostly on the basis of United Nations maritime conventions (discussed in Chapter 13). The map shows the extent of these conflicting claims. Which pairs of countries are engaged in the greatest territorial conflicts? Are there any conflicts that do not involve China? What would your position be on these conflicts if you were the government of Singapore?

Access your WileyPLUS course to interact with a dynamic version of this map and to engage with online exercises and questions.

The disputes and conflicts over territorial waters in the South China Sea are numerous and complicated. The most important concern: (1) the *Paracel Islands*, claimed by China, Taiwan, and Vietnam; (2) the *Spratly Islands*, claimed by China, Vietnam, Malaysia, and Brunei; and (3) the *Scarborough Shoal*, claimed by China, Taiwan, and the Philippines (Fig. 11-10). ASEAN has been hopelessly divided on these geopolitical controversies, with Vietnam, Malaysia, and the Philippines strongly countering Chinese claims; Cambodia and Laos (tacitly) supporting China; and Singapore and Indonesia trying to steer a more neutral, diplomatic course.

China's behavior and strategies in the South China Sea are an example of what some political geographers call **hybrid warfare [22]**: a deliberate mix of tactics set out to achieve aggressive geopolitical objectives while avoiding conventional (large-scale) war. This mix involves diplomacy, intelligence, trade and investments, technology and cyberspace, intellectual property theft, military muscle flexing, and small-scale or incremental deployments. While a number of these tactics have always been part of conflict or war, today it is the avoidance of all-out war that is key, by maneuvering simultaneously in different arenas, and technology is essential in this regard. Russia is also often said to engage in hybrid warfare, not only through its activities in cyberspace but also in the ways it has gone about the carefully planned annexation of the Crimea (see the chapters on Europe and Russia/Central Asia). Thus, hybrid warfare still plays out in real space with real geopolitical ramifications.

When satellite imagery in mid-2015 confirmed that China was dredging enormous quantities of sand to reclaim and enlarge several major reefs (especially Fiery Cross Reef and Mischief Reef; photos below) in the southern part of the sea to construct runways and naval bases, Vietnam, Thailand, and the Philippines began to pursue closer ties to the United States, clearly in search of an appropriate counterweight to Chinese leverage. All three quickly concluded military agreements with the United States that include joint military exercises as well as flyovers near Chinese construction sites by the U.S. Air Force. Significantly, the Philippines took its case to the International Court of Justice, which ruled against the PRC in July 2016, saying that China's claims had no legal basis and that it had violated international law. The Chinese quickly declared the ruling "invalid" and with "no binding force." Nevertheless, this ruling is likely to strengthen the non-Chinese claims.

Whereas the United States has declared its neutrality on the matter of who owns which islands and shoals, it has emphasized the importance of free access to the waters of the South China Sea. Why? Because this is one of the most crucial trade corridors in the world. Think not only of major shipping routes within the Southeast Asian realm but also of all the intercontinental trade coming from the west—from Europe and Africa as well as South and Southwest Asia—that passes through the Strait of Malacca, around Singapore, and then across the South China Sea on the way to Taiwan, South Korea, Japan, and China itself—and vice versa. Nearly half of all international oceanic trade passes through the South China Sea, and the stakes could not be higher.

But China's economic and military tactics and its overall strategy of hybrid warfare in the South China Sea appear to have paid off, and the U.S. has been maneuvered to a peripheral position. The observation of a high-ranking U.S. military official in 2018 said it all: "China is now capable of controlling the South China Sea in all scenarios short of war with the U.S."

DigitalGlobe/ScapeWare3d/Getty Images

Before-and-after (ca. 2013/September 22, 2015) satellite images of Fiery Cross Reef, located in the western part of the Spratly Islands group in the South China Sea (Fig. 11-10). The 2015 image shows the nearly completed construction of a Chinese military base, replete with a 10,000-foot runway.

Regions of Southeast Asia

Southeast Asia's first-order regionalization recognizes the mainland–island dichotomy that is obvious from any map of this realm. But things are quite a bit more complicated than that because the southern part of the Malay Peninsula, occupied by the states of Malaysia and Singapore, exhibits physiographic, historical, and cultural characteristics to justify its inclusion in the *Insular* (island) rather than the *Mainland* region. Using the political framework as our grid, we map the broadest regional division of Southeast Asia in **Figure 11-11**:

Mainland region: Vietnam, Cambodia, Laos, Thailand, and Myanmar (Burma)

Insular region: Malaysia, Brunei, Singapore, Indonesia, East Timor (Timor-Leste), and the Philippines

Special note should also be taken of the discordance between the political matrix and the cultural-geographic reality along the eastern periphery of the Insular region. Here, even though Indonesia controls the western half of the island of New Guinea, all of New Guinea belongs to the neighboring Pacific realm.

Region Mainland Southeast Asia

Five countries form the Mainland region of Southeast Asia (Fig. 11-11): the three remnants of French Indochina—Vietnam, Cambodia, and Laos—in the east; Thailand in the center; and Myanmar (formerly Burma) in the west. Although a single religion, Buddhism, dominates cultural landscapes, this is a multicultural, multiethnic region (see Fig. 11-5). This may be one of the less urbanized regions of the world, yet the Mainland contains several major cities, and the pace of urbanization is quickening today. Moreover, as Figure 11-11 shows, two countries (Vietnam and Myanmar) possess dual core areas.

Vietnam

Vietnam officially became a unified state in 1976, and since then the contrasts between north and south have diminished. The northern capital, Hanoi, has long lagged behind bustling Saigon

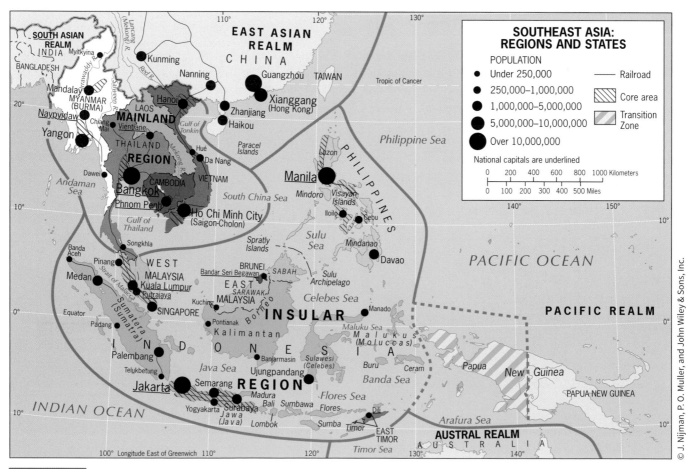

FIGURE 11-11

© J. Nijman, P. O. Muller, and John Wiley & Sons, Inc.

(now renamed Ho Chi Minh City to honor North Vietnam's revolutionary leader), but today its growing skyline reflects modernization and the overland links to its outport of Haiphong have been upgraded. With more than 8 million residents, Hanoi anchors the northern core area of Vietnam (**Fig. 11-12**), the lower basin of the Red River (its agricultural hinterland).

A leading development focus in Vietnam since 2000 has been the expansion of agricultural production. One result is that the country is now the world's second-largest coffee producer after Brazil, with most production taking place in the Central Highlands (Fig. 11-12). However, because Vietnam's mass-produced bulk coffee is raised plantation-style in the sun, the diffusion of this crop into the interior highlands has sparked substantial deforestation, worsening the annual monsoon-related floods that plague this now-unstable countryside. Other major Vietnamese crops, led by rice, include cashew nuts, rubber, tea, sugar, and spices—all with increasingly favorable export opportunities. Vietnam has also in recent years become an oil exporter.

About 80 percent of Vietnam's land is mountainous; the other 20 percent is made up of the major river deltas in the far north and the far south. Most of this country between the deltas has high relief and is forested, and is not so easily accessible. Not surprisingly, when Vietnam was a single state in historic times, the rulers preferred to have the government capital near its topographic center, about halfway north and south. During the entire nineteenth century and the first half of the twentieth century (even during the French colonial times) that royal capital was located in the city of Hué (Fig. 11-12; photo right).

The south has experienced more significant change and faster economic growth (**Box 11-7**) as this country of 97 million has opened up to the global economy. Today, Ho Chi Minh City is a burgeoning metropolis of almost 11 million, propelled by the rapid development of the east bank of the Saigon River, which includes a Chinese-style Special Economic Zone as well as the booming New Saigon business/residential district. In fact, the Saigon area now accounts for more than a quarter of Vietnam's industrial output and about a third of its tax revenues.

Somewhat ironically, Vietnam these days has become one of the most pro-American countries in Southeast Asia despite its persistence as a communist state. Undoubtedly,

The city of Hué is located at the waist of the country, on the coast, almost equidistant from Hanoi and Saigon. It was the royal capital of Vietnam from 1802 to 1945. The photo shows the entrance to the Citadel, the historic royal grounds that are now a UNESCO World Heritage Site. Not surprisingly, given its location near the dividing line between North and South, this area witnessed one of the longest and bloodiest battles of the Vietnam War.

Ho Ngoc Binh/Moment/Getty Images

Box 11-7 From the Field Notes . . .

© J. Nijman

"On a walking tour of Ho Chi Minh City (many here still call it Saigon), I encounter two contrasting images of present-day Vietnam, just minutes apart: a statue of the still-revered former communist leader Ho Chi Minh in front of City Hall; and, just three blocks away, the country's first Louis Vuitton store. The latter caters to the expanding base of higher-income Vietnamese consumers. So much for communism!"

FIGURE 11-12

these good feelings toward their former enemy in the "American War" at least partially reflect the growing perception of dominance by China, against whom the United States can provide a counterbalance. As noted above, American–Vietnamese relations warmed considerably during the 2010s, with Vietnam even opening its strategically located port at Cam Ranh Bay to visiting U.S. naval forces—the

original builder of these facilities during the their war a half-century earlier (Fig. 11-12).

Cambodia

Compact Cambodia is heir to the ancient Khmer Empire whose capital was Angkor and whose legacy is a vast landscape of imposing monuments, including the majestic Buddhist-inspired temple complex of Angkor Wat (see **Box 11-8**). Today, more than 85 percent of Cambodia's 16.5 million inhabitants are ethnic Khmers, with most of the remainder divided between Vietnamese and Chinese. The current capital, Phnom Penh, lies on the Mekong River, which bisects Cambodia before it enters and forms its massive delta in Vietnam (Fig. 11-12).

From 1970 to 1976, Cambodia suffered one of the most brutal and murderous regimes of the twentieth century. The so-called *Khmer Rouge* (*Red Khmer*, the name being a reference to its self-declared communist convictions and to the glorious past of the Khmer people) embarked on a malevolent reign of terror and destruction to reconstruct *Kampuchea*—as leader Pol Pot referred to Cambodia—as a rural society. They drove townspeople into the countryside where they had no place to live or work, emptied hospitals to send the sick and dying into the streets, and outlawed both religion and family. By the time they were ousted from power by neighboring Vietnamese forces, as many as 2 million Cambodians (out of a population of 8 million at the time) had lost their lives in what forever will be known as the Cambodian Genocide.

Even five decades later, the country's postwar trauma still lingers. Many young children of the 1970s, now in their forties or fifties, became orphans and faced an uphill struggle in life. Almost every family can tell of parents and/or close relatives who were murdered by the regime. The long-lasting effects of social and economic dislocation, especially in the countryside, have still not been overcome (see **Box 11-9**). The forced migration in the 1970s of people from the cities to rural areas effectively destroyed the economy of whatever urban centers there were. Today, the urban share of the population is still one of the lowest in the world, at a mere 24 percent.

Today, despite Cambodia's continuing status as one of Southeast Asia's poorest countries, it has finally begun to achieve some meaningful economic growth, especially in the garment manufacturing sector. Some offshore oil reserves were discovered about a decade ago, but exploration has yet to gather momentum. Other investments from abroad are multiplying, mostly by producers seeking to relocate their operations out of China (where wages are steadily rising) to take advantage of Cambodia's cheaper labor. The Cambodian government is also optimistic that tourism, largely focused on

---

Box 11-8 Technology & Geography

Making the Invisible Visible—LiDAR at Cambodia's Angkor Wat

Remote sensing [23] refers to the indirect observation and measurement of the Earth with cameras, sensors, and scientific instruments mounted on unmanned aerial vehicles (UAVs, or "drones"), aircraft, or satellites. A special type of remote sensing technology known as Light Detection and Ranging or *LiDAR* is now ushering in a new era of geographic exploration and discovery. Originally developed to measure such atmospheric phenomena as clouds and wind, LiDAR is now used to survey and map the surface of the Earth. LiDAR sensors actively transmit thousands of pulses of light per second with a laser and then record the length of time it takes for those pulses to be reflected back from the surface. When combined with the position, direction, and orientation of the aircraft and sensor, an incredibly rich elevation dataset is produced. What makes LiDAR imagery so special is its high resolution and accuracy. The density of LiDAR data also permits features such as forest and other vegetation cover to be filtered out, thereby making invisible objects on the Earth's surface visible.

LiDAR technology today is used and applied in an ever-increasing variety of situations such as agriculture and farming, humanitarian response and disaster preparedness, climate change research, and cultural heritage preservation. Here at the famous site of Angkor Wat in Cambodia, LiDAR is used by archeologists to obtain insights into the Khmer Empire and its preindustrial urban complexes. A recent expedition that mapped the area surrounding Angkor Wat using LiDAR revealed a series of previously unknown roads, canals, and even temple structures that were obscured by dense vegetation—and made even more difficult to survey due to

land mines buried during decades of conflict and war. It turned out that Angkor's urban region was far more extensive than previously thought. LiDAR can play an important role in the preservation and management of such cultural heritage sites by revealing their true geographic extent; providing detailed inventories of structures, temples, and buildings; and even detecting illegal excavations, looting, or encroachment.

2.4 km

The Angkor Wat temple complex and surroundings. Conventional high-resolution satellite image (top) showing mainly vegetation, compared to the LiDAR image (bottom) that reveals elevation details and terrain features such as roads, mounds, and canals.

Damian H. Evans et al. PNAS 2013;1110:31:12595-12600; ©2019 National Academy of Sciences

"Along one of the access roads leading to the monumental temple complex of Angkor Wat, I come across a group of middle-aged men playing traditional Cambodian music. When I get closer, I notice that all are amputees, a legacy of the vicious Pol Pot regime of the 1970s. During that upheaval and the war that followed, which resulted in the regime's ouster, millions of land mines were deployed across the country. Ominously, a large number remain buried in place, and the records indicating their location are poor or nonexistent. Not surprisingly, they continue to inflict casualties to this day, maiming and killing civilian men, women, and children. Most of the men performing here lost their limbs when they were adolescents or young adults. They now earn their livelihood from performing and selling their music—while actively campaigning for better government efforts to detect and disarm the mines. 'It is 40 years since Pol Pot,' one of the men tells me during a performance break, 'but our country is still recovering.'"

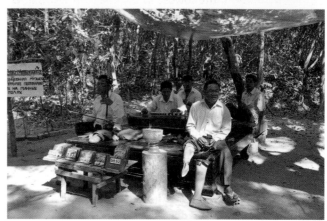

the Angkor temple complex in the northwest, will continue to expand and perhaps soon become a mainstay of the economy. For the time being, a great many people make a living in the primary sector, much as Cambodians have done for centuries, in rice farming and fishing (**Box 11-10**).

Laos

Landlocked Laos has five neighbors, one of which is China (Fig. 11-12). The Mekong River forms a long stretch of its western boundary, and the important sensitive border with Vietnam to the east lies in mountainous terrain. With only 7 million people—more than half of them ethnic Lao who are related to the Thai of Thailand—the country has a woefully inadequate infrastructure and hardly any industry, and less than 5 percent of its land is suitable for agriculture. Laos is only about one-third urbanized, but the capital and largest city, Vientiane, has a fast-growing population of just over one million.

Communist Laos has been very slow to follow the lead of its neighbors in opening up its economy, but today momentum is finally beginning to build. The country's first stock market began operating in 2011. Around the same time, a major new hydropower project on the central Laotian Nam Theun River, a tributary of the Mekong, began producing electricity for much of central Laos as well as selling surplus power to Thailand. And we have already discussed the much bigger Xayaburi Dam project on the Mekong itself (see page 400), which despite drawing wide regional criticism opened in 2019. Moreover, the new high-speed rail line from Kunming to Vientiane will open in 2021, and is certain to pull Laos even more tightly into China's orbit (see Box 11-5).

Thailand

In virtually every way, centrally positioned Thailand is the leading Mainland state. In contrast to its awakening neighbors, Thailand has been a strong participant in the realm's

"This boat ride on Lake Tonle Sap gives me a good impression of life on the lake. It is a huge freshwater lake (see Fig. 11-12), the largest in this realm, used for fishing, living, and as a reservoir for irrigation during the dry season. For many centuries, rice and fish have been the main staple of people here. It is a busy part of the lake, with floating homes, stilt houses, stores, and lots of boat traffic. The photo shows an NGO-supported charity, in the blue floating house to the left, that provides cheap food to locals—it is not easy to make a living around here, and many people are quite poor. Most of Cambodia's fresh fish comes from here, and the fish industry overall contributes about 15 percent to the national economy. As Cambodia slowly modernizes, this share is sure to come down, but it is hard to see how it will affect the people who have lived here for many generations."

economic development for decades. This country of 70 million exhibits some impressive demographic indicators: its annual growth rate is the lowest in the realm (even below Singapore's), and it leads the region in literacy and life expectancy. The capital, Bangkok, is one of the world's most prominent **primate cities [24]** and the largest urban agglomeration in the Mainland region—a megacity containing 16 million residents today (see photo).

As Figure 11-11 indicates, Thailand occupies the heart of the Mainland region. Even though it has no Red, Mekong, or Irrawaddy Delta, this country's central lowland is watered by a network of streams that flow off the northern highlands and the Khorat Plateau to the east. One of these waterways, the Chao Phraya, is the Mississippi of Thailand and forms the axis of the country's core area (**Fig. 11-13**). From the head of the Gulf of Thailand to Nakhon Sawan, this river functions as a highway network for boat traffic. Barge trains loaded with rice head toward the coast, ferries move upstream, and freighters transport tin and tungsten (of which Thailand is among the world's leading producers). Bangkok sprawls along both sides of the lower Chao Phraya floodplain, here flanked by skyscrapers, pagodas, modern factories, boat-sheds, ferry landings, luxury hotels, and myriad modest dwellings—all crammed together in crowded confusion on the swampy terrain where the central lowland meets the Gulf of Thailand.

At least some of Thailand's internal political instability of recent years can be attributed to the overreaching dominance of Bangkok and its elites as well as this primate city's considerable distance from the substantial electorates of the more peripheral rural areas. Marked by frequent government turnovers and episodes of military intervention, Thailand has yet to face up to the challenge of legitimizing national politics across this sprawling, unevenly developed country. And given Bangkok's location, particularly in the context of Thailand's protruded territorial configuration, the task of national governance is not made any easier by the presence

of a large Malaysian—and overwhelmingly Islamic—minority on the Malay Peninsula, Thailand's most remote corner that abuts the Malaysian border in the far south (Fig. 11-13).

As noted earlier in this chapter, Thailand is a classical protruded state, with its southward-pointing land corridor extending away from the more compactly shaped main body of the state. Thailand's protrusion extends for hundreds of kilometers into the upper Malay Peninsula, narrowing to less than 32 kilometers (20 mi) in the vicinity of the Kra Isthmus (Fig. 11-13). Figure 11-5 shows the dominance of ethnic Malays in the southernmost periphery of the protrusion—a population that is about 90 percent Muslim (the figure for Thailand as a whole is less than 6 percent). Moreover, the international border between Thailand and Malaysia is quite porous and can be crossed practically at will.

For more than a century, these southernmost provinces of Thailand have had closer ties with Malaysia across the border than with the Thai capital 1000 kilometers (600 mi) to the north. Today, rising Islamic militancy and violence here pose a mounting challenge to the Bangkok government (an estimated 7000 people have been killed and thousands more maimed by local terrorist attacks since 2004). These events also have economic consequences: Phuket and other coastal resorts facing the Andaman Sea in this area—key components of Thailand's important tourist industry—are located just northwest of the troubled Malaysian border zone and are regarded by many as increasingly vulnerable to a spillover of continuing hostilities (see Fig. 11-13, inset map).

Myanmar

Thailand's western neighbor, Myanmar (often still referred to as Burma, its former name), has a population of 55 million and is one of Asia's poorest countries. Long oppressed by one of the world's most corrupt and brutal military dictatorships, its government turned a corner in 2011 and soon surprised

A fish-eye, panoramic view of Central Bangkok's business district, seen from across the Chao Phraya River. With 16 million inhabitants, this is the biggest city in Mainland Southeast Asia, and the third biggest in the realm.

Frank Bienewald/LightRocket/Getty Images

FIGURE 11-13

the world by taking bold steps forward. Grinding poverty and crippling isolation along with sanctions (imposed by much of the world because of human rights violations) must finally have played a part in pushing Myanmar's leadership in a new direction. When the West embargoed and worked to isolate Myanmar's military regime in years past, China took the opportunity to invest here, and the Chinese have further ramped up their involvement in Myanmar in recent years.

Myanmar's Geographic Challenges

Myanmar's challenges are in some ways comparable to those of Thailand: both are protruded states, and in both cases the primary urban region is quite distant from the outermost rural peripheries (**Fig. 11-14**). Yangon (previously named Rangoon) is the country's main port, economic center, and primate city, and has a population of nearly 6 million. In 2006, the military junta capriciously relocated the capital from Yangon 300 kilometers

FIGURE 11-14

(200 mi) northward to a newly constructed headquarters named Naypyidaw, with the government asserting that its more central location would yield greater administrative efficiency as well as better protection against (unspecified) enemies. After a slow start, the new capital city now houses a million people in the wider urban area, not far behind the population size of the second city, the precolonial hub of Mandalay in central Myanmar (Fig. 11-14). Here, we discuss three important challenges to Myanmar: ethnic divisions, illicit drug production, and the impacts of global climate change (**Box 11-11**).

Myanmar's Ethnic Divisions. As Figure 11-14 reveals, the peripheral peoples of Myanmar occupy significant parts of the country that are designated as seven different States. The majority Burman-dominated areas, encompassing 68 percent of the national population, are designated as Divisions. Among the leading minorities, the Shan of the northeast and far north, who are related to the neighboring Thai, account for 9 percent of the population, or 5 million. The Kayin (Karen), who constitute 7 percent, live in the neck of Myanmar's protrusion and have proclaimed their desire to create an autonomous territory within a federal Myanmar. The Rohingyas of Rakhine State on the northwestern coast comprise the besieged Muslim minority discussed earlier in this chapter. The Mon (2 percent) were in what is today Myanmar long before the Burmans, and even introduced Buddhism to the national heartland of the Irrawaddy Basin; in fact, they continue to seek the return of ancestral lands from which they were ousted.

When Myanmar finally shook off the excesses of dictatorship in the early 2010s and embraced long-awaited reforms, it was hoped that this change would usher in a period of national progress and a more democratic society. To a certain extent, advances were made—political prisoners were freed, press censorship was largely ended, and long-overdue elections took place. But a byproduct of the ending of that lengthy era of repression was the resurfacing of ugly, long-simmering ethnic tensions. Besides the appalling ethnic cleansing of the Rohingya community in Rakhine State, violent clashes have taken place in the northernmost State of Kachin between government forces and the Kachin Independence Army. This strife also has potential international ramifications because the Kachin people, constituting 1.5 percent of Myanmar's population, are part of a more extensive ethnic community that spills across both the Chinese and Indian borders.

Illicit Drugs in the Golden Triangle. Yet another challenge from the periphery continues to plague the mountainous eastern margins of Shan State, which has long been part of the infamous opium-producing region known as the **Golden Triangle [25]**. That name refers to the huge profitability of opium (the raw material of heroin, harvested from the sap of poppies) on the global market, together with the triangular shape of this territory that encompasses northwestern Laos and northernmost Thailand in addition to eastern Shan State (Fig. 11-14). Although poppy cultivation has almost been wiped out in Thailand, it is strong in Laos and especially Myanmar. Today, with only Afghanistan producing more opium than Myanmar, it is hard to discern if the Naypyidaw government is truly interested in clamping down on the opium trade: this illicit narcotic remains the source of enormous foreign revenues that, through elusive business connections, tend to find their way into the formal economy—including real estate transactions and infrastructural works in Myanmar's cities. Moreover, in the last few years the area around Kutkai (Fig. 11-14) has become one of the world's largest hubs in the manufacturing of methamphetamine, a chemical drug that has seen rapidly rising demand in Asia. To some extent, in the marketplace, meth has taken over from heroin, and the organized traffickers and manufacturers (militias, ethnic insurgents, and the military are all probably involved) in the Golden Triangle have diversified their business.

Box 11-11 Global Climate Change in Myanmar

According to various listings of countries most at risk due to global climate change, Myanmar ranks near the top (other highly vulnerable countries are Haiti, the Philippines, and Bangladesh). Myanmar is a special case because it comprises highly diverse subregions that are very differently exposed to the impacts of climate change. The low-lying Irrawaddy delta is prone to flooding and destructive winds: from the monsoon water carried down by the river, but also by cyclones—which are expected to increase in frequency and severity with the warming of ocean water and rising sea levels. In 2018 alone, about 150,000 people were displaced because of flooding.

The problems are quite the reverse in Myanmar's north-central Dry Zone (Fig. 11-14). This is one of Myanmar's most climate-sensitive and poor areas, but it contains about a third of the country's population. It is situated in the **rain shadow [26]** of the Arakan Mountains (see Fig. 11-1) and therefore does not get sufficient rainfall. But things have deteriorated in recent decades: the monsoon has been getting shorter, from around 140 days in 1998 to around 125 today. Across the Dry Zone, water is scarce, vegetation cover is thin, and in places the soils are severely eroded. Irrigation systems are poorly developed. Most of the people here make a living in **subsistence farming [27]**, growing rice, sesame, and groundnuts (photo to right). There are reports that increasing heatwaves, droughts, and dry spells are forcing growing numbers of people to migrate elsewhere in search of better and more predictable livelihoods.

© J. Nijman

A small subsistence farm in Myanmar's Dry Zone. This farmer has a couple of cows, chickens, and a plot of land to grow crops. But it's late February 2019, well into the dry season, and nothing much seems to be growing at this time. The Irrawaddy River is about seven miles from here, but it is running low, and irrigation systems are poorly developed. The intensifying droughts in this part of the country are attributed to global climate change.

Myanmar's New Opportunities Geography is all about constraints and opportunities: while Myanmar faces some major challenges that we just outlined, it also has some important new opportunities after it opened up to the world. We focus here on three major infrastructural projects that are presently underway, involving the neighboring countries of India, Thailand, and China. Each of these projects would strengthen Myanmar's external economic relations, inside and beyond the realm.

The first is a joint venture with India to renovate and expand port facilities in Sittwe, the strategically located coastal city facing the northern end of the Bay of Bengal (Fig. 11-14). India's overriding interest here is to increase the accessibility of its landlocked easternmost States, a deeply impoverished highland periphery difficult to reach via domestic overland routes. A fully developed port at Sittwe would allow the opening of a shorter, easier southern routeway: ships sailing around Bangladeshi waters from nearby Kolkata can offload their goods onto smaller vessels at Sittwe; the smaller vessels could then travel up the Kaladan River to the Indian border, from where trucks using upgraded roads would cover the final leg (Fig. 11-14).

The second infrastructural project concerns China's completion of a dual set of oil and gas pipelines from a new port terminal at Kyaukphyu, south of Sittwe, running 1150 kilometers (700 mi) northeastward to Kunming, China's thriving gateway city for Southeast Asia (Fig. 11-14). These linkages are hugely important to the PRC because they directly supply interior southern China with urgently needed fuels, thereby circumventing the much longer maritime route through the Strait of Malacca and the South China Sea. And Myanmar benefits as well: the pipelines cross the center of the country, providing a series of delivery points in such key urban centers as Mandalay.

The third project, located in the far south, is perhaps the most significant enterprise from Myanmar's point of view—the so-called Dawei Development Project, a joint initiative with neighboring Thailand to convert the small southern city of Dawei on the Andaman Sea into a world-class port facility and regional trade hub—bolstered by new overland connections to Bangkok in the heart of Thailand's burgeoning core area.

Region Insular Southeast Asia

On the peninsulas and islands of Southeast Asia's southern and eastern periphery lie 6 of the realm's 11 states (Fig. 11-11). Few regions in the world contain so diverse a set of countries. Malaysia, the former British colony, consists of two major territories separated by hundreds of kilometers of South China Sea. The realm's southernmost state, Indonesia, sprawls across thousands of islands from Sumatera in the west to New Guinea in the east. North of the Indonesian archipelago lies the Philippines, another island-chain country that once was

a U.S. colony. Territorially, these are three of the most acutely fragmented states on Earth, and each has faced the challenges that such politico-geographical division presents. This Insular region of Southeast Asia also contains three small but significant sovereign entities: a city-state, a sultanate, and a ministate. The city-state is Singapore, once a part of Malaysia (and one instance in which internal centrifugal forces were too powerful to be overcome). The sultanate is Brunei, an oil-rich Muslim country on the island of Borneo that seems to be transplanted from the Persian Gulf. The ministate is severely impoverished East Timor, which only achieved independence (from Indonesia) in 2002.

Malaysia

One portion of Malaysia's national territory lies on the Asian mainland and the other on one or more islands. The country is a colonial political artifice that combines a pair of decidedly disparate components into a single state: the southern end of the Malay Peninsula and the northern sector of the island of Borneo. These are referred to, respectively, as *West Malaysia* and *East Malaysia* (Fig. 11-11). The appellation "Malaysia" came into use in 1963, when the original Federation of Malaya, on the Malay Peninsula, was expanded to incorporate the territories of Sarawak and Sabah in Borneo. When the name *Malaya* is used, it refers to the peninsular part of the Federation, whereas *Malaysia* refers to the total political entity.

The Malays of the peninsula constitute 50 percent of the country's population of 32 million. They possess a compelling cultural identity expressed in adherence to the Muslim faith, a common language, and a strong sense of territoriality. The Chinese came to the Malay Peninsula as well as to northern Borneo in substantial numbers during the colonial era, and now they account for roughly 23 percent of Malaysia's population. Hindu South Asians were in this area long before the Europeans, and for that matter before the Arabs and Islam arrived on these shores; today they still form a substantial minority of almost 7 percent of the population, clustered, like the Chinese, on the western side of the peninsula (see Fig. 11-5 and **Fig. 11-15**).

Malaysia's ethnic and racial groups have long coexisted in relative harmony, but relations today are marked by rising social tensions. As Malays increasingly assert claims based on "majority rights," the Chinese find themselves in much the same minority position that marks the overseas Chinese across the realm. The Indian population is much smaller and feels increasingly crowded out. Add religious tensions to the mix (that occasionally include Muslim demands for a separate *Sharia* legal system), and the notion of a common Malaysian identity begins to look rather fragile.

West Malaysia. The populous peninsular component of Malaysia remains the country's dominant sector, containing 11 of its 13 States and approximately 80 percent of its

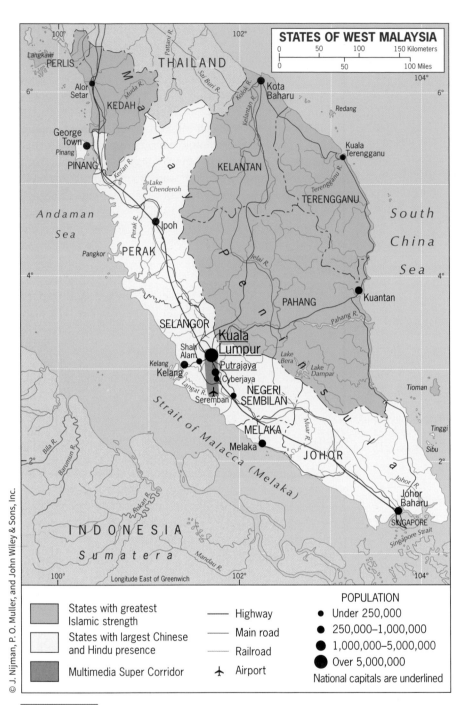

© J. Nijman, P. O. Muller, and John Wiley & Sons, Inc.

FIGURE 11-15

that annually transit through it constitute fully one-third of the world's shipping and carry one-quarter of its oil.

Malaysia, despite the loss of Singapore and notwithstanding its recurrent ethnic troubles, has become a leading player in contemporary Southeast Asia. Its new *Multimedia Super Corridor* not only anchors the national core area but also showcases the country's commitment to twenty-first-century technology—symbolized by Putrajaya, the recently built high-tech administrative capital. Elsewhere, the strong skills and modest wages of the local workforce have attracted many multinational corporations, and the government has capitalized on its opportunities, such as encouraging the creation of an ultramodern manufacturing complex on the far northwestern island of Pinang, where Chinese outnumber Malays by two to one.

East Malaysia. The decision to combine the 11 sultanates of Malaya with the States of Sabah and Sarawak on Borneo (Fig. 11-11), creating the country we now call Malaysia, had far-reaching consequences. Those two offshore States may constitute 60 percent of Malaysia's territory, but they are home to barely 20 percent of the population. They endowed Malaysia with abundant energy resources, huge stands of timber, and vast tracts of land suitable for palm oil plantations (see Box 11-3). But they also complicated Malaysia's ethnic complexion because each State contains more than two dozen indigenous groups (for example, immigrant Chinese form Sarawak's largest ethnic minority). These locals complain that the federal government in Kuala Lumpur treats Malaysian Borneo as a colony, and politics here are contentious and fractious. It is therefore quite likely that Malaysia will confront intensifying devolutionary forces in East Malaysia.

Brunei

Also located on Borneo—near the midpoint of the northwestern coast, adjacent to where Sarawak and Sabah meet—is Brunei (see inset map in Fig. 11-17 on page 426), a rich, oil-exporting, Islamic sultanate far from the Persian Gulf. Brunei, which became independent from the United Kingdom in 1984,

population. Here, the government has tightly controlled economic and social policies as it pursues nationwide modernization. The Strait of Malacca (Meloka) to the west continues to be one of the world's busiest and most strategic waterways, and forms yet another crucial *choke point* in the seaborne flow of resources and goods between major geographic realms. A longstanding problem in the Strait, which markedly resurfaced in the mid-2010s, is piracy. The spatial configuration of this narrow, 1000-kilometer- (625-mi)-long waterway makes it highly susceptible to such attacks. Moreover, the Strait is always a most inviting target because the 50,000-plus vessels is only slightly larger than Delaware and has a population of just 440,000. Thus, the sultanate is a mere ministate—but a most prominent one since the discovery of oil and natural gas, which turned it into Southeast Asia's richest country. The Sultan of Brunei rules as an absolute monarch; his palace in the capital, Bandar Seri Begawan, is reputed to be the world's largest. And he will have no difficulty finding customers for his oil in energy-poor eastern Asia. Even though Brunei's oilfields may become exhausted by mid-century, major new supplies have recently been discovered beneath the seafloor of Brunei's territorial waters. Not taking any chances, and with an eye clearly on the future, the government has been investing heavily in overseas assets such as real estate in London and elsewhere—working to develop alternative revenue sources should oil income begin to decline.

Brunei's expanding population—in which immigration plays a larger role than internal natural increase—is ethnically two-thirds Malay and about 12 percent Chinese. Most inhabitants live and work near the offshore oilfields in the northern district of Brunei Muara that contains the capital city. Evidence of profligate spending is everywhere, ranging from sumptuous palaces to magnificent mosques to the most luxurious of resort hotels. In this respect, Brunei offers a stark contrast to surrounding East Malaysia, but even within Brunei there are significant differences. The country's interior, where a small minority of indigenous groups

survive, remains an area of subsistence agriculture and rural isolation, its villages a world apart from the modern splendor of Bandar Seri Begawan.

Singapore

In 1965, a fateful event occurred in Southeast Asia. Singapore, the crown jewel of British colonialism in this realm, seceded from the recently independent (1957) Malaysian Federation and became a sovereign state, albeit a tiny city-state (**Fig. 11-16**). With its unparalleled relative location, its multiethnic and well-educated population, and its no-nonsense government, Singapore then overcame the severe limitations of space and the absence of raw materials to become one of the four original Asian Tigers in the late 1970s.

With a mere 619 square kilometers (239 sq mi) of territory and 5.9 million people, space is at a premium in Singapore, and this is a constant challenge for the government. The highest priority is to develop cutting-edge but space-conserving manufacturing and service industries. Benefiting from its relative location, the old port of Singapore had become one of the world's busiest even before independence. It thrived as an *entrepôt* **[28]** between the Malay Peninsula, the rest of Southeast Asia, Japan, and other emerging economic powers on the Asian Pacific Rim. Crude oil from Southeast Asia still is unloaded and refined in Singapore, then shipped to East Asian

FIGURE 11-16

© J. Nijman, P. O. Muller, and John Wiley & Sons, Inc.

FIGURE 11-17

destinations. Raw rubber from the adjacent Malay Peninsula and from Indonesia's nearby island of Sumatera is shipped to China, Japan, the United States, and many other countries. Timber from Malaysia, rice, spices, and other foodstuffs are processed and forwarded through Singapore. In return, automobiles, machinery, and equipment are imported into Southeast Asia and distributed almost exclusively via Singapore.

Singapore's current development strategy stresses two primary objectives. The first is to focus its industries more tightly on leading-edge information technology, automation, and biotechnology. The second is to forge an ever stronger **Growth Triangle [29]** involving Singapore's developing neighbors, Malaysia and Indonesia. Accordingly, Malaysia and Indonesia would supply the necessary raw materials and relatively inexpensive labor, and Singapore would supply the capital and technological expertise. The ethnic composition of Singapore's population today is just over 75 percent Chinese, 15 percent Malay, and 7 percent South Asian. The government is Chinese-dominated, and its policies have served to sustain ethnic-Chinese control. Indeed, the government of Singapore has encouraged immigration from the People's Republic of China to stabilize the ethnic mosaic of the city-state, where natural increase rates, especially among its native Chinese, have for some time been well below the replacement level. But, as noted earlier, this may not be a long-term solution because tensions are mounting between the resident and newly arrived immigrant Chinese communities.

© J. Nijman

Singapore is in many ways at the center of the Southeast Asian realm. This is a highly developed, rich, sophisticated city-state that owes much to its prime location at the tip of the Malay Peninsula overlooking the Strait of Malacca, one of the most important shipping lanes in the world.

Indonesia

The fourth-largest country in the world in terms of human numbers is also the globe's most expansive archipelago (**Fig. 11-17**). Spread across a chain of more than 17,000 mostly volcanic islands, Indonesia's 274 million people live both separated and clustered—separated by water and clustered on islands large and small. With more than 300 discrete ethnic clusters, over 250 languages, and just about every religion practiced on Earth, Indonesia's survival as a unified state is remarkable. Indeed, Indonesia's national motto is *bhinneka tunggal ika*—diversity in unity. Nevertheless, Indonesia

nominally is the largest Muslim country in the world: 88 percent of the people adhere to Islam, and in the cities the silver domes and minarets of neighborhood mosques rise like a dense forest above the townscape. Although until recently Indonesian Islam has been relatively moderate, as noted earlier more assertive Islamization has been on the rise, with new laws now in effect to ban public displays of interpersonal affection and to dictate the kinds of clothing women may wear outside their homes.

Indonesia's development challenges are complicated by spatial fragmentation and a particularly uneven population distribution (see Fig. 11-4). From 1974 to 2001, Indonesia's central government pursued a policy known as *transmigration*, inducing millions from the densely populated inner islands (especially Jawa, Bali, and Madura) to relocate to such sparsely inhabited, peripheral islands as Kalimantan and Sulawesi. As many as 8 million Jawanese were relocated to outer islands, but a majority of them experienced a decline in their standard of living; many were even reduced to bare subsistence to scratch out a living in the tropical woodlands confiscated from indigenous inhabitants—which turned out to be unsuitable for the farming methods used by the settlers. In the end, cultural conflict, ecological havoc, and rampant deforestation finally led the Indonesian government to admit failure and terminate this grandiose social program in 2001.

Today, Indonesia's stature within Southeast Asia is steadily rising. This is not just the biggest country in the realm: it is also striving to achieve a commensurate level of economic development and prosperity. The economy expanded vigorously during the 2000s; global recession notwithstanding, per capita incomes here now exceed those of Laos, Cambodia, and Myanmar—yet they still rank well below those of Thailand or Malaysia. At the same time, centrifugal forces are intensifying (growing Islamic militancy in Jawa is an overriding concern), and persistent decentralization and devolution throughout this immense archipelago is certain to continue and challenge the Indonesian government well into the future.

Indonesia's Major Islands
With more than 150 million inhabitants, **Jawa** is one of the world's most densely peopled territories (see Figs. 1-7 and 11-4) as well as one of the most agriculturally productive, with its terraced rice-growing paddies rising up the highly fertile flanks of myriad active volcanoes. Jawa also is the most highly urbanized component of a country in which almost half the people still live off the land. The primate city of Jakarta has today become the heart of the much larger Greater Jakarta conurbation that now exceeds 32 million, nearly 25 percent of Indonesia's urban population. Thousands of factories, their owners taking advantage of low prevailing wages, have been built in this sprawling conurbation (which goes by the acronym *Jabodetabek*, after its constituent cities), badly straining its infrastructure and overburdening the port of Jakarta.

Sumatera, spelled *Sumatra* in the colonial era, is Indonesia's westernmost island and forms the west shore of the critical Strait of Malacca; Singapore lies across the strait from approximately the middle of this island. Although much larger than Jawa, Sumatera has only about one-third as many people (roughly 50 million). In colonial times, the island became a base for rubber and palm oil production; its high relief makes possible the cultivation of a wide variety of crops, and neighboring Bangka and Belitung yield petroleum and natural gas. Palembang is the key urban center in the south. In recent years, the northeastern coastlands of Sumatera, along with parts of mainland Malaysia and Kalimantan, have been blanketed by huge and continually expanding palm oil plantations to drive up export revenues—but at enormous cost to the regional environment (as noted earlier in the chapter and underscored below in the satellite image of Sumatera's smoke plumes).

In Sumatera's far north lies Aceh, the land area closest to the epicenter of the colossal 2004 undersea earthquake that triggered a disastrous tsunami that claimed about 230,000 lives in the Indian Ocean Basin. Prior to this catastrophe, the province of Aceh was designated a Special Territory within Indonesia because of the separatist movement of its indigenous Acehnese population. However, that crisis became defused thanks to the massive post-tsunami relief effort and follow-up of the central government. As a result, relations between the province and Jakarta today are much improved.

Kalimantan is the Indonesian sector that comprises 73 percent of the island of Borneo, a slab of the Earth's crust whose mountainous backbone is of erosional, not plate-tectonic, origin. Larger than Texas, Borneo has a deep, densely forested interior that is a last refuge for some 20,000 orangutans as well as dwindling numbers of Asian elephants, rhinoceroses, and tigers. Along with a number of indigenous peoples, these wildlife species survive even as loggers, plantation builders, and small

In June 2013, human-ignited forest fires on Sumatera (lower left) generated enormous plumes of thick smoke that drifted eastward to engulf Singapore and surrounding areas of Malaysia and Indonesia. The Singaporean authorities warned people to stay indoors, and the city-state's government loudly insisted to its Indonesian counterpart that a greater effort must be made to prevent plantation owners and smaller-plot farmers from deliberately burning vast stands of trees to clear land for additional palm oil production.

farmers relentlessly penetrate their shrinking habitat. Borneo is home to a comparatively small human population (roughly 20 million on the Indonesian side, less than 8 percent of the country's total) on poor tropical soils. Aboriginal peoples, principally the Dayak clans, have traditionally had a less negative impact on the natural environment than the Indonesian and Chinese immigrants as well as the multinational corporations that log the rainforests and clear space for new farms. As Figure 11-17 shows, the only towns of any size in Kalimantan lie on or near the coast; access routes into the interior are still few and far between.

Sulawesi consists of a set of intersecting, volcanic mountain ranges rising above sea level; the 800-kilometer (500-mi) Minahasa Peninsula, propelled by active volcanism, continues to build itself up from the floor of the Celebes Sea. This northern peninsula is the most developed part of an otherwise rugged and remote island, with Manado its relatively prosperous urban focus. Seven major ethnic groups inhabit the valleys and basins scattered among the mountains, but the population of about 18 million also includes many immigrants from Jawa, especially in and around the southern center of Ujungpandang. Subsistence farming is the leading mode of life, although logging, scattered mining, and fishing augment the economy. Clashes between Muslims and Christians occur intermittently in remote areas.

Papua, the Indonesian name for the western half of the island of New Guinea, has become an issue in national politics. Papua is bordered on the east by a superimposed, mostly straight-line boundary that separates it from the state of Papua New Guinea on the eastern half of the island (Fig. 11-17). This territory was seized by Indonesia from the Dutch in 1969, the last remnant of the Netherlands East Indies colony. Papua contains about 22 percent of Indonesia's territory, but its (fast-growing) population totals only about 3.8 million—just under 1.5 percent of the country's total. The indigenous inhabitants of this territory—which is essentially an Indonesian colony—are Papuan, with most living in the remote reaches of this mountainous, densely forested island. Papua is economically important to Indonesia because it contains what is reputed to be the world's richest gold mine as well as its third-largest open-pit copper mine. But political consciousness has now reached the Papuans: the Free Papua Movement has been active for the past quarter-century, mostly holding small demonstrations in the capital (Jayapura), displaying a Papuan flag, and demanding recognition.

East Timor

As Figure 11-17 shows, Timor is the easternmost of the sizeable Lesser Sunda Islands, and throughout the colonial Dutch East Indies period the Portuguese maintained their own colony on the eastern half of it. In 1975, Indonesian forces overran that colony and formally annexed it the following year. Indonesian rule, however, was even less benign than that of the Portuguese, and soon a bitter struggle for independence was under way. Eventually, in 1999, under UN supervision, the people of East Timor voted overwhelmingly for independence, but Indonesia still refused to let go and unleashed a brutal military occupation. It took another three years of armed conflict and foreign intervention (led by Australia, which previously had supported Indonesia) until independence was finally realized in 2002. By then, the violence had devastated most of this ministate and its infrastructure. East Timor's leaders proclaimed the official name of their country to be *Timor-Leste*—Portuguese for "Timor East." Not surprisingly, nation-building in this Connecticut-sized country still remains a big challenge.

As Figure 11-17 indicates, East Timor is a fragmented state, with a dominant east (where the coastal capital, Dili, is located) and a tiny **exclave [30]**, on the northern coast of Indonesian West Timor, named Ocussi. Even though East Timor has always been overwhelmingly agricultural, farming was badly neglected during the turmoil unleashed under the Indonesian occupation. To make matters worse, the country's population of 1.4 million continues to grow rapidly.

Most recently, the economic growth rate has been more encouraging, and there is the prospect of billions of dollars to be earned from future oil revenues, though the small country has faced an uphill battle negotiating with Australia (which shares the rights to the offshore oil reserves in the Timor Sea) and with the big oil companies.

The Philippines

North of Indonesia lies a lengthy archipelago of more than 7000 islands inhabited by 110 million people. This island chain constitutes the Philippines, and it can be subdivided into three groups: (1) Luzon, largest island of all, and Mindoro in the north; (2) the Visayan group in the center; and (3) Mindanao, second-largest, located in the south (**Fig. 11-18**). The Philippines' population,

FIGURE 11-18

© J. Nijman, P. O. Muller, and John Wiley & Sons, Inc.

AP Images/Girlie Linao

May 8, 2018: A woman passes by a bombed mosque in the southern city of Marawi. The Battle of Marawi was fought a year earlier, but there had not been a cleanup. The government claimed to have killed a thousand ISIS-affiliated terrorists in this battle. For the innocent Muslim civilians in this area, living conditions have only gotten worse, and tensions remain high.

concentrated where productive farmlands lie, is densest in three general areas: the northwestern and south-central components of Luzon; the southeastern extension of Luzon; and the islands of the Visayan Sea between Luzon and Mindanao (Fig. 11-4). Luzon is the site of the capital, Manila (with nearly 25 million people, or 23 percent of the entire population), a sprawling megacity.

Today, the Philippines, lying adjacent to the world's largest Muslim state (Indonesia), is 81 percent Roman Catholic, 9 percent Protestant, and only 5 percent Muslim. Out of the Philippines melting pot, where Malay, Arab, Chinese, Japanese, Spanish, and American elements met and mixed, has emerged the distinctive Filipino culture. It is not a homogeneous or unified culture, as is reflected by the dozens of Malay languages in use in the islands, but it is in several ways unique. At independence in 1946, the largest of the Malay languages, Tagalog (also known as Pilipino), became the country's official language. But English is widely learned as a second language, and a Tagalog–English hybrid, "Taglish," is increasingly heard today. The Chinese component of the population is small (1.6 percent) but dominates numerous local business communities.

The Philippines' small Muslim population, concentrated in the southernmost periphery of the archipelago, especially in southern and western Mindanao (Fig. 11-18), has long decried its marginalization in this predominantly Christian country. Over the past generation, a half-dozen Islamic organizations have promoted the Muslim cause here through tactics that have ranged from peaceful negotiation with the government to violent insurgency. In 2017, the violence escalated into a

five-month military conflict, known as the Battle of Marawi, named for a city in the southern province of Lanao del Sur (Fig. 11-18, and photo left). Government forces bombed the city for several months, allegedly killing nearly a thousand militants and ISIS-affiliated terrorists.

Agriculture continues to dominate the Philippines' economy. Alluvial as well as volcanic soils, together with ample moisture in this mostly humid-tropical environment, enable self-sufficiency in rice and other staples and allow the Philippines to be a net exporter of farm products. However, unemployment is rampant, trade linkages are insufficient, and additional land reform is badly needed, as is social restructuring to reduce the controlling influence of a comparatively small group of families over national affairs.

Perhaps more than any other people, Filipinos take jobs in foreign countries in massive numbers, proving their capacity to succeed in jobs they cannot find at home. More than 10 million Filipinos, over one-quarter of the total workforce, are now employed abroad. The global merchant marine would not exist without Filipino sailors, and Filipina nurses and domestic workers can be found from Dubai to Dublin to Dubuque. Funds sent home to family members by the emigrants make the Philippines a world leader in remittances, a monetary inflow that is the basis for at least one-tenth of the country's GDP in any given year.

Overall, the Philippines sustains a lower-middle-income economy, and given a longer period of political stability and success in mitigating the problems outlined above, it should be able to rise to the next level and take its long overdue place among the Asia Pacific's growth poles.

Since 2000, the Philippines has established itself in a pair of niches in the global economy: international call centers and the outsourcing of digital services. The latter operate through such Internet websites as *oDesk*—one of several that serve as a global marketplace for digital services by freelancers who perform data entry and other routine back-office work. Although such outsourcing is not a promising growth industry and compensation is low by worldwide standards, the **business process outsourcing [31]** sector today employs about one million full-time workers and is the second-largest contributor (after remittances) to the Philippines' foreign exchange earnings.

In its foreign relations, this country is changing course. Nearly 30 years ago, it terminated the lease for the huge U.S. Subic Bay Naval Base—then the largest of all American foreign military installations. But these days the Philippine government is once again seeking closer relations with the United States, and in 2018 it resumed joint military exercises. As noted earlier, this is in response to China's growing maritime-territorial assertiveness in Southeast Asia—not only regarding the South China Sea in general but also longstanding specific disputes between China and the Philippines concerning the Spratly Islands and Scarborough Shoal.

Points to Ponder

- The two smallest countries of the realm are also by far the wealthiest.

- In your daily life, how many products do you use that contain palm oil? How can you find out? As a consumer, do you thereby contribute to deforestation?

- With the deepening involvement of India, China, and Thailand in Myanmar over the past several years, it is hard to imagine this country could ever return to its isolationist recent past.

- In 2016, the International Court of Justice ruled that China's claims in the South China Sea were unlawful. Beijing retorted that the ruling was invalid and not binding.

- Most of the countries considered to be at greatest risk from global climate change also rank among the poorest in the world.

FIGURE 12-1

© J. Nijman, P. O. Muller, and John Wiley & Sons, Inc.

Vineyards in the Barossa Valley in South Australia, about 40 miles (60 km) northeast of the city of Adelaide, especially well-known for its Shiraz red wines. Australia and New Zealand are major wine producers, located in the same wine-growing latitudinal belt as Chile, Argentina, and South Africa (mapped in Box 4-8).

CHAPTER 12

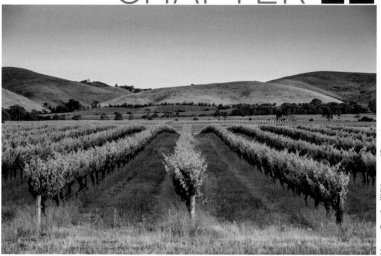

Marco Bottigelli/Moment/Getty Images

The Austral Realm

IN THIS CHAPTER

- The Asian turn of Australia and New Zealand
- Challenges of climate change in Australia
- China covets Australian commodities
- Aboriginal rights to land and resources

CONCEPTS, IDEAS, AND TERMS

Austral	[1]	Unitary state	[9]
Southern Ocean	[2]	Outback	[10]
Subtropical Convergence	[3]	Environmental degradation	[11]
West Wind Drift	[4]	El Niño	[12]
Biogeography	[5]	Desalination	[13]
Wallace's Line	[6]	Aboriginal land issue	[14]
Aboriginal population	[7]	APEC	[15]
Federation	[8]	Peripheral development	[16]

The Austral realm is the only one that lies entirely in the Southern Hemisphere. It is also unique in that it has no land link of any kind to a neighboring realm and is thus completely surrounded by oceans and seas (**Fig. 12-1**). It is second only to the Pacific as the world's least populous geographic realm. Appropriately, its name refers to its location (the word **austral [1]** derives from the Latin for "south")—a location far from the sources of its dominant cultural heritage, but relatively close to its newfound economic partners on the Asian Pacific Rim.

Defining the Realm

Two countries constitute the Austral realm: Australia, in every way the dominant national entity, and New Zealand, physiographically more varied but demographically much smaller (**Fig. 12-2**). Between them lies the South Pacific Ocean's Tasman Sea. To the realm's west lies the Indian Ocean, to its east the Pacific, and to the south the frigid Southern Ocean (**Box 12-1**).

Today, this southern realm is at a crossroads. On the doorstep of populous eastern Asia, the realm's Anglo-European legacies have been diversified accordingly by the infusion of many Asian cultural strains. Markets throughout the Asia Pacific are buying ever larger quantities of raw materials. Chinese and other Asian tourists fill hotels and resorts. The streets of Sydney and Melbourne display a multicultural panorama unimagined just two generations ago. Indigenous minorities—the Polynesian Maori in New Zealand and Aboriginal communities in Australia—are demanding greater rights and wider acknowledgment of their cultural heritage. All these changes have stirred political debate. Issues ranging from immigration quotas to indigenous land rights dominate, exposing social fault lines (city versus Outback in Australia; North Island versus South Island in New Zealand). Aboriginals and Maori first settled this realm, then the Europeans arrived, and now Asians are an increasingly significant economic and cultural element.

Physiography of the Austral Realm

The physiographic contrasts between massive, compact Australia and elongated, fragmented New Zealand are related to their locations with respect to the Earth's tectonic plates (see Fig. 1-5, Chapter 1). Australia, carrying some of the geologically most ancient rocks on the planet, lies at the center of its own Australian Plate. New Zealand, younger and less stable, lies at the convulsive convergence of the Australian and Pacific plates. Earthquakes are rare in Australia, and volcanic eruptions are unknown; New Zealand has plenty of both. This locational contrast is also reflected by differences in relief (Fig. 12-1). Australia's highest relief occurs in what Australians call the Great Dividing Range, the mountains that line the east coast from the Cape York Peninsula in the north to southern Victoria State, with an outlier in Tasmania. The highest point along these old, now eroding mountains is Mount Kosciusko, 2230 meters (7316 ft) tall. In New Zealand, entire ranges are higher—Mount Cook in the Southern Alps, for example, reaches 3754 meters (12,316 ft). West of Australia's Great Dividing Range, the physical landscape mostly exhibits low relief, with some local exceptions such as the Macdonnell Ranges near the

---

Box 12-1 Major Geographic Features of the Austral Realm

1. Australia and New Zealand constitute a geographic realm by virtue of relative location, territorial dimensions, and dominant cultural landscape.

2. Despite their inclusion in a single geographic realm, the two countries differ significantly. Australia has a vast, dry, low-relief interior; New Zealand is mountainous and has a humid temperate climate.

3. Australia and New Zealand are marked by development near their territorial edges—Australia because of its aridity, New Zealand because of its topography.

4. The populations of each are not only peripherally distributed but also highly clustered in urban centers.

5. The economic geography of Australia and New Zealand is dominated by the exporting of livestock products and specialty goods such as wine. Australia also is a significant producer of wheat and possesses a rich, diverse base of mineral resources.

6. Australia and New Zealand are now integrated into the economic framework of the Asian Pacific Rim, principally as suppliers of raw materials. Mineral-rich Australia's newly discovered and exploited gas reserves are adding to its export base.

7. As immigration from the western Pacific Rim increases, Australia has, in a cultural sense, "returned" to Asia.

8. Both Australia and New Zealand face challenges with the integration of aboriginal populations in mainstream society.

9. Australia is faced with serious environmental challenges due to climate change.

POPULATION DISTRIBUTION: AUSTRAL REALM

Population per Square Kilometer

0 to 5	
5 to 10	
10 to 50	
50 to 100	
100 to 500	
Over 500	

© J. Nijman, M. Shin, P. O. Muller, and John Wiley & Sons, Inc.

FIGURE 12-2 **Interactive**

Western Plateau and Margins in the lower map of Figure 12-3 contains much of Australia's mineral wealth.

Climates

The world climate map (Fig. 1-6) reveals the effects of latitudinal position and interior isolation on Australia's climates. In this respect, Australia is far more varied than New Zealand, its climates ranging from tropical in the far north, where rainforests flourish (see **Box 12-2**), to Mediterranean and humid subtropical, respectively, in the two corners of the south. The interior is dominated by desert and steppe conditions, the semiarid steppes providing the grasslands that sustain tens of millions of livestock. Only in the extreme east does Australia possess a zone of humid temperate climate, and here lies most of the country's core area. New Zealand, by contrast, is totally under the influence of the Southern and Pacific oceans, creating moderate, moist conditions, temperate in the north and progressively colder in the south.

continental center; plateaus and plains dominate (photo below; **Fig. 12-3**, top map). The Great Artesian Basin is a key physiographic region, providing underground water sources in what is otherwise desert country. To its south lies the continent's predominant Murray–Darling river system. The area mapped as

The Southern Ocean

Twice now we have referred to the **Southern Ocean** [2]. Many maps actually don't make mention of the Southern Ocean, and they suggest that the Atlantic, Pacific, and Indian oceans reach all the way to the shores of Antarctica. Australians and New Zealanders know better. They experience the frigid waters and persistent winds of this great weather-maker on a daily basis.

For geographers, and everybody else, it is a good exercise to turn the globe upside down now and then. After all, the usual orientation is quite arbitrary. Modern mapmaking started in the Northern Hemisphere, and the cartographers put their hemisphere on top and the other at the bottom. That is now the norm, and it can distort our view of the world. In souvenir shops in Australia and New Zealand, you see upside-down maps showing Australia and Argentina at the top, and Europe and Canada at the bottom (**Fig. 12-4**). Among other things, an inverted view of the globe shows us the vastness of the ocean encircling Antarctica.

Where do the northward limits of the Southern Ocean lie? This ocean is bounded not by land but by a marine transition known as the **Subtropical Convergence** [3]. Here the very cold, extremely dense waters of the Southern Ocean meet the warmer waters of the Atlantic, Pacific, and Indian oceans. It is quite sharply defined by changes in temperature, chemistry, salinity, and marine fauna. Flying over it, you can actually observe it in the changing colors of the water: the Antarctic side is a deep gray, the northern side a greenish blue. Even though the

walesjacqueline/Pixabay

Ayers Rock, located in the heart of the immense, interior Outback near the geographic center of Australia, is the most famous landform in the Austral Realm. This miniature tableland is all that is left of a much larger surrounding plateau that has been eroded away to form the flat plain that frames this iconic formation of red sandstone rock. Geomorphologists use the term *inselberg* (German for "island mountain") to describe this kind of geologic remnant, which consists of rocks that are far more resistant to the local forces of erosion than those that once surrounded it. Ayers Rock—called Uluru by the Aboriginal people who first settled Australia—stands about 350 meters (1150 ft) above the plain and has a circumference of nearly 10 kilometers (5.8 mi). Despite its extreme remoteness from where most Australians live, about 300,000 visitors annually make the long trek out here, and the tourist infrastructure has kept pace.

AUSTRALIA: PHYSIOGRAPHY

Desert and semi-desert

Tropical and sub-tropical grassland and forest

Mediterranean scrub

PHYSICAL LANDSCAPES

Eastern Uplands

Central Lowlands

Western Plateau & Margins

FIGURE 12-3

Box 12-2 From the Field Notes . . .

"The northeastern coastal strip of the Australian State of Queensland contains tropical rainforest. It is not that far from here to Papua New Guinea (Figs. 12-1, 12-3). I am visiting a unique ethnobotanical garden designated 'The Botanical Ark.' It is run by a couple from the United States, Susan and Alan Carle, who moved here over 25 years ago. They traveled the tropical habitats of the world and collected an enormous variety of wild plants that yield either fruits or nuts, with the purpose not only of expanding our knowledge of many species but also of preserving them. They grow them on their sprawling estate here, not far from the coastal town of Port Douglas. Of course, plants are crucial to our diets and as medicines; myriad species are still unknown or remain poorly researched; and many are probably disappearing before we can even discover them. Here, standing before a colorful arrangement of edibles, our host holds up breadfruit, a highly nutritious staple food in many tropical areas, but relatively unknown in the United States and other mid-latitude countries. Cut it in slices, fry the slices briefly, and sprinkle on a little salt, and they make for a delicious snack that goes well with a late afternoon beer—a perfect combination in the hot tropical weather of this part of Australia."

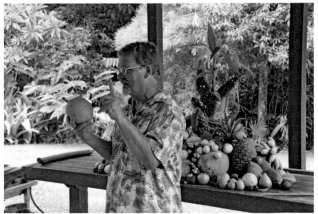

© J. Nijman

Subtropical Convergence shifts seasonally, its position does not vary far from latitude 40° South, which also is the approximate northern limit of Antarctic icebergs. Defined this way, the great Southern Ocean is a huge body of water that circulates clockwise (from west to east) around Antarctica, which is why we also call it the **West Wind Drift** [4].

Biogeography

One of the Austral Realm's defining characteristics is its wildlife. Australia is the land of kangaroos and koalas, wallabies

William Chopart/AGE Fotostock

The breathtakingly beautiful Great Barrier Reef, off the coast of the State of Queensland in northeastern Australia (Fig. 12-3, top map). This is the largest coral reef on the planet, sometimes described as the only living thing on Earth that is visible from space. It was designated a World Heritage Site in 1981. Today, it ranks as one of the planet's most prominent ecotourism destinations, despite the threat of global warming effects (even very slightly higher ocean temperatures can cause massive "bleaching" of coral).

and wombats, possums and platypuses. These and numerous other *marsupials* (animals whose young are born very early in their development and then are carried in an abdominal pouch) owe their survival to Australia's early isolation during the breakup of Gondwana (see Fig. 8-3). Before more advanced mammals could enter Australia and replace the marsupials, as happened in every other part of the world, this landmass was separated from Antarctica and India, and today it contains the world's largest assemblage of marsupial fauna. Australia's vegetation has distinctive qualities as well, notably the hundreds of species of eucalyptus trees native to this geographic realm. The study of fauna and flora in spatial perspective integrates the disciplines of biology and geography in a field known as **biogeography** [5], and Australia serves as a gigantic research laboratory for its practitioners.

Biogeographers are especially interested in the distribution of plant and animal species, as well as the relationships between plant and animal communities and their natural environments. (The spatial analysis of plant life is called *phytogeography*, and that of animal life is known as *zoogeography*.) In 1876, one of the founders of biogeography, the great British naturalist Alfred Russel Wallace, posited that the zoogeographic boundary of Australia's fauna was located beyond Australia in the Sunda island chain to the northwest, between Borneo and Sulawesi, and just east of Bali (**Fig. 12-5**).

Wallace's Line [6] soon was challenged by other researchers, who found species Wallace had missed and who visited islands Wallace had not. There was no question that Australia's zoogeographic realm terminated somewhere in Indonesia's Sunda archipelago, but where? Western Indonesia was the habitat of nonmarsupial animals such as tigers, rhinoceroses, and elephants in addition to primates; New Guinea clearly was part of the realm of the marsupials. How far had the more advanced mammals progressed eastward along the

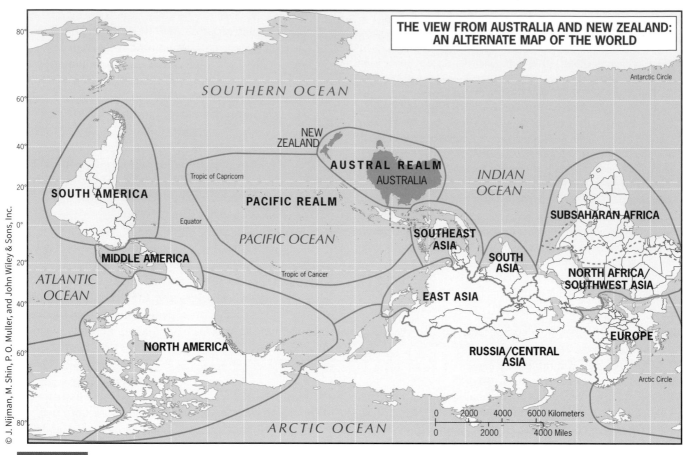

FIGURE 12-4

island stepping stones toward New Guinea? The zoogeographer Max Weber discovered evidence that led him to postulate his own *Weber's Line*, which, as Figure 12-5 shows, lay very close to New Guinea.

In Australia, the arrival of the **Aboriginal population** [7] about 50,000 years ago appears to have triggered an ecosystem collapse. Widespread burning of the existing forest, shrub, and grassland vegetation all across Australia probably led to the spread of desert scrub and to the rapid extinction of most of the continent's large mammals soon after the human invasion occurred. Those species that survived faced a second crisis tens of thousands of years later when the European colonizers introduced their livestock, leading to the further destruction of remaining wildlife habitats. Survivors include marsupials such as the koala bear and the wombat, but the list of extinctions is much longer.

The Realm's Asian Turn

Since the late eighteenth century, with the arrival of successive waves of settlers from the United Kingdom and other European countries, both Australia and New Zealand became decidedly Western-style countries in terms of their political systems and institutions, the predominance of the English language, their cultural fabric, and their religion. The two countries could not possibly be located any further away from Europe, but they very much maintained their Western character for about two hundred years.

Toward the end of the twentieth century, things started to change for two important reasons. First, as migration from Europe declined, the influx of migrants from nearby Asian countries started to increase. This was not a gradual process and, especially in the case of Australia, it required overcoming initial resistance against growing cultural diversity and ethnicity. In 2019, about 30 percent of Australia's population was foreign born, and more than half the Australian population were immigrants or children of immigrants (by comparison, foreign-born residents of the U.S. that year comprised 14 percent of the population). Some 200,000 new migrants arrived in 2019, about 60 percent of them from Asia (especially China, India, Indonesia, Vietnam, and the Philippines). In New Zealand, an estimated 23 percent of the population in 2019 was foreign born, and nearly half of the immigrants came from Asia. Most who enter Australia and New Zealand are well educated and skilled. Today, there is strong overall support for immigration, with a realization that low birthrates will not suffice in maintaining a stable population and workforce. Arguably, like other graying populations, both Australia and New Zealand actually

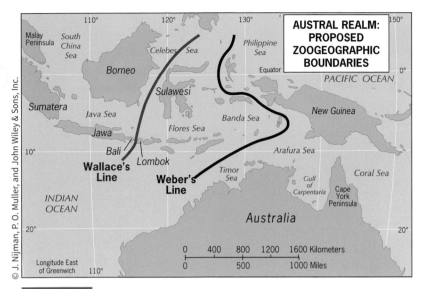

AUSTRAL REALM: PROPOSED ZOOGEOGRAPHIC BOUNDARIES

FIGURE 12-5

most recently several Southeast Asian countries), Australia and New Zealand have become increasingly integrated in Asian trade and investment flows. Thus, Australia's top four export destinations in 2019 were all in Asia: China (30 percent), Japan (10 percent), South Korea (5 percent), and India (4 percent)—the United States ranked fifth, at 3.5 percent. For New Zealand, China was also by far the biggest export destination, at 25 percent, followed by Australia (15 percent), the United States (9 percent), Japan (6 percent), and South Korea (3 percent).

Both Australia and New Zealand still maintain strong political ties with the United States and "the West" in general, including a formal military alliance with the United States (ANZUS). However, there can be little doubt that the steadily advancing economic and cultural integration with Asia, and particularly the intensification of economic relations with China, are likely to give Australia and New Zealand a different perspective on global political issues across these adjacent realms.

need (young) immigrants to compensate for low birthrates and an increasing *dependency ratio*.

Second, due to the rapid economic rise of Asian economies (first Japan and the Asian Tigers, then China, and

Regions of the Austral Realm

Region Australia

Australia is the predominant component of the Austral realm, a continent-scale country in a size category that also includes China, Canada, the United States, and Brazil. For two reasons, however, Australia has fewer regional divisions than these other countries: its comparatively uncomplicated physiography and its modest human population of 25.4 million.

Positioned on the Pacific Rim; almost as large as the 48 contiguous U.S. States; rich with farmlands and vast pastures, rivers, groundwater supplies, mineral deposits, and energy resources; and served by good natural harbors, Australia is one of the most naturally endowed countries in the world.

Administrative Framework

From the late eighteenth century on, the Europeanization of Australia doomed the continent's Aboriginal societies. The first to suffer were those situated in the path of British settlement on the coasts, where penal colonies and free towns were founded. Distance protected the Aboriginal communities of the northern interior longer than elsewhere; but in Tasmania, the indigenous Australians died off in just decades after having lived there for perhaps 45,000 years. Eventually, the major coastal settlements became the centers of seven different colonies, each with its

own hinterland; by 1861, Australia was internally delimited by its now-familiar pattern of straight-line State boundaries (**Fig. 12-6**).

On January 1, 1901, following years of difficult negotiations, the Australia we know today finally emerged: the Commonwealth of Australia, consisting of six States and two Federal Territories. The two Federal Territories are the *Northern Territory*, assigned to protect the interests of the substantial Aboriginal population concentrated there and agitating for Statehood, and the *Australian Capital Territory*, carved from southern New South Wales to accommodate the federal capital of Canberra that was completed in 1927.

Australia's six States are New South Wales (capital Sydney), at 8.1 million the most populous and politically powerful; Victoria (Melbourne), small but populous by Australian standards with 6.6 million residents; Queensland (Brisbane), at 5.1 million, with the Great Barrier Reef offshore and tropical rainforests in its north; South Australia (Adelaide), with 1.8 million, where the Murray–Darling river system reaches the sea; Western Australia (Perth) with barely more than 2.6 million people in an area of more than 2.5 million square kilometers (nearly 1 million sq mi); Tasmania (Hobart), containing 533,000, the island across the Bass Strait from the mainland's southeastern corner that lies in the path of Southern Ocean storms; and the Northern Territory (Darwin), with 246,000 people, and the largest share of Australia's aboriginal population.

In earlier chapters, we referred to the concept of federalism, which is a form of politico-territorial organization. The word "federal" comes from the Latin *foederis*, implying alliance

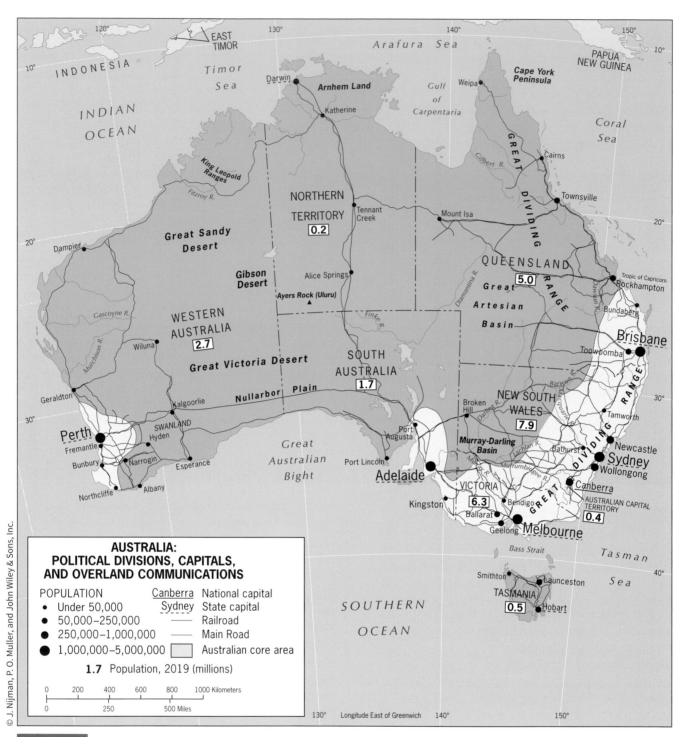

© J. Nijman, P. O. Muller, and John Wiley & Sons, Inc.

AUSTRALIA:
POLITICAL DIVISIONS, CAPITALS,
AND OVERLAND COMMUNICATIONS

POPULATION
· Under 50,000
● 50,000–250,000
● 250,000–1,000,000
⬤ 1,000,000–5,000,000

Canberra National capital
Sydney State capital
— Railroad
— Main Road
☐ Australian core area

1.7 Population, 2019 (millions)

0   200   400   600   800   1000 Kilometers
0   250   500 Miles

FIGURE 12-6

and coexistence, a union of consensus and common interest—a **federation [8]**. It stands in contrast to the idea of a centralized or **unitary state [9]**. For this, too, the ancient Romans devised a term: *unitas*, meaning "unity." Most of Europe's states are unitary, including the United Kingdom. Although most Australians came from that tradition, they managed to overcome their differences and forge a Commonwealth that was, in effect, a federation of States with different viewpoints, economies, and objectives, separated by enormous distances along the rim of a particularly remote island-continent.

Spatial Organization

Australians often talk about distance. One of their leading historians, Geoffrey Blainey, labeled it a tyranny—an imposed remoteness from most of the rest of the world; and, within, remoteness from much of the rest of the country and a divisive part of life. Even today, Australia is far from nearly everywhere on Earth. A trans-Pacific jet flight from Los Angeles to Sydney takes about 14 hours nonstop and is correspondingly expensive. You may have had the fortune of traveling abroad several times, yet never made it to

Australia. Freighters carrying products to European markets take ten days to two weeks to get there. Within Australia, distances are of continental proportions as well, and Australians pay the price—literally. Australians tend to pay more per mile for their domestic flights than air passengers in most other parts of the world.

As Figure 12-2 shows, Australia is a large landmass, but its population is almost entirely concentrated in a (discontinuous) core area that lies in the east and southeast, most of which faces the Pacific Ocean's Tasman Sea. Figure 12-6 shows that this crescent-shaped Australian heartland extends from north of the city of Brisbane to the vicinity of Adelaide and includes the largest city, Sydney; the capital, Canberra; and the second-largest city, Melbourne. A secondary core area has developed in the far southwest, centered on Perth. Beyond the core areas lies the enormous periphery that Australians call the **Outback [10]**.

To better understand the evolution of this spatial arrangement, it helps to refer again to the map of world climates (Fig. 1-6). Environmentally, Australia's most favored strips face the Pacific and Southern oceans as well as a sector of the Indian Ocean, and they are not large. We can describe the country as a coastal rimland, with cities, towns, farms, and forested slopes giving way to the immense, arid, interior Outback. On the western flanks of the east coast's Great Dividing Range lie the extensive grassland pastures that catapulted Australia into its first commercial age—and on which still graze one of the largest sheep herds in the world (more than 70 million sheep in 2018, producing a major share of all the apparel wool sold in the world). Where it is moister, to the north and east, cattle by the millions graze on ranchlands. This is frontier Australia, over which livestock have ranged for nearly two centuries.

Despite the vast open spaces and romantic notions of frontier and Outback, Australia is an urban country, with 86 percent of all Australians living in cities and towns. The large cities lie along the coast, the centers of manufacturing complexes as well as the foci of agricultural areas. For all its vastness and youth, Australia has developed a remarkable cultural identity, a sameness of urban and rural landscapes that persists from one end of the continent to the other.

Sydney, often called the New York of Australia, lies on a spectacular estuarine site, its compact, high-rise central business district overlooking a port bustling with ferry and freighter traffic (photo this page). Sydney today is a widely spread out metropolis of 4.6 million, with multiple outlying centers dominating its burgeoning suburban ring; brash modernity and reserved British ways somehow blend here. Melbourne (4.3 million), sometimes regarded as the Boston of Australia, prides itself on its more interesting architecture and more cultured ways. Brisbane (2.2 million), the capital of Queensland, which also anchors Australia's Gold Coast and adjoins the Great Barrier Reef, is the Miami of Australia; unlike Miami, however, its residents can find nearby relief from the summer heat in the mountains of its immediate hinterland (as well as at its beaches). Perth (1.9 million) is separated from its eastern Australian neighbors by three-quarters of a continent, and from Southeast Asia and Africa by thousands of kilometers of ocean—but due to the ever-expanding mining activities of Western Australia it is increasingly drawn into the global economy.

An aerial panorama of central Sydney, showing the high-rise CBD adjoining the Royal Botanic Garden (center left), with the famously designed Sydney Opera House on the waterfront, the iconic Sydney Harbour Bridge connecting downtown to North Sydney, and the upscale suburb of Kirribilli (bottom right).

And yet, each of these cities—as well as the capitals of South Australia (Adelaide), Tasmania (Hobart), and, to a lesser extent, the Northern Territory (Darwin)—exhibits an Australian character of unmistakable quality. Life is both orderly and unhurried, streets are clean, slums are uncommon, and graffiti rarely seen. Standards of public transport, urban schools, and healthcare provision are high. Spacious parks, pleasing waterfronts, and plentiful sunshine make Australia's urban life more agreeable than just about anywhere else on Earth.

Australia's Economic Geography

Australia today ranks tenth among the world's countries in GDP per capita and, perhaps more important, its economy has been remarkably stable over the last two decades. For the vast majority of Australians life is quite comfortable. In terms of key development indicators, Australia is far ahead of all its western Pacific Rim competitors except Japan and Singapore. The country's initial prosperity was achieved in the mines and on the farms, not in the cities. Australia has material assets that other Asia Pacific countries can only dream of. But today the economy is highly diversified.

In agriculture, sheep-raising was the earliest commercial venture, but it was the technology of refrigeration that brought world markets within the reach of Australian beef producers. Wool, meat, and wheat have long been the country's big-three income earners; **Figure 12-7** displays the immense pastures in the east, north, and west that constitute the ranges of Australia's huge herds. The zone of commercial grain farming forms a broad crescent extending from northeastern New South Wales through Victoria into South Australia, with a major outlier covering much of the hinterland of Perth.

And keep in mind the scale of this map: Australia is only slightly smaller than the continental United States. Commercial grain farming in Australia is big business. As the world climate

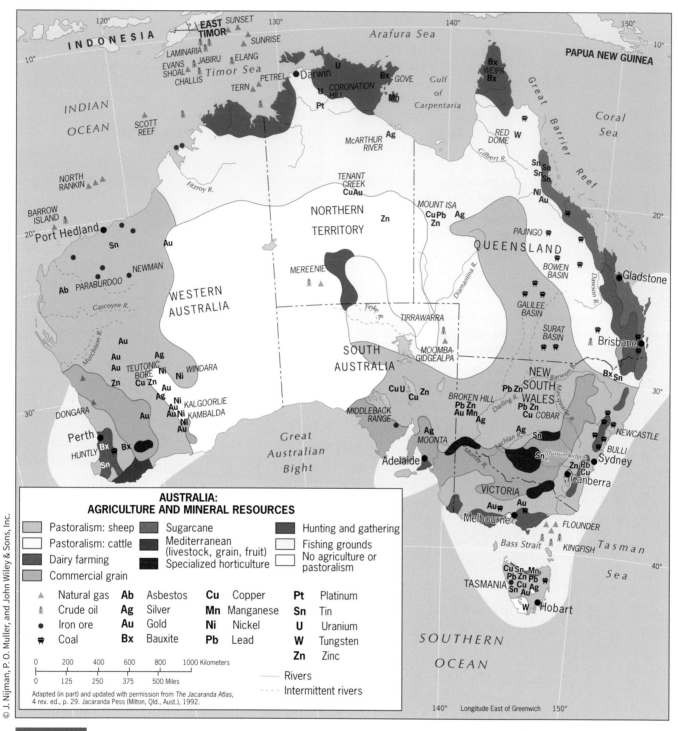

**AUSTRALIA:
AGRICULTURE AND MINERAL RESOURCES**

Pastoralism: sheep
Pastoralism: cattle
Dairy farming
Commercial grain
Sugarcane
Mediterranean
(livestock, grain, fruit)
Specialized horticulture
Hunting and gathering
Fishing grounds
No agriculture or
pastoralism

▲ Natural gas   **Ab** Asbestos   **Cu** Copper   **Pt** Platinum
⚒ Crude oil   **Ag** Silver   **Mn** Manganese   **Sn** Tin
● Iron ore   **Au** Gold   **Ni** Nickel   **U** Uranium
⛏ Coal   **Bx** Bauxite   **Pb** Lead   **W** Tungsten
   Zn Zinc

0   200   400   600   800   1000 Kilometers
0   125   250   375   500 Miles

——— Rivers
– – – Intermittent rivers

Adapted (in part) and updated with permission from *The Jacaranda Atlas*,
4 rev. ed., p. 29. Jacaranda Pess (Milton, Qld., Aust.), 1992.

© J. Nijman, P. O. Muller, and John Wiley & Sons, Inc.

FIGURE 12-7

map (Fig. 1-6) suggests, sugarcane grows along most of the humid, subtropical, coastal-lowland strip of Queensland, and Mediterranean crops (including grapes for Australia's highly prosperous wine industry) cluster in the hinterlands of Sydney, Adelaide, and Perth. Mixed horticulture concentrates in the Murray River Basin, including rice, grapes, and citrus fruits, all under irrigation. In addition, as elsewhere across the world, dairying has developed close to the large metropolitan areas. With its considerable range of environments, Australia yields a great variety of crops.

Australia's mineral resources, as Figure 12-7 shows, also are diverse. Major gold discoveries in Victoria and New South Wales produced a ten-year gold rush starting in 1851 and ushered in a new economic era. By the middle years of that decade, Australia was producing 40 percent of the world's gold. Subsequently, the search for more gold led to the discoveries of additional minerals, including nickel, copper, iron ore, uranium, tungsten, bauxite, and more. Since the 1990s, the enormous demand for raw materials from China as well as emerging India has marked a long-running commodity boom in Australia.

About 320 kilometers (200 mi) south-southeast of Port Hedland in the Pilbara region of Western Australia, Mount Whaleback near the town of Newman is one of the world's leading sources of high-grade iron ore. China's insatiable demand for this ore has generated the planet's largest "open-cut" iron mine, a huge and growing gash in the natural landscape—just one of many such negative impacts resulting from Australia's role as raw-material supplier to industrializing economies (to appreciate the enormous size of this mine, spot the truck in the center of the picture).

New finds are still being made today, and even oil and natural gas have been discovered both on land and offshore (Fig. 12-7). Most promising are the undersea deposits off the northwest coast, which has sparked an energy boom that is transforming Darwin. This northernmost city of Australia directly faces the fossil-fuel riches of the Timor Sea and is becoming a major processor of natural gas, liquefying it for export to East and Southeast Asia. Another liquid natural gas (LNG) boomtown is the east-coast, central Queensland port of Gladstone: here, an ultramodern processing-plant complex produces LNG from piped-in coal-seam gas, which is extracted by new hydraulic fracturing technologies from coal deposits in interior Queensland's Bowen and Surat Basins (Fig. 12-7).

Australian manufacturing caters mainly to the domestic market and is quite diversified, producing a wide array of industrial goods ranging from machinery made of locally produced steel to chemicals to textiles to paper products. These industries cluster in and near the major urban areas where consumers are located. The domestic market in Australia is not large, but it remains relatively affluent. This makes it attractive to foreign producers, and Australia's retail outlets overflow with high-priced goods from China, Japan, South Korea, the United States, and Europe.

Australia's Environmental Problems

Unfortunately, **environmental degradation [11]** is practically synonymous with Australia. First the Aboriginals, then the Europeans and their livestock, inflicted catastrophic damage on Australia's natural environments and ecologies. Great stands of magnificent forest were ravaged. In Western Australia,

centuries-old trees were simply ringed and left to die so that the sun could penetrate through their leafless crowns to nurture the grass below for pasture for the introduced livestock. In island Tasmania, where Australia's native eucalyptus tree reaches its greatest concentrations (comparable to California's redwood stands), tens of thousands of hectares of this irreplaceable treasure have been lost to chain saws and pulp mills. Many of Australia's unique marsupial species have been driven to extinction, and many more are endangered or threatened. "Never have so few people wreaked so much havoc on the ecology of so large an area in so short a time," observed one prominent Australian geographer not long ago.

More recently, the intensification of mining activities as well as oil and gas exploration have raised ecological stress to the next level. The energy-related development of Gladstone referred to above offers a telling example. Environmental activists are sounding alarms concerning the new hydraulic fracturing techniques for coal-seam gas production because they threaten Queensland's agricultural resources. Moreover, these same groups are vehemently protesting the expansion of port facilities because the coastal zone surrounding Gladstone lies within the supposedly protected confines of the Great Barrier Reef just to the north.

Another environmental problem involves Australia's wide and long-term climatic variability. In a predominantly arid continent, droughts in the moister fringes are the worst enemy, and Australia's history is replete with devastating dry spells. Australia is vulnerable to **El Niño [12]** events (periodic anomalies of sea-surface warming in the equatorial Pacific that have far-reaching effects on skewing regional weather patterns), increasingly reinforced by global processes of climate change (see **Box 12-3**).

These problems are exacerbated by rising demands for water in Australia's burgeoning urban areas in addition to excessive damming, well-drilling, and water diversion in the valleys of upstream tributaries of the vital Murray-Darling river system. The Australian government responded by initiating the first fully coordinated drainage-basin-control program in the country, but all stakeholders recognize that it will take many years (and delays caused by future dry spells) as well as agricultural adjustments for the full benefits to materialize.

Western Australia, too, faces colossal challenges in the provision of water for its urban, agricultural, and extractive activities. Here, much like in the western United States, water supplies in this dry environment are increasingly stressed by population growth—and especially all those massive new mining projects. The largest population cluster here is centered on west-coast Perth, and some scientists have warned that it could become the first "ghost metropolis," abandoned because of acute water shortages. To stave off that scenario, Perth's prospects have improved through the use of innovative **desalination [13]** technologies that convert ocean water into a viable source of supply for households and businesses (also see Box 7-4 on page 232). Today, two desalination plants south of Perth provide about half of this metropolitan area's water needs. This technology is now beginning to be applied to Australia's southeastern coast, and a number of plants have already come online (**Fig. 12-8**).

Box 12-3 Global Climate Change in Australia

Few rich countries in the world are as exposed to climate change as Australia: the tropical fringe of the north is experiencing increasing frequency and intensity of cyclones and flooding, whereas much of the rest of the country is increasingly plagued by droughts and wildfires. The droughts have been more frequent and harsher in recent decades, due to overall higher temperatures and less reliable rainfall. The Murray-Darling river basin, Australia's breadbasket, has been drying out, and southeastern Australia has seen increased wildfires.

Figure 12-8 shows what are believed to be the effects of climate change on Australia between 1999 and 2018: whereas the northern (tropical) zones have experienced enhanced rainfall, the populated southwestern and southeastern coastal regions received much less precipitation and have been more vulnerable to drought (see photos). Serious droughts, including the worst ever recorded, have occurred since the turn of the century and have been especially damaging in the Murray–Darling river basin.

Source: Adapted from *State of the Climate 2018*, Bureau of Meteorology, Australian Government, p. 6.

AUSTRALIA: CLIMATE CHANGE AND WATER SHORTAGES
GROWING SEASON (APRIL-OCT) RAINFALL PATTERN, 1999–2018

- Highest on record
- Very much above average
- Above average
- Average
- Below average
- Very much below average
- Lowest on record
- ■ Major desalination plant

FIGURE 12-8

Extreme weather events have plagued Australia in recent years. The photo to the left is of a raging wildfire in January 2020, about 60 miles west of the national capital, Canberra. Since November of 2019, hundreds of fires ravaged the drought-stricken southeastern part of the country (Fig. 12-8). By January 12, two dozen people had died (most from smoke inhalation) and thousands of homes were destroyed. To the right, one month earlier, a demonstrator with a mask attends a climate protest rally in Sydney on December 11, 2019. The protesters demanded urgent climate action from Australia's government.

Australia's Aboriginal Challenge

Despite Australia's generally positive fortunes, not everyone shares sufficiently in the national wealth. The indigenous (Aboriginal) population, though a small minority today of less than 900,000, remains disproportionately disadvantaged in almost every way—shorter life expectancies, higher infant mortality rates, elevated unemployment, lower high school graduation rates, and much higher imprisonment rates (**Box 12-4**).

For the past several decades, the Aboriginal challenge has focused on two questions: (1) official acknowledgment, by the government as well as the Australian majority, of mistreatment of the Aboriginal minority accompanied by appropriate apologies and reparations; and (2) land ownership. The first question was resolved in 2008 when Prime Minister Kevin Rudd offered the long-awaited formal apology for the historic mistreatment of the Aboriginals. The second question has particular geographic implications. Although comprising only about 3 percent of the total population, the Aboriginal population has been gaining influence in national affairs over the past three decades: since as far back as the 1980s, Aboriginal leaders have been waging a campaign to obstruct mineral exploration on what they regard as ancestral and sacred lands.

Prior to 1992, Australians had taken it for granted that Aboriginals had no right to land ownership, but in that year,

---

Box 12-4 Geographic Information Analysis

The Aboriginal Population Distribution

The Aboriginal population totals more than 800,000 today and remains marginalized within Australian society. Not unlike the situation with Native Americans in the United States, land ownership and entitlement are important and contentious issues. **Figure 12-9** shows the spatial distribution of both the Aboriginal and overall Australian populations across the country's States and Territories. Where are the Aboriginals most heavily concentrated,

and how does that compare with the distribution of the total national population? And what would you say are some of the implications of the clustering of Aboriginals in relation to the distribution of Australia's largest cities (Fig. 12-6) and its geography of agricultural and mineral resources (Fig. 12-7)?

Access your WileyPLUS course to interact with a dynamic version of this map and to engage with online exercises and questions.

A) ABORIGINAL POPULATION

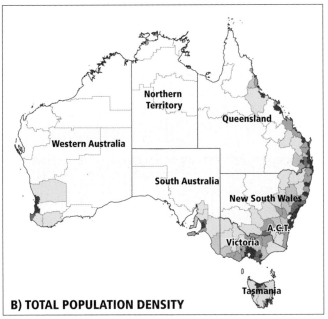

B) TOTAL POPULATION DENSITY

Aboriginal population (percentage of total), 2016

| No Data | < 5% | 5% - 10% | 10% - 20% | 20% - 30% | 30% - 50% | > 50% |

Population density (people per square kilometer), 2016

| No Data | < 1 | 1 - 3 | 3 - 5 | 5 - 10 | 10 - 15 | > 15 |

FIGURE 12-9 Interactive

the Australian High Court rendered the first of a series of landmark rulings in favor of Aboriginal claimants. A subsequent court decision implied that enormous areas (perhaps as much as 78 percent of the entire continent) could potentially be subject to Aboriginal claims. Today, the **Aboriginal land issue [14]** remains mostly (though not exclusively) an Outback issue, but it has the potential to overwhelm Australia's court system and to constrain economic growth.

Australia's Place in the World

Since World War II, as its historic ties to the United Kingdom have faded, Australia has become ever more embedded in the Asia Pacific—economically, culturally, and politically. As noted earlier in this chapter, the country's strengthening orientation toward its Asian neighbors is reflected in economic policies and patterns of trade and investment. Australia is now a key member of the Asia Pacific Economic Cooperation forum, **APEC [15]**; China and Japan, by far, have become its most important trading partners. Australia itself, as the result of changing immigration patterns, is becoming culturally more heterogeneous and increasingly Asian (photo below). And, politically, Australia has become more involved with the geographic realms to its north, especially the countries of Indonesia, East Timor, and Papua New Guinea. Multiculturalism is sure to remain a long-term challenge for Australia.

Geostrategically, Australia finds itself in a relatively secure location. However, security issues in the South China Sea and the Indian Ocean have become a growing concern precisely because Australia's interdependence with Asia is intensifying. Also of much interest is Australia's intent to forge closer relations with the United States as a counterweight to China's rapidly increasing influence across these realms, especially Southeast Asia. To nobody's surprise, the United States is pleased to reciprocate: since 2012, several thousand American troops have been stationed at Australian military bases, and the U.S. Air Force has been given access to Australian airfields in the Northern Territory that are within easy flying distance of the South China Sea.

Territorial dimensions, relative location, and raw-material wealth have helped determine Australia's place in the world and, more specifically, on the western Pacific Rim. Australia's population may be just over 25 million, but the country's importance in the international community greatly exceeds its human numbers.

New Zealand

Physiography

Twenty-four hundred kilometers (1500 mi) east-southeast of Australia, in the Pacific Ocean across the Tasman Sea, lies New Zealand, also known as *Aotearoa* in Maori (meaning "land of the long white cloud"). In an earlier age, New Zealand would have been part of the Pacific geographic realm because its population was entirely Maori, a people with Polynesian roots (see Chapter 13). But New Zealand, like Australia, was invaded and occupied by Europeans. Today, its population of 4.8 million is about two-thirds European, and the Maori comprise a substantial minority that numbers about 800,000, with many of mixed Euro-Polynesian ancestry (including Pacific Islanders).

New Zealand consists of two large mountainous islands and many scattered smaller islands (**Fig. 12-10**). The two main islands, with the South Island somewhat larger than the North Island, look diminutive in the great Pacific Basin, but together they are bigger than Britain. In sharp contrast to generally low-relief Australia, the more rugged terrain of the two large islands contains several peaks rising far higher than any on the Australian landmass (**Box 12-5**). Along its western margin, the South Island has a spectacular snowcapped range appropriately called the Southern Alps, with a number of summits reaching beyond 3300 meters (10,000 ft). The smaller North Island has proportionately more land under low relief, but it also has an area of central highlands along whose lower slopes lie the pastures of New Zealand's chief dairying district. The convergence of the Australian and Pacific tectonic plates

Cabramatta, one of Sydney's western suburbs, is seen as the "multicultural capital of Australia," a tourist attraction and proof of Australia's capacity to accommodate non-Europeans and especially Asians. This "Freedom Gate" in the Vietnamese community is flanked by a Ming horse and a replica of a Forbidden City lion— all reflecting Australia's growing ties to the Asia Pacific.

FIGURE 12-10

underlies much of New Zealand, and it makes the country vulnerable to volcanic eruptions and earthquakes.

Social Geography

The most promising areas for habitation, therefore, are the lower-lying slopes and lowland fringes on both islands. On the North Island, the largest urban area, Auckland, occupies a comparatively low-lying peninsula. On the South Island, the largest lowland is the agriculture-dominated Canterbury Plain, focused on the city of Christchurch. What makes these lower-elevation zones so attractive, apart from their availability as cropland, is their magnificent pastures. The range of soils and pasture plants allows both summer and winter grazing. Moreover, the Canterbury Plain, the chief farming region, also produces a wide variety of vegetables, cereals, and fruits. About half of all New Zealand is pasture land, and much of the farming provides fodder for the pastoral industry. About 29.5 million sheep, 6.5 million dairy cattle, and 3.6 million beef cattle dominate these livestock-raising activities, with wool, milk products, and meat providing about 75 percent of the islands' export revenues.

Despite their contrasts in size, shape, physiography, and history, New Zealand and Australia share a number of characteristics. Apart from their joint British heritage, they share a

Box 12-5 From the Field Notes . . .

"The drive from Christchurch to Arthur's Pass on the South Island of New Zealand was a lesson in physiography and biogeography. Here, on the eastern side of the Southern Alps, you leave the Canterbury Plain and its agriculture and climb into the rugged terrain of the glacier-cut, snowcapped mountains. A vast pasture lies on a patch of flatland in the foreground; in the background is the unmistakable wall of a U-shaped valley sculpted by moving ice. Natural vegetation ranges from pines to ferns, becoming even more luxuriant as you approach the moister western side of the island."

© H.J. de Blij

sizeable pastoral economy with growth in specialty goods such as wines, a small local market, the problem of great distances to world markets, and a desire to stimulate (through protectionist policies) domestic manufacturing. The very high degree of urbanization in New Zealand (87 percent of the total population) once again brings up the comparison with Australia: substantial employment in city-based industries—mostly the processing and packing of livestock and farm products—as well as government jobs.

Spatially, New Zealand further shares with Australia its pattern of **peripheral development [16]** and population distribution (Fig. 12-2), imposed not by deserts but by high, rugged mountains and the fragmented layout of the country. The major cities—Auckland and the capital city, Wellington, on the North Island; Christchurch and Dunedin on the South Island—are all situated on the coast, and the rail and highway networks are therefore dominantly peripheral in their configuration. Furthermore, the two main islands are separated by Cook Strait, a windswept waterway that can only be crossed by ferry or plane (Fig. 12-10). On the South Island, the Southern Alps are the country's most formidable barrier to surface communications.

The Maori Factor and New Zealand's Future

Like Australia, New Zealand has had a history of difficult relations with its indigenous population. The Maori, who today account for 16.7 percent of the country's total population, appear to have reached the islands during the tenth century AD. By the time the European colonists arrived, the Maori had also had a tremendous impact on the islands' ecosystems, especially on the North Island, where most of them lived. In 1840, the Maori and the British signed a treaty at Waitangi that granted the colonizers sovereignty over New Zealand but guaranteed the Maori rights covering established tribal lands. Although the British abrogated parts of the treaty in 1862, the Maori still had reason to believe that vast reaches of New Zealand, as well as offshore waters, were theirs in perpetuity.

As in Australia, judicial rulings beginning in the 1990s supported the Maori position, which led to expanded claims and growing demands. Culturally, the declaration of Maori as an official language in New Zealand and its teaching throughout the school system are seen as significant progress toward an acceptance of the Maori cultural heritage. But the most persistent Maori complaint concerns their slow pace of integration into modern New Zealand society. Although Maori land claims encompass much of rural New Zealand, they also cover prominent sites in the major cities. Today, the Maori question continues to be the foremost domestic issue.

At the same time, as noted earlier, New Zealand's population, like that of Australia, is becoming ever more multicultural. Nearly a quarter of New Zealanders today are foreign born, and immigration from Asian countries has been on the rise. While there is overall strong support for immigration, there are some reactionary elements. This came to the fore in a very painful manner in March 2019, when a white supremacist attacked two mosques in Christchurch, killing 51 Islamic worshippers. It was a shocking event, highly uncharacteristic for this otherwise tolerant and peaceful society—one that is likely to become more international, more multicultural, and more Asian in the years ahead.

Points to Ponder

- In this era of globalization, Australia is becoming more strongly tied to neighboring geographic realms, whereas its links to faraway Europe steadily diminish.

- Australia's advancing desalination technologies provide relief for the country's immediate water needs but seem to be delaying much-needed long-term conservation strategies.

- For countries like Australia, where ecological conditions (such as a desert environment) demand careful stewardship, global climate change can be enormously consequential.

- New Zealand experiences more than 10,000 earthquakes each year; as many as 150 can be felt, and about 20 are strong enough to cause damage on the human landscape.

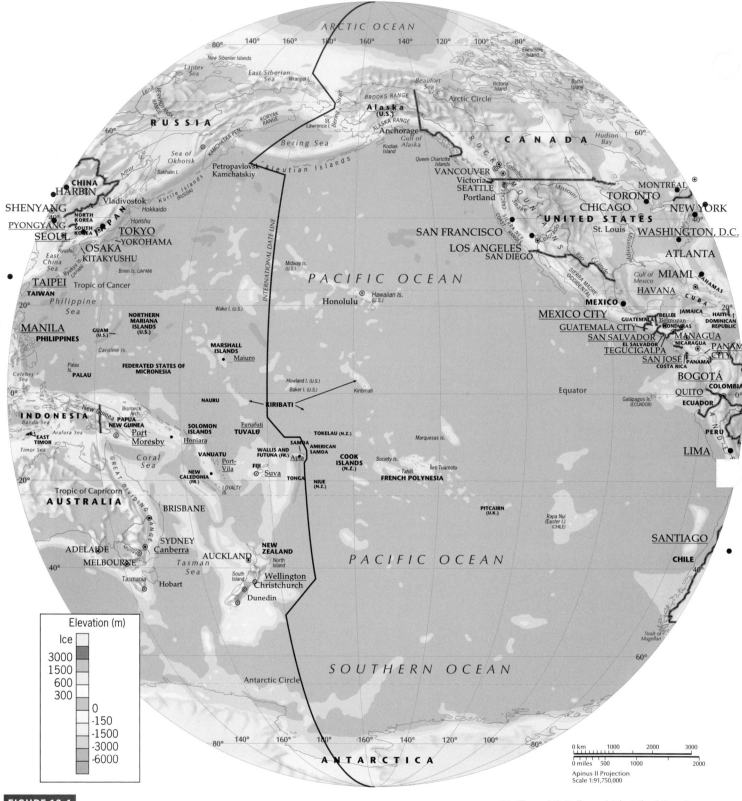

FIGURE 13-1

©J. Nijman, P. O. Muller, and John Wiley & Sons, Inc.

CHAPTER 13

In the Weddell Sea near the Antarctic Peninsula (see Fig. 13-6, inset map), a pair of Gentoo penguins waddle across the face of a huge flat iceberg that has detached from the Antarctic Ice Sheet. This largest ice mass on our planet covers about 98 percent of the continent, its areal size roughly one-and-a-half times that of the United States. Through the recent past, the Ice Sheet has held stable, but in the last several years a number of massive ice losses have been reported along its outer edges.

Jay Dickman/Corbis Documentary/Getty Images

The Pacific Realm and Polar Regions

IN THIS CHAPTER

- Water, water everywhere—but who owns it?
- Low islands threatened by rising sea level
- The partitioning of Antarctica
- Geopolitics in the warming Arctic Basin
- Up close with Greenland's Ilulissat Glacier

CONCEPTS, IDEAS, AND TERMS

High island **[1]**

Low island **[2]**

Atoll **[3]**

Tsunami **[4]**

Tsunami Warning System (TWS) **[5]**

Small-island developing economies **[6]**

Economies of scale **[7]**

Connectivity **[8]**

Marine geography **[9]**

Territorial sea **[10]**

High seas **[11]**

Continental shelf **[12]**

Exclusive Economic Zone (EEZ) **[13]**

Maritime boundary **[14]**

Median-line boundary **[15]**

Global climate change **[16]**

Rising sea level **[17]**

Arctic Circle **[18]**

Antarctic Circle **[19]**

Antarctic Treaty **[20]**

Maritime claims **[21]**

Northeast Passage **[22]**

Northwest Passage **[23]**

In this final chapter, we focus on three of the world's gigantic expanses, so different from those discussed in previous chapters that we need a different perspective. Water and ice, not land and soil, dominate the physiography of these areas. Of the utmost importance is that environmental change occurring in all three has consequences for the entire planet.

The largest by far of these expanses, the *Pacific Realm*, covers almost half the globe from the Bering Sea to the Southern Ocean and from Central America to Indonesia (**Fig. 13-1**). The Pacific Ocean, larger than all the world's landmasses combined, is studded with tens of thousands of islands, of which New Guinea is by far the largest. The two polar regions to which it is linked could hardly be more different. The *Arctic* consists entirely of water, frozen at the surface, but you can reach the North Pole in a submarine. The *Antarctic* is mainly land, weighed down by the planet's largest accumulation of permanent ice; the surface journey to the South Pole, first accomplished more than a century ago, has cost many an explorer's life. The Arctic and Antarctic cannot be considered realms in the same way as the others in this book because of the near absence of human populations and overall physiographic homogeneity. We treat them, instead, as distinct polar regions.

Climate change is altering geography everywhere, as is the enhanced technology of mineral (especially oil and gas) exploration and extraction. Rising sea levels would imperil thousands of populated low-elevation islands. Areas of ocean floor once beyond the scope of human intervention are now coming within reach. Melting ice may be clearing new maritime shipping routes. At the same time, as we will see, international rules to govern who owns what in these remote frontiers are subject to dispute. When land boundaries jut outward into the sea, who owns what depends on who draws the maps. We should be aware of the potential for competition, even conflict.

Defining the Pacific Realm

Our survey begins with the Pacific Realm, which covers just about an entire hemisphere, the one sometimes called the Sea Hemisphere (Fig. 13-1). This Sea Hemisphere meets the Russian/Central Asian and North American realms in the far north and merges into the Southern Ocean in the south. Despite the preponderance of water, this fragmented, culturally complex realm does possess regional identities. It includes the Hawaiian Islands, Tahiti, Tonga, Fiji, and Samoa—fabled names in a world apart.

In terms of modern cultural and political geography, Indonesia and the Philippines are not part of the Pacific Realm, though Indonesia's political system reaches into it; nor do Australia and New Zealand belong to it. Before the European invasion and colonization, Australia would have been included (because of its Aboriginal population) as well as New Zealand (its Maori population has Polynesian affinities). But the Europeanization of their countries (and, more recently, Asianization) has engulfed indigenous Australians and Maori New Zealanders, and the regional geography of the Austral Realm today is decidedly not Pacific. In New Guinea, by contrast, Pacific peoples, numerically and culturally, remain the dominant element (**Box 13-1**).

In examining the Pacific Realm, it is important to keep in mind the dimensional contrasts this part of the world presents. The realm may be enormous, but its total land area is a mere 975,000 square kilometers (377,000 sq mi), about the

---

Box 13-1 Major Geographic Features of the Pacific Realm

1. The Pacific Realm's areal extent is the largest of all geographic realms. Its total land area, however, is the smallest, as is its population.

2. The highly fragmented Pacific Realm consists of three island regions: Melanesia (including New Guinea), Micronesia, and Polynesia.

3. The Pacific Realm, with its huge expanses of water and numerous islands, has been strongly affected by United Nations Law of the Sea provisions regarding states' rights over economic assets in their adjacent waters.

4. Papua New Guinea (the eastern, Pacific half of the island of New Guinea), with 9 million people, alone contains almost three-quarters of the Pacific Realm's population.

5. The Pacific Realm's islands and cultures may be divided into volcanic high-island cultures and coral-based low-island cultures.

6. Rising sea level is threatening the low islands in the Pacific, which are desperately trying to adapt to increased flooding.

7. In Micronesia, U.S. influence has been particularly strong and continues to affect local societies.

8. In Polynesia, indigenous culture exhibits remarkable consistency and uniformity throughout the region, its enormous dimensions and dispersal notwithstanding. Yet at the same time, local cultures nearly everywhere are severely stressed by external influences. In Hawai'i, as in New Zealand, indigenous culture has been largely submerged by Westernization.

© J. Nijman

Aerial view of a circular atoll in the Fijian archipelago, which marks the eastern edge of Melanesia. This atoll is almost completely submerged except for the tiny sliver of sand visible at the extreme left. From the plane, atolls can often be detected first by the clouds clustered above them: the shallow water atop an atoll (for that matter, any low island) heats up more than the deeper ocean around it—causing the air to rise, condense into clouds, and occasionally trigger a local rainstorm.

Colonization and Independence in the Pacific Realm

The Pacific islands were colonized by the French, British, and Americans—an indigenous Polynesian kingdom in the Hawaiian Islands was annexed by the United States in 1898, and since 1959 has functioned as the 50th U.S. State. Because political geography evolves slowly within the Pacific Basin, today's map is still an assemblage of colonial territories and independent microstates (Fig. 13-1). France still controls New Caledonia and French Polynesia. The United States administers Guam as well as American Samoa, the Line Islands, Wake Island, the Midway Islands, and several smaller island groups; the United States also has special relationships with other territories, former dependencies that are now nominally independent. The British, through New Zealand, have responsibility for the Pitcairn group of islands, and New Zealand administers and supports the Cook, Tokelau, and Niue Islands. Easter Island, that storied speck of land in the southeastern Pacific, is part of Chile (see Box 13-4). And, as discussed in Chapter 11, Indonesia rules Papua, the western half of the island of New Guinea.

Other island groups have now become independent states. The largest are Fiji, once a British dependency; the Solomon Islands, also formerly British; and Vanuatu, until 1980 ruled jointly by France and Britain. Also on the current map are such impoverished microstates as Tuvalu, Kiribati, Nauru, and Palau, which must depend on foreign aid for their survival. Tuvalu, for instance, has a total area of 26 square kilometers (10 sq mi), a population of barely 12,000, and a per capita GNI under U.S. $6,000—derived from fishing, sales of copra (coconut meat used to make coconut oil), and some tourism. But what really keeps Tuvalu going is an international trust fund set up by Australia, New Zealand, the United Kingdom, Japan, and South Korea. Annual grants from that fund, as well as remittances sent back to families by workers who have left for New Zealand and elsewhere, allow Tuvalu to survive.

In some ways, the Pacific islands can be compared to those in the Caribbean (Chapter 3) in that they face similar challenges of **small-island economies [6]**: their modest size severely limits potential benefits from **economies of scale [7]** and their lack of **connectivity [8]** hampers external economic relations. Argu-ably, the challenges for the Pacific islands are even more severe because the islands' minuscule sizes and isolated locations are more extreme than in the Caribbean. It also explains, at least in part, the continued strong political involvement of outside powers. Another indication of the comparable vulnerability of these small islands in the Caribbean and Pacific is found in recent reports about the illicit drug trade outflow (mainly cocaine and methamphetamines) from South America to Australia and New Zealand, in sailing boats, via Pacific islands such as French Polynesia and Fiji that are used as stopovers and storage hubs.

size of Texas plus New Mexico, with nearly half constituted by Papua New Guinea. Indeed, the realm's inhabitants are so widely and thinly scattered that a population distribution map would be impractical; moreover, the population totals only around 16 million (less than for the compact megacity of Dhaka, Bangladesh—see Box 9-5).

Physiographically, the main landforms of the Pacific can be categorized as high islands, low islands, and atolls. **High islands [1]** are seafloor-based volcanic peaks and ridges that protrude above the ocean surface and can reach thousands of meters in elevation. **Low islands [2]** are far more common, consist mainly of coral, and sometimes barely rise above sea level. Livelihoods tend to vary among these different environments: farming prevails on high islands (thanks to fertile soils and orographic rainfall), whereas fishing dominates the economy of low islands (where freshwater supplies are limited).

The third landform category is the **atoll [3]**, a ring-shaped coral reef that encircles a lagoon, lying more or less at sea level (photo above). The circular shape results from the reef growing on the rim of an underwater volcano (which may long have eroded away); typically, atolls are uninhabitable.

Whether high or low, the Pacific islands are a reminder of the geological nature of the ocean floor, with its volcanic peaks, ridges, and tectonic activity. Earthquakes in the Pacific, wherever they may occur, can cause huge seismic sea waves known as **tsunamis [4]** (also see p. 378 in Chapter 10). Because the low islands and low-lying shores of the Pacific are so vulnerable, an internationally organized **Tsunami Warning System (TWS) [5]** has been established to constantly monitor sea-surface conditions (**Box 13-2**).

Box 13-2 Technology & Geography

The Tsunami Warning System in the Pacific

After a major earthquake strikes near or under a large body of water, a series of long, fast-moving waves is sometimes generated. Earthquakes are quite common in the Pacific and Atlantic oceans due to tectonic-plate motion (see Fig. 1-5, Chapter 1). Such a seismic sea wave is called a *tsunami*. As a tsunami crosses the ocean surface, it can travel at speeds of up to 1000 kilometers per hour (625 mph) and be more than 200 kilometers (125 mi) wide. As the wave approaches shallow water, its forward speed decreases and its height increases. When tsunamis make landfall, at times cresting at over 10 meters (35 ft), low-lying coastal areas and their inhabitants are imperiled. A colossal, magnitude-9.0 quake off Japan in 2011 created a gigantic tsunami that destroyed dozens of coastal settlements, killed almost 20,000 people, and caused disastrous damage at an oceanside nuclear-power generation facility (photo p. 377; Fig. 10-2).

During the mid-twentieth century, to warn coastal populations and minimize the loss of life from tsunamis, several Pacific Basin countries joined forces to establish the *Tsunami Warning System (TWS) in the Pacific*. Consisting of an extensive network of buoys armed with sensors and instruments specifically designed to measure and record the depth of the water column across the Pacific Ocean, the TWS is designed to detect unexpected shifts in water depth and ocean-wave irregularities. Global positioning systems (GPS) satellites record the exact location and movement of these buoys, from which the speed and direction of a tsunami event can be calculated. Should a tsunami be detected, its direction, speed, and potential strength can be estimated, and member-states are swiftly alerted to take appropriate safety measures. **Figure 13-2** shows the locations of coastal seismographic stations (that measure both tectonic activity as well as tidal variations), deep-ocean *tsunameters* (which gauge seafloor pressures and water movement), and average tsunami travel times from Honolulu, Hawai'i.

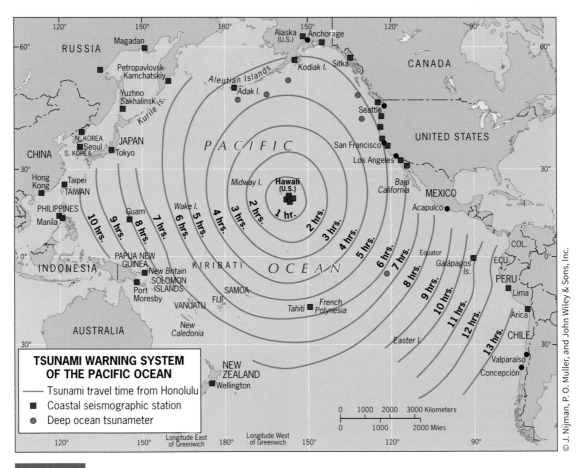

FIGURE 13-2

© J. Nijman, P. O. Muller, and John Wiley & Sons, Inc.

The Pacific Realm and Its Marine Geography

Certain land areas may not be part of the Pacific Realm (we mentioned the Philippines and New Zealand), but the Pacific Ocean extends from the shores of North and South America to mainland East and Southeast Asia and from the Bering Sea off Alaska southward to the Subtropical Convergence around latitude 40°S. This means that several seas, including the South China Sea, the East China Sea, the Sea of Japan (East Sea), the Java Sea, and the Tasman Sea, are all part of the Pacific Ocean. As we are about to discover, this relationship matters. Pacific coastal countries, large and small, mainland as well as island, compete for jurisdiction over the waters that bound them.

The ocean-focused Pacific Realm, therefore, provides an ideal setting for introducing some basic principles of **marine geography [9]**. This field encompasses a variety of approaches to the study of oceans and seas; some marine geographers concentrate on the biogeography of coral formations, others on the geomorphology of coasts and beaches, and still others on the circulation of ocean currents. A most fascinating specialty within marine geography has to do with the definition and delimitation of political boundaries at sea. Here, geography intersects both political science and maritime law.

The State at Sea

Littoral (coastal) states do not end where atlas maps suggest they do. States have claimed various forms of jurisdiction over coastal waters for centuries, closing off bays and estuaries and ordering foreign fishing fleets to stay away from nearby fishing grounds. Thereby arose the notion of the **territorial sea [10]**, where all the rights of a coastal state would prevail. Beyond lay the **high seas [11]**, free, open, and unfettered by national interests—so-called international waters.

It was in the interest of colonizing and/or trading states to keep territorial seas narrow and high seas wide, thereby interfering as little as possible with their commercial fleets. In the seventeenth and eighteenth centuries, the territorial sea was 3, 4, or at most 6 nautical miles wide, and the colonizing powers claimed the identical widths for their colonies (1 nautical mile = 1.85 kilometers or 1.15 statute miles).*

During the twentieth century, these constraints weakened. States without trading fleets saw no reason to limit their territorial seas. States with nearby fishing grounds traditionally exploited by their own fleets wanted to keep the increasing number of foreign trawlers away. States with shallow

*Here and in the rest of this section all distances involving maritime boundaries are given in nautical miles only. Throughout the modern history of maritime law, this unit of measurement has been the only one used for such boundary-making. Any distance stated in nautical miles can be converted to its metric equivalent by multiplying it by 1.85.

continental shelves [12] (offshore continuations of coastal plains) wished to control their resources on and below the seafloor, made more accessible by improved technology. States also disagreed on the methods by which offshore boundaries, whatever their width, should be defined. Early efforts by the League of Nations (the precursor to the United Nations) during the 1920s to resolve these issues met with only partial success, mainly in the technical sphere of boundary delimitation.

Scramble for the Oceans   In 1945, the United States helped precipitate what has become known as "the scramble for the oceans." President Truman issued a proclamation that claimed U.S. jurisdiction and control over all the resources "in and on" the continental shelf down to its margin, around 100 fathoms (183 meters/600 ft) deep. In some areas, the shallow continental shelf of the United States extends more than 300 kilometers (200 mi) offshore, and Washington did not want foreign countries drilling for oil just beyond the 3-mile territorial sea.

Few observers foresaw the impact the Truman Proclamation would have, not only on U.S. waters but on oceans everywhere, including the Pacific. It set off a rush of additional claims. In 1952, a group of South American countries, some with little continental shelf to claim, issued the Declaration of Santiago, claiming exclusive fishing rights up to a distance of 200 nautical miles off their coasts. Meanwhile, as part of the Cold War competition, the Soviet Union claimed a 12-mile territorial sea and urged its allies to do likewise.

UNCLOS Intervention

At this point, the United Nations intervened, and a series of United Nations Conference on the Law of the Sea (UNCLOS) meetings began. These meetings addressed issues ranging from the closure of bays to the width and delimitation of the territorial sea, and after three decades of negotiations they achieved a convention that changed the political and economic geography of the world's oceans forever. Among its key provisions were the authorization of a 12-mile territorial sea for all countries as well as the establishment of a 200-mile (230-statute-mi/370-km)-wide **Exclusive Economic Zone (EEZ) [13]** over which a coastal state would have total economic rights. Resources in and under this EEZ (fish, minerals, oil, and gas) belong to the coastal state, which could either exploit them or lease, sell, or share them as it saw fit. These zones were noted in the discussion of the new competition for the Arctic Ocean in Chapter 6, but as we shall see, EEZs take on even greater importance in the Pacific Realm.

These new provisions had a far-reaching impact on the world's oceans and seas (**Fig. 13-3**), especially the Pacific. Unlike the Atlantic Ocean, the Pacific is studded with myriad islands, so that a microstate consisting of a single small island suddenly acquired an EEZ covering 166,000 square nautical miles. European colonial powers that still

© J. Nijman, P. O. Muller, and John Wiley & Sons, Inc.

WORLD MARITIME CLAIMS

◯ Area within 200-Nautical-Mile Limit
(Arctic claims in Fig. 13-7;
Antarctic claims in Fig. 13-6)

FIGURE 13-3

controlled minor Pacific possessions (most notably France) saw their maritime jurisdictions vastly expanded. Small, low-income archipelagos could now bargain with large, rich fishing nations over fishing rights in their now huge EEZs. And for all the UNCLOS Convention's provisions for the "right of innocent passage" of shipping through EEZs and via narrow straits, the geographic extent of the world's high seas has been diminished accordingly.

Maritime Boundaries

The extension of the territorial sea to 12 nautical miles and the EEZ to an additional 188 nautical miles created new **maritime boundary [14]** problems. Waters less than 24 nautical miles wide separate many countries all over the world, so that **median-line boundaries [15]**, equidistant from opposite shores, have been delimited to establish their territorial seas. And even more countries lie closer than 400 nautical miles apart, requiring further maritime-boundary delimitation to determine their EEZs. In such maritime arenas as the North Sea, the Caribbean Sea, and the East and South China seas (and more recently the Arctic Ocean), a maze of maritime boundaries emerged, some of them subject to dispute. Furthermore, political changes on land can also lead to concomitant modifications at sea.

EEZ Implications

The UNCLOS provisions created new opportunities for some states to expand their spheres of influence. Wider territorial-sea and EEZ allocations raised the stakes: claiming an island now entailed potential control over a huge maritime area. In Chapter 6, we mentioned the impending need to further carve up the Arctic Basin as the ice melts and untapped resources beckon. In Chapters 10 and 11, we referred to several island disputes off mainland East and Southeast Asia, which involve Japan and Russia, Japan and South Korea, Japan and China, China and Vietnam, and China and the Philippines. Ownership of several offshore Asian islands is uncertain, and small specks of island territory have become large stakes in today's scramble for the oceans. For instance, in the case of the Spratly Islands, several countries claim ownership, including both Taiwan and China (see Box 11-6 on p. 412). China's island claims in the South China Sea (within its alleged "Nine-Dashed Line" maritime boundary) support Beijing's contention that this body of water is part and parcel of the Chinese state—a position that increasingly troubles other states with coasts facing it.

Figure 13-3 indicates what EEZ regulations have meant to Pacific Realm countries such as Tuvalu, Kiribati, and Fiji, with

nearly circular EEZs now surrounding the clusters of islands in this vast oceanic space. Japan, Taiwan, and other nations with fisheries have purchased fishing rights in these EEZs from the island governments. Nonetheless, violations of EEZ rights do occur; not long ago, Vanuatu and the Philippines were at odds over unauthorized Filipino fishing in Vanuatu's EEZ. The process of boundary delimitation continues to evolve, and the Pacific (and world) map of maritime boundaries remains very much a work in progress.

That point is underscored by a development that occurred at the end of the 1990s, when the implications of provisions in Article 76 of the 1982 UNCLOS Convention were reexamined. Whenever a continental shelf extends beyond the 200-mile limit of the EEZ, the Convention apparently permits a coastal state to claim that extension as a natural prolongation of its landmass. Although there is as yet no regulation that will also allow the state to extend its EEZ to the edge of the continental shelf, it is not difficult to foresee such an amendment to the Convention. As it stands, states are now delimiting their proposed natural prolongations under a current deadline, which puts poorer states at a disadvantage since the required marine surveys are quite expensive. Today, this issue has become quite heated, especially regarding the Arctic Ocean, as countries scramble to claim prolongations (most notably Russia, as we shall see at the end of this chapter). The big winners are likely to include Russia, the United States, Canada, Australia, New Zealand, and India—although a total of 60 coastal countries could ultimately benefit to some extent from the provision.

Another complicating factor is the role of new technologies that are accelerating deep-sea exploration and the exploitation of seabed resources. Today, the quest to discover additional mineral deposits is increasingly an underwater endeavor, led by the search for so-called massive sulfides—enormous concentrations of gold, silver, copper, and other metallic minerals formed below the ocean floor in sulfide-rich volcanic environments. The technological breakthroughs entail sensors, advanced robotic devices, and other state-of-the-art gear (such as "sea gliders") developed by the offshore oil and gas industry. The growing use of sea gliders is of special interest because these small, slow-moving, unmanned vessels can track enormous distances as they collect detailed information on everything below—from marine life to seabed resources to the movements of submarines. And so the scramble for the oceans marches ever onward, and with it the constriction of the open waters of the world's high seas.

Regions of the Pacific Realm

Sail across the Pacific Ocean, and one spectacular vista follows another. Dormant and extinct volcanoes, sculpted by erosion into basalt spires draped by luxuriant tropical vegetation and encircled by reefs and lagoons, tower over azure waters. Low

atolls with nearly snow-white beaches, crowned by stands of palm trees, seem to float in the water. Pacific islanders, where foreign influences have not overtaken them, appear to take life with enviable ease.

PACIFIC REGIONS

ATLANTIC OCEAN

PACIFIC OCEAN

POLYNESIA

MICRONESIA

MELANESIA

INTERNATIONAL DATE LINE

FIGURE 13-4

More intensive investigations, however, reveal that such Pacific sameness is more apparent than real. Even the Pacific Realm, with its centuries-old sailing traditions, its still-diffusing populations, and its historic migrations, has a durable regional framework. **Figure 13-4** outlines the three regions of the Pacific that can be distinguished mainly on the basis of culture and language groups:

Melanesia: Papua (province of Indonesia), Papua New Guinea, Solomon Islands, Vanuatu, New Caledonia (France), Fiji

Micronesia: Palau, Federated States of Micronesia, Northern Mariana Islands, Republic of the Marshall Islands, Nauru, western Kiribati, Guam (United States)

Polynesia: Hawaiian Islands (U.S.), Samoa, American Samoa, Tuvalu, Tonga, eastern Kiribati, Cook and other New Zealand–administered islands, French Polynesia, Easter Island (Chile)

Region Melanesia

New Guinea island lies at the western end of a Pacific region that extends eastward to Fiji and includes the Solomon Islands, Vanuatu, and New Caledonia (Fig. 13-4). The human mosaic here is complex, both culturally and ethnically. Most of the 12.7 million people of New Guinea (divided between the Indonesian province of Papua and the independent state of Papua New Guinea) are Papuans, but there are many minorities. Altogether there are more than 800 communities speaking different languages; the Papuans are most numerous in the densely forested highland interior and in the lowland south. In total, Melanesia contains around 15 million inhabitants, making this the most populous Pacific region by far.

With almost 9 million people today, **Papua New Guinea (PNG)** became a sovereign state in 1975 after nearly a century of British and Australian administration. Almost all of PNG's limited development is taking place along its coasts, whereas most of the interior remains hardly touched by the changes that transformed neighboring Australia. Perhaps four-fifths of the population is part of the self-sufficient subsistence economy, cultivating root crops, raising pigs, hunting wildlife, and gathering forest products. Age-old traditions of the kind lost in Australia continue on here, protected by remoteness and the rugged terrain.

Welding this disparate population into a nation is a task hardly begun, and PNG faces numerous obstacles in addition to its cultural complexity. Not only are hundreds of languages in use, but more than one-third of the population is estimated to be illiterate. English, the official language, is used by the educated minority but is of little use beyond the coastal zone and its towns. The capital, Port Moresby, contains just under 400,000 residents, reflecting the very low level of urbanization (around 13 percent) in this developing economy and generally across the realm (it is the second biggest city in the realm after Honolulu).

Yet Papua New Guinea is not without economic opportunities. Oil was discovered in the 1980s, and by 2000 crude oil was PNG's largest export by value. More recently, newly found gas deposits have propelled the growth of the energy sector, and Port Moresby's major new PNG-LNG liquid natural gas facility began export operations in 2014 (even larger gas production projects are now being developed). Other important exports include gold, copper, silver, timber, and several agricultural products that include coffee and cocoa, reflecting the country's diversity of environments and resources. And PNG has not been immune to Asia Pacific influences: although the biggest share of exports still goes to its nearest neighbor, Australia, Japan ranks second, and fast-rising China is close behind.

Turning eastward, it is a measure of Melanesia's cultural fragmentation that as many as 120 languages are spoken in the approximately 1000 islands that make up the **Solomon Islands** (about 80 of them support almost all the inhabitants, who number almost 670,000). Inter-island, historic animosities among the islanders were worsened by the events of World War II, when U.S. forces moved thousands of Malaitans to Guadalcanal. This started a postwar cycle of violence that finally led Australia to intervene in 2003.

New Caledonia, still under French rule, is in a very different situation. Only around 40 percent of its population of roughly 280,000 are Melanesian; not that far behind are the 29 percent of European ancestry (known as the Caldoches), many of them descended from the inhabitants of the penal colony France established here during the nineteenth century. Nickel mines, based on reserves that rank among the world's largest, dominate New Caledonia's export economy. The mining industry attracted additional French settlers, and social problems soon arose. Most of the French population lives in or near the capital city of Nouméa, steeped in French cultural landscapes, located in the southeastern quadrant of the island. Melanesian demands for an end to colonial rule have led to violence, and the two communities remain engaged in a long-running process of coming to terms.

On its eastern margins, Melanesia includes one of the Pacific realm's most interesting countries, **Fiji**. On two larger and over 100 smaller islands live around 900,000 Fijians, of whom about 55 percent are Melanesians and 35 percent South Asians, the latter brought to Fiji from India during the British colonial occupation to work on the sugar plantations. When Fiji achieved independence in 1970, the indigenous Fijians owned most of the land and held political control, whereas the Indians were concentrated in the towns (chiefly Suva, the capital) and dominated commercial life. It was a recipe for trouble, which surfaced as soon as the British had departed. Nearly a half-century of ethnic conflict and political malfunctioning has cast a long shadow across this country. In 2009, after some 40 years of strife and violations of democratic rule, Fiji was suspended from the British Commonwealth. The sugar industry was damaged by the nonrenewal of Indian-held leases by Fijian landowners, which also resulted in a substantial movement of Indians from the countryside to the towns, where unemployment is already high. Only after democratic elections were finally held in 2014 was Commonwealth status reinstated; more important, at the same time both Australia and New Zealand restored diplomatic relations. While Fiji was shunned by these other countries,

China stepped in to fill the void. Chinese foreign aid to Fiji has increased sevenfold since the last military coup in 2006, and Chinese investments in infrastructural projects have increased commensurately. This, of course, is yet another sign that China's influence is now being projected globally—and not always in a manner reassuring to the international community.

Region Micronesia

North of Melanesia and to the east of the Philippines lie the islands that constitute the second Pacific region, Micronesia (Fig. 13-4). The name refers to the size of the islands: the 2000-plus islands of Micronesia are not only tiny (many of them are no larger than one square kilometer [half a square mile]), but they are also much lower-lying, on average, than those of Melanesia. Some are volcanic high islands, but they are outnumbered by low coral islands and atolls that barely reach above sea level. Guam, with 550 square kilometers (210 sq mi), is Micronesia's largest island, and no elevation anywhere in Micronesia reaches 1000 meters (3300 ft).

Take, for example, the country of **Kiribati**, consisting of three main island groups that lie across Micronesia and Polynesia (Fig. 13-4). The average elevation of Kiribati's 33 islands is 2 meters (6.5 ft), and, like all other low islands in this realm, they are particularly susceptible to rising sea level driven by global warming (**Box 13-3**).

The high-island/low-island dichotomy, introduced in the earlier overview of Pacific physiography, is particularly applicable in Micronesia. The high islands wrest substantial orographic rainfall from the moist maritime air, tend to be well watered, and are usually endowed with productive volcanic soils. As a result, agriculture can thrive, life is reasonably secure, and populations tend to be larger than those of low islands. On those low islands, drought is the rule, and fishing and the coconut palm are the mainstays of life; most of the islands can only support small communities, and over time many have died out. Thus the major precolonial, indigenous migrations, which sent fleets to populate islands from Hawai'i to New Zealand, originated mainly in the high islands.

From 1945 until the mid-1980s, Micronesia was largely a United States Trust Territory (the last of the post–World War II trusteeships supervised by the United Nations). But, as Figure 13-4 shows, Micronesia is now divided into countries bearing the names of independent states. These days the **Marshall Islands**, where the United States used to test nuclear weapons (giving prominence to the name *Bikini*), is a republic in "free association" with the U.S., having the same status as Palau and the **Federated States of Micronesia**. The **Northern Mariana Islands** comprise a commonwealth "in political union" with the United States, which in effect provides billions of dollars in assistance to these countries, in return for which they commit themselves to avoid foreign policy actions that are contrary to American interests.

The Republic of **Palau** is another fascinating example of how some of these small Pacific nations make a living. This group of islands some 800 kilometers (500 mi) east of the Philippines was granted independence in 1994 and has a total population of around 18,000. Palau has no military but entered into an agreement with the United States to provide security in return for a 50-year lease on a military base. Though highly dependent on American foreign assistance, Palau also receives funds and investments from Taiwan—a reward for Palau's official recognition of Taiwan's status as the Republic of China as well as for withholding recognition of the PRC.

Also part of Micronesia is the U.S. territory of **Guam**, where independence is not in sight and where U.S. military installations and tourism provide most of its income. And then there is the remarkable Republic of **Nauru**, with a population of barely 10,000 and just 20 square kilometers (8 sq mi) of land. Nauru got rich by selling its phosphate deposits to Australia and New Zealand, where they are used as fertilizer. In 2013, GDP per capita had risen above U.S. $15,000, making Nauru one of the realm's high-income societies (!); but the main surface deposits of phosphate have now been depleted, and great uncertainty prevails as the mining of a secondary layer proceeds. By 2019, GDP per capita had declined to around $9000.

In this region of tiny islands, almost all inhabitants subsist on farming or fishing, and virtually every country needs infusions of foreign aid to survive. The natural economic complementarity between the high-island farming cultures and the low-island fishing communities is all too often negated by distance, spatial as well as cultural. Life here may seem idyllic to the casual visitor, but for too many Micronesians it is an arduous daily struggle.

Region Polynesia

To the east of Micronesia and Melanesia lies the heart of the Pacific, enclosed by a great triangle stretching from the Hawaiian Islands to New Zealand to Chile's Easter Island (see **Box 13-4**). This third Pacific region is Polynesia (Fig. 13-4), which consists of numerous islands (*poly* means many) ranging from volcanic mountains rising far above the Pacific's waters (Mauna Kea on Hawai'i reaches over 4200 meters [nearly 13,800 ft]), clothed by luxuriant tropical forests and drenched by well over 250 centimeters (100 in) of rainfall each year, to low coral atolls where, if there is just enough elevation, a few palm trees form the only vegetation and drought is a persistent problem (see photo p. 453).

Despite its vast extent (it is about twice the areal size of Melanesia and Micronesia combined), Polynesia clearly constitutes a distinct Pacific region. Polynesian culture, though spatially fragmented, exhibits a remarkable consistency and uniformity from one island to the next, from one end of this widely dispersed region to the other. This consistency is particularly expressed in vocabularies, technologies, housing, and art forms. The Polynesians are uniquely

Box 13-3 Global Climate Change in the Pacific

Adaptation to Rising Sea Level in Kiribati

Kiribati (pronounced "KEER-uh-bahss"), an elongated archipelago straddling the regional boundary between Micronesia and Polynesia (Fig. 13-4), consists of 33 low-lying coral atolls that are home to just 120,000 people. About half of them reside on the main island of Tarawa Atoll and depend on the surrounding ocean for nutrition and livelihoods. Surprisingly, overcrowded South Tawara has a population density similar to Tokyo or Hong Kong, and its freshwater supply, sanitation facilities, and infrastructure are under immense strain.

Tarawa and the other atolls lie less than 3 meters (10 feet) above sea level, and find themselves directly in the crosshairs of **global climate change [16]** impacts—particularly ocean warming and **rising sea level [17]**. Heightened flooding, coastal erosion, and damage to infrastructure are already scarring the landscape, and saltwater intrusion into groundwater and soils increasingly endangers the raising of crops. Offshore, ocean warming and acidification are simultaneously degrading coral reefs that serve as feeding grounds for the fish that are critical to the dietary needs of the islanders.

Adaptation planning has now become a key component of the toolkit to mitigate climate change impacts. The Kiribati Adaptation Program (KAP) is an international effort aimed at reducing the country's vulnerability to sea-level rise. The project entails measures such as mangrove planting, construction of seawalls, strengthening laws to reduce coastal erosion, and population settlement planning. It also involves the promotion of skills and capacity building so that local communities can provide safe drinking water and maintain resilient coastal infrastructure.

Unfortunately, such measures can only postpone the inevitable: most studies forecast that Kiribati and certain other Pacific island-countries will become uninhabitable by 2050. Debates continue about the responsibilities of the world's leading emitters of greenhouse gases and the rights of climate-change refugees, and about ways of slowing the effects of global warming on sea-level rise. In the meantime, daily life in Kiribati goes on as usual, with most of the increasingly inconvenienced islanders seeming to accept the higher tides, more frequent floods, and incessant seawall construction (photos below).

Jonas Gratzer/Getty Images

The AGE/Getty Images

Kiribati and myriad other low islands in the Pacific Realm face the dire consequences of rising sea levels. In the left photo, taken in late 2015, a family wades through the latest invasion of seawater that flooded their village to collect seabed stones. These stones are then used to build seawalls higher to deflect future floodwaters away from homes (right).

Box 13-4 From the Field Notes ...

"Easter Island, known as Rapa Nui in the native language, is a typical high island that is constituted by three extinct volcanoes protruding above the ocean surface. In the foreground is the spectacular crater and lake named Rano Kau, its rim reaching about 330 meters (1000 ft) above sea level. We have just taken off from the airport runway aligned left to right behind the crater where the slopes of Rano Kau end. Our pilot executed a 180° right turn to let us have this splendid view of the island. Just beyond the left end of the runway is the town of about 5000 people. Most high islands of the Pacific carry abundant vegetation, and Easter Island was indeed once blanketed by palm trees. However, during the seventeenth century the local population effectively cut down all the trees (for reasons that are still not understood), eliminating an essential resource and triggering massive soil erosion. Today, some scholars consider Easter Island to be a symbol of environmental destruction and societal collapse. The trees now found on Easter Island were all replanted. Most food and supplies for the islanders—who total about 6000—have to be brought in from Chile, 3500 kilometers (2200 mi) away."

© J. Nijman

adapted to their marine environment, and long before European sailing ships began to arrive in their waters, Polynesian seafarers had learned to navigate their enormous expanses of ocean in huge double canoes as long as 45 meters (150 ft). They traveled hundreds of kilometers to favorite fishing zones and engaged in inter-island barter trade, using maps constructed from bamboo sticks and cowrie shells and navigating by the stars. However, modern descriptions of a Pacific Polynesian paradise of emerald seas, lush landscapes, and gentle people distort harsh realities. Polynesian society was forced to accept much loss of life at sea when storms claimed their boats; families were ripped apart by accidents as well as by migration; hunger and starvation afflicted the inhabitants of smaller islands; and the island communities were often embroiled in violent conflicts and cruel retributions.

The political geography of Polynesia is complex. In 1959, the **Hawaiian Islands** became the 50th State to join the United States. The State's population is now 1.5 million, with more than 70 percent living on the island of Oahu. There, the superimposition of cultures is symbolized by the panorama of Honolulu's seaside skyscrapers framed against the famous extinct volcano at nearby Diamond Head. The Kingdom of **Tonga** became an independent country in 1970 after seven decades as a British protectorate; the British-administered Ellice Islands

were renamed **Tuvalu** and received independence from the UK in 1978. Other islands continue under French control (including the Marquesas Islands and Tahiti), under New Zealand's administration (Cook Islands), as well as under British, U.S., and Chilean flags.

In the process of politico-geographical fragmentation, Polynesian culture has suffered severe blows. Land developers, hotel builders, and tourist dollars have set **Tahiti** on a course along which Hawai'i has already traveled far. The Americanization of eastern **Samoa** has created a new society different from the old. Today, Polynesia has lost much of its ancient cultural consistency, and the region has become a patchwork of new and old—the new often bleak and barren, with the old under ever intensifying pressure.

The countries and cultures of the Pacific Realm lie in an enormous ocean basin along whose rim a great drama of economic and political transformation will play itself out during the twenty-first century. Already, the realm's own former margins—in Hawai'i in the north and in New Zealand in the south—have been so thoroughly reshaped by foreign intervention that very little remains of the kingdoms and cultures that once prevailed. In the 2020s, the world's largest country and next superpower (China) faces the richest and most powerful (the United States) across the Pacific. How will the microstates of the Pacific Realm fare?

The Polar Regions

We can define the Polar Regions as lying poleward of the **Arctic Circle [18]** and **Antarctic Circle [19]**, respectively located at latitudes 66°34' North and 66°34' South. Why these precise latitudes? On the longest day of the year in the Northern Hemisphere (which varies between June 20 and 22), the sun never sets at or poleward of the Arctic Circle; on the same day, the sun never rises at or poleward of the Antarctic Circle. On the longest day in the Southern Hemisphere (varying between December 20 and 23), everything is reversed: the sun never sets at or poleward of the Antarctic Circle, and it never rises at or poleward of the Arctic Circle. This situation results from the constant tilt of the Earth's axis (at an angle of 66°34') with respect to the plane of its orbit as it revolves yearly around the sun, which is diagrammed in **Figure 13-5**.

Partitioning the Antarctic

South of the Pacific Realm lies Antarctica and its encircling Southern Ocean (which is also discussed in Chapter 12). The combined area of these two gargantuan geographic expanses constitutes 40 percent of the entire planet—two-fifths of the Earth's surface containing a mere one one-thousandth of the world's population.

We define Antarctica as a region, but it is also a huge continent, the size of which is rarely acknowledged because most world map projections barely show it (but see Fig. 12-4, in Chapter 12, for a different perspective). It is practically twice the size of Australia, but virtually all of it is covered by a dome-shaped ice sheet nearly 3.2 kilometers (2 mi) thick near its center. The continent frequently is referred to as "the white desert" because, despite all of its ice and snow, annual precipitation is extremely low—averaging less than 15 centimeters (6 in) per year. Temperatures are extremely

frigid, with winds so powerful that Antarctica is also called the home of the blizzard. For all its size, no functional regions have developed here, nor have any towns or transport networks except the supply lines of research stations. Antarctica still is a frontier, even a scientific frontier only grudgingly giving up its secrets. Underneath all that ice lie some 70 lakes, of which Lake Vostok is the largest at over 14,000 square kilometers (5400 sq mi), and it may be as much as 600 meters (2000 ft) deep.

Like virtually all frontiers, Antarctica has long attracted explorers and pioneers. Whale and seal hunters destroyed huge populations of Southern Ocean fauna during the eighteenth and nineteenth centuries, and explorers planted the flags of their countries on Antarctic shores. By the first decade of the twentieth century, the quest for the South Pole had become an international obsession; Roald Amundsen, the Norwegian, reached it first in 1911. All this activity soon led to national claims in Antarctica during the interwar era (1918–1939).

The geographic effect was the partitioning of Antarctica into pie-shaped sectors centered on the South Pole (**Fig. 13-6**). In the least frigid area of the continent, the Antarctic Peninsula, British, Argentinian, and Chilean claims overlapped—and they still do. Note that one sector, Marie Byrd Land (shown in neutral beige on the main map), was never formally claimed by any country.

Why should states be interested in territorial claims in so remote and difficult an area? It is because these remote areas near the poles contain substantial resources and are attractive for possible future use, despite the formidable challenges of access. Antarctica is no different, as both land and sea contain raw materials that may someday become crucial: proteins in the waters, and fossil fuels as well as mineral deposits beneath the huge ice sheet. As noted earlier, Antarctica

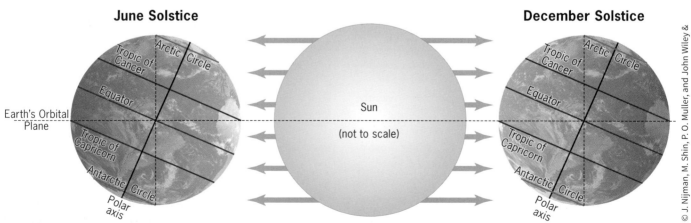

FIGURE 13-5 **The Position of the Polar Regions in Relation to the Sun on June 21 and December 21**

FIGURE 13-6

at 14.2 million square kilometers (5.5 million sq mi) is nearly twice the size of Australia, and the Southern Ocean is almost as large as the North and South Atlantic combined. However distant actual future exploitation may be, countries want to hang onto their stakes here.

But the states claiming Antarctic territory also recognize the need for cooperation. During the late 1950s, they participated in the International Geophysical Year (IGY) that launched major scientific programs and established a number of permanent research stations across the continent. This spirit of cooperation led to the 1961 signing of the **Antarctic Treaty [20]**, which ensures continued scientific collaboration, prohibits military activities, safeguards the environment, and holds national claims in abeyance. In 1991, when the treaty was extended under the terms of the Wellington Agreement, concerns were raised that it does not do enough to control future resource exploitation.

Notice, in Figure 13-6, that this map shows no **maritime claims [21]** off the pie-shaped sectors that blanket most of Antarctica. In fact, some of the claimant states did draw maps that extended their claims into the Southern Ocean, but the 1991 treaty extension terminated that initiative and restricted the existing claims to the landmass. It is, of course, possible that those claims could be reinstated should the Wellington Agreement disintegrate: national self-interest has abrogated many an international treaty in the past. But delimiting

maritime claims off Antarctica is particularly challenging for practical reasons. As Figure 13-6 indicates, Antarctica in a number of places is flanked by *ice shelves* attached to its coast, permanent slabs of floating ice (such as the Ronne and the Ross). These ice shelves are replenished on the landward side and break off (calve) on the seaward side, where huge icebergs float northward into the Southern Ocean to become part of the wide belt of *pack ice* that encircles the continent. From where would any territorial sea or EEZ be measured? From the inner edge of the ice shelf? That would put the territorial "sea" boundary on the ice shelf! From the ever-changing outer edge of the ice shelf? Measuring a territorial sea or EEZ from such an unstable perimeter would not work either. No UNCLOS regulations would be applicable in situations like this, so no matter what the maps showed, such claims are legally unsupportable. That is just as well: the last thing the world needs is a scramble for Antarctica.

In an age of growing national self-interest and increasing raw material consumption, the possibility exists nonetheless that Antarctica and its offshore waters may yet become an arena for international rivalry. Until now, its remoteness and its forbidding environments have saved it from that fate. The entire world benefits from this because evidence is mounting that Antarctica plays a critical role in the global environmental system, so that human modifications are likely to have worldwide (and unpredictable) consequences.

Geopolitics and Global Climate Change in the Arctic Basin

A Polar Sea Surrounded by Northern Countries

To observe how different the physiographic as well as the political situation is in the Arctic, consider the implications of **Figure 13-7**. Not only does the North Pole lie on the floor of a relatively small body of water (grandiosely labeled the Arctic Ocean), but the entire Arctic is ringed by countries whose

EEZs, delimited under UNCLOS rules, would allocate much of that ocean floor (the *subsoil*, to use its legal designation) to those states. Moreover, again under UNCLOS regulations, states can expand their rights to the seafloor even farther than the 200-mile EEZ if they are able to prove their continental shelves continue beyond that limit—in fact, up to 350 nautical miles offshore.

As Figure 13-7 shows, the Siberian Continental Shelf is by far the largest in the Arctic; indeed, it is the largest in the world. It extends from Russia's north coast beneath the waters of the Arctic Ocean and under the floating icecap at the ocean's center. Because several island groups off the mainland (such as Franz Josef Land, North Land, and the New Siberian Islands) belong to Russia, the Russians can claim virtually all of Eurasia's northern continental shelf under existing regulations. But even that is not enough: the Russians want to extend their

COMPETITION IN THE ARCTIC
- —·—·— EEZ Boundary
- —·—·— Treaty Boundaries
- · · · · Disputed Boundary
- ☐ Continental Shelf

ARCTIC CLAIMS BEYOND THE EEZ
- ☐ Russia
- ☐ Canada
- ☐ United States
- ▨ Denmark

FIGURE 13-7

claim all the way to the North Pole itself, where in 2007 they sent a submarine to plant a Russian flag on the seafloor nearly 4000 meters (13,000 ft) below the Pole (photo p. 468).

To bolster their seafloor claim, the Russians assert that the Lomonosov Ridge, a submerged mountain range, is a "natural extension" of their Siberian Continental Shelf. Few, if any, neutral observers might agree that this ridge, or several others rising from the deep floor of the Arctic Ocean, is a continental-shelf landform. But in 2014, Denmark made a new claim on the same grounds: following "ten years of scientific research," their conclusion was that the Lomonosov Ridge is actually connected to Greenland's shelf and therefore lies in Danish territory.

Take another look at Figure 13-7 and observe how far the Danes have extended their claims, which now overlap with those of Canada and, more significantly, Russia (the North Pole is included in the Danish claim as well). The stakes are high: estimates of the quantity of oil and natural gas reserves to be found beneath the Arctic seafloor run as high as 25 percent of the world's remaining total. In the end, whatever method is used to finally resolve the Lomonosov and other claims, Russia will be the biggest winner.

Disputation and Navigation

Even as a dispute over Arctic maritime boundaries looms, another key issue that has emerged in recent years involves the effects of sustained global warming. The Greenland Ice Sheet, in many ways a smaller version of Antarctica's, is now in a melting phase, and the zone of seasonally expanding and contracting Arctic sea ice has also shown significant attrition in both surface extent and thickness (see **Box 13-5, Figure 13-8,** and **Box 13-6**). Researchers with the Greenland Ice Core Project have reported in *Science* that during a previous warming episode, more than 450,000 years ago, the southernmost part of Greenland was covered by boreal forest—suggesting that both the ice sheet and the sea ice may disappear entirely if the ongoing global warming trend continues. Indeed, this process seems well underway: the polar icecap is now less than half of what it was in 1979, and Figure 13-8 highlights the current shrinkage phase. The ecological consequences are far-reaching, threatening the habitats of polar bears, whales, walruses, seals, and other species. But the economic and politico-geographical implications are going to be significant as well.

When the Arctic's sea ice melts in summer and does not fully recover in winter, once-blocked waterways begin to open up and accessibilities can be transformed. The **Northeast Passage [17]**, that legendary high-latitude route between the Far East and Europe, has been the object of hope and despair for centuries, causing much loss of human life in the quest to forge a viable sea route. But in 2007, for the first time in recorded history, it was completely ice-free for a brief period during the late-summer peak of seasonal melting. Since then, the Northeast Passage has been ice-free every year during that August/September window; by 2013, a total of 71 specially equipped vessels—carrying well over a million tons of cargo—were able to successfully navigate the route (mapped in Fig. 13-8). That map also shows the presence of a similar route around the northern perimeter of North America, the **Northwest Passage [18]**. This other icon of (unsuccessful) pre-twentieth-century polar exploration would connect the Pacific and Atlantic oceans, following an eastward watercourse along Alaska's northern coast, through Canada's massive Arctic Archipelago, and then a final southward traverse via Baffin Bay and the Davis Strait between Canada and Greenland.

If either or both of these waterways can be activated, the impact on global shipping will be momentous. Consider what it would mean if ships bound for Asia's Pacific Rim from the U.S. eastern seaboard, and too large to fit through the upgraded Panama Canal, could shorten their voyages by several days and thousands of kilometers by rounding North America instead of South America. No wonder countries like China, India, Japan, and even Singapore have applied for "observer status" at the meetings of the Arctic Council, whose regular members include the United States, Russia, Canada, Norway, Sweden, Denmark, Finland, and Iceland.

But who really owns the waterways vacated by the ice? Canada regards the Northwest Passage as a domestic waterway, giving it the right to control all shipping. The United States holds that the Northwest Passage is an international waterway, and American and Russian ships have sailed through it without Canada's permission. The Canadian government has affirmed its position by starting work on a deepwater port at Nanisivik on Nunavut's Baffin Island at the eastern entrance to the Northwest Passage, and by building a military base at Resolute Bay to the west; it has also announced the construction of eight Arctic patrol vessels to police the route. In the meantime, the United States is opening a new Coast Guard station to patrol the narrow Bering Strait between Alaska and Russia. The lines, if not yet battle lines, are already being drawn in the warming Arctic.

Humankind's growing numbers, steadily escalating demands, environmental impacts, and technological capacities are transforming even the most remote recesses of our resilient planet. No international resolution is more critical than the avoidance of destructive conflict in fragile polar environments.

Box 13-5 Geographic Information Analysis

The Shrinking North Polar Icecap

One of the most visible effects of global warming is the shrinking of the polar icecap in the Arctic. As the maps and diagram in Figure 13-8 show, since 1979 the polar icecap has lost over two million square kilometers of surface area (more than the size of Mexico). Despite growing concerns surrounding climate change, the receding levels of sea ice also create opportunities. For instance, Arctic ports may be able to stay open longer, and new transportation and trade routes may be available. Which countries with polar access stand to benefit the most from the decrease in polar sea ice? What other opportunities and concerns does a melted polar icecap present to these countries and the world?

Access your WileyPLUS course to interact with a dynamic version of this map and to engage with online exercises and questions.

A) 1980 (SEPTEMBER)

B) 2018 (SEPTEMBER)

C) SURFACE AREA OF SEA ICE IN SEPTEMBER, 1979 TO 2018

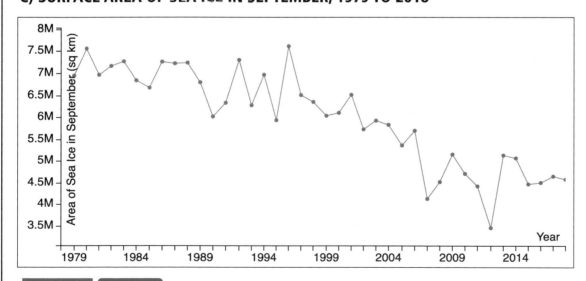

FIGURE 13-8 Interactive

© J. Nijman, M Shin, P. O. Muller, and John Wiley & Sons, Inc.

Box 13-6 From The Field Notes …

"The small town of Ilulissat lies on the west coast of Greenland, inside the Arctic Circle. It is at the outlet of Ilulissat Glacier, the fastest-moving and most 'productive' in the Northern Hemisphere, carrying about 35 billion tons of ice to the North Atlantic every year. When the glacier reaches the fjord at lower elevations, the ice breaks into different pieces (a process called 'calving'). It then moves through the ice fjord to the sea. The ice fjord is about 16 miles long and 3 miles wide. In this photo, I am looking south across the fjord, watching this enormous mass of ice. It moves left to right, about 25 meters a day (not quite fast enough to actually see it moving). Once dumped in the bay, some of the big icebergs will sit on the ground for years, until they are light enough to float, and then they drift south. When the doomed *Titanic* hit an iceberg in 1912, about 400 miles east of Newfoundland, that iceberg very likely came from Ilulissat Glacier. Greenland, and this particular area, are important sites in observations and measurements of climate change and global warming. In the late spring and early summer of 2019, scientists reported historical records for high temperatures and melting. For decades, the mouth of the glacier (where it calves, into the ice fjord) has been retreating inland, indicating progressively higher temperatures inland, in the heart of the Greenland Ice Sheet. Today is June 18, 2019; temperatures have been soaring into the 70s Fahrenheit, and I have to swat away mosquitoes as I write this note. I hope this glacier will be here forever."

© J. Nijman

The Ilulissat ice fjord. The ice moves left to right (westward), toward Disko Bay, an arm of the North Atlantic Ocean.

AP Images/TV

On August 3, 2007, tens of millions of Russian television viewers witnessed live the scene pictured here: the operator of a Russian mini-submarine planting a metal Russian flag at the North Pole, on the seafloor of the Arctic Ocean beneath the polar icecap. As Figure 13-7 shows, the North Pole lies far beyond Russia's continental shelf. Yet in future international negotiations to settle national claims in the Arctic Basin, the Russians will claim that the Lomonosov Ridge actually constitutes an extension of their continental shelf, thereby entitling Russia to draw international boundaries up to, and even beyond, the North Pole. Other countries with actual and pending Arctic claims are making less dramatic but equally assertive moves in the run-up to negotiations that have the potential to transform the map of the Arctic region. With new sea routes as well as newly exploitable oil and gas reserves in prospect, diplomacy will be difficult but crucial.

Points to Ponder

- Governments of low-lying island-states in the Pacific and Indian oceans are demanding international assistance as they are confronted by the threat of rising sea levels.

- Fiji receives support from Beijing for officially recognizing the People's Republic of China (PRC) but not Taiwan; Palau gets support from the Taiwanese for officially recognizing Taiwan but not the PRC.

- Singapore, located almost directly on the equator, has requested and been granted "observer status" to monitor the meetings of the Arctic Council because of the impact the possible opening of ice-free far northern passages will have on global shipping.

- If all the ice on Greenland melted, global sea levels would rise by an estimated 6 meters (20 ft).

Metric (Standard International [SI]) and Customary Units and Their Conversions

Length

Metric Measure

1 kilometer (km)	= 1000 meters (m)
1 meter (m)	= 100 centimeters (cm)
1 centimeter (cm)	= 10 millimeters (mm)

Nonmetric Measure

1 mile (mi)	= 5280 feet (ft)
	= 1760 yards (yd)
1 yard (yd)	= 3 feet (ft)
1 foot (ft)	= 12 inches (in)
1 fathom (fath)	= 6 feet (ft)

Conversions

1 kilometer (km)	= 0.6214 mile (mi)
1 meter (m)	= 3.281 feet (ft)
	= 1.094 yards (yd)
1 centimeter (cm)	= 0.3937 inch (in)
1 millimeter (mm)	= 0.0394 inch (in)
1 mile (mi)	= 1.609 kilometers (km)
1 foot (ft)	= 0.3048 meter (m)
1 inch (in)	= 2.54 centimeters (cm)
	= 25.4 millimeters (mm)

Area

Metric Measure

1 square kilometer (km^2)	= 1,000,000 square meters (m^2)
	= 100 hectares (ha)
1 square meter (m^2)	= 10,000 square centimeters (cm^2)
1 hectare (ha)	= 10,000 square meters (m^2)

Nonmetric Measure

1 square mile (mi^2)	= 640 acres (ac)
1 acre (ac)	= 4840 square yards (yd^2)
1 square foot (ft^2)	= 144 square inches (in^2)

Conversions

1 square kilometer (km^2)	= 0.386 square mile (mi^2)
1 hectare (ha)	= 2.471 acres (ac)
1 square meter (m^2)	= 10.764 square feet (ft^2)
	= 1.196 square yards (yd^2)
1 square centimeter (cm^2)	= 0.155 square inch (in^2)
1 square mile (mi^2)	= 2.59 square kilometers (km^2)
1 acre (ac)	= 0.4047 hectare (ha)
1 square foot (ft^2)	= 0.0929 square meter (m^2)
1 square inch (in^2)	= 6.4516 square centimeters (cm^2)

Temperature

To change from Fahrenheit (F) to Celsius (C)

$$°C = \frac{°F - 32}{1.8}$$

To change from Celsius (C) to Fahrenheit (F)

$$°F = °C \times 1.8 + 32$$

Aboriginal land issue The legal campaign in which Australia's **indigenous peoples** have claimed title to traditional land in several parts of that country. The courts have upheld certain claims, fueling Aboriginal activism that has raised broader issues of indigenous rights.

Aboriginal population Native or *aboriginal* peoples; often used to designate the inhabitants of areas that were conquered and subsequently colonized by the imperial powers of Europe.

Absolute location The position or place of a certain item on the surface of the Earth as expressed in degrees, minutes, and seconds of latitude and longitude.

Acculturation Cultural modification resulting from intercultural borrowing. In **cultural geography**, the term refers to the change that occurs in the **culture** of **indigenous peoples** when contact is made with a society that is technologically superior.

AFTA The ASEAN Free Trade Agreement that since 1992, through lowered tariffs and other incentives, has fostered increased trade within Southeast Asia. This is the economic centerpiece of the **Association of Southeast Asian Nations (ASEAN)**, a **supranational** organization whose members include 10 of that realm's 11 states (only East Timor does not participate).

Agribusiness The agricultural operations of large, often multinational, corporations.

Al Qaeda The terrorist organization that evolved into an expanding global network under the directorship of Osama bin Laden between the mid-1990s and his elimination by the U.S. in 2011. It sought to coordinate the efforts of once loosely allied Muslim revolutionary movements, and unleash a *jihad* aimed at what it perceived to be Islam's enemies in the West.

Altiplano High-elevation plateau, basin, or valley between even higher mountain ranges, especially in the Andes of South America.

Altitudinal zonation Vertical regions defined by physical-environmental zones at various elevations (see Fig. 3-5), particularly in the highlands of South and Middle America.

American Manufacturing Belt North America's near-rectangular **core area**, whose corners are Boston, Milwaukee, St. Louis, and Baltimore.

Antarctic Circle The parallel (circle) at 66.34 degrees South that defines the Antarctic.

Antarctic Treaty International cooperative agreement on the use of Antarctic territory. Originally signed in 1961, it was extended in 1991 under the Wellington Agreement.

Anthropogenic Environmental impacts emanating from human activities, particularly the production of pollutants associated with atmospheric warming.

Apartheid Literally, *apartness*. The Afrikaans term for South Africa's pre-1994 policies of racial separation, a system that produced highly segregated socio-geographical patterns.

APEC The Asia Pacific Economic Cooperation forum. Its 21 Pacific Rim member-economies promote free trade throughout the Asia Pacific region.

Aquifer An underground reservoir of water contained within a porous, water-bearing rock layer.

Arab Spring A series of anti-government, pro-democracy, protests and uprisings, both peaceful and violent, that spread across the realm of North Africa and Southwest Asia (NASWA) in the early 2010s.

Archipelago A set of islands grouped closely together, usually elongated into a *chain*.

Arctic Circle The parallel (circle) at 66.34 degrees North that defines the Arctic.

ASEAN The **Association of Southeast Asian Nations (ASEAN)**, a **supranational** organization whose members include 10 of that realm's 11 states: Brunei, Cambodia, Indonesia, Laos, Malaysia, Myanmar (Burma), the Philippines, Singapore, Thailand, and Vietnam (only minuscule East Timor does not participate, but is currently seeking to join).

Asylum Legally protected residency status; usually granted by a host country to immigrants fleeing political oppression in their former homeland.

Atoll A ring-like coral reef surrounding an empty lagoon that probably formed around the rim of a now-completely-eroded volcanic cone standing on the seafloor. They are common in certain tropical areas of the Pacific Ocean where they are classified among that realm's **low islands**.

Austral South.

Balkanization The fragmentation of a **region** into smaller, often hostile political units. Named after the historically contentious Balkan Peninsula of southeastern Europe.

Barrio Term meaning "neighborhood" in Spanish. Usually refers to an urban community in a Middle or South American city.

Basin irrigation Farming technique devised by ancient Egyptians to trap the Nile's annual floodwaters and their fertile silt by building fields with earthen ridges.

Belt and Road Initiative China's global development strategy, announced in 2013, involving over a hundred infrastructure-investment projects. Some serve to connect China to surrounding countries (e.g., Kazakhstan or Laos), others support infrastructure developments within or between countries around the world. The BRI has significantly increased China's global presence.

Biodiversity Shorthand for *biological diversity*; the total variety of plant and animal species that exists in a given area.

Biodiversity hot spot A much higher than usual, world-class geographic concentration of natural plant and/or animal species. Tropical rainforest environments have dominated, but their recent ravaging by deforestation has had catastrophic results.

Biofuel Renewable fuel produced from contemporary (not geological) biomass sources, mainly plants.

Biogeography The study of *flora* (plant life) and *fauna* (animal life) in spatial perspective.

Boko Haram The violent, jihadist, terrorist organization that is based in—and controls parts of—northeastern Nigeria. It engages in extreme terrorist attacks, increasingly beyond its territory as it seeks to expand into other parts of the country (see Fig. 8-16).

Borderland General term for a linear zone that parallels a political boundary. The most dynamic of these areas, such as those lining the U.S.-Mexico border, are marked by significant cultural and economic interaction across the boundary that separates them.

Boreal forest The subarctic, mostly coniferous snowforest that blankets Canada south of the **tundra** that lines the Arctic shore; known as the **taiga** in Russia.

Break-of-bulk A location along a transport route where goods must be transferred from one carrier to another. In a port, the cargoes of oceangoing ships are unloaded and put on trains, trucks, or perhaps smaller river boats for inland distribution. An *entrepôt*.

BRICs Acronym for the four biggest emerging national markets in the world today—**B**razil, **R**ussia, **I**ndia, and **C**hina.

Buffer state A country or set of countries separating ideological or political adversaries. In southern Asia, Afghanistan, Nepal, and Bhutan were parts of a buffer zone set up between British and Russian-Chinese imperial spheres. Thailand was a *buffer state* between British and French colonial domains in mainland Southeast Asia.

Business-process outsourcing The subcontracting of specific business tasks, such as subscriptions or payroll, to another, specialized, company (sometimes abroad), at lower cost.

Caliphate An imperial-scale Islamic government led by a caliph, considered a direct successor to the Prophet Muhammad, who rules and exerts moral authority over Muslims worldwide.

Cartography The branch of Geography that specializes in map-making. This specialization has a long history and has been enhanced to new levels through the digital revolution.

Cash economy An economic system in which all transactions are made in cash.

Caste system The strict **social stratification** and residential segregation of people—

*Words in **boldface type** within an entry are defined elsewhere in this Glossary.

G-1

specifically in India's Hindu society—on the basis of ancestry and occupation.

Central business district (CBD) The downtown heart of a central city; marked by high land values, a concentration of business and commerce, and the clustering of the tallest buildings.

Centrality The strength of an urban center in its capacity to attract producers and consumers to its facilities; a city's "reach" into the surrounding region.

Centrifugal forces A term employed to designate forces that tend to divide a country—such as internal religious, linguistic, ethnic, or ideological differences.

Centripetal forces Forces that unite and bind a country together—such as a strong national culture, shared ideological objectives, and a common faith.

Cerrado Regional term referring to the fertile savannas of Brazil's interior Central-West that make it one of the world's most promising agricultural frontiers. Soybeans are the leading crop, and other grains and cotton are expanding. Inadequate transport links to the outside world remain a problem.

China-Pakistan Economic Corridor The 2000-kilometer (1250-mi), northeast-southwest development axis stretching between the westernmost Chinese city of Kashgar and Pakistan's new Indian Ocean port of Gwadar. A major future trade route that aligns with China's **Belt and Road Initiative**, its centerpiece is a bundled routeway of ultramodern rail, road, and pipeline connections expected to spawn various development nodes within the corridor, mostly in Pakistan (see Fig. 9-17).

Choke point A narrowing of an international waterway causing marine-traffic congestion, requiring reduced speeds and/or sharp turns, and increasing the risk of collision as well as vulnerability to attack. When the waterway narrows to a distance of less than 38 kilometers (24 mi), this necessitates the drawing of a **median-line (maritime) boundary**.

City-state An independent political entity consisting of a single city with (and sometimes without) an immediate **hinterland**.

Climate The long-term conditions (over at least 30 years) of aggregate weather over a region, summarized by averages and measures of variability; a synthesis of the succession of weather events we have learned to expect at any given location.

Cold War The rivalry and geopolitical conflict between the United States and the Soviet Union during most of the second half of the 20th century.

Command economy The tightly controlled economic system of the former Soviet Union, whereby central planners in Moscow assigned the production of particular goods to particular places, often guided more by socialist ideology than the principles of **economic geography**.

Commercial agriculture For-profit agriculture.

Communal tension Persistent stress among a country's sociocultural groups that can often erupt into communal violence, particularly in India.

Compact state A politico-geographical term to describe a **state** that possesses a roughly circular, oval, or rectangular territory in which the distance from the geometric center to any point on the boundary exhibits little variance.

Complementarity Exists when two regions, through an exchange of raw materials and/or finished products, can specifically satisfy each other's demands.

Connectivity The degree of direct linkage between a particular location and other locations within a regional, national, or global transportation network.

Containment The U.S. geopolitical strategy during the **Cold War** of containing the influence of the Soviet Union within its own borders or nearby regions.

Continental drift The slow movement of continents controlled by the processes associated with **plate tectonics**.

Continentality The variation of the continental effect on air temperatures in the interior portions of the world's landmasses. The greater the distance from the moderating influence of an ocean, the greater the extreme in summer and winter temperatures. Continental interiors also tend to be dry when the distance from oceanic moisture sources becomes considerable.

Continental shelf Beyond the coastlines of many landmasses, the ocean floor declines very gently until the depth of about 600 feet (183 m). Beyond the 600-foot line the sea bottom usually drops off sharply, along the *continental slope*, toward the much deeper mid-oceanic basin. The submerged continental margin is called the continental shelf, and it extends from the shoreline to the upper edge of the continental slope.

Conurbation General term used to identify a large multimetropolitan complex formed by the coalescence of two or more major urban areas.

Core area In geography, a term with several connotations. *Core* refers to the center, heart, or focus. The core area of a **nation-state** is constituted by the national heartland, the largest population cluster, the most productive region, and the part of the country with the greatest **centrality** and accessibility probably containing the capital city as well.

Core-periphery relationship The contrasting spatial characteristics of, and linkages between, the *have* (core) and *have-not* (periphery) components of a national, regional, or the global system.

Cultural diffusion The process of spreading and adopting a cultural element from its place of origin across a wider area.

Cultural geography The wide-ranging and comprehensive field of geography that studies spatial aspects of human **cultures**.

Cultural landscape The forms and artifacts sequentially placed on the **natural landscape** by the activities of various human occupants. By this progressive imprinting of the human presence, the physical (natural) landscape is modified into the cultural landscape, forming an interacting unity between the two.

Cultural pluralism A society in which two or more population groups, each practicing its own **culture**, live adjacent to one another without mixing inside a single **state**.

Culture hearth Heartland, source area, or innovation center; place of origin of a major **culture.**

Dalits Term (meaning "the oppressed") used for the lowest caste in India's Hindu caste system. The official government rubric for Dalits is Scheduled Castes.

Deindustrialization Process by which companies relocate manufacturing jobs to other regions or countries with cheaper labor, leaving the newly-deindustrialized region to convert to a service economy while struggling with the accompanying effects of increased unemployment and meeting the retraining needs of its workforce.

Demographic transition Multi-stage model, based on western Europe's experience, of changes in population growth exhibited by countries undergoing industrialization. High *birth rates* and *death rates* are followed by plunging death rates, producing a huge net population gain; birth and death rates then converge at a low overall level.

Dependencia **theory** Originating in South America during the 1960s, it was a new way of thinking about economic development and underdevelopment that explained the persistent poverty of certain countries in terms of their unequal relations with other (i.e., rich) countries.

Dependency ratio An indicator of the pressure on a country's workers, the age-population ratio of (dependent) people who are not in the labor force to those (productive) people who are in the labor force.

Desalination The process of removing dissolved salts from water, thereby producing fresh (drinking) water from seawater or brackish water.

Development The economic, social, and institutional growth of national **states**.

Devolution The process whereby regions within a **state** demand and gain political strength and growing autonomy at the expense of the central government.

Digital elevation model A representation of a unit of terrain obtained from **remote sensing** imagery. The technique is discussed in Box 5-3.

Distance decay The various degenerative effects of distance on human spatial structures and interactions.

Distribution center A centralized focus of economic activity specializing in the distribution of goods, situated as a major hub on its regional transportation network. Atlanta, Georgia, with its outstanding highway, rail, and air-freight connections to the surrounding southeastern United States, is a classic example.

Domino effect The belief that political destabilization in one **state** can result in the collapse of order in a neighboring state, triggering a chain of events that, in turn, can affect a series of contiguous states.

Domino theory The belief that political destabilization in one **state** can result in the collapse of order in a neighboring state, triggering a chain of events that, in turn, can affect a series of contiguous states.

Double delta South Asia's combined delta formed by the Ganges and Brahmaputra rivers. All of Bangladesh lies on this enormous deltaic plain, which also encompasses surrounding parts of eastern India. Well over 200 million people live here, attracted by the fertility of its soils that are constantly replenished by the silt transported and deposited by these two of Asia's largest river systems. Natural hazards abound here as well, ranging from the flooding caused by excessive **monsoonal** rains to the intermittent storm surges of powerful cyclones (hurricanes) that come from the Bay of Bengal to the south.

Dravidian languages The language family, indigenous to the South Asian realm, that dominates southern India today; as opposed to the Indo-European languages, whose tongues dominate northern India.

Drone warfare The use of remote-controlled unmanned aerial vehicles (UAVs) as delivery systems to conduct military attacks. Tactical advantages include highly precise targeting, long flight times without refueling, difficulty of detection from the ground, and advanced imaging capabilities that allow drones to often "see" what is invisible to human pilots.

Dynasty A succession of Chinese rulers that came from the same line of male descent, sometimes enduring for centuries. Dynastic rule in China lasted for thousands of years, only coming to an end in 1911.

Economic geography The field of geography that focuses on the diverse ways in which people earn a living and on how the goods and services they produce are expressed and organized spatially.

Economic integration The economic benefits of forging **supranational** partnerships among three or more countries. The European Union (EU) is the prototype; NAFTA and CAR-ICOM are examples in the Middle American realm.

Economies of scale The savings that accrue from large-scale production wherein the unit cost of manufacturing decreases as the level of operation enlarges. Supermarkets operate on this principle and are able to charge lower prices than small grocery stores.

Ejidos Mexican farmlands redistributed to peasant communities after the Revolution of 1910–1917. The government holds title to the land, but user rights are parceled out to village communities and then to individuals for cultivation.

Electoral geography The spatial distribution of political preferences as expressed in voting behavior for political parties and/or candidates. The mapping of election results (see Box 2-6) is the foundation of electoral geography.

El Niño A periodic, abnormal warming of the sea surface in the eastern equatorial Pacific Ocean. Disturbs weather patterns across much of the world, especially in northwestern and northeastern South America.

Elongated state A **state** whose territory is decidedly long and narrow in that its length is at least six times greater than its average width.

Elongation In political geography, refers to the territorial configuration of a state that is at least six times longer than its average width. Chile is the most prominent example of this shape on the world map.

Emerging market The world's fastest-growing national market economies as measured by economic growth rates, attraction of **foreign direct investment**, and other key indicators. Led by the **BRICs** (Brazil, Russia, India, and China), but this club is now expanding to include many other countries.

Endemic Refers to a disease in a host population that affects many people in a kind of equilibrium without causing rapid and widespread deaths.

Entrepôt A place, usually a port city, where goods are imported, stored, and transshipped; a **break-of-bulk point**.

Environmental degradation The accumulated human abuse of a region's **natural landscape** that, among other things, can involve air and water pollution, threats to plant and animal ecosystems, misuse of natural resources, and generally upsetting the balance between people and their habitat.

Environmental determinism A geographic school of thought, popular in the first quarter of the twentieth century, that maintains the physical environment determines human behavior and/or social outcomes.

Epidemic A local or regional outbreak of a disease.

Estuary The widening mouth of a river as it reaches the sea; land subsidence or a rise in sea level has overcome the tendency to form a delta.

Ethnicity The combination of a people's **culture** (traditions, customs, language, and religion) and racial ancestry.

Eurasian Customs Union A **supranational** organization created by Russia in 2010 to maintain economic ties with the friendliest countries in the **Near Abroad**. The three charter members (Russia, Belarus, and Kazakhstan) were joined in 2015 by the two smaller countries of Armenia and Kyrgyzstan.

European state model A **state** consisting of a legally defined territory inhabited by a population governed from a capital city by a representative government.

Euro Zone The 19 countries (as of late 2019) whose official currency is the euro. They are mapped in Figure 5-8.

Exclave A bounded (non-island) piece of territory that is part of a particular **state** but lies separated from it by the territory of another state.

Exclusive Economic Zone (EEZ) An oceanic zone extending up to 200 nautical miles (370 km) from a shoreline, within which the coastal **state** can control fishing, mineral exploitation, and additional activities by all other countries.

Expansion diffusion The spreading of an innovation or an idea through a fixed population in such a way that the number of those adopting it grows continuously larger, resulting in an expanding area of dissemination.

Externality effect A negative consequence of an action, which countries try to minimize. Ethiopia's new dam on the Blue Nile is deliberately situated just upstream from the Sudan border in order to minimize reduced water flow and silt entrapment within Ethiopian territory.

Extreme weather events Unprecedented, record-breaking departures from the longer-term weather patterns of a certain area. In India, such events have tripled since 1980 in the form of severe heat waves, droughts, and non-**monsoonal** rainstorms that trigger massive flooding and widespread landslides.

Failed state A country whose institutions have collapsed and in which anarchy prevails.

Favela Shantytown on the outskirts or even well within an urban area in Brazil.

Federation A country adhering to a political framework wherein a central government represents the various subnational entities within a **nation-state** where they have common interests—defense, foreign affairs, and the like—yet allows these various entities to retain their own identities and to have their own laws, policies, and customs in certain spheres.

Fertility rate More precisely the Total Fertility Rate, it is the average number of children born to women of childbearing age in a given population.

First Nations Name given Canada's **indigenous peoples** of American descent, whose U.S. counterparts are called Native Americans.

Floating population China's huge mass of mobile workers who respond to shifting employment needs within the country. Most are temporary urban dwellers with restricted residency rights, whose movements are controlled by the *hukou* system.

Foreign direct investment A key indicator of the success of an **emerging market** economy, whose growth is accelerated by the infusion of foreign funds to supplement domestic sources of investment capital.

Formal economy The part of a national economy that is registered with government agencies and complies with laws and regulations, especially taxation. The complement to a country's **informal economy**.

Formal region A type of **region** marked by a certain degree of homogeneity in one or more phenomena; also called *uniform region* or *homogeneous region*.

Forward capital Capital city positioned in actually or potentially contested territory, usually near an international border; it confirms the **state's** determination to maintain its presence in the area of contention.

Fossil fuels The energy resources of coal, natural gas, and petroleum (oil), so named collectively because they were formed by the geologic compression and transformation of tiny plant and animal organisms.

Four Motors of Europe *Rhône-Alpes* (France), *Baden-Württemberg* (Germany), *Catalonia* (Spain), and *Lombardy* (Italy). Each is a high-technology-driven region marked by exceptional industrial vitality and economic success not only within Europe but on the global scene as well.

Fragmented modernization A checkerboard-like spatial pattern of modernization in an **emerging-market** economy wherein a few localized regions of a country experience most of the development while the rest are largely unaffected.

Fragmented state A **state** whose territory consists of several separated parts, not a contiguous whole. The individual parts may be isolated from each other by the land area of other states or by international waters.

Functional region A **region** marked less by its sameness than by its dynamic internal structure; because it usually focuses on a central **node**, also called *nodal region* or *focal region*.

Gender imbalance The demographic imbalance of males outnumbering females resulting from selective birth control. In China, this is an outcome of the **One-Child Policy**.

Gender pay gap The (large) difference in pay between males and females.

Gentrification The upgrading of an older residential area through private reinvestment, usually in the downtown area of a central city. Frequently, this involves the displacement of established lower-income residents, who cannot afford the heightened costs of living, and conflicts are not uncommon as such neighborhood change takes place.

Geographic information system (GIS) A form of spatial analysis that integrates computer hardware, mapping software, and such specialized tools as models and algorithms. A versatile technique that is constantly being expanded in its applications (e.g., digital terrain mapping, as shown in Box 5-3).

Geographic realm The basic spatial unit in our world regionalization scheme. Each realm is defined in terms of a synthesis of its total human geography—a composite of its leading cultural, economic, historical, political, and appropriate environmental features.

Geopolitics Political relations among states or regions that are strongly influenced by their geographical setting, including proximity, accessibility, sovereign boundaries, natural resources, **population distribution**, and the like.

Geospatial data Data pertaining to a particular location on or near the Earth's surface.

Gini Index A measure of inequality within a given area, ranging from 0 to 100. A value of 0 indicates that income is equally distributed across an area's population; a value of 100 indicates that all income is concentrated in the hands of a single recipient.

Global climate change The shift in the characteristics and spatial distribution of Earth's climates in response to a long-term trend in atmospheric warming.

Global core The constellation of countries with the most highly developed and influential economies (anchored by the United States; the 27 member-countries of the European Union; East Asia's China and Japan; and Australia/New Zealand). Although the **global core** (mapped in Fig. 1-11) is not a contiguous geographic region, its countries as a whole dominate international trade and investment flows.

Globalization The gradual reduction of regional differences at the world **scale**, resulting from increasing international cultural, economic, and political interaction.

Global periphery All of the countries that lie outside the **global core** (mapped in Fig. 1-11). Economically, these countries are subordinate to those of the global core in terms of development and international influence. Geographically, the global periphery is constituted by the 8 realms that lie outside North America, Europe, East Asia, and the Austral Realm.

Golden Triangle Opium-producing area encompassing northwestern Laos, northernmost Thailand, and east-central Myanmar's Shan State.

GPS (Global Positioning System) The orbiting-satellite-based navigation system that provides locational and time information, anywhere on or near the Earth's surface where there is an unobstructed line of sight to four or more GPS satellites.

Great Recession The economic downturn from 2008–2011 that was in large part based on the so-called "mortgage meltdown," leading to foreclosures and other socio-economic distress.

Green Revolution The successful late-twentieth-century development of higher-yield, fast-growing varieties of rice and other cereals in certain developing countries.

Gross National Happiness Index A set of measures of well-being designed initially by the country of Bhutan in 1972, giving more or less equal weight to economic and social indicators of well-being (e.g., education or environmental conditions).

Ground subsidence Sinking land, caused by long-term geological processes or short-term human activity such as groundwater extraction.

Growth-pole concept An urban center with a number of attributes that, if augmented by investment support, will stimulate regional economic **development** in its **hinterland**.

Growth Triangle An increasingly popular economic **development** concept along the western **Pacific Rim**, especially in Southeast Asia. It involves the linking of production in growth centers of three countries to achieve benefits for all.

Hacienda Literally, a large estate in a Spanish-speaking country. Sometimes equated with the **plantation**, but there are important differences between these two types of agricultural enterprise.

Hanification Imparting a cultural imprint by the ethnic Chinese (the "people of Han"). Within China often refers to the steadily increasing migration of Han Chinese into the country's **periphery**, especially Xinjiang and Xizang (Tibet). **Overseas Chinese** imprints, more generally referred to as **Sinicization**, have been significant as well, most important in the Southeast Asian realm.

High-value-added goods Products of improved net worth.

High island Cultures associated with volcanic islands of the Pacific Realm that are high enough in elevation to wrest substantial moisture from the tropical ocean air (see **rain shadow effect**). They tend to be well watered, their volcanic soils enable productive agriculture, and they support larger populations than **low islands.**

High seas Areas of the oceans away from land, beyond national jurisdiction, open and free for all to use.

High-speed rail Railway infrastructure for high-speed passenger trains, typically exceeding maximum speeds of 200km (125m) per hour.

Hindutva "Hinduness" as expressed through Hindu nationalism, Hindu heritage, and/or Hindu patriotism. The cornerstone of a fundamentalist movement that has been gaining strength since the late twentieth century that seeks to remake India as a society dominated by Hindu principles. It has been the guiding agenda of the Bharatiya Janata Party (BJP), which has emerged a powerful force in national politics and in big States like Maharashtra, Gujarat, and Madhya Pradesh.

Hinterland Literally "country behind," a term that applies to a surrounding area served by an urban center. That center is the focus of goods and services produced for its hinterland and is its dominant urban influence as well.

Holocene The current *interglacial* epoch (the warm period of glacial contraction between the glacial expansions of an ice age); extends from 10,000 years ago to the present. Also known as the *Recent Epoch*.

Hukou **system** A long-standing Chinese system whereby all inhabitants must obtain and carry with them residency permits that indicate where an individual is from and where they may exercise particular rights such as education, health care, housing, and the like.

Human Development Index A UN index that is a composite measure of life expectancy, education, and income per capita. It is used to rank countries within a four-level classification under this name.

Human evolution Long-term biological maturation of the human species. Geographically, all evidence points toward East Africa as the source of humankind. Our species, *Homo sapiens*, emigrated from this hearth to eventually populate the rest of the habitable world.

Hurricane Alley The most frequent pathway followed by tropical storms and hurricanes over the past 150 years in their generally westward movement across the Caribbean Basin. Historically, hurricane tracks have bundled most tightly in the center of this route, most often affecting the Lesser Antilles between Antigua and the Virgin Islands, Puerto Rico, Hispaniola (Haiti/Dominican Republic), Jamaica, Cuba, southernmost Florida, Mexico's Yucatán, and the Gulf of Mexico.

Hybrid warfare A calculated mix of tactics to achieve aggressive geopolitical objectives while avoiding direct (large-scale) war. The tactics include military, diplomatic, economic, and intelligence means.

Hydraulic civilization theory The theory that cities that managed to control irrigated farming over large **hinterlands** held political power over other cities. Particularly applies to early Asian civilizations based in such river valleys as the Chang (Yangzi), the Indus, and those of Mesopotamia.

Inclusive development The extent of equal economic (and social) development opportunities for different population groups, especially minorities and the poor.

Indigenous Aboriginal or native; an example would be the pre-Columbian inhabitants of the Americas.

Indirect rule British colonial practice that kept indigenous power structures in place, co-opting individual rulers. The purpose was to minimize armed conflict and maximize profits.

Indo-European language family The major world language family that dominates the European **geographic realm.** This language family is also the most widely dispersed globally (Fig. 1-8), and about half of humankind speaks one of its languages.

Indo-European languages The major world language family that dominates the European **geographic realm.** This language family is also the most widely dispersed globally (Fig. 1-8), and about half of humankind speaks one of its languages.

Indochina Broadly speaking, the region of Southeast Asia that has historically been influenced by both India and China; more narrowly, the part of Southeast Asia under French control during colonial times (Vietnam, Cambodia, and Laos).

Industrial Revolution The term applied to the social and economic changes in agriculture, commerce, and especially manufacturing and urbanization that resulted from technological innovations and greater specialization in late-eighteenth-century Europe.

Informal economy The part of a national economy that is not registered with the government, and for which reliable statistics are rarely available. The complement to a country's **formal economy**.

Informal sector Dominated by unlicensed sellers of homemade goods and services, the primitive form of capitalism found in many developing countries that takes place beyond the control of government. The complement to a country's **formal economic** sector.

Information economy The new, increasingly dominant, postindustrial economy that is maturing in the most highly advanced countries of North America, Europe, and the Pacific Rim. Here, traditional industry is being eclipsed by a higher-technology productive complex focused on information-related activities.

Insurgent state Territorial embodiment of a successful guerrilla movement. The establishment by antigovernment insurgents of a territorial base in which they exercise full control; thus a **state** within a state.

Intermodal connections Facilities and activities related to the transfer of goods in transit from one transportation mode to another (e.g., the loading of containers from a ship directly onto a truck or railcar).

Intermodal transport system One that smoothly integrates different surface transportation modes. The shipping of cargo containers depends on fast and efficient transfers: they can be stacked on the decks and in the holds of ships as well as attaching to flatbed railcars and trailer trucks.

Irredentism A policy of cultural extension and potential political expansion by a **state** aimed at a community of its nationals living in a neighboring state.

Islamic Frontier The southern border of the African Transition Zone that marks the religious frontier of the Muslim faith in its southward penetration of Subsaharan Africa (see Fig. 7-20).

Jihad A doctrine within Islam. Commonly translated as *holy war*, it entails a personal or collective struggle on the part of Muslims to live up to the religious standards prescribed by the *Quran* (Koran).

Land alienation One society or culture group taking land from another.

Land bridge A narrow isthmian link between two large landmasses. They are temporary features—at least when measured in geologic time—subject to appearance and disappearance as the land or sea level rises and falls.

Land Hemisphere The half of the globe containing the greatest amount of land surface, centered on western Europe.

Landlocked location An interior **state** wholly surrounded by land. Without coasts, such a country is disadvantaged in terms of accessibility to international trade routes, and in the scramble for possession of areas of the **continental shelf** and control of the **Exclusive Economic Zone** beyond.

Land tenure The way people own, occupy, and use land.

Liberation theology A powerful religious movement that arose in South America during the 1950s, and subsequently gained followers throughout the **global periphery**. At its heart is a belief system, based on a blend of Christian faith and socialist thinking, that interprets the teachings of Christ as a quest to liberate the impoverished masses from oppression.

Lingua franca A "common language" prevalent in a given area; a second language that can be spoken and understood by many peoples, although they speak other languages at home.

Local functional specialization A hallmark of Europe's **economic geography** that later spread to many other parts of the world, whereby particular people in particular places concentrate on the production of particular goods and services.

Low island Cultures associated with low-lying coral islands of the Pacific Realm that cannot wrest sufficient moisture from the tropical maritime air to avoid chronic drought. Thus productive agriculture is impossible, and their modest populations must rely on fishing and the coconut palm for survival. Moreover, they are now threatened by rising sea level.

Maquiladora The term given to modern industrial plants in Mexico's U.S. border zone. These foreign-owned factories assemble imported components and/or raw materials, and then export finished manufactures, mainly to the United States. Import duties are disappearing under **NAFTA**, bringing jobs to Mexico and the advantages of low wage rates to the foreign entrepreneurs.

Marine geography The geographic study of oceans and seas. Its practitioners investigate both the physical (e.g., coral-reef **biogeography**, ocean–atmosphere interactions, coastal geomorphology) as well as human (e.g., **maritime boundary**-making, fisheries, beachside development) aspects of oceanic environments.

Maritime boundary An international boundary that lies in the ocean. Like all boundaries, it is a vertical plane, extending from the seafloor to the upper limit of the air space in the atmosphere above the water.

Maritime claims Intended legal assertions made by **states** concerning **sovereignty** over an oceanic space as parts of their **territorial seas**, or as rights to an **Exclusive Economic Zone** (or as extensions into waters adjacent to land claims, such as in Antarctica [Fig. 13-6]).

Median-line boundary An international **maritime boundary** drawn where the width of a sea is less than 400 nautical miles. Because the **states** on either side of that sea claim **Exclusive Economic Zones** of 200 nautical miles, it is necessary to reduce those claims to a (median) distance equidistant from each shoreline. Delimitation on the map almost always appears as a set of straight-line segments that reflect the configurations of the coastlines involved.

Medical geography The study of health and disease within a geographic context and from a spatial perspective. Among other things, this geographic field examines the sources, **diffusion** routes, and distributions of diseases.

Megacity Informal term referring to the world's most heavily populated cities; in this book, the term refers to a **metropolis** containing a population of greater than 10 million.

Megacity-region An unusually large urban region in China, territorially and population-wise, comprised of multiple coalescing metropolitan areas, connected through newly developed transportation infrastructures and with internal functional connections (e.g., the Pearl River Delta).

Melting pot Traditional characterization of American society as a blend of numerous immigrant **ethnic** groups that over time were assimilated into a single societal mainstream. This notion always had its challengers among social scientists, and is now increasingly difficult to sustain given the increasing complexity and sheer scale of the U.S. ethnic mosaic in the twenty-first century.

Mental maps Maps that individuals carry around in their minds that reflect their constantly evolving perception of how geographic space (ranging from their everyday activity space to the entire world) is organized around them.

Mestizo Derived from the Latin word for *mixed*, refers to a person of mixed European (white) and Amerindian ancestry.

Micro-credit Small loans extended to poverty-stricken borrowers who would not otherwise qualify for them. The aim is to help combat poverty, encourage entrepreneurship, and to empower poor communities—especially their women.

Microstate A sovereign **state** that contains a minuscule land area and population. They do not have the attributes of "complete" states, but are on the map as tiny yet independent entities nonetheless.

Migration A change in residence intended to be permanent.

Mobile money The dominant system of text-messaged money exchange in Subsaharan Africa that is attributable to the realm's general lack of a formal banking infrastructure. Elaborated in Box 8-9.

Monocentric geographic realm A world geographic realm dominated—territorially and/or demographically—by a single country. Russia in Russia/Central Asia is a prime example; others are the United States (North America), India (South Asia), and China (East Asia).

Monsoon Refers to the seasonal reversal of wind and moisture flows in certain parts of the subtropics and lower-middle latitudes. The *dry monsoon* occurs during the cool season when dry offshore winds prevail. The *wet monsoon* occurs in the hot summer months, which produce onshore winds that bring large amounts of rainfall. The air-pressure differential over land and sea is the triggering mechanism, with wind-flows always moving from areas of relatively higher pressure toward areas of relatively lower pressure. Monsoons make their greatest regional impact in the coastal and near-coastal zones of South Asia, Southeast Asia, and East Asia.

Multilingualism A society marked by a mosaic of local languages. Constitutes a **centrifugal force** because it impedes communication within the larger population. Often a *lingua franca* is used as a "common language," as in many countries of Subsaharan Africa.

NAFTA The free-trade area launched in 1994 involving the United States, Canada, and Mexico.

Nation Legally a term encompassing all the citizens of a **state**, it also has other connotations. Most definitions now tend to refer to a group of tightly knit people possessing bonds of language, **ethnicity**, religion, and other shared **cultural** attributes. Such homogeneity actually prevails within very few states.

Nation-state A country whose population possesses a substantial degree of **cultural** homogeneity and unity. The ideal form to which most **nations** and **states** aspire—a political unit wherein the territorial state coincides with the area settled by a certain national group or people.

NATO Established in 1950 at the height of the Cold War as a U.S.-led **supranational** defense pact to shield postwar Europe against the Soviet military threat. NATO is now in transition, expanding its membership while modifying its objectives in the post-Soviet era. Its 28 member-states (as of late 2019) are: Albania, Belgium, Bulgaria, Canada, Croatia, Czech Republic, Denmark, Estonia, France, Germany, Greece, Hungary, Iceland, Italy, Latvia, Lithuania, Luxembourg, the Netherlands, Norway, Poland, Portugal, Romania, Slovakia, Slovenia, Spain, Turkey, the United Kingdom, and the United States (see Fig. 5-11).

Natural landscape The array of landforms that constitutes the Earth's surface (mountains, hills, plains, and plateaus) and the physical features that mark them (such as water bodies, soils, and vegetation). Each **geographic realm** has its distinctive combination of natural landscapes.

Near Abroad The 14 former Soviet republics that, in combination with the dominant Russian Republic, constituted the USSR. Since the 1991 breakup of the Soviet Union, Russia has asserted a sphere of influence in these now-independent countries, based on its proclaimed right to protect the interests of ethnic Russians who were settled there in substantial numbers during Soviet times. These 14 countries include Armenia, Azerbaijan, Belarus, Estonia, Georgia, Kazakhstan, Kyrgyzstan, Latvia, Lithuania, Moldova, Tajikistan, Turkmenistan, Ukraine, and Uzbekistan.

Neighborhood effect The impact of one's neighborhood on an individual's outlook, aspirations, socialization, and life chances.

Neoliberal policies Policies adhering to an ideology or development strategy that advocates the privatization of state-run companies, lowering of international trade tariffs, reduction of government subsidies, cutting of corporate taxes, and overall deregulation of business activity.

Neoliberalism A national or regional development strategy based on the privatization of state-run companies, lowering of international trade tariffs, reduction of government subsidies, cutting of corporate taxes, and overall deregulation of business activity.

Nine-Dashed-Line map A map used by Chinese authorities to indicate Chinese claims to the South China Sea.

Node A center that functions as a point of **connectivity** within a regional network or system. All urban settlements possess this function, and the higher the position of a settlement in its **urban system** or hierarchy, the greater its nodality.

Non-governmental organization (NGO) A legitimate organization that operates independently from any form of government and does not function as a for-profit business. Mostly seeks to improve social conditions, but is not affiliated with political organizations.

Northeast Passage The high-latitude sea route of the Arctic Ocean that follows the entire north coast of Eurasia from northern Norway in the west to the northeasternmost corner of Russia where it meets the Bering Strait. Increased seasonal melting of the Arctic ice cap in recent years has begun to open up this waterway as a summer route for shipping between Europe and East Asia.

Northwest Passage The high-latitude, Arctic Ocean sea route around North America extending from Alaska's Bering Strait in the west to the Davis Strait between Canada and Greenland in the east. Heightened summertime melting of the Arctic ice cap in recent years has increased the likelihood of opening up this waterway for shipping between Asia's Pacific Rim and the Atlantic seaboard of the Americas.

Offshore banking Term referring to financial havens for foreign companies and individuals, who channel their earnings to accounts in such a country (usually an "offshore" island-state) to avoid paying taxes in their home countries.

One-Child Policy Chinese population control policy initiated in the late 1970s that proscribed (and enforced) a limit of one child per family of most population groups (mainly urban, **Han** populations). This policy, ended in 2016, had become increasingly controversial over the past decade.

Outback The name given by Australians to the vast, peripheral, sparsely settled interior of their country.

Overseas Chinese The more than 50 million ethnic Chinese who live outside China. About two-thirds live in Southeast Asia, and many have become quite successful. A large number maintain links to China and as investors played a major economic role in stimulating the growth of **SEZs** and Open Cities in China's **Pacific Rim**.

Pacific Rim A far-flung group of countries and components of countries (extending clockwise on the map from New Zealand to Chile) sharing the following criteria: they face the Pacific Ocean; they exhibit relatively high levels of economic development, industrialization, and urbanization; their imports and exports mainly move across Pacific waters.

Pacific Ring of Fire Zone of crustal instability along **tectonic plate** boundaries, marked by earthquakes and volcanic activity, that ring the Pacific Ocean Basin.

Pandemic An outbreak of a disease that spreads worldwide.

Partition The subdivision of the British Indian Empire into India and Pakistan at the end of colonial rule on August 15, 1947.

Peripheral development Spatial pattern in which a country's or region's development (and population) is most heavily concentrated along its outer edges rather than in its interior.

Permafrost Permanently frozen water in the near-surface soil and bedrock of cold environments, producing the effect of completely frozen ground. Surface can thaw during brief warm season.

Physiographic region A **region** within which there prevails substantial **natural-landscape** homogeneity, expressed by a certain degree of uniformity in surface relief, **climate**, vegetation, and soils.

Physiologic density The number of people per unit area of arable land.

Plantation A large estate owned by an individual, family, or corporation and organized to produce a cash crop. Almost all plantations were established within the tropics; in recent decades, many have been divided into smaller holdings or reorganized as cooperatives.

Polycentric urban region An urban region with multiple central cities and/or first-order suburban activity **nodes**.

Population density The number of people per unit area. Also see arithmetic density and physiologic density measures.

Population distribution The way people have arranged themselves in geographic space. One of human geography's most essential expressions because it represents the sum total of the adjustments that a population has made to its natural, cultural, and economic environments. A population distribution map is included in every chapter in this book.

Population geography The field of geography that focuses on the spatial aspects of demography and the influences of demographic change on particular countries and regions.

Population implosion The opposite of population explosion; refers to the declining populations of many European countries and Russia in which the death rate exceeds the birth rate and immigration rate.

Population pyramid Graphic representation or *profile* of a national population according to age and gender. Such a diagram of age-sex structure typically displays the percentage of each age group (commonly in five-year increments) as a horizontal bar, whose length represents its relationship to the total population (see Fig. 9-13).

Primate city A country's largest city—ranking atop its urban hierarchy—most expressive of the national culture and usually (but not in every case) the capital city as well.

Protruded state Territorial shape of a **state** that exhibits a narrow, elongated land extension (or *protrusion*) leading away from the main body of territory.

Rain shadow effect The relative dryness in areas downwind of mountain ranges resulting from orographic precipitation, wherein moist air masses are forced to deposit most of their water content as they cross the highlands.

Region A commonly used term and a geographic concept of paramount importance. An area on the Earth's surface marked by specific criteria, which are discussed in the Introduction.

Regional complementarity Exists when a pair of regions, through an exchange of raw materials and/or finished products, can specifically satisfy each other's demands.

Regional growth pole A center of growth (an existing town/city or planned economic center) intended to spur development in the wider region.

Relative location The regional position or **situation** of a place relative to the position of other places. Distance, accessibility, and **connectivity** affect relative location.

Religious revivalism Religious movement whose objectives are to return to the foundations of that faith and to influence state policy. Often called *religious fundamentalism*; but in the case of Islam, Muslims prefer the term *revivalism*.

Relocation diffusion Sequential diffusion process in which the items being diffused are transmitted by their carrier agents as they relocate to new areas. The most common form of relocation diffusion involves the spreading of innovations by a **migrating** population.

Remittances Money earned by emigrants that is sent back to family and friends in their home country, mostly in cash; forms an important part of the economy in poorer countries.

Remote sensing The indirect capture of images by specially equipped, Earth-orbiting satellites.

Residential geography The spatial distribution of a residential population. The term is most often used by urban geographers to describe the clustering of various social groups into the neighborhoods that form the residential fabric of cities and suburbs.

Rift valley The trough or trench that forms when a thinning strip of the Earth's crust sinks between two parallel faults (surface fractures).

Rural-to-urban migration The dominant **migration** flow from countryside to city that continues to transform the world's population, most notably in the less advantaged geographic realms.

Russification Demographic resettlement policies pursued by the central planners of the Soviet Empire (1922–1991), whereby ethnic Russians were encouraged to emigrate from the Russian Republic to the 14 non-Russian republics of the USSR.

Satellite state The countries of eastern Europe under Soviet hegemony between 1945 and 1989. This tier of countries—the "satellites" captured in Moscow's "orbit" following World War II—was bordered on the west by the Iron Curtain and on the east by the USSR. Using the names then in force, they included Bulgaria, Czechoslovakia, East Germany, Hungary, Poland, and Romania.

Scale Representation of a real-world phenomenon at a certain level of reduction or generalization. In **cartography**, the ratio of map distance to ground distance; indicated on a map as a bar graph, representative fraction, and/or verbal statement. *Macroscale* refers to a large area of national proportions; *microscale* refers to a local area no bigger than a county.

Schengen Area The territory constituted by most of Europe's countries within which people are free to cross international boundaries without formal border checks. Certain EU members do not fully participate: Bulgaria, Croatia, Cyprus, Ireland, and Romania. Four non-EU countries do participate: Iceland, Norway, Switzerland, and the microstate of Liechtenstein. Another non-participant is the United Kingdom, which voted to leave the EU in 2016.

Sex ratio A demographic indicator showing the ratio of males to females in a given population.

Sharia **law** The strict criminal code based in Islamic law that prescribes corporal punishment, amputations, stonings, and lashings for both major and minor offenses. Its occurrence today is associated with the spread of **religious revivalism** in Muslim societies.

Sharing economy A recent development in urban economies based on the digital revolution that allows for sharing of heretofore private resources ranging from cars and scooters to overnight accommodation, from work space to power tools.

Shi'ite Islam The smaller of the two main Islamic sects, comprising about 10 percent of Muslims overall, but in the majority in both Iran and Iraq. The origin of Shi'ism dates back to the death of the Prophet Muhammad, whose adherents believe that only a blood relative of Muhammad (his cousin) could be considered his legitimate successor.

Sinicization Giving a Chinese cultural imprint; Chinese **acculturation**. See also **Hanification**.

Site The internal locational attributes of an urban center, including its local spatial organization and physical setting.

Situation The external locational attributes of an urban center; its **relative location** or regional position with reference to other non-local places.

Small-island developing economies The additional disadvantages faced by lower-income island-states because of their often small territorial size and populations as well as overland inaccessibility. Limited resources require expensive importing of many goods and services; the cost of government operations per capita are higher; and local production is unable to benefit from **economies of scale**.

Social stratification In a layered or stratified society, the population is divided into a hierarchy of social classes. In an industrialized society, the working class is at the lower end; elites that possess capital and control the means of production are at the upper level.

South-to-North Water Diversion Project (SNWDP) The PRC's inter-basin water transfer scheme (the world's largest) to deliver massive quantities of fresh water from the Huang He and Chiang Jiang river systems to the burgeoning urban areas of northern China that face severe water shortages.

Southern Ocean The ocean that surrounds Antarctica (discussed in Chapter 12).

Sovereignty Controlling power and influence over a territory, especially by the government of an autonomous **state** over the people it rules.

Spatial diffusion The spatial spreading or dissemination of a **culture** element (such as a technological innovation) or some other phenomenon (e.g., a disease outbreak).

Spatial perspective Broadly, the geographic dimension or expression of any phenomenon; more specifically, anything related to the organization of space on the Earth' surface.

Spatial system The components and interactions of a **functional region**, which is defined by the areal extent of those interactions.

Special Economic Zone (SEZ) Manufacturing and export center in China, created after 1980 to attract foreign investment and technology transfers. Seven SEZs—all located on China's Pacific coast—currently operate: Shenzhen, adjacent to Hong Kong; Zhuhai; Shantou; Xiamen; Hainan Island, in the far south; Pudong, across the river from Shanghai; and Binhai New Area, next to the port of Tianjin.

State A politically organized territory that is administered by a sovereign government and is recognized by a significant portion of the international community. A state must also contain a permanent resident population, an organized economy, and a functioning internal circulation system.

State capitalism Government-controlled corporations competing under free-market conditions, usually in a tightly regimented society.

State capture A systematic form of political corruption whereby certain private interest groups significantly influence (capture) government decision-making through bribery and other means.

State formation The creation of a state, exemplifying traditions of human territoriality that go back thousands of years.

State territorial configuration The particular geographic layout of sovereign territory of a state, that influences its internal functioning and external relations.

Stateless nation A national group that aspires to become a an independent **state** but lacks the territorial means to do so.

Suburbanization Urban growth away from the center, typically in the edges of urban areas.

Subsistence agriculture Farmers who eke out a living on a small plot of land on which they are only able to grow enough food to support their families or at best a small community.

Subtropical Convergence A narrow marine **transition zone**, girdling the globe at approximately latitude 40°S, that marks the equatorward limit of the frigid **Southern Ocean** and the poleward limits of the warmer Atlantic, Pacific, and Indian oceans to the north.

Sunbelt The popular name given to the southern tier of the United States, which is anchored by the mega-States of California, Texas, and Florida. Its warmer climate, superior recreational opportunities, and other amenities have been attracting large numbers of relocating people and activities since the 1960s; broader definitions of the Sunbelt also include much of the western United States, even Colorado and the coastal Pacific Northwest.

Sunni Islam The larger of Islam's two main sects (encompassing) roughly 90 percent of all Muslims) who adhere to the conviction that any devout follower of the Prophet Muhammad is eligible to be his legitimate successor.

Supranationalism A venture involving three or more **states**—political, economic, and/or cultural cooperation to promote shared objectives.

Sustainable development Viable, long-term development that does not deplete resources and/or cause significantly negative side-effects.

Suwalki Gap The 65-km (40-mile) southern stretch of border between Lithuania and Poland, the only overland connection between the Baltic states and other NATO countries (the other borders are with Russia and Belarus).

Taiga The subarctic, mostly coniferous snowforest that blankets northern Russia and Canada south of the **tundra** that lines the Arctic shore. Known as the **boreal forest** in North America.

Taliban The term means "students" or "seekers of religion." Specifically, refers to the Islamist militia group that emerged from madrassas in Pakistan and ruled neighboring Afghanistan between 1996 and 2001; it has been trying to regain control of that country in its continuing conflict with U.S.-led NATO troops. Taliban rule, in adherence with an extremist interpretation of *Sharia* law, was marked by one of the most virulent forms of militant Islam ever seen.

Tar sands The main source of oil from nonliquid petroleum reserves. The oil is mixed with sand and requires massive open-pit mining as well as a costly, complicated process to extract it. The largest-known deposits are located in the northeast of Canada's province of Alberta, and by most estimates these Athabasca Tar Sands constitute one of the largest oil reserves in the world. The high oil prices of recent years have led to greatly expanded production here, but the accompanying environmental degradation caused by strip-mining and waste disposal has triggered a widening protest movement that may limit the exploitation of this resource.

Technopole A planned techno-industrial complex (such as California's Silicon Valley) that innovates, promotes, and manufactures the products of the postindustrial **information economy**.

Tectonic plates The slabs of heavier rock on which the lighter rocks of the continents rest. The plates are in motion, propelled by gigantic circulation cells in the red-hot, molten rock below. Most earthquakes and volcanic eruptions are associated with collisions of the mobile plates, as is the building of mountain chains.

Territorial sea Zone of seawater adjacent to a country's coast, held to be part of the national territory and treated as a component of the sovereign **state**.

Transculturation Cultural borrowing and two-way exchanges that occur when different **cultures** of approximately equal complexity and technological level come into close contact.

Transferability The capacity to move a good from one place to another at a bearable cost; the ease with which a commodity may be transported.

Transition zone An area of spatial change where the **peripheries** of two adjacent **realms** or **regions** join; marked by a gradual shift (rather than a sharp break) in the characteristics that distinguish these neighboring geographic entities from one another.

Triple Frontier The turbulent and chaotic area in southern South America that surrounds the convergence of Brazil, Argentina, and Paraguay. Lawlessness pervades this haven for criminal elements, which is notorious for money laundering, arms and other smuggling, drug trafficking, and links to terrorist organizations, including money flows to the Middle East.

Tropical deforestation The clearing and destruction of tropical rainforests in order to make way for expanding settlement frontiers and the exploitation of new economic opportunities.

Tsunami A seismic (earthquake-generated) sea wave that can attain gigantic proportions and cause coastal devastation. The tsunami of December 26, 2004, centered in the Indian Ocean near the Indonesian island of Sumatera (Sumatra), produced the first great natural disaster of the twenty-first century. Our new century's second major tsunami disaster occurred along the coast of Japan's northeastern Honshu Island on March 11, 2011.

Tsunami Warning System (TWS) The ocean-basin-scale network of instrumented buoys and other sensors to detect and warn of Pacific **tsunamis** following coastal and seafloor earthquakes (Fig. 13-2). The Pacific TWS is the most advanced and has proven its worth many times. Similar systems are or soon will be in place in other maritime arenas, including the Indian Ocean, the Northeast Atlantic/Mediterranean, and the Caribbean Sea.

Tundra The treeless plain that lies along the Arctic shore in northernmost Russia and Canada, whose vegetation consists of mosses, lichens, and certain hardy grasses.

Uneven development The notion that economic development varies spatially, a central tenet of **core-periphery relationships** in realms, regions, and lesser geographic entities.

Unitary state A **nation-state** that has a centralized government and administration that exercises power equally over all parts of the **state**.

Unity of place The great German natural scientist Alexander von Humboldt's notion that in a particular locale or region intricate connections exist among climate, geology, biology, and human cultures. This laid the foundation for modern geography as an *integrative discipline* marked by a spatial perspective.

Urbanization A term with a variety of connotations. The proportion of a country's population living in urban places is its level of urbanization. The process of urbanization involves the movement to, and the clustering of, people in towns and cities—a major force in every geographic realm today. Another kind of urbanization occurs when an expanding city absorbs rural countryside and transforms it into suburbs; in the case of cities in disadvantaged countries, this also generates peripheral shantytowns.

Urban primacy Refers to a country's largest city—ranking atop its urban hierarchy—most expressive of the national culture and usually (but not in every case) the capital city as well.

Urban-rural divide The divide between cities and rural areas in terms of well-being, employment, services, and consumption; sometimes also in terms of culture and politics.

Urban system A hierarchical network or grouping of urban areas within a finite geographic area, such as a country.

Viticulture The growing of grapes for the production of wine.

Wallace's Line The zoogeographical boundary proposed by Alfred Russel Wallace that separates the marsupial fauna of Australia and New Guinea from the non-marsupial fauna of Indonesia (see Figure 12-5).

West Wind Drift The clockwise movement of water as a current that circles around Antarctica in the **Southern Ocean**.

World-City A large city with particularly significant international (economic) linkages that also has a high ranking in the global urban system. Leading world-cities include London, New York, Tokyo, Shanghai, Singapore, and Paris.

Xenophobia Extreme dislike and fear of foreigners, sometimes fueled by populist politicians.

Index

A page number followed by "f" indicates the entry is within a figure; a page number followed by "t" indicates the entry is within a table.

www.wiley.com/go/nijman/theworldtoday8e